U.S. Department of Defense
Law of War Manual

June 2015
(Updated December 2016)

Office of General Counsel
Department of Defense

Abridged Primary Text Edition.

This edition is published with just the primary text to provide a portable paperback version. The Kindle edition of this work, including all the references as endnotes, is free with the purchase of this edition. The original Electronically issued DOD document is also available at the Department of Defense website as well as on the Internet Archive. The footnote numbers by section have been retained in the text to facilitate reference to the Kindle version by text location. Footnote text can also be referenced by Section note number in the original 1,200 page Department of Defense Electronic Edition PDF.

First Abridged Text Only Paperback 1.0 (2018)
Specifications: Manual Text, 415 Pages, 287k Words.
ISBN-13: 978-1-7320124-2-4
Imprint: Payton Fireman Attorney at Law

Contents

Foreword

The law of war is of fundamental importance to the Armed Forces of the United States. The law of war is part of who we are. George Washington, as Commander in Chief of the Continental Army, agreed with his British adversary that the Revolutionary War would be "carried on agreeable to the rules which humanity formed" and "to prevent or punish every breach of the rules of war within the sphere of our respective commands." During the Civil War, President Lincoln approved a set of "Instructions for the Government of the Armies of the United States in the Field," which inspired other countries to adopt similar codes for their armed forces, and which served as a template for international codifications of the law of war. After World War II, U.S. military lawyers, trying thousands of defendants before military commissions did, in the words of Justice Robert Jackson, "stay the hand of vengeance and voluntarily submit their captive enemies to the judgment of law" in "one of the most significant tributes that Power has ever paid to Reason." Reflecting on this distinctive history, one Chairman of the Joint Chiefs of Staff observed that "[t]he laws of war have a peculiarly American cast." And it is also true that the laws of war have shaped the U.S. Armed Forces as much as they have shaped any other armed force in the world. The law of war is a part of our military heritage, and obeying it is the right thing to do. But we also know that the law of war poses no obstacle to fighting well and prevailing. Nations have developed the law of war to be fundamentally consistent with the military doctrines that are the basis for effective combat operations. For example, the self-control needed to refrain from violations of the law of war under the stresses of combat is the same good order and discipline necessary to operate cohesively and victoriously in battle. Similarly, the law of war's prohibitions on torture and unnecessary destruction are consistent with the practical insight that such actions ultimately frustrate rather than accomplish the mission.

This manual reflects many years of labor and expertise, on the part of civilian and military lawyers from every Military Service. It reflects the experience of this Department in applying the law of war in actual military operations, and it will help us remember the hardlearned lessons from the past. Understanding our duties imposed by the law of war and our rights under it is essential to our service in the nation's defense.

Stephen W. Preston
General Counsel of the Department of Defense

Preface

This manual is a Department of Defense (DoD)-wide resource for DoD personnel – including commanders, legal practitioners, and other military and civilian personnel – on the law of war.

This manual has many distinguished antecedents that have provided important guidance to the U.S. Armed Forces. For example, General Order No. 100, the Instructions for the Government of Armies of the United States in the Field, commonly known as the Lieber Code, was prepared by Professor Francis Lieber and approved by President Abraham Lincoln during the Civil War in 1863.[1] A similar code related to naval warfare titled The Law and Usages of War at Sea: A Naval War Code was prepared by then-Captain Charles H. Stockton and approved by President William McKinley in 1900.[2] The War Department published instructions for the armed land forces of the United States in a 1914 manual titled Rules of Land Warfare, which was updated in 1917, 1934, and 1940.[3]

After World War II, in connection with U.S. ratification of the 1949 Geneva Conventions, the Department of the Navy published Naval Warfare Information Publication 10-2, Law of Naval Warfare, in 1955,[4] and the Department of the Army published Field Manual 27-10, The Law of Land Warfare, in 1956, which was updated in 1976.[5] The Department of the Army also published pamphlets on international law applicable in peace and war in the 1960s, and, in 1979, an updated version of the pamphlet on the law of peace.[6] The Department of the Air Force published in 1976 Air Force Pamphlet 110-31, International Law – The Conduct of Armed Conflict and Air Operations, which was updated in 1980.[7] More recently, the Judge Advocate General of the Air Force's School has published a manual titled Air Force Operations and the Law in 2002, with new editions in 2009 and 2014.[8] The Navy, Marine Corps, and Coast Guard have published several editions of The Commander's Handbook on the Law of Naval Operations starting in 1987 and most recently in 2007.[9] Helpful annotated supplements have also been published.[10]

In addition to these major publications, DoD components have produced many other publications that have supported DoD lawyers in giving advice on the law of war. For example, since 1895, the Naval War College has published its International Law Studies journal.[11] The Judge Advocate General of the Army's Legal Center & School has published many editions of a Law of Armed Conflict Deskbook, a Law of Armed Conflict Documentary Supplement, and an Operational Law Handbook.[12]

The preparation of this manual also has benefited greatly from consulting foreign experts and resources – for example, the 2004 edition of the Manual of the Law of Armed Conflict by United Kingdom Ministry of Defence.[13] In this way, the preparation of this manual is no different from its predecessors. For example, the 1956 Army Field manual benefited from considering a draft of what ultimately became the 1958 United Kingdom law of war manual, and the preparation of the 1914 War Department manual benefited from the Rules of Land Warfare prepared by officers of the English Army and Professor Lassa Oppenheim.[14] The law of war manuals of Germany, Australia, and Canada were also helpful resources in the preparation of this manual.[15]

The preparation of this manual has also benefited from the participation of officers from the United Kingdom's Royal Air Force and the Royal Australian Air Force on exchange assignments with the U.S. Air Force. In addition, military lawyers from Canada, the United Kingdom, New Zealand, and Australia reviewed and commented on a draft of the manual in 2009 as part of review that also included comments from distinguished scholars. Promulgating a DoD-wide manual on the law of war has been a long-standing goal of DoD lawyers. Memoranda and meeting notes from the 1970s reflect that the international law offices of the Department of the Army's Office of the Judge Advocate General and the Department of the Navy's Office of the Judge Advocate General generally agreed on a concept plan for a new all-Services law of war manual that would be a resource for implementing the 1977 Additional Protocols to the 1949 Geneva Conventions.[16] At the time, it was anticipated that the United States would ratify the Protocols, which has not occurred.

The origin of this manual may be traced to work in the late 1980s to update Department of the Army Field Manual 27-10, The Law of Land Warfare.[17] Then, in the mid-1990s, work began on an all-Services law of war manual to reflect the views of all DoD components. It was envisioned that the manual would provide not only the black letter rules, but also discussion, examples of State practice, and references to past manuals, treatises, and other documents to provide explanation, clarification, and elaboration. The present manual has sought to realize that vision and thus it falls within the tradition of the 1914 War Department manual, as well as the 1989 and 1997 Commander's Handbook on the Law of Naval Operations, which also adopted this general approach of an annotated manual.

This manual is an institutional publication and reflects the views of the Department of Defense, rather than the views of any particular person or DoD component. An effort has been made to reflect in this manual sound legal positions based on relevant authoritative sources of the law, including as developed by the DoD or the U.S. Government under such sources, and to show in the cited sources the past practice of DoD or the United States in applying the law of war.

This manual primarily has been prepared by the DoD Law of War Working Group, which is chaired by a representative of the DoD General Counsel and includes representatives of the Judge Advocates General of the Army, Navy, and Air Force; the Staff Judge Advocate to the Commandant of the Marine Corps; the offices of the General Counsels of the Military Departments; and the Legal Counsel to the Chairman of the Joint Chiefs of Staff.[18] This manual has been reviewed by principals of these offices. The preparation of this manual has also benefited significantly from the participation of experts from the Department of State, Office of the Legal Adviser, and the Department of Justice, Office of Legal Counsel, although the views in this manual do not necessarily reflect the views of those Departments or the U.S. Government as a whole.

Comments and suggestions from users of the DoD Law of War Manual are invited. All such correspondence should be addressed by email to: osd.pentagon.ogc.mbx.ia-law-of-war-manual-comments@mail.mil.

Summary of Important Abbreviations

Editor Note: For a full list of abbreviations and citations see the full original List of Abbreviations Section at the end of the text. Also, the description of changes dated May 2016 and December 2016 are not reproduced in this work. Accordingly, only the final 2016 version is reproduced in its entirety.

1899 Hague II: Convention with Respect to the Laws and Customs of War on Land, Jul. 29, 1899.

1909 Declaration of London: Declaration Concerning the Laws of Maritime War, Feb. 26, 1909).

1914 Rules of Land Warfare: War Department, Office of the Chief of Staff, *Rules of Land Warfare* (Apr. 25, 1914)

1925 Geneva Gas and Bacteriological Protocol: Protocol for the Prohibition of the Use in War of Asphyxiating, Poisonous, or Other Gases, and of Bacteriological Methods of Warfare, Jun. 17, 1925, 94 LNTS 65

1929 GPW: Geneva Convention Relative to the Treatment of Prisoners of War, Jul. 27, 1929, 47 Stat. 2021

1929 GWS Geneva Convention for the Amelioration of the Condition of the Wounded and Sick of Armies in the Field, Jul. 27, 1929, 47 Stat. 2074

1940 Rules of Land Warfare: War Department, Field Manual 27-10, *Rules of Land Warfare* (1940)

1944 Chicago Convention: Convention on International Civil Aviation, Dec. 7, 1944, 61 Stat. 1180

1955 NWIP 10-2 : Department of the Navy, Office of the Chief of Naval Operations, Naval Warfare Information Publication 10-2, *Law of Naval Warfare* (Sept. 1955),

1956 FM 27-10 (Change No. 11976): Department of the Army Field Manual 27-10, *The Law of Land Warfare* (Jul. 18, 1956 with Change 1, Jul. 15, 1976)

AP I :Protocol (I) Additional to the Geneva Conventions of August 12, 1949, and Relating to the Protection of Victims of International Armed Conflicts, Jun. 8, 1977, 1125 UNTS 3

CCW: Convention on Prohibitions or Restrictions on the Use of Certain Conventional Weapons Which May be Deemed to be Excessively Injurious or to Have Indiscriminate Effects, Oct. 10, 1980, 1342 UNTS 137

Chemical Weapons Convention: Convention on the Prohibition of the Development, Production, Stockpiling and Use of Chemical Weapons and on Their Destruction, Jan. 13, 1993, 1974 UNTS 317

ENMOD Convention: Convention on the Prohibition of Military or Any Other Hostile Use of Environmental Modification Techniques, Dec. 10, 1976, 1108 UNTS 151

GC : Geneva Convention Relative to the Protection of Civilian Persons in Time of War, Aug. 12, 1949, 75 UNTS 287

GPW:Geneva Convention Relative to the Treatment of Prisoners of War, Aug. 12, 1949, 75 UNTS 135

Grotius, Law of War & Peace Hugo Grotius, *The Law of War and Peace: De Jure Belli ac Pacis Libri Tres* (translated by Francis W. Kelsey, 1925)

WS : Geneva Convention for the Amelioration of the Condition of the Wounded and Sick in Armed Forces in the Field, Aug. 12, 1949, 75 UNTS 31

GWS-Sea: Geneva Convention for the Amelioration of the Condition of the Wounded, Sick and Shipwrecked Members of Armed Forces at Sea, Aug. 12, 1949, 75 UNTS 85

Hague IV: Convention (IV) Respecting the Laws and Customs of War on Land, Oct. 18, 1907, 36 Stat. 2277

Hague IV Reg.: Regulations Respecting the Laws and Customs of War on Land, Annex to Convention (IV) Respecting the Laws and Customs of War on Land, Oct. 18, 1907, 36 Stat. 2295

Hague IX: Convention (IX) Concerning Bombardment by Naval Forces in Time of War, Oct. 18, 1907, 36 Stat. 2351

Hague V:Convention (V) Respecting the Rights and Duties of Neutral Powers and Persons in Case of War on Land, Oct. 18, 1907, 36 Stat. 2310

Hague VIII: Convention (VIII) Relative to the Laying of Automatic Submarine Contact Mines, Oct. 18, 1907, 36 Stat. 2332

Hague X: Convention (X) for the Adaptation to Maritime Warfare of the Principles of the Geneva Convention, Oct. 18, 1907, 36 Stat. 2371

Hague XI: Convention (XI) Relative to Certain Restrictions with Regard to the Exercise of the Right of Capture in Naval War, Oct. 18, 1907, 36 Stat. 2396

Hague XIII: Convention (XIII) Concerning the Rights and Duties of Neutral Powers in Naval War, Oct. 18, 1907, 36 Stat. 2415

ICJ Statute: Statute of the International Court of Justice (Annex to the Charter of the United Nations), Jun. 26, 1945, 59 Stat. 1055

ICTR: International Criminal Tribunal for the Prosecution of Persons Responsible for Genocide and Other Serious Violations of International Humanitarian Law Committed in the Territory of Rwanda and Rwandan Citizens Responsible for Genocide and Other Such Violations Committed in the Territory of Neighbouring States, between 1 January 1994 and 31 December 1994

ICTY: International Tribunal for the Prosecution of Persons Responsible for Serious Violations of International Humanitarian Law Committed in the Territory of the Former Yugoslavia since 1991

Lieber Code: E. D. Townsend, Assistant Adjutant General, General Orders No. 100, *Instructions for the Government of Armies of the United States in the Field*, Apr. 24, 1863, *reprinted in Instructions for the Government of Armies of the United States in the Field* (Government Printing Office, 1898).

LNTS League of Nations Treaty Series

LOS Convention: United Nations Convention on the Law of the Sea, Dec.10, 1982, 1833 UNTS 396.

Roerich Pact: Treaty between the United States of America and other American Republics on the Protection of Artistic and Scientific Institutions and Historic Monuments, Apr. 15, 1935, 49 Stat. 3267

Rome Statute: Rome Statute of the International Criminal Court, July 17, 1998, 2187 UNTS 90

U.S.C.: United States Code (as of the date of publication of this manual)

UNTS: United Nations Treaty Series

VCLT: Vienna Convention on the Law of Treaties, May 23, 1969, 1155 UNTS 331

I – General Background

Chapter Contents

1.1 PURPOSE AND SCOPE OF THIS MANUAL

1.1.1Purpose. The purpose of this manual is to provide information on the law of war to DoD personnel responsible for implementing the law of war and executing military operations.[1]

This manual represents the legal views of the Department of Defense. This manual does not, however, preclude the Department from subsequently changing its interpretation of the law. Although the preparation of this manual has benefited from the participation of lawyers from the Department of State and the Department of Justice, this manual does not necessarily reflect the views of any other department or agency of the U.S. Government or the views of the U.S. Government as a whole. This manual is not intended to, and does not, create any right or benefit, substantive or procedural, enforceable at law or in equity against the United States, its departments, agencies, or other entities, its officers or employees, or any other person.

1.1.2 Scope. This manual is not a definitive explanation of all law of war issues. This manual focuses on *jus in bello* – law relating to the conduct of hostilities and the protection of war victims.

This manual seeks to address the law of war that is applicable to the United States, including treaties to which the United States is a Party, and applicable customary international law. It provides legal rules, principles, and discussion, particularly with respect to DoD practice. Although the views of other States may be referenced in this manual, it is not a purpose of this manual to describe the views of other States, which may differ from views expressed in this manual.

This manual is not a substitute for the careful practice of law. As specific legal issues arise, legal advisers should consider relevant legal and policy materials (*e.g.*, treaty provisions, judicial decisions, past U.S. practice, regulations, and doctrine), and should apply the law to the specific factual circumstances.

This manual is intended to be a description of the law as of the date of the manual's promulgation. In this vein, much of this manual has been written in the past tense to help ensure that the text remains accurate, even after subsequent developments have occurred. Every effort has been made to ensure the accuracy of the manual, but it must be read in the light of later developments in the law.

1.2 USE OF FOOTNOTES, SOURCES, CROSS-REFERENCES, AND SIGNALS IN THIS MANUAL

1.2.1Use of Footnotes in This Manual. This manual uses footnotes to provide sources or cross-references to other sections of the manual in order to clarify, elaborate on, or support the main text.

An effort has been made to avoid introducing discussion in the footnotes that addresses different propositions than those discussed in the main text. Although providing tangential information in footnotes is common in academic legal writing, this practice has been avoided to the extent possible for principally two reasons. First, it was desirable that this manual's main text convey as much information as possible without the reader needing to read the footnotes. For example, it was desirable to avoid the possibility that a reader might misunderstand a legal rule addressed in the main text because a notable exception to that rule was addressed only in a footnote accompanying the text. Second, tangential discussion on a given issue in footnotes would have made it much more difficult to keep the manual's treatment of that issue consistent from section to section and to allow the reader to find all the relevant information about a single topic. Thus, tangential discussion in footnotes has been avoided, to the extent possible, in favor of cross-references to the appropriate section of the manual that addresses that topic in more detail.

1.2.2 Use of Sources in This Manual. This manual cites sources in the footnotes to support or elaborate upon propositions in the main text. These sources are cited in the footnotes to help practitioners research particular topics discussed in the main text. Reviewing the cited sources in their entirety may provide additional contextual information, especially where sources are only partially quoted in the footnotes.

1.2.2.1 *Selection of Sources*. The sources cited in the footnotes have been chosen for a variety of reasons. For example, a source may contain a particularly helpful explanation or illustration. A source may have been chosen to illustrate U.S. practice or legal interpretation. A source may have been selected because its author was a particularly influential and respected international lawyer. For example, the 1956 Department of Army Field Manual 27-10, The Law of Land Warfare, has been a source of legal guidance for the U.S. armed forces for more than 50 years, and was published in connection with the U.S. ratification of the 1949 Geneva Conventions. One of the persons who helped prepare the 1956 manual was Richard Baxter, a highly respected DoD lawyer, who later became a judge on the International Court of Justice. Citation to a particular source should not be interpreted to mean that the cited source represents an official DoD position, or to be an endorsement of the source in its entirety. For example, parts of a source, such as an opinion by the International Court of Justice or a commentary published by the International Committee of the Red Cross, may reflect the DoD legal interpretation, while other parts of the source may not. Similarly, the citation of the jurisprudence of the Inter-American Commission on Human Rights should not be understood to indicate that the United States has accepted the competence of the Inter-American Commission on Human Rights to apply the law of war.[2]

1.2.2.2 *Use of Older Sources*. Older sources are sometimes cited: (1) because that source is particularly influential; (2) to demonstrate the origin of a legal proposition; or (3) to illustrate that a particular rule or formulation has a long history. The citation of an older source should not necessarily be interpreted as an endorsement that every aspect of that source remains current law. For example, the Lieber Code is a canonical law of war document for the United States, but parts of it no longer reflect current law.[3]Moreover, an older document produced by a State does not necessarily reflect its current legal views. For example, the 1958 UK Manual, although a particularly influential law of war manual prepared by distinguished experts Hersch Lauterpacht and Gerald Draper, has been superseded by subsequent UK Manuals, which reflect more recent developments in the law for the United Kingdom (*e.g.*, its ratification of AP I).

1.2.2.3 *Quotes Provided From Sources*. Quotes from sources are sometimes given in parentheticals within footnotes. These parentheticals are provided to help practitioners, such as by facilitating comparison between the main text of the manual and the language used in the sources. Every effort has been made to quote sources accurately. Practitioners, however, should verify quotations using the original source.

Certain formatting rules have been followed for quoted material. Two spaces have been placed after each period ending a sentence. Footnote numbers and carriage returns have been omitted from quoted text. Otherwise, quotes have not been changed unless noted through the use of ellipses, brackets, or parentheticals after the quotes indicating the changes made.

1.2.2.4 *Citation of Policies and Regulations*. Policies and regulations of the U.S. Government or particular DoD components are sometimes cited as examples of past practice. This manual, however, seeks primarily to address the law and not to address applicable U.S. Government or DoD policies or regulations. Many policies and regulations are not addressed in this manual, and the discussion of some policies, where relevant, should not be understood to indicate that other pertinent policies or regulations do not exist. Moreover, policies and regulations are constantly updated, so practitioners are advised to ascertain whether more recent versions of cited policies and regulations have been issued.

In some cases, cancelled issuances or superseded policies or regulations are cited to show the past practice, and, at times, a series of issuances are cited to illustrate a continuity in practice. Policies and regulations often exceed the requirements of applicable law, and the mere citation of a policy or regulation in this manual should not be understood to reflect the view that the policy or regulation's requirements have been promulgated out of a sense of legal obligation for the purposes of assessing customary international law or otherwise intending to reflect legal requirements.

1.2.2.5 *Citation Forms*. An effort has been made to make citations forms consistent throughout the manual, and to provide enough information about each cited source to reflect its significance and to enable readers to find it.[4]This manual has not strictly adhered to an established system of citation. Although certain citation systems were consulted, modifications were made as deemed appropriate for this type of resource, to make the citation forms straightforward and simple and relatively easy for readers to understand. In regard to abbreviations, for example, this manual generally does not abbreviate the names of academic journals. Moreover, it is hoped that the quotations from the cited sources that have been included in footnotes will help readers find the cited sources electronically.

1.2.3 Use of Cross-References in This Manual. This manual uses cross-references in the footnotes to point the reader to other sections of the manual containing relevant discussion of a particular topic. In particular, an effort has been made to use cross-references rather than to repeat discussion of a recurring issue or duplicate citation of legal sources. In sections in which a law of war rule is only mentioned tangentially or as an example, a cross-reference is used to direct the reader to the section of the manual in which a more in-depth discussion of that rule and supporting sources are provided.[5]

Cross-references are linked to enable the reader to access the referenced section quickly.

1.2.4 Use of Signals in This Manual. This manual uses signals to introduce the sources and cross-references in the footnotes. The table below identifies the signals used in this manual, describes their function, and provides examples of their use.

[no signal] Directly states the proposition

If a person joins a *levée en masse*, he or she may be held as a POW even if he or she actually took no part in fighting.1

1 1958 UK MANUAL ¶100 ("If it is shown that they joined the *levée en masse*, but took no part in the defence, they may be held as prisoners of war.").

Identifies the source of a quotation

As the Supreme Court has explained: "Lawful combatants are subject to capture and detention as prisoners of war by opposing military forces."2

2 Ex parte Quirin, 317 U.S. 1, 31 (1942).

Identifies an authority referred to in the text

There are additional provisions of the CCW Amended Mines Protocol addressing international exchanges of information and cooperation in this respect.3

3 CCW AMENDED MINES PROTOCOL art. 11.

See Clearly supports the proposition but does not directly state it

In addition, observers on military reconnaissance aircraft have not been regarded as acting clandestinely or under false pretenses.4

4 *See* HAGUE IV REG. art. 29 ("Persons sent in balloons for the purpose of carrying despatches and, generally, of maintaining communications between the different parts of an army or a territory" are not considered spies.).

See also Elaborates on the proposition

This means that a combatant's "killing, wounding, or other warlike acts are not individual crimes or offenses."5

5 LIEBER CODE art. 57. *See also* WINTHROP, MILITARY LAW & PRECEDENTS 778 ("The State is represented in active war by its contending army, and the laws of war justify the killing or disabling of members of the one army by those of the other in battle or hostile operations.").

Cf. Supports the proposition by analogy, *i.e.*, discusses a different proposition that is sufficiently similar to support the original proposition

A person must engage in acts of espionage in the zone of operations of a belligerent to be considered a spy. "Zone of operations" has been construed broadly to include areas supporting the war effort.6

6 *Cf.* Ex parte Quirin, 317 U.S. 1, 37 (1942) ("The law of war cannot rightly treat those agents of enemy armies who enter our territory, armed with explosives intended for the destruction of war industries and supplies, as any

the less belligerent enemies than are agents similarly entering for the purpose of destroying fortified places or our Armed Forces.").

Refer to Refers to another manual section that supports or elaborates on the proposition

Certain categories of persons are not members of the armed forces, but are nonetheless authorized to support the armed forces in the fighting:

• persons authorized to accompany the armed forces, but who are not members thereof;7

7 *Refer to* § 4.15 (Persons Authorized to Accompany the Armed Forces).

Compare Refers to another manual section that is analogous to the proposition

Persons authorized to accompany the armed forces who provide security against criminal elements generally would not be viewed as taking a direct part in hostilities (and do not forfeit their protection from being made the object of attack).269

269 *Compare* § 4.23.1 (Police as Civilians).

Consider Identifies a treaty that relates to the proposition but to which the United States is not a Party (*e.g.*, AP I)

Under international law, every treaty in force is binding upon the Parties to it and must be performed by them in good faith.10

10 *Consider* VCLT art. 26 ("Every treaty in force is binding upon the parties to it and must be performed by them in good faith.").

For example, Illustrates the proposition with an example drawn from historical practice

Adjusting the timing of an attack may reduce the risk of incidental harm. For example, attacking a military objective when civilians are less likely to be present may be appropriate.11

11 *For example,* Jennifer O'Connor, General Counsel, Department of Defense, *Remarks at New York University School of Law: Applying the Law of Targeting to the Modern Battlefield*, Nov. 28, 2016 ("Given the facts of this specific situation, including the intelligence that the cash was being used to pay ISIL fighters and conduct terrorist operations, the targeting authority determined that the bulk cash was a legitimate military objective. The next step then for the targeting authority before approving the strike was to conduct a proportionality analysis. This bulk cash storage site happened to be in an area where civilians were often present—as you might expect from a building that used to be a civilian bank before ISIL turned it into a cash storage site. As a result of the high number of civilians in the area, the targeting authority took specific steps to minimize potential harm to civilians by ordering the attack to occur at a time when the potential for any civilian casualties was deemed to be the lowest."); FINAL REPORT ON THE PERSIAN GULF WAR 100 (noting that during Operation DESERT STORM "attacks on known dual (i.e., military and civilian) use facilities normally were scheduled at night, because fewer people would be inside or on the streets outside.").

e.g., Added to any of the other signals when the cited authority is one of several authorities (some of which remain uncited) that stand for the same proposition

International humanitarian law is an alternative term for the law of war that may be understood to have the same substantive meaning as the law of war.12

12 *See, e.g., Overview of the Amendment to the Convention on the Physical Protection of Nuclear Material*, 6, Enclosure to Condoleezza Rice, Letter of Submittal, Jun. 11, 2007, MESSAGE FROM THE PRESIDENT OF THE UNITED STATES TRANSMITTING AMENDMENT TO THE CONVENTION ON THE PHYSICAL PROTECTION OF NUCLEAR MATERIAL (THE "AMENDMENT"). A CONFERENCE OF STATES PARTIES TO THE CONVENTION ON THE PHYSICAL PROTECTION OF NUCLEAR MATERIAL, ADOPTED ON OCTOBER 28, 1979, ADOPTED THE AMENDMENT ON JULY 8, 2005, AT THE INTERNATIONAL ATOMIC ENERGY AGENCY IN VIENNA, TREATY DOC. 110-6, 6 (2007) ("(2) The United States of America understands that the term 'international humanitarian law' in Paragraph 5 of the Amendment (Article 2 of the Convention on the Physical Protection of Nuclear Material, as amended) has the same substantive meaning as the law of war.").

1.3 DEFINITION OF THE LAW OF WAR

For the purposes of this manual, the *law of war* is that part of international law that regulates the resort to armed force; the conduct of hostilities and the protection of war victims in both international and non-international armed conflict; belligerent occupation; and the relationships between belligerent, neutral, and non-belligerent States.[6]

For the purposes of this manual, the law of war comprises treaties and customary international law applicable to the United States.[7]

1.3.1 Law of War – Notes on Terminology.

1.3.1.1 *Different Definitions of the Law of War*. The law of war may be defined slightly differently in other publications. For example, DoD issuances have defined the law of war more narrowly than the definition discussed in this section (*e.g.*, by omitting reference to that part of international law that regulates the resort to armed force).[8]

1.3.1.2 *Law of War Versus International Humanitarian Law and Law of Armed Conflict*. The *law of war* is often called the *law of armed conflict*. Both terms can be found in DoD directives and training materials. *International humanitarian law* is an alternative term for the law of war that may be understood to have the same substantive meaning as the law of war.[9]In other cases, international humanitarian law is understood more narrowly than the law of war (*e.g.*, by understanding international humanitarian law not to include the law of neutrality).[10]

1.3.2 The Law of War's Relationship to Other Bodies of Law.

An issue that often confronts law of war practitioners is the relationship of the law of war to other bodies of law, especially when rules in those bodies of law may appear to conflict with rules reflected in the law of war. These apparent conflicts are often resolved by considering the principle that the law of war is the *lex specialis* governing armed conflict.[11]How a law of war rule relates to a particular rule that is not grounded in the law of war may depend on the specific legal rule in question.

In general, the law of war may relate to other bodies of law through: (1) law of war rules superseding rules in other bodies of law with respect to armed conflict; (2) construing the rules in other bodies of law to avoid conflict with law of war rules; (3) law of war rules informing the content of general standards in other bodies of law, should such standards be construed to apply to armed conflict; and (4) law of war treaties explicitly incorporating concepts from other bodies of law.

In some cases, it may be difficult to distinguish between these approaches, and different entities may apply different approaches to achieve the same result.[12]Although there are different approaches and although the ultimate resolution may depend on the specific rules and context, the law of war, as the *lex specialis* of armed conflict, is the controlling body of law with regard to the conduct of hostilities and the protection of war victims.[13]

1.3.2.1 *The Law of War as the Lex Specialis Governing Armed Conflict*. The maxim *lex specialis derogat legi generali* means that "[a]s a rule the special rule overrides the general law."[14]The rule that is more specifically directed towards the action receives priority because it takes better account of the particular features of the context in which the law is to be applied, thus creating a more equitable result and better reflecting the intent of the authorities that have made the law.[15]

The law of war has been developed with special consideration of the circumstances of war and the challenges inherent in its regulation by law. Thus, for example, the exigencies of armed conflict cannot justify violating the law of war.[16]Moreover, lawmakers sometimes have considered peacetime rules appropriate to apply during armed conflict, and in certain of these cases, they have explicitly incorporated such concepts into the law of war.[17]

Thus, traditionally, the law of war has been described as the only "authoritative rules of action between hostile armies," or as superseding ordinary law in the actual theater of military operations.[18]Similarly, law of war treaties have been viewed as a clear example of a *lex specialis* in relation to treaties providing peacetime norms concerning the same subjects.[19]

1.3.2.2 *Construing Other Laws to Avoid Conflict With the Law of War*. Potential conflicts between the law of war and other law may be resolved by construing such other law to avoid conflict with law of war rules.

Underlying this approach is the fact that the law of war is firmly established in customary international law as a well-developed body of law that is separate from the principles of law generally applicable in peace.[20]Lawmakers have been understood not to amend that well developed body of law, absent affirmative evidence of an intention to do so.[21]In a similar fashion, for comparison, the GC deliberately excludes from its application the nationals of certain States in order to avoid creating complications or inconsistencies in procedures should both the GC and the law applicable to normal diplomatic representation apply.[22]

In some cases, treaties explicitly clarify that they do not affect States' rights under the law of war.[23]For example, the 1944 Chicago Convention on civil aviation explicitly provides that it does not affect the freedom of action of States during armed conflict.[24]However, even when not explicitly stated, infringements on the law of war through international agreements that primarily address situations other than armed conflict are not to be presumed.[25]For example, the LOS Convention has been interpreted not to impair a State's rights during armed conflict, even though this principle is not explicitly stated in the treaty.[26]In addition, the International Convention for the Suppression of the Financing of Terrorism has been understood not to preclude any State Party to the Convention from conducting any legitimate activity against any lawful target in accordance with the law of armed conflict.[27]

In addition to treaties, domestic statutes have also been construed not to violate international law, including the law of war, if any other construction remains possible.[28]Certain domestic statutes have been interpreted not to apply to situations addressed by the law of war because such intention was not made clear and unequivocal.[29]

1.3.2.3 *Using the Law of War to Determine the Content of General Standards if Applied to Armed Conflict.* Another way in which the law of war has been applied as *lex specialis* is to determine the content of a more general standard with respect to the situation of armed conflict. For example, the law of war has been used to inform the content of general authorizations to conduct military operations.[30]

As another example, to the extent that the concept of "due regard" for the safety of civil aircraft may be deemed to apply during armed conflict, what regard is due would be understood in terms of the requirements of the law of war.[31]Similarly, to the extent that the concept of "due regard" for the rights of other States under the law of the sea may be deemed to apply during armed conflict, what regard is due would be understood in terms of the requirements of the law of war.[32]

Lastly, even where international courts or commissions have characterized human rights obligations as applicable during armed conflict, they generally have characterized the content of those obligations as determined by standards and tests drawn from the law of war.[33]

1.3.2.4 *Explicit Incorporation of Concepts From Other Bodies of Law Into the Law of War.* In some cases, law of war treaties explicitly incorporate concepts from other bodies of law. For example, the peacetime property law concept of usufruct is made applicable to the duties of the Occupying States.[34]Similarly, the GC explicitly applies a peacetime rule with respect to the nationals of the Occupying Power who, before the outbreak of hostilities, have sought refuge in the occupied territory.[35]And as another example, Common Article 3 of the 1949 Geneva Conventions incorporates by reference those judicial guarantees that are recognized as indispensable by civilized peoples.[36]

1.3.3 Restrictive and Permissive Character of the Law of War. In certain respects, the law of war may be viewed as prohibitive; in other respects, the law of war may be viewed as permissive.[37]

1.3.3.1 *Law of War as Prohibitive Law.* The law of war that relates to the conduct of hostilities has generally been viewed as "prohibitive law," in the sense that it forbids rather than authorizes certain uses of force.[38]For example, the lawfulness of the use of a type of weapon does not depend on an absence of authorization, but, on the contrary, on whether the weapon is prohibited.[39]

One rationale for this view is that the rules binding upon States in treaties and customary law reflect restrictions that they have accepted, and that States are otherwise independent entities with freedom to act.[40]Thus, the authority to take actions under the law of war would be viewed as emanating from the State's rights as a sovereign entity rather than from any particular instrument of international law.

The prohibitive character of the law of war that relates to the conduct of hostilities is also consistent with the view that *jus in bello* applies to aggressors and defenders alike. The fact that an aggressor complies with *jus in bello* does not justify the legality of its military operations under *jus ad bellum.*[41]

The lack of an express prohibition in treaty law, however, does not necessarily mean that an action is lawful under *jus in bello*. When no specific rule applies, the principles of the law of war form the general guide for conduct during war.[42]

1.3.3.2 *Law of War as Permissive Law.* Although the law of war is generally viewed as "prohibitive law," in some respects, especially in the context of domestic law, the law of war may be viewed as permissive or even as a source of authority.[43]

For example, the principle of military necessity in the customary law of war may be viewed as justifying or permitting certain acts.[44]Similarly, under the law of belligerent occupation, the fact of occupation is the basis for the Occupying Power to exercise authority over the occupied territory.[45]In addition, law of war treaties also sometimes recognize States' authorities in war.[46]

1.3.4 Purposes of the Law of War. The main purposes of the law of war are:

• protecting combatants, noncombatants, and civilians from unnecessary suffering;[47]

• providing certain fundamental protections for persons who fall into the hands of the enemy, particularly prisoners of war, civilians, and military wounded, sick, and shipwrecked;[48]

• facilitating the restoration of peace;[49]

• assisting military commanders in ensuring the disciplined and efficient use of military force;[50]and

• preserving the professionalism and humanity of combatants.[51]

1.4 OBJECT AND NATURE OF WAR

Understanding the object and nature of war is important in understanding and applying the law of war.[52]

1.4.1 Object of War. The object of war has been understood to be the submission of the enemy as quickly and efficiently as possible.[53]The military defeat of the enemy in war is intended to advance political objectives.[54]Even where those political objectives are limited, the object of war is nonetheless to ensure the submission of the enemy as quickly and efficiently as possible.[55]

The object of war informs the principle of *military necessity* and what uses of force may be justified in war.[56]Nevertheless, the law of war limits what uses of force the object of war may justify.[57]

1.4.2 Nature of War.

1.4.2.1 *Nature of War – Violence and Suffering.* War has been described as a violent clash of interests characterized by the use of force.[58]The fact that violence is an essential element of war has been viewed as important in understanding the nature of war.[59]The violent nature of war has also meant that suffering has been an unfortunate and tragic, but unavoidable consequence of war.[60]

Law of war treaties such as the Hague and Geneva Conventions have been negotiated with the understanding that suffering and destruction are unavoidably part of war.[61]But these treaties and the principle of humanity seek to reduce unnecessary suffering and destruction.[62]

1.4.2.2 *Nature of War – Limited and Unreliable Information – "Fog of War".* During war, information is often limited and unreliable.[63]The uncertainty of information in war results from the chaotic nature of combat and from the opposing sides' efforts to deceive one another, which generally is not prohibited by the law of war.[64]

The limited and unreliable nature of information available during war has influenced the development of the law of war. For example, it affects how the principle of military necessity is applied.[65]The limited and unreliable nature of information available during war also is recognized in the law of war's standards for how persons are to assess information.[66]

1.5 "WAR" AS A LEGAL CONCEPT

"War" is sometimes used as a legal concept, *i.e.*, the application or operation of a legal rule may depend on the existence of a "war," "armed conflict," or "hostilities." As a legal concept, "war" has traditionally been viewed as a condition in which a State is prosecuting its rights by military force, usually against another State. However, the precise definition of "war" often depends on the specific legal context in which it is used.

1.5.1 Traditional Conception of War Under International Law. As international law began to regulate "war," "hostilities," and "armed conflict," it became necessary to determine what "war" is for the purpose of triggering those legal obligations.[67]

As a legal concept, war has usually been described as a condition or state that applies more broadly than only the mere employment of force or the mere commission of acts of violence.[68]

When treated as a legal concept, "war" has been associated with a State's use of force to vindicate its rights (principally, its inherent right of self-defense) under international law.[69]

Traditionally, war has often been described as a legal condition between two or more States.[70] However, certain law of war rules apply to non-international armed conflicts (such as intrastate conflicts or conflicts between a State and a non-State armed group).[71]

1.5.2 Different Definitions of "War" for Different Legal Purposes. There is no single legal definition of "war," "hostilities," or "armed conflict," and the definition of these terms has varied in both domestic and international law. In domestic law, "war," "hostilities," and "armed conflict" have been interpreted differently depending on the specific legal context at issue.[72] For example, under the Constitution, Congress has the power to "declare war."[73] Thus, "war" might be interpreted to determine whether a military operation constitutes "war" in this sense.[74] Similarly, the War Powers Resolution states certain requirements that are triggered when U.S. forces are introduced into "hostilities."[75] Other statutes may require a determination that conduct has occurred "[w]hen the United States is at war" or during "time of war."[76]

Under international law, "war," "hostilities," and "armed conflict" may also be interpreted with different purposes in mind.[77] A state of "war" can affect what duties States that are not participating in the conflict have under the law of neutrality.[78] A state of "war" can affect whether peacetime treaties between two States continue to apply. Most importantly for the purposes of this manual, the terms "war" and "armed conflict" are used to describe when *jus in bello* rules apply.[79]

1.6 LAW OF WAR DISTINGUISHED FROM CERTAIN TOPICS

The law of war may be distinguished from the following topics: (1) operational law; (2) arms control; (3) human rights treaties; (4) the Just War Tradition; (5) rules of engagement; and (6) the Code of Conduct for U.S. Armed Forces.

1.6.1 Operational Law. The law of war is an important part of, but not the entirety of, operational law. Operational law consists of that body of domestic, foreign, and international law that specifically pertains to the activities of military forces across the entire conflict spectrum. Operational law includes diverse legal disciplines, such as military justice, administrative and civil law, legal assistance, claims, procurement law, national security law, fiscal law, and the law of war.[80]

1.6.2 Arms Control. Arms control is a broad term that includes a variety of efforts to reduce the numbers, types, performance characteristics, proliferation, testing, or other aspects of certain categories of weapons. Arms control usually proceeds through bilateral or multilateral treaties. Arms control can also include non-binding political commitments, as well as reciprocal unilateral statements of intention or policy. The overall goals of arms control are to reduce: (1) the likelihood of war; (2) the consequences of war, should it occur; and (3) the costs of preparing for war.

Arms control is closely related to other concepts. For example, "non-proliferation" refers specifically to efforts to restrict the spread of weapons (in particular, weapons of mass destruction). "Disarmament" refers to efforts to eliminate entirely, rather than to restrict, a particular category of weapon. And sometimes States accept "confidence-building measures" (or confidence, security, and transparency-building measures) that do not directly reduce the quantity or quality of armaments, but rather increase States' certainty that ambiguous activities by other States are not secret actions in violation of arms control obligations. Arms control and the law of war frequently overlap in treaties. For example, the CCW Protocol IV on Blinding Laser Weapons has both arms control and law of war provisions.[81] Similarly, the Chemical Weapons Convention prohibits, *inter alia*, the development and stockpiling of chemical weapons, but it is also directly relevant to the law of war because it prohibits the use of chemical weapons in all circumstances.[82]

1.6.3 Human Rights Treaties.[83] Human rights treaties address primarily the obligations of governments with respect to the rights of individuals, including their own nationals.[84] For example, governments must refrain from subjecting individuals to arbitrary detention, to arbitrary deprivation of life, or to cruel, inhuman, or degrading treatment or punishment.[85] As a general matter, human rights treaties have been described as primarily applicable to the relationship between a State and individuals in peacetime.[86] Some human rights treaties also provide for derogation from certain provisions in emergency situations.[87]

Law of war treaties have been described as chiefly concerned with the conditions particular to armed conflict and the relationship between a State and nationals of the enemy State.[88] Law of war treaties generally do not provide for derogation because necessity is not a basis for derogating from law of war rules.[89]

1.6.3.1 *Relationship Between Human Rights Treaties and the Law of War*. In some circumstances, the rules in the law of war and the rules in human rights treaties may appear to conflict; these apparent conflicts may be resolved by the principle that the law of war is the *lex specialis* during situations of armed conflict, and, as such, is the controlling body of law with regard to the conduct of hostilities and the protection of war victims.[90]

For example, the right to challenge the lawfulness of an arrest before a court provided in Article 9 of the International Covenant on Civil and Political Rights (ICCPR) would appear to conflict with the authority under the law of war to detain certain persons without judicial process or criminal charge.[91] However, the United States has understood Article 9 of the ICCPR not to affect a State's authorities under the law of war, including a State's authority in both international and non-international armed conflicts to detain enemy combatants until the end of hostilities.[92] Some international courts or commissions have interpreted the rights conveyed by human rights treaties in light of the rules of the law of war, as the applicable *lex specialis,* when assessing situations in armed conflict.[93]

On the other hand, during armed conflict, human rights treaties would clearly be controlling with respect to matters that are within their scope of application and that are not addressed by the law of war. For example, a time of war does not suspend the operation of the ICCPR with respect to matters within its scope of application. Therefore, as an illustration, participation in a war would in no way excuse a State Party to the ICCPR from respecting and ensuring the right and opportunity of every citizen to vote and to be elected at genuine periodic elections.[94]

1.6.3.2 *Different Views on the Applicability of Human Rights Treaties*. In conducting operations with coalition partners, it may be important to consider that some States may have different perspectives on the applicability of human rights treaties. Such differences may result from different legal interpretations or from the fact that the other State is a Party to different human rights treaties than the United States. For example, the European Court of Human Rights – as well as some European States – have construed certain obligations under the European Convention on Human Rights (ECHR) as applicable to their military forces abroad during occupation.[95]

1.6.3.3 *International Covenant on Civil and Political Rights (ICCPR)*. The United States is a Party to the International Covenant on Civil and Political Rights (ICCPR). The ICCPR creates obligations for a State with respect to persons within its territory and subject to its jurisdiction.[96] The United States has long interpreted the ICCPR not to apply abroad.[97] The inclusion of the reference to "within its territory" in Article 2(1) of the ICCPR was adopted as a result of a proposal made by U.S. delegate Eleanor Roosevelt – specifically to ensure that a State Party's obligations would not apply to persons outside its territories, such as in occupied territory and leased territory.[98]

1.6.3.4 *Convention Against Torture*. The United States is a Party to the Convention Against Torture and Other Cruel, Inhuman or Degrading Treatment or Punishment.[99]

The Convention against Torture was not intended to supersede the prohibitions against torture already contained in customary international law and the 1949 Geneva Conventions or its Additional Protocols.[100] The law of war is the controlling body of law with respect to the conduct of hostilities and the protection of war victims. Nevertheless, a time of war does not suspend operation of the Convention Against Torture. The Convention Against Torture continues to apply even when a State is engaged in armed conflict.[101] For example, a state of war could not justify a State's torture of individuals during armed conflict.[102]

In addition, where the text of the Convention Against Torture provides that obligations apply to a State Party in "any territory under its jurisdiction," such obligations, including the obligations in Articles 2 and 16 to prevent torture and cruel, inhuman, or degrading treatment or punishment, extend to certain areas beyond the sovereign territory of the State Party, and more specifically to "all places that the State Party controls as a governmental authority."[103]

1.6.4 Just War Tradition. The Just War Tradition describes customs, ethical codes, and moral teachings associated with warfare that military thinkers and philosophers have developed over centuries to seek the moral justification of and the limitations to war.[104]

The Just War Tradition provides part of the philosophical foundation for the modern law of war and has considered both *jus ad bellum* and *jus in bello*. The Just War Tradition developed criteria or principles that have provided the foundation for modern *jus ad bellum* rules.[105]

Similarly, law of war treaties that provide *jus in bello* rules, such as the 1949 Geneva Conventions, are also rooted in the Just War Tradition. The Just War Tradition remains relevant for decisions to employ U.S. military forces and in warfighting.[106]

1.6.5 Rules of Engagement (ROE). *Rules of engagement* (ROE) have been defined as "[d]irectives issued by competent military authority that delineate the circumstances and limitations under which United States forces will initiate and/or continue combat engagement with other forces encountered."[107] ROE are used by States to tailor the rules for the use of force to the circumstances of a particular operation.[108]

ROE reflect legal, policy, and operational considerations, and are consistent with the international law obligations of the United States, including the law of war.[109] ROE may restrict actions that would be lawful under the law of war, but may not permit actions prohibited by the law of war. States have used ROE as part of the implementation of their law of war obligations during military operations.[110]

1.6.6 Code of Conduct for U.S. Armed Forces. The Code of Conduct is a moral guide for U.S. forces to govern their conduct in resisting capture and their actions in the event they fall into hostile hands.[111] The Code of Conduct was developed after the Korean War and was promulgated by Executive Order.[112] The Code of Conduct is consistent with the law of war obligations of the United States, including obligations in the GPW.[113]

1.7 TREATIES

Treaties are generally defined as international agreements concluded between States in written form and governed by international law.[114] Under international law, a treaty is binding upon States that are Parties to it.[115]

The United States is a Party to a number of law of war treaties.[116] For many years, the Department of State has published annually a listing of treaties and other international agreements in force for the United States. This publication has provided helpful information about such treaties, including the date of U.S. ratification and a listing of other Parties to each treaty.

1.7.1 Treaties – Notes on Terminology. Treaties may be titled or referred to by several other terms in addition to "treaty" – including convention, protocol, or agreement. In the context of the law of war, "protocol" often refers to an agreement that supplements or updates an existing agreement.

1.7.1.1 *"Treaties" Under U.S. Constitutional Law.* Under the Constitution, a "treaty" must receive the advice and consent of the Senate before U.S. ratification or accession. Certain international agreements, such as Executive agreements, are not classified as "treaties" for the purposes of this requirement, although they may be characterized as "treaties" for the purposes of international law and impose obligations upon the United States.[117]

1.7.2 Reservations to Treaties. A State may limit the application of provisions of a treaty by reservation upon ratification of the treaty as long as the treaty does not prohibit such reservations and the reservation is compatible with the object and purpose of the treaty.[118] For example, the United States has taken a reservation to certain provisions of CCW Protocol III on Incendiary Weapons.[119] On the other hand, for example, the Chemical Weapons Convention expressly prohibits reservations to the Convention and prohibits reservations to the Convention's Annexes that are incompatible with its object and purpose.[120]

1.7.3 Withdrawal From Treaties. Under certain circumstances, States may withdraw from treaties.[121] Some law of war and arms control treaties specify the conditions under which Parties may withdraw from them.[122] Even upon denunciation of a treaty, States remain bound by customary international law, including law of war principles.[123]

1.7.4 Use of Certain Subsequent Practice in Treaty Interpretation. Certain subsequent State practice in the application of a treaty provision may be taken into account when interpreting that provision.[124] Subsequent State practice is important as an element of interpretation when it constitutes objective evidence of the understanding of the Parties as to the meaning of the treaty.[125] For example, the subsequent practice of States in the application of the GWS-Sea's requirements for hospital ships has clarified that States may use hospital ships with the capability to conduct encrypted communications.[126]

1.7.5 Treaties and Domestic Implementing Legislation. States may enact domestic legislation to implement treaty provisions. Although such implementing legislation is not international law, it may reflect a State's interpretation of those provisions.[127]

A State's domestic implementing legislation, or lack of such legislation, however, does not justify that State's noncompliance with an international obligation as a matter of international law.[128]

1.8 CUSTOMARY INTERNATIONAL LAW

Customary international law results from a general and consistent practice of States that is followed by them from a sense of legal obligation (*opinio juris*).[129] Customary international law is an unwritten form of law in the sense that it is not created through a written agreement by States. Customary international law is generally binding on all States, but States that have been persistent objectors to a customary international law rule during its development are not bound by that rule.[130]

Assessing whether State practice and *opinio juris* have resulted in a rule of customary international law may be a difficult inquiry.[131]

1.8.1 Relationship Between Treaties and Customary International Law. Treaty provisions may, *inter alia*: (1) not reflect customary international law; (2) reflect customary international law; or (3) be based on customary law, but not precisely reflect it. In most cases, treaty provisions do not reflect customary international law. For example, AP I's provisions changing which persons would be entitled to the privileges of combatant status were viewed as novel at the time of the adoption of AP I and as not reflecting customary international law.[132]

In some cases, a treaty provision may reflect customary international law. The rule reflected in the treaty would thus be understood to be binding, even if the treaty provision was not applicable, because the rule maintains a separate existence as a customary norm.[133]For example, provisions of Hague IV and the Hague IV Regulations have been found to reflect customary international law.[134]Law of war treaties have specified that customary law and principles continue to apply even if the treaty is not applicable.[135]

A treaty provision may be based on an underlying principle that is an accepted part of customary law, but the precise language of the treaty provision may not reflect customary international law because there may be considerable disagreement as to the precise statement of that underlying principle.[136]For example, the United States has expressed support for the customary principle on which Article 51(3) of AP I is based, but has noted that Article 51(3) of AP I, as drafted, does not reflect customary international law.[137]

1.8.2 State Practice. One part of determining whether a purported rule is customary international law is to analyze whether there is a general and consistent practice of States that supports the purported rule. An analysis of State practice to determine whether a purported rule reflects the customary international law of war should include consideration of, *inter alia*: (1) whether the State practice is extensive and virtually uniform; (2) actual operational practice; (3) the practice of specially affected States; and (4) contrary practice.

1.8.2.1 *Extensive and Virtually Uniform*. State practice should be sufficiently dense and consistent to meet the "extensive and virtually uniform" standard generally required for existence of a customary rule.[138]

1.8.2.2 *Actual Operational Practice*. An analysis of State practice should include an analysis of actual operational practice by States during armed conflict. Although manuals or other official statements may provide important indications of State behavior, they cannot replace a meaningful assessment of operational State practice.[139]

1.8.2.3 *Specially Affected States*. The practice of "States whose interests are specially affected," *e.g.*, States with a distinctive history of participation in the relevant matter, must support the purported rule.[140]States that have had a wealth of experience, or that have otherwise had significant opportunities to develop a carefully considered military doctrine, may be expected to have contributed a greater quantity and quality of State practice relevant to the law of war than States that have not.

For example, "specially affected States" could include, depending upon the relevant matter, the nuclear powers, other major military powers, and occupying or occupied States.[141]As a case in point, the United Kingdom has been viewed as a specially affected State with respect to the law of the sea.[142]

1.8.2.4 *Contrary Practice*. Evidence of contrary practice, *i.e.*, the practice of States that does not support the purported rule, must be considered in assessing whether that rule exists as a rule of customary international law.[143]

In addition, the persistent objection of States may be relevant after the formation of that rule by preventing the application of that rule to States that have objected to that rule during its development.[144]

1.8.3 *Opinio Juris*. In addition to analyzing State practice, one must determine whether the State practice results from a sense of legal obligation (*opinio juris*) or merely reflects States' policy or practical interests. *Opinio juris* cannot simply be inferred from consistent State practice, which may exist for reasons other than *opinio juris*.[145]For example, the fact that nuclear weapons have not been used to conduct attacks during armed conflict since 1945 does not reflect a prohibition in customary international law against their use because such lack of use has not resulted from *opinio juris*.[146]

1.8.3.1 *Potential Sources of* Opinio Juris. It may be difficult to find evidence of *opinio juris*, and care should be exercised in assessing whether a source reflects *opinio juris* on the part of a State. For example, treaty provisions do not necessarily reflect *opinio juris*.[147]

Similarly, rather than indicating a position expressed out of a sense of a customary legal obligation, a State's military manual often recites requirements applicable to that State under treaties to which it is a Party, or provides guidance to its military forces for reasons of national policy.[148]

1.8.4 Objection During Development. Even if a rule otherwise reflects customary international law, the rule is not binding upon a State that has persistently objected to that rule during its development.[149] This principle is an accepted application of the traditional principle that international law essentially depends on the consent of States.[150]

1.9 SUBSIDIARY MEANS OF DETERMINING INTERNATIONAL LAW

As a subsidiary means, it may be helpful to consult judicial decisions and the teachings of the most highly qualified publicists of various nations in determining the applicable rules of international law. These means are subsidiary in the sense that they do not, in themselves, constitute sources of treaty or customary international law.

Discretion must be exercised in weighing sources, however, because sources vary significantly in their probative value. For example, the United States has said that it is not in a position to accept without further analysis the conclusions in a study on customary international humanitarian law published by the ICRC.[151]

1.9.1 Judicial Decisions. Judicial decisions have sometimes been used as a subsidiary means of determining the rules of international law.[152]

Judicial decisions are generally consulted as only persuasive authority because a judgment rendered by an international court generally binds only the parties to the case in respect of that particular case.[153] The legal reasoning underlying the decisions of the International Court of Justice is not binding on States.[154] Similarly, the decisions of the International Criminal Tribunal for the Former Yugoslavia and the International Criminal Tribunal for Rwanda cannot, as a strictly legal matter, "bind" other courts.[155]

The legal principle of *stare decisis* does not generally apply between international tribunals, *i.e.*, customary international law does not require that one international tribunal follow the judicial precedent of another tribunal in dealing with questions of international law.[156] Moreover, depending on the international tribunal, a tribunal may not be bound by its prior decisions. Some international courts, however, may adhere to their own prior decisions in resolving a case absent a sufficiently persuasive reason to reconsider the point of law.[157]

1.9.2 Legal Writings of Highly Qualified Publicists. The writings "of the most highly qualified publicists" have sometimes been used as a subsidiary means of determining the rules of international law.[158] For example, classical publicists, such as Hugo Grotius and Emmerich de Vattel, and recognized scholars, such as Francis Lieber and Hersch Lauterpacht, have been widely cited and relied upon as practitioners have sought to interpret and apply the law of war. The standard for whose writings should be relied upon is high, and writings are only as authoritative as the evidence upon which they are based. The writings should only be relied upon to the degree they accurately reflect existing law, rather than the author's views about what the law should be.[159]

1.10 LEGAL FORCE OF THE LAW OF WAR

This section addresses the technical legal force of the law of war under international and U.S. domestic law. As a matter of policy, DoD personnel may be required to adhere to law of war rules, even where the rules do not technically apply as a matter of law.[160]

1.10.1 Legal Force of the Law of War Under International Law. The technical force of a law of war rule depends on whether it takes the form of a treaty or customary international law.

1.10.1.1 *Legal Force of Treaties Among States*. Under international law, every treaty in force is binding upon the Parties to it and must be performed by them in good faith.[161]

A treaty enters into force for a State after, *inter alia*, it has provided its consent to be bound by the treaty.[162] In some cases, the terms of a treaty may cause it to expire, and in other cases, States may withdraw from a treaty.[163] In some cases, a reservation may also modify the obligations imposed by a treaty on that State.[164]

A treaty does not create either obligations or rights for a third State without its consent.[165]

Thus, a treaty generally would not be binding on non-Parties to the treaty or create rights or obligations for a non-Party to the treaty with respect to a Party to the treaty. Instead, a treaty only creates law (*i.e.*, rights that may be invoked) as between the States that are Parties to it.[166]

1.10.1.2 *Legal Force of Customary International Law Among States.* The customary law of war generally binds all States. However, States that have objected to a customary international law rule during its development are not bound by that rule.[167]

1.10.1.3 *Predominately Inter-State Nature of International Obligations.* International obligations are generally viewed as running to other States, although individuals may have responsibility under international law. Traditionally, international law has governed relations between States, although over time it has increasingly regulated the relationships between States and persons. Under the traditional view, a State's international law obligations run to other States, even when the obligations relate to an individual (*e.g.*, by protecting that individual), such that individuals' "place in international life depends largely on their status as nationals of states."[168]For example, the 1949 Geneva Conventions and the customary law of war do not provide a private right for individuals to claim compensation directly from a State for violations of the law of war; rather, such claims are made by other States.[169]

International law has long prescribed certain rules regulating the conduct of individuals.[170]Under international law, there may be responsibility for individuals, apart from State responsibility.[171]

1.10.1.4 *Force of International Law Notwithstanding a State's Domestic Law.* A State's domestic law does not justify that State's noncompliance with an international obligation as a matter of international law.[172]Similarly, the fact that a State's domestic law does not provide for a penalty with respect to a violation of international law does not relieve a person from responsibility for that act under international law.[173]

1.10.2 Force of the Law of War Under U.S. Domestic Law. The specific legal force of a law of war rule under U.S. domestic law may depend on whether that rule takes the form of a self-executing treaty, non-self-executing treaty, or customary international law. Longstanding DoD policy has been to require DoD personnel to comply with the law of war obligations of the United States.[174]

Even if a violation of a rule is not directly punishable under U.S. law, a variety of tools in U.S. domestic law may be used to enforce a law of war obligation of the United States. For example, a violation of a law of war obligation may be made punishable through implementation of the obligation in military instructions, regulations, and procedures.[175]

1.10.2.1 *Force of Self-Executing and Non-Self-Executing Treaties Under U.S. Domestic Law.* Under domestic law, treaties to which the United States is a Party are part of U.S. law.[176]

The terms "self-executing" and "non-self-executing" may be used to explain how a treaty is to take effect in U.S. domestic law. A treaty may be classified as a self-executing treaty that "operates of itself, without the aid of any legislative provision," or as a non-self-executing treaty that would require "that the Legislature must execute the contract before it can become a rule for the Court."[177]

1.10.2.2 *Force of Customary International Law Under U.S. Domestic Law.* The customary law of war is part of U.S. law insofar as it is not inconsistent with any treaty to which the United States is a Party, or a controlling executive or legislative act.[178]

1.11 *JUS AD BELLUM* The law of war has been categorized into *jus ad bellum* (law concerning the resort to force) and *jus in bello* (law concerning conduct during war).[179]Although *jus ad bellum* is an essential part of the law of war to consider in the political process of whether to resort to military force, this manual focuses on *jus in bello*.[180]Although *jus in bello* rules generally operate independently of whether a side has comported with *jus ad bellum* in the resort to force, parts of *jus ad bellum* are relevant to *jus in bello*.[181]

This section provides a brief overview of some basic aspects of *jus ad bellum*. *Jus ad bellum* issues might raise questions of national policy that, in the Executive Branch, would be decided by the President. In U.S. practice, legal advice provided to national-level principal officials on such issues generally would need to be addressed through interagency discussions coordinated by the legal adviser to the National Security Council, including consultation and coordination among senior counsel of relevant U.S. departments and agencies.

1.11.1 *Jus ad Bellum* Criteria. Certain *jus ad bellum* criteria have, at their philosophical roots, drawn from principles that have been developed as part of the Just War Tradition.[182]

These principles have included:

- a competent authority to order the war for a public purpose;
- a just cause (such as self-defense);
- the means must be proportionate to the just cause;
- all peaceful alternatives must have been exhausted; and
- a right intention on the part of the just belligerent.[183]

These principles may be reflected in modern law of war rules. For example, the Charter of the United Nations recognizes the inherent right of States to use force in individual or collective self-defense—a just cause for military action.[184]

These principles have also been incorporated into military doctrine.[185]

1.11.1.1 *Competent Authority (Right Authority) to Wage War for a Public Purpose*. One longstanding criterion for a just war is that war must be ordered by a competent authority for a public purpose. This *jus ad bellum* principle (sometimes called *right authority*) acknowledges that the resort to military force is a prerogative of the State.[186]

The criterion that war must be ordered by a competent authority for a public purpose is reflected in the requirement that armed groups must belong to a State to receive the privileges of combatant status.[187]This criterion is also reflected in the general denial to private persons of the entitlement to the privileges of combatant status.[188]This criterion is also reflected in the condemnation under international law of certain types of private acts of hostility (such as piracy or terrorism) outside the context of patriotic resistance against an enemy State during international armed conflict.[189]

1.11.1.2 *The Means Must Be Proportionate to the Just Cause (Proportionality – Jus ad Bellum). Proportionality* involves a weighing of the contemplated actions with the justification for taking action.[190]For example, the proportionality of the measures taken in self-defense is to be judged according to the nature of the threat being addressed.[191]Force may be used in self-defense, but only to the extent that it is required to repel the armed attack and to restore the security of the party attacked.[192]As an illustration, assessing the proportionality of measures taken in self-defense may involve considerations of whether an actual or imminent attack is part of an ongoing pattern of attacks or what force is reasonably necessary to discourage future armed attacks or threats thereof.[193]

The *jus ad bellum* criterion of proportionality is different from the *jus in bello* principle of proportionality in conducting attacks.[194]These concepts should not be confused with one another.[195]

1.11.1.3 *All Peaceful Alternatives Must Have Been Exhausted (Necessity – Jus ad Bellum)*. The *jus ad bellum* condition of necessity requires that no reasonable alternative means of redress are available.[196]For example, in exercising the right of self-defense, diplomatic means must be exhausted or provide no reasonable prospect of stopping the armed attack or threat thereof.[197]

The *jus ad bellum* criterion of *necessity* is different from the *jus in bello* concept of *military necessity*.[198]

1.11.2 U.N. Charter Framework and the U.N. Security Council. The Charter of the United Nations provides the modern treaty framework for *jus ad bellum*. Under the Charter of the United Nations, the U.N. Security Council has primary responsibility for the maintenance of international peace and security.[199]The U.N. Security Council may determine the existence of any threat to the peace, breach of the peace, or act of aggression, and may decide what measures shall be taken under the Charter to maintain or restore international peace and security.[200]For example, the U.N. Security Council may recognize that a State is acting lawfully in self-defense or that another State is the aggressor in an armed conflict.[201]In addition, the U.N. Security Council may authorize the use of military force.[202]

1.11.2.1 *U.N. Member State Obligations With Respect to U.N. Security Council Decisions*. Members of the United Nations have agreed to give the United Nations every assistance in any action it takes in accordance with the Charter and to refrain from giving assistance to any State against which the United Nations is taking preventive or enforcement action.[203]

Members of the United Nations have agreed to accept and carry out the decisions of the U.N. Security Council in accordance with the Charter.[204]They have also agreed to join in affording mutual assistance in carrying out the measures decided upon by the U.N. Security Council.[205]

Moreover, in the event of a conflict between the obligations of the Members of the United Nations under the Charter and their obligations under any other international agreement, their obligations under the Charter prevail.[206]

1.11.3 Prohibition on Certain Uses of Force. Under Article 2(4) of the Charter of the United Nations, "[a]ll Members shall refrain in their international relations from the threat or use of force against the territorial integrity or political independence of any State, or in any other manner inconsistent with the Purposes of the United Nations.[207]Numerous other treaties also reflect these prohibitions on the threat or use of force.[208]

The resort to force must have a legal basis in order not to violate these prohibitions. The legality of the use of force must be assessed in light of the particular facts and circumstances at issue.[209]

1.11.3.1 *Aggression*. Aggression is the most serious and dangerous form of the illegal use of force.[210]Not every act of illegal use of force prohibited by Article 2(4) of the Charter constitutes aggression.[211]Initiating a war of aggression is a serious international crime.[212]

U.N. General Assembly Resolution 3314 suggested considerations that the Security Council should bear in mind in determining whether an act of aggression had occurred.[213]

Although this resolution states basic principles as guidance for such determinations, it recognizes that whether an act of aggression has been committed must be considered in light of all the circumstances of each particular case.[214]

The United States has expressed the view that the definition of the act of aggression in the Kampala amendments to the Rome Statute does not reflect customary international law.[215]

1.11.4 Rationales for the Resort to Force.

1.11.4.1 *Use of Force in Self-Defense*. The right to use force in self-defense is an inherent right of States.[216]

1.11.4.2 *Use of Force Authorized by the U.N. Security Council Acting Under Chapter VII of the Charter of the United Nations*. Chapter VII of the Charter of the United Nations provides that the U.N. Security Council may take such action by air, sea, or land forces as may be necessary to maintain or restore international peace or security, including demonstrations, blockade, and other military operations.[217]

1.11.4.3 *Use of Force With the Consent of the Territorial State*. Military action in the territory of another State is not a violation of Article 2(4)'s prohibition against the use of force against that State where it consents to such military action.218[218]

1.11.4.4 *Humanitarian Intervention*. Violations of law of war treaties applicable to non-international armed conflict generally have not been understood to provide an independent basis for intervening in a State.219[219]

Although the United Kingdom and certain other States have argued that intervention for humanitarian reasons may be a legal basis for the resort to force, the United States has not adopted this legal rationale.[220]Consistent with this view, the United States did not adopt this theory as a legal rationale for NATO's military action to address the humanitarian catastrophe in Kosovo in 1999, but rather expressed the view that such action was justified on the basis of a number of factors.[221]

Military action for humanitarian reasons may, however, be authorized by the U.N. Security Council.[222]

1.11.5 Use of Force in Self-Defense. Article 51 of the Charter of the United Nations provides that "[n]othing in the present Charter shall impair the inherent right of individual or collective self-defence if an armed attack occurs against a Member of the United Nations, until the Security Council has taken measures necessary to maintain international peace and security."[223]

The Charter of the United Nations was not intended to supersede a State's inherent right of individual or collective self-defense in customary international law.[224]

To constitute legitimate self-defense under customary international law, it is generally understood that the defending State's actions must be necessary.[225]For example, reasonably available peaceful alternatives must be exhausted.[226]In addition, the measures taken in self-defense must be proportionate to the nature of the threat being addressed.[227]

1.11.5.1 *Responding to an Imminent Threat of an Attack*. The text of Article 51 of the Charter of the United Nations refers to the right of self-defense "if an armed attack occurs against a Member of the United Nations."[228]Under customary international law, States had, and continue to have, the right to take measures in response to imminent attacks.[229]

1.11.5.2 *Use of Force Versus Armed Attack.* The United States has long taken the position that the inherent right of self-defense potentially applies against any illegal use of force.[230]Others, however, would be inclined to draw more of a distinction between "armed attacks" and uses of force that do not give rise to the right to use force in self-defense.[231]

1.11.5.3 *Use of Force to Protect Nationals Abroad.* A State's right to use force in self-defense may be understood to include the right to use force to protect its nationals abroad.[232]The United States has taken action to protect U.S. nationals abroad when the government of the territory in which they are located was unwilling or unable to protect them.[233]A State need not await actual violence against its nationals before taking such action if an attack against them is imminent.[234]

1.11.5.4 *Right of Self-Defense Against Non-State Actors.* The inherent right of self-defense, recognized in Article 51 of the Charter of the United Nations, applies in response to any "armed attack," not just attacks that originate with States.[235]As with any other exercise of the right of self-defense, actions taken in self-defense against non-State actors must comply with applicable international law.[236]

1.11.5.5 *Right of Collective Self-Defense.* Article 51 of the Charter of the United Nations also recognizes a right of States to engage in collective self-defense with a State that can legitimately invoke its own right of national self-defense. Collective self-defense of a State must proceed with that State's consent, although this consent need not necessarily be expressed in the form of an explicit request.[237]

Some treaties include commitments by States to assist one another in collective self-defense. [238]

1.11.5.6 *Reporting to the U.N. Security Council.* Measures taken in the exercise

of national self-defense shall be immediately reported to the U.N. Security Council.239[239]

II – Principles

Chapter Contents

2.1 INTRODUCTION

Three interdependent principles – *military necessity*, *humanity*, and *honor* – provide the foundation for other law of war principles, such as *proportionality* and *distinction*, and most of the treaty and customary rules of the law of war.

This Chapter briefly addresses certain specific rules to illustrate these foundational principles. For more information about a specific rule, practitioners should refer to the cross-referenced section that addresses that rule.[1]

2.1.1 Legal Principles as Part of International Law. General principles of law common to the major legal systems of the world are a recognized part of international law.[2]Law of war principles have been understood to be included in this category of international law.[3]

2.1.2 Uses of Law of War Principles. Law of war principles provide the foundation for the specific law of war rules. Legal principles, however, are not as specific as rules, and thus interpretations of how principles apply to a given situation may vary.

Law of war principles: (1) help practitioners interpret and apply specific treaty or customary rules; (2) provide a general guide for conduct during war when no specific rule applies; and (3) work as interdependent and reinforcing parts of a coherent system.

2.1.2.1 *Law of War Principles as an Aid in Interpreting and Applying Law of War Rules*. Understanding law of war principles helps practitioners interpret and apply specific law of war rules. For example, the principle of *military necessity* has been incorporated into specific law of war rules.[4]Similarly, the principle of *humanity* can assist in the proper interpretation and application of law of war rules that are based on humanitarian considerations.[5]

2.1.2.2 *Law of War Principles as a General Guide*. When no specific rule applies, the principles of the law of war form the general guide for conduct during war. States have reflected this idea in certain treaty provisions, including the "Martens Clause," which make clear that situations not covered by the treaty remain governed by principles of international law.[6]

The considerable progress States have made in developing specific law of war rules, however, has lessened the need to rely solely on these principles to guide conduct during war.[7]

2.1.2.3 *Law of War Principles as a Coherent System*. Law of war principles work as interdependent and reinforcing parts of a coherent system.

Military necessity justifies certain actions necessary to defeat the enemy as quickly and efficiently as possible.[8]Conversely, *humanity* forbids actions unnecessary to achieve that object.[9]*Proportionality* requires that even when actions may be justified by *military necessity*, such actions not be unreasonable or excessive.[10]*Distinction* underpins the parties' responsibility to comport their behavior with *military necessity*, *humanity*, and *proportionality* by requiring parties to a conflict to apply certain legal categories, principally the distinction between the armed forces and the civilian population.[11]Lastly, *honor* supports the entire system and gives parties confidence in it.[12]

2.2 MILITARY NECESSITY

Military necessity may be defined as the principle that justifies the use of all measures needed to defeat the enemy as quickly and efficiently as possible that are not prohibited by the law of war. *Military necessity* has been defined in military publications,[13] judicial decisions,[14] and scholarly works.[15]

2.2.1 Military Necessity as a Justification. *Military necessity* justifies actions, such as destroying and seizing persons and property.[16] Thus, *military necessity* underlies law of war concepts that explain when persons and property may be the object of attack, *e.g.*, the concepts of "taking a direct part in hostilities"[17] and "military objective."[18]

Military necessity may justify not only violence and destruction, but also alternative means of subduing the enemy.[19] For example, military necessity may justify the capture of enemy persons,[20] or non-forcible measures, such as propaganda and intelligence-gathering.[21]

Military necessity also justifies certain incidental harms that inevitably result from the actions it justifies.[22] The extent to which *military necessity* justifies such harms is addressed by the principle of proportionality.[23]

2.2.2 Military Necessity and Law of War Rules.

2.2.2.1 *Military Necessity Does Not Justify Actions Prohibited by the Law of War. Military necessity* does not justify actions that are prohibited by the law of war. From the late 19th Century through World War II, Germany asserted that *military necessity* could override specific law of war rules (*Kriegsraeson geht vor Kriegsmanier* – "necessity in war overrules the manner of warfare").[24] This view was strongly criticized.[25] Post World War II war crimes tribunals rejected it as well.[26]

Military necessity cannot justify departures from the law of war because States have crafted the law of war specifically with war's exigencies in mind.[27] In devising law of war rules, States considered military requirements.[28] Thus, prohibitions on conduct in the law of war may be understood to reflect States' determinations that such conduct is militarily unnecessary *per se*. The fact that law of war rules are formulated specifically with military requirements in mind has played an important part in the doctrine that the law of war is the *lex specialis* governing armed conflict.[29]

2.2.2.2 *Incorporation of Military Necessity Into Law of War Rules*. Although *military necessity* cannot justify actions that have been prohibited by the law of war, some law of war rules expressly incorporate *military necessity*. For example, certain law of war rules specify that departures from what would otherwise be the rule are permissible when *absolutely* or *imperatively* necessary. In these cases, *military necessity* must not be conflated with mere convenience.[30] Examples of rules incorporating the concept of *absolute* or *imperative* necessity include the following:

• The activities of the representatives or delegates of the Protecting Powers shall only be restricted as an exceptional and temporary measure when this is rendered necessary by imperative military necessities.[31]

• The internment or placing in assigned residence of protected persons may be ordered only if the security of the Detaining Power makes it absolutely necessary.[32]

• If the Occupying Power considers it necessary, for imperative reasons of security, to take safety measures concerning protected persons, it may, at the most, subject them to assigned residence or to internment.[33]

• The seizure or destruction of enemy property must be imperatively demanded by the necessities of war.[34]

Certain law of war rules may direct that persons comply with an obligation, but only to the extent feasible or consistent with *military necessity*. Examples of rules incorporating the concept of *feasibility* or *necessity* include the following:

• Certain affirmative duties to take feasible precautions to reduce the risk of harm to the civilian population and other protected persons and objects.[35]

• Military medical and religious personnel, if their retention is not indispensable, are to be returned to the party to the conflict to which they belong as soon as a road is open for their return and military requirements permit.[36]

• Whenever military considerations permit, POW camps shall be indicated in the day time by the letters PW or PG, placed so as to be clearly visible from the air.[37]

• Should military necessity require the quantity of relief shipments to civilian internees to be limited, due notice thereof shall be given to the Protecting Power and to the International Committee of the Red Cross, or to any other organization giving assistance to the internees and responsible for the forwarding of such shipments.[38]

2.2.3 Applying Military Necessity. *Military necessity* is a difficult concept to define and apply.[39]What is necessary in war may depend closely on the specific facts and circumstances of a given situation, and different people often assess *military necessity* differently. The limited and unreliable nature of information available during war compounds this difficulty in evaluating what is necessary.[40]This difficulty runs throughout the law of war, since *military necessity* is itself important and is an element of many other principles and rules.

The law of war seeks to ameliorate these difficulties in applying *military necessity* by: (1) permitting consideration of the broader imperatives of winning the war as quickly and efficiently as possible; (2) recognizing that certain types of actions are, as a general matter, inherently militarily necessary; and (3) recognizing that persons must assess the *military necessity* of an action in good faith based on the information available to them at the relevant time and that they cannot be judged based on information that subsequently comes to light.

2.2.3.1 *Consideration of the Broader Imperatives of Winning the War*. In evaluating *military necessity*, one may consider the broader imperatives of winning the war as quickly and efficiently as possible and is not restricted to considering only the demands of the specific situation.

This is the case because *military necessity* justifies those measures necessary to achieve the object of war, and the object of war is not simply to prevail, but to prevail as quickly and efficiently as possible.[41]Thus, *military necessity* may consider the broader imperatives of winning the war and not only the demands of the immediate situation. For example, in assessing the military advantage of attacking an object, one may consider the entire war strategy rather than only the potential tactical gains from attacking that object.[42]An interpretation of *military necessity* that only permitted consideration of the immediate situation could prolong the fighting and increase the overall suffering caused by the war.[43]

Some commentators have argued that *military necessity* should be interpreted so as to permit only what is actually necessary in the prevailing circumstances, such as by requiring commanders, if possible, to seek to capture or wound enemy combatants rather than to make them the object of attack.[44]This interpretation, however, does not reflect customary international law or treaty law applicable to DoD personnel.[45]For example, the law of war does not require that enemy combatants be warned before being made the object of attack, nor does the law of war require that enemy combatants be given an opportunity to surrender before being made the object of attack.[46]Moreover, the law of war may justify the use of overwhelming force against enemy military objectives.[47]

2.2.3.2 *Recognizing Certain Types of Actions as Generally Inherently Militarily Necessary*. The law of war recognizes that certain types of actions are, as a general matter, inherently militarily necessary.[48]For example, attacking enemy combatants is generally lawful.[49]Similarly, the internment of enemy POWs is generally lawful.[50]Such rules may be viewed as an example of how, when specific rules are applicable, there is less need to resort to fundamental law of war principles as a general guide for conduct during war.[51]

2.2.3.3 *Good Faith Evaluation of Military Necessity Based on the Available Information*. In what is sometimes called the "Rendulic Rule," the law of war recognizes that persons must assess the *military necessity* of an action based on the information available to them at that time; they cannot be judged based on information that subsequently comes to light.[52]

2.3 HUMANITY

Humanity may be defined as the principle that forbids the infliction of suffering, injury, or destruction unnecessary to accomplish a legitimate military purpose.[53]

2.3.1 Humanity as a Prohibition. Although *military necessity* justifies certain actions necessary to defeat the enemy as quickly and efficiently as possible, *military necessity* cannot justify actions not necessary to achieve this purpose, such as cruelty or wanton violence.[54]

Moreover, once a military purpose has been achieved, inflicting more suffering is unnecessary and should be avoided. For example, if an enemy combatant has been placed *hors de combat* (*e.g.*, incapacitated by being severely wounded or captured), no military purpose is served by continuing to attack him or her.[55]Thus, the principle of *humanity* forbids making enemy combatants who have been placed *hors de combat* the object of attack.[56]Similarly, the principle of *humanity* has been viewed as the source of the civilian population's immunity from being made the object of attack because their inoffensive and harmless character means that there is no military purpose served by attacking them.[57]

2.3.1.1 *Relationship Between the Principles of Humanity and Military Necessity*. Humanity is related to military necessity, and these principles logically complement one another.

Humanity may be viewed as the logical inverse of the principle of military necessity. If certain necessary actions are justified, then certain unnecessary actions are prohibited. The principle of humanity is an example of how the concept of necessity can function as a limitation as well as a justification.[58]

Because humanity forbids those actions that are unnecessary, the principle of humanity is not in tension with military effectiveness, but instead reinforces military effectiveness.[59]

2.3.2 Humanity and Law of War Rules. *Humanity* animates certain law of war rules, including:

- fundamental safeguards for persons who fall into the hands of the enemy;[60]
- protections for the civilian population and civilian objects;[61]
- protections for military medical personnel, units, and transports;[62]
- prohibitions on weapons that are calculated to cause superfluous injury;[63]and
- prohibitions on weapons that are inherently indiscriminate.[64]

The principle of *humanity* may help interpret or apply these and other law of war rules. For example, the requirement that POWs be interned only in premises located on land has been understood not to prohibit POW detention aboard ships pending the establishment of suitable facilities on land, if detention aboard ships provides the most appropriate living conditions for POWs.[65]Similarly, the U.S. reservation to CCW Protocol III on Incendiary Weapons makes clear that U.S. forces may use incendiary weapons against military objectives located in concentrations of civilians where it is judged that such use would cause fewer casualties and/or less collateral damage than alternative weapons.[66]

2.4 PROPORTIONALITY

Proportionality may be defined as the principle that even where one is justified in acting, one must not act in a way that is unreasonable or excessive.[67]

Proportionality has also been viewed as a legal restatement of the military concept of economy of force.[68]

2.4.1 Proportionality as a Limit on the Exercise of a Right.

2.4.1.1 *Justification in Acting*. *Proportionality* is implicated in cases in which one is justified in acting. In *jus in bello*, for example, *military necessity* justifies making military objectives, such as enemy combatants, the object of attack. So, an attack on enemy combatants that incidentally damages civilian property would raise *proportionality* considerations. On the other hand, where there is no justification for acting, such as unlawful attacks directed against the civilian population, a *proportionality* analysis would not be necessary to reach the conclusion that the attack would be unlawful.

2.4.1.2 *Unreasonable or Excessive*. *Proportionality* generally weighs the justification for acting against the expected harms to determine whether the latter are disproportionate in comparison to the former. In war, incidental damage to the civilian population and civilian objects is unfortunate and tragic, but inevitable.[69]Thus, applying the principle of proportionality in conducting attacks does not require that no incidental damage result from attacks.[70]Rather, this principle creates obligations to refrain from attacks in which the expected harm incidental to such attacks would be excessive in relation to the concrete and direct military advantage anticipated to be gained and to take feasible precautions in planning and conducting attacks to reduce the risk of harm to civilians and other persons and objects protected from being made the object of attack.[71]

Under the law of war, judgments of *proportionality* often involve difficult and subjective comparisons.[72]Recognizing these difficulties, States have declined to use the term "proportionality" in law of war treaties because it could incorrectly imply an equilibrium between considerations or suggest that a precise comparison between them is possible.[73]

2.4.2 Examples Where Proportionality Is Reflected in Law of War Rules. The principle of proportionality is reflected in many areas in the law of war. *Proportionality* is often used to refer to the *jus in bello* prohibition on attacks expected to cause excessive incidental harm.[74]However, the principle of proportionality described in a more general sense in this section also underlies the affirmative duties to take feasible precautions for the protection of civilians and other protected persons and objects.[75]These obligations apply both in conducting attacks and to the party subject to attack.[76]*Proportionality* also plays a role in assessing whether weapons are prohibited because they are calculated to cause unnecessary suffering.[77]

In *jus ad bellum*, *proportionality* requires that the State's means in resorting to force be proportionate to its just cause in using force, such as the threat that the State seeks to address.[78] *Proportionality* is also a requirement for reprisals, which must respond in a proportionate manner to the preceding illegal act by the party against which they are taken.[79]

2.5 DISTINCTION

Distinction, sometimes called *discrimination*, obliges parties to a conflict to distinguish principally between the armed forces and the civilian population, and between unprotected and protected objects.[80]

Distinction may be understood as encompassing two sets of reinforcing duties. Parties to a conflict must apply a framework of legal classes for persons and objects by: (1) discriminating in conducting attacks against the enemy; and (2) distinguishing a party's own persons and objects.

2.5.1 Distinction as a Framework of Legal Classes. *Distinction* requires parties to a conflict to apply a framework of legal classes for persons and objects, each class characterized by different rights, duties, and liabilities.

Principally, *distinction* seeks to separate the armed forces and the civilian population.[81] There are, however, certain special cases, such as military medical and religious personnel, who may be treated like a combatant for one purpose (*e.g.*, detention) but like a civilian for another purpose (*e.g.*, not being made the object of attack).[82] However, for any particular legal purpose, a person may not claim the distinct rights afforded to both combatants and civilians at the same time.[83]

2.5.2 Discriminating in Conducting Attacks Against the Enemy. *Distinction* requires parties to a conflict to discriminate in conducting attacks against the enemy.[84] On the one hand, consistent with *military necessity*, parties may make enemy combatants and other military objectives the object of attack.[85] On the other hand, consistent with *humanity*, parties may not make the civilian population and other protected persons and objects the object of attack.[86]

Moreover, persons using force must discriminate between legitimate and illegitimate objects of attack in good faith based on the information available to them at the time.[87]

2.5.3 Distinguishing a Party's Own Persons and Objects. *Distinction* enjoins the party controlling the population[88] to use its best efforts to distinguish or separate its military forces and war-making activities from members of the civilian population to the maximum extent feasible so that civilian casualties and damage to civilian objects incidental to attacks on military objectives will be minimized as much as possible.[89]

Parties to a conflict must: (1) take certain measures to help ensure that military forces and civilians can be visually distinguished from one another; (2) physically separate, as feasible, their military objectives from the civilian population and other protected persons and objects; and (3) refrain from the misuse of protected persons and objects to shield military objectives.

2.5.3.1 *Measures to Help Ensure That Military Forces and Civilians Are Visually Distinguishable From One Another*. Parties to a conflict must take certain measures, in offense or defense, to help ensure that military forces and civilians can be visually distinguished from one another.

First, parties to a conflict must not disguise their armed forces as civilians or as other protected categories of persons in order to kill or wound opposing forces.[90] Second, other rules obligate parties to mark protected persons and objects to help ensure they receive the protections of that status.[91] Third, certain rules encourage parties to a conflict to identify certain persons and objects as unprotected. For example, during international armed conflict, members of organized resistance movements must, *inter alia*, wear fixed distinctive signs visible at a distance and carry arms openly to distinguish themselves from the civilian population in order for members of their group to receive POW status.[92]

2.5.3.2 *Feasible Measures to Separate Physically a Party's Own Military Objectives From the Civilian Population and Other Protected Persons and Objects*. *Distinction* also creates obligations for parties to a conflict to take feasible measures to separate physically their own military objectives from the civilian population and other protected persons and objects.[93]

For example, it may be appropriate to evacuate civilians from danger areas.[94] Similarly, if feasible, military commanders should avoid placing military objectives in densely populated areas.[95] In addition, it may be appropriate to establish zones where civilians and other protected persons may seek refuge.[96]

2.5.3.3 *Refrain From the Misuse of Protected Persons and Objects to Shield Military Objectives*. Parties to a conflict must refrain from the misuse of civilians and other protected persons and objects to shield their own military objectives.[97] For example, it is prohibited to take hostages or otherwise to endanger deliberately protected persons or objects for the purpose of deterring enemy military operations. Misusing protected persons and objects to shield military objectives also offends *honor* because it constitutes a breach of trust with the enemy and thus undermines respect for the law of war.[98]

2.5.4 Misconceptions About Distinction. *Distinction* seeks to ensure that protected and unprotected categories are distinct from one another, not distinct in the abstract. For example, using camouflage is consistent with *distinction* because foliage is not a protected category and because civilians generally do not wear camouflage.[99] Similarly, U.S. forces have worn nonstandard uniforms to blend with local forces while remaining distinct from the civilian population.[100]

Distinction addresses the different rights, duties, and liabilities of the categories; it does not require that a particular person or object fall within a particular category. For example, the principle of *distinction* does not prohibit an otherwise civilian object from being used for military purposes, thereby turning it into a military objective.[101] However, if such an object were seized from the enemy, such seizure would have to have been imperatively demanded by the necessities of war.[102] Similarly, persons with medical training or who provide medical care on the battlefield are not necessarily military medical personnel and need not be identified as such.[103] Rather, a State may reserve the ability to use these persons as combatants by refraining from designating them as exclusively engaged in medical activities.

2.5.5 Reinforcing Duties – Discriminating in Conducting Attacks and Distinguishing a Party's Own Persons and Objects. Discriminating in conducting attacks against the enemy and distinguishing a party's own persons and objects reinforce one another. A party is not relieved of its obligations to discriminate in conducting attacks by the failures of its adversary to distinguish its military objectives from protected persons and objects.[104] Nonetheless, the practical ability of a party to a conflict to discriminate in conducting attacks often depends on the degree to which its enemy has distinguished its military objectives from its protected persons and objects. For example, if enemy forces intermingle with civilians, then a party may be less able to avoid incidentally harming the civilian population. In addition, the degree to which an enemy force in fact discriminates in conducting attacks may affect whether a party distinguishes its protected persons and objects from its military objectives. For example, if enemy forces do not respect the red cross emblem, but instead specifically attack persons wearing it, then the party receiving these attacks is less likely to distinguish its medical personnel and transports.[105] Similarly, if insurgents seek to attack civilians in a non-international armed conflict, positioning military forces near the civilian population may be essential to protecting the civilian population.[106]

2.6 HONOR

Honor demands a certain amount of fairness in offense and defense and a certain mutual respect between opposing military forces.[107]

2.6.1 Honor – Background and Notes on Terminology. *Honor* has been vital to the development of the law of war, which was preceded by warriors' codes of ethical behavior.[108] U.S. military canons of personal conduct continue to emphasize the importance of honor as a core value. Honor as a core value and other ethical norms continue to be vital as a medium for the implementation of the law of war.[109]

Honor has also been called *chivalry*. *Chivalry* is often associated with a specific historical context—a code of behavior for knights in Europe during the Middle Ages. Rather than refer to this specific context, the term *honor* is used in this manual to indicate more clearly that the law of war principle of *honor* draws from warriors' codes from a variety of cultures and time periods.

2.6.2 Certain Amount of Fairness in Offense and Defense. *Honor* requires a certain amount of fairness in offense and defense. *Honor* forbids resort to means, expedients, or conduct that would constitute a breach of trust with the enemy.[110]

2.6.2.1 *Acceptance That Belligerent Rights Are Not Unlimited.* In requiring a certain amount of fairness in offense and defense, *honor* reflects the principle that parties to a conflict must accept that certain limits exist on their ability to conduct hostilities. For example, parties to the conflict must accept that the right of belligerents to adopt means of injuring the enemy is not unlimited.[111] Here, *honor* does not address what those limits are so much as requires that parties accept that there are legal limits that govern their conduct of hostilities.[112] This acceptance is a prerequisite for the existence and operation of the law of war in the way that the principle of *pacta sunt servanda* (treaties are binding on parties and must be performed by them in good faith) provides a necessary foundation for treaties to exist and operate as instruments that are legally binding on States.[113]

Thus, *honor* may be understood to provide a foundation for obligations that help enforce and implement the law of war or special agreements between belligerents during armed conflict.[114]For example, *honor* may be understood to provide the foundation for the requirement for persons to comply with the law of war in good faith.[115]Similarly, POWs are bound to adhere to paroles on their personal honor.[116]

2.6.2.2 *Prohibition on Conduct That Breaches Trust With the Enemy*. In requiring a certain amount of fairness in offense and defense, *honor* also forbids resort to means, expedients, or conduct that would constitute a breach of trust with the enemy. In particular, *honor* requires a party to a conflict to refrain from taking advantage of its opponent's adherence to the law by falsely claiming the law's protections. *Honor* forbids such conduct because it may: (1) undermine the protections afforded by the law of war; (2) impair non-hostile relations between opposing belligerents; and (3) damage the basis for the restoration of peace short of complete annihilation of one belligerent by another.[117]

For example, enemies must deal with one another in good faith in their non-hostile relations.[118]And even in the conduct of hostilities, good faith prohibits: (1) killing or wounding enemy persons by resort to perfidy; (2) misusing certain signs; (3) fighting in the enemy's uniform; (4) feigning non-hostile relations in order to seek a military advantage; and (5) compelling nationals of a hostile party to take part in the operations of war directed against their own country.[119]*Honor*, however, does not forbid parties from using ruses and other lawful deceptions against which the enemy ought to take measures to protect itself.[120]

2.6.3 Certain Mutual Respect Between Opposing Military Forces. *Honor* demands a certain mutual respect between opposing military forces.

Opposing military forces should respect one another outside of the fighting because they share a profession and they fight one another on behalf of their respective States and not out of personal hostility.[121]For example, this principle is reflected in the rule that POWs are legally in the hands of the enemy State and not the individuals or military units who have captured them.[122]

2.6.3.1 *Honor and Rules for POW Captivity*. In demanding a certain mutual respect between opposing military forces, *honor* animates rules that relate to the treatment of POWs. For example, *honor* is one of the foundations for the humane treatment of POWs.[123]

The principle of *honor* is also reflected in rules that require POWs and their captors to treat one another with respect. For example, POWs must be treated with respect for their honor and person.[124]As another example, POWs, with the exception of officers, must salute and show to all officers of the Detaining Power the external marks of respect provided for by the regulations applying in their own forces.[125]In addition, capitulations agreed upon between belligerents must take into account the rules of military honor.[126]

2.6.3.2 *Honor and the Class of Military Forces*. In demanding a certain mutual respect between opposing military forces, *honor* also reflects the premise that military forces are a common class of professionals who have undertaken to comport themselves honorably.[127]

Honor thus animates the rules that determine who qualifies for privileges of combatant status. For example, an armed group must, *inter alia*, be organized under a responsible command and conduct its operations in accordance with the law of war in order for its members to be entitled to POW status during international armed conflict.[128]On the other hand, private persons are generally denied the privileges of combatant status because they do not belong to this class of combatants.[129]

The principle that combatants share a common class has also been a foundation for the trial of enemy combatants by military tribunals.[130]For example, the GPW expresses a preference for POWs to be tried by military courts rather than civilian courts.[131]

III – Application of the Law of War

Chapter Contents

3.1 INTRODUCTION

Whether a particular law of war rule applies to a situation may depend on a variety of issues, such as (1) whether a state of "war," "hostilities," or "armed conflict" exists;[1] (2) whether a party is recognized as a belligerent or as a State; [2] or (3) whether an enemy State has accepted that law of war rule.[3] This Chapter addresses these and other broad issues underlying the application of law of war rules. Later chapters address issues in applying specific rules, including considerations specific to those rules. For example, whether a specific provision of the GPW applies to a particular person may depend on whether that person is entitled to POW status under the GPW.[4]

Many of the legal issues underlying the application of law of war rules may be confusing because they are complex and may appear to result in contradictory legal positions. For example, a State may be at "war" for some purposes, but not for other purposes.[5] The legal character of U.S. military operations may change rapidly. For example, some operations may begin as military operations other than war and later evolve into war, or an international armed conflict may change into a non-international armed conflict.

Although this Chapter addresses how rules apply as a matter of law, DoD practice has often been to act consistently with law of war rules, even in certain cases where these rules might not technically be applicable as a matter of law.

3.1.1 DoD Practice of Applying Law of War Rules Even When Not Technically Applicable. DoD policy and doctrine make clear the importance of compliance with, implementation of, and enforcement of the law of war.[6]

In addition, DoD practice has often been to act consistently with law of war rules, even in certain cases where these rules might not technically be applicable as a matter of law. This means, for example, that DoD practice has often been to act consistently with a particular law of war treaty rule, even if that rule might not apply as a matter of treaty law. Moreover, DoD practice has sometimes been to adhere to the standards in the law of war, even in situations that do not constitute "war" or "armed conflict," because these standards in the law of war reflect legal standards that must be adhered to in all circumstances (*i.e.*, whether there is a state of armed conflict or not).

3.1.1.1 *Reasons for Acting Consistent With a Treaty Rule, Even Though the Treaty Does Not Apply*. DoD practice has often been to act consistently with a treaty rule, even if that rule might not apply as a matter of treaty law. First, it may be appropriate to act consistently with the terms of a treaty (even as applied in dealings with a non-Party to a treaty) because the general principles of the treaty have been determined to be declaratory of customary international law.[7] In such cases, practice that is consistent with the treaty's terms with regard to a particular matter likewise would be in compliance with applicable customary international law. In addition, it may be important to act consistently with the terms of the treaty because the treaty represents "modern international public opinion" as to how military operations should be conducted.[8] Other policy considerations, including efficacious training standards or close relations with coalition partners, may lead to a policy decision that DoD practice should be consistent with a particular law of war treaty rule, even if that rule does not apply to U.S. forces as a matter of law.

3.1.1.2 *Applying Law of War Standards as Reflecting Minimum Legal Standards*. DoD practice also has been to adhere to certain standards in the law of war, even in situations that do not constitute "war" or "armed conflict," because these law of war rules reflect standards that must be adhered to in all circumstances. Applying these standards provides assurance that the standards adhered to equal or exceed those required. This practice has been reflected in DoD policies to comply with the law of war, or to comply with the spirit and principles of the law of war, during military operations outside the context of armed conflict.[9]

Certain prohibitions and certain other rules in the law of war that reflect customary international law have been described as reflecting "elementary considerations of humanity."[10]

These "elementary considerations of humanity" have been understood to be "even more exacting in peace than in war."[11] Thus, these legal standards, at a minimum, must be adhered to in all circumstances.

In particular, Common Article 3 of the 1949 Geneva Conventions reflects a minimum yardstick of humane treatment protections for all persons taking no active part in hostilities, including members of armed forces who have laid down their arms and those placed *hors de combat* by sickness, wounds, detention, or any other cause.[12] The standards in Common Article 3 of the 1949 Geneva Conventions may be understood to reflect minimum humane treatment standards for detainees in any military operation, including those during international or non-international armed conflict or occupation, and those military operations that are not addressed by the law of war. Additional rules will apply depending on the particular context. For example, the United States has supported adherence to the guarantees in Article 75 of AP I during international armed conflict.[13]

3.2 SITUATIONS TO WHICH THE LAW OF WAR APPLIES

Although the law of war is commonly understood as applying to the conduct of hostilities and the protection of war victims, the law of war addresses other situations as well. The law of war establishes:

- rules governing the resort to force (*jus ad bellum*);[14]
- rules between enemies for the conduct of hostilities and the protection of war victims in international and non-international armed conflict;[15]
- rules between belligerents and neutrals;[16]
- rules for military occupation;[17] and
- duties during peacetime that help implement the above rules.[18]

In addition, these rules in the law of war can sometimes be applied by analogy to other contexts.[19]

3.3 STATUS OF THE PARTIES AND THE LAW OF WAR

The law of war distinguishes between: (1) international armed conflicts, *i.e.*, conflicts between States; and (2) other armed conflicts, typically called non-international armed conflicts.

Three situations—unrecognized governments, recognition of belligerency, and national liberation movements—merit further discussion because they may affect whether law of war rules relating to international armed conflict apply.

3.3.1 International Armed Conflict and Non-International Armed Conflict. The law of war treats situations of "war," "hostilities," or "armed conflict" differently based on the legal status of parties to the conflict. If two or more States oppose one another, then this type of armed conflict is known as an "international armed conflict" because it takes place between States. However, a state of war can exist when States are not on opposite sides of the conflict.[20]

These other types of conflict are described as "not of an international character" or "noninternational armed conflict."[21] For example, two non-State armed groups warring against one another or States warring against non-State armed groups may be described as "non-international armed conflict," even if international borders are crossed in the fighting.[22]

The law of war rules applicable to non-international armed conflict are addressed in Chapter XVII.[23] There are a number of important differences between the law applicable to international armed conflict and the law applicable to non-international armed conflict.[24]

3.3.1.1 *Application of* Jus in Bello *Rules Does Not Necessarily Affect the Legal Status of Parties*. Although the legal status of an opponent affects the character of the conflict and what rules apply as a matter of law, the application of *jus in bello* rules does not necessarily affect the legal status of parties to a conflict. For example, a belligerent may, as a policy matter, afford a person POW protections and treatment without affording that person legal status as a POW.[25] Similarly, the application of humanitarian rules, such as those reflected in Common Article 3 of the 1949 Geneva Conventions, towards enemy non-State armed groups does not affect their legal status (*e.g.*, such application does not amount to recognizing the group as lawful belligerents or as the legitimate government of a State).[26]

3.3.1.2 *Mixed Conflicts Between Opposing States and Non-State Armed Groups.*

Rather than viewing a situation as either an international armed conflict or a non-international armed conflict, it may be possible to characterize parts of a conflict as international in character, while other parts of that armed conflict may be regarded as non-international in character.[27] For example, under this view, during a situation involving conflict between a variety of States and non-State armed groups, as between the States, the rules of international armed conflict would apply, while as between the States and non-State armed groups, the rules of non-international armed conflicts would apply.[28]

3.3.2 Unrecognized Governments. Even if a State does not recognize an opponent as the legitimate government of a State, under certain circumstances, rules of international armed conflict may apply to a conflict between a State and a government that it does not recognize. For example, members of the regular armed forces who profess allegiance to a government or authority not recognized by the Detaining Power nonetheless would be entitled to POW status if they fall into the power of the enemy during international armed conflict.[29]

3.3.3 State Recognition of Armed Groups as Belligerents. In certain cases, States have recognized armed groups as belligerents for certain legal purposes.

For the purpose of applying humanitarian rules, recognition of the armed group as having belligerent rights is neither a prerequisite for nor a result of applying humanitarian rules.[30]

3.3.3.1 *Recognition by Outside States of a Rebel Faction as a Belligerent in a Civil War.* In the past, in cases of a major civil war in a State, other States have recognized the rebel faction as a belligerent with the effect of treating the rebels as though they were a State with belligerent rights under the law of neutrality.[31]

Historically, certain conditions had to be satisfied before outside States could recognize a non-State armed group as a belligerent:

- a general state of armed conflict within a territory;
- the armed group occupies and administers a significant portion of national territory;
- the armed group acts under responsible chain of command and respects the laws of war; and
- circumstances exist that make it necessary for outside States to define their attitude toward the conflict.[32]

This doctrine has not been invoked often, especially after the adoption of the Charter of the United Nations, and the unwarranted recognition of an insurgent group likely would, at least, be considered an unfriendly act against a State engaged in hostilities against that group.[33]

3.3.3.2 *Assertion of War Powers by a State Engaged in Hostilities Against a Non- State Armed Group.* Occasionally, a State that has been engaged in hostilities against a non- State armed group has taken actions that have recognized the belligerency of the non-State armed group, at least for certain purposes. For example, President Lincoln's proclamation of a blockade during the U.S. Civil War was viewed as recognizing the existence of a state of war, at least for the purposes of imposing the blockade on foreign vessels seeking to trade with the Confederacy.[34]

3.3.3.3 *Recognition of Friendly Armed Groups as Lawful Belligerents.* In some cases, a State has recognized an armed group that fights alongside it as a belligerent.[35] One reason for such recognition might be to seek to persuade other States that have not recognized the armed group as a belligerent, such as an enemy State, to grant the privileges of combatant status to members of those armed groups.[36] An armed group that has been recognized in this way by a friendly State may be viewed as analogous to an organized resistance movement that belongs to a State that is a party to a conflict.[37]

3.3.4 AP I Provision on National Liberation Movements. AP I treats as international armed conflicts "armed conflicts in which peoples are fighting against colonial domination and alien occupation and against racist regimes in the exercise of their right of self-determination."[38]

The United States has strongly objected to this provision as making the applicability of the rules of international armed conflict turn on subjective and politicized criteria that would eliminate the distinction between international and non-international conflicts.[39] The United States has understood these types of conflicts to be non-international armed conflicts.[40]

Moreover, the United States has not accepted this provision in the context of the CCW.[41] The United States has expressed the view that it would not be appropriate to treat this provision of AP I as customary international law.[42]

3.4 WHEN *JUS IN BELLO* RULES APPLY

Jus in bello treaties often provide that they apply in cases of "declared war or of any other armed conflict," even if the state of war is not recognized by them.[43] This standard has also been understood to result in the application of the customary law of war.[44]

A case of "declared war or any other armed conflict" for the purpose of determining whether parties must comply with *jus in bello* rules may be understood as arising in two ways:

(1) when a party intends to conduct hostilities; or (2) when parties are actually conducting hostilities. "War," "hostilities," and "armed conflict" may be defined differently for other legal purposes.[45] It must be emphasized that the discussion in this section is for the purpose of assessing whether *jus in bello* restrictions apply and not necessarily for other purposes. For example, the fact that *jus in bello* restrictions apply is not determinative of whether a State's actions are lawful under *jus ad bellum*.[46] Similarly, the fact that *jus in bello* restrictions apply is not determinative of whether the permissions that are sometimes viewed as inherent in *jus in bello* rules may be relied upon by a State or non-State actor.[47]

3.4.1 Intent-Based Test for Applying *Jus in Bello* Rules. *Jus in bello* rules apply when a party intends to conduct hostilities.

If a State chooses to go to war, then it is bound by *jus in bello* rules for the conduct of those hostilities.[48] For example, if a State considers it necessary to respond to attacks with military force, then those military operations must comply with *jus in bello* rules.[49]

The fact that the intention to conduct hostilities gives rise to obligations to comply with the law of war is important because law of war obligations must be taken into account even before the fighting actually begins, such as in the planning of military operations.[50] Similarly, certain obligations under the GPW and the GC are triggered by the onset of hostilities, and it may be necessary to implement these obligations even before the fighting actually begins.[51] As another example, the party that is subject to attack is often in a position to take feasible precautions for the protection of the civilian population even before the fighting begins.[52]

3.4.1.1 *Declarations of War or Other Official Recognition by States of a State of Hostilities*. The application of the 1949 Geneva Conventions and other law of war treaties to cases of "declared war" is an example of how *jus in bello* restrictions apply when a party intends to conduct hostilities.[53] Traditionally, a State could create a state of hostilities with another State simply by providing objective evidence of its decision to resort to force through formal declarations that hostilities exist between them.[54]

Although States generally no longer file formal declarations of war with one another, officials may make public statements that are like declarations of war in that they provide notice of a state of hostilities.[55] For example, States may make statements that indicate their view that they are engaged in armed conflict in the context of reporting measures taken in the exercise of their inherent right of self-defense to the U.N. Security Council.[56] Similarly, the authorization by Congress of the use of military force has been interpreted as triggering the application of certain parts of the law of war.[57]

These types of statements concerning *jus ad bellum* may be probative of the applicability of *jus in bello* restrictions. For example, a statement by a State indicating that it had suffered a wrongful attack under *jus ad bellum* would also indicate that the State viewed *jus in bello* restrictions as applicable to its adversary's operations against it and its own military operations against its adversary.[58]

Similarly, statements by States that justify the legality of their actions or assert authority under *jus in bello* rules may also provide evidence that States have the intention of conducting hostilities and that *jus in bello* restrictions apply to the activities that will effectuate those intentions.[59]

3.4.1.2 *Non-State Armed Groups With the Intention of Conducting Hostilities*. A non-State armed group, such as a rebel group, might also intend to conduct hostilities.[60] Non- State armed groups are similarly bound by the restrictions in the law of war for the conduct of hostilities when they intend to conduct hostilities.[61] However, in contrast to States, non-State armed groups lack *competent authority*.[62] Thus, there would not be a basis for non-State armed groups to claim the permissions that may be viewed as inherent in parts of the law of war.[63] For example, members of a non-State group may be subject to prosecution under a State's domestic law for their participation in hostilities against it.[64]

3.4.2 Act-Based Test for Applying *Jus in Bello* Rules. *Jus in bello* rules apply when parties are actually conducting hostilities, even if the war is not declared or if the state of war is not recognized by them.[65] The *de facto* existence of an armed conflict is sufficient to trigger obligations for the conduct of hostilities.[66]

The United States has interpreted "armed conflict" in Common Article 2 of the 1949 Geneva Conventions to include "any situation in which there is hostile action between the armed forces of two parties, regardless of the duration, intensity or scope of the fighting."[67]

3.4.2.1 *Reasons for States to Seek to Deny the Existence of Hostilities*. States have specified that *jus in bello* rules apply even if a state of hostilities is not recognized by them because States have frequently denied that they are at "war." And, in some cases, this denial has resulted in a refusal to comply with *jus in bello* obligations.[68]

There are a variety of reasons why States might deny that they are at "war." For example, government officials may deny that an armed conflict exists: (1) to avoid an escalation in fighting and to facilitate a diplomatic resolution; (2) for reasons of domestic law; (3) to avoid economic costs (such as discouraging commerce and foreign investment); or (4) to avoid appearing to acknowledge the military effectiveness of an opposing force.

In particular, States have often been reluctant to acknowledge that operations by non- State armed groups have triggered the application of Common Article 3 of the 1949 Geneva Conventions.[69] This reluctance often stems from an unwillingness to take any action that could enhance the perceived status of rebels or give any appearance of legitimacy to their actions.[70]

Nonetheless, the application of the humanitarian rules does not affect the legal status of parties to a conflict.[71]

3.4.2.2 *Distinguishing Armed Conflict From Internal Disturbances and Tensions*. In assessing whether *de facto* hostilities exist for the purpose of applying *jus in bello* restrictions, situations of internal disturbances and tensions, such as riots, isolated and sporadic acts of violence, and other acts of a similar nature do not amount to armed conflict.[72]

Any hostile action between the armed forces of different States (*i.e.*, international armed conflict) may readily be distinguished from an "internal disturbance or tension." However, it has been more difficult to distinguish "armed conflict not of an international character" from "internal disturbances or tensions."

"Armed conflict not of an international character" for the purpose of applying the obligations in Common Article 3 of the 1949 Geneva Conventions was not specifically defined in those conventions.[73] There has been a range of views on what constitutes an "armed conflict not of an international character" for this purpose.[74] The intensity of the conflict and the organization of the parties are criteria that have been assessed to distinguish between noninternational armed conflict and "internal disturbances and tensions."[75] A variety of factors have been considered in assessing these criteria and in seeking to distinguish between armed conflict and internal disturbances and tensions.[76]

A helpful rule of thumb may be that where parties are, in fact, engaged in activities that the law of war contemplates (*e.g.*, detention of enemy military personnel without criminal charge, bombardment of military objectives), those activities are subject to the law of war.[77]

3.4.2.3 *Responding to Ordinary Crimes, Including Acts of Terrorism*. States are not required to apply law of war rules when using domestic law enforcement tools to respond to ordinary crimes, including acts of terrorism.[78] For example, States may apply their domestic law and prosecute acts of terrorism by non-State armed groups.[79] States, however, have at times decided to resort to military force to counter a terrorist or similar threat that is beyond the capabilities of ordinary law enforcement to address.[80] If States intend to conduct hostilities, then they are also bound by the applicable *jus in bello* restrictions in the law of war.[81]

Acts of terrorism during armed conflict are prohibited by the law of war.[82]

3.5 RELATIONSHIP BETWEEN *JUS IN BELLO* AND *JUS AD BELLUM*

3.5.1 General Distinction Between *Jus in Bello* and *Jus ad Bellum*. As a general matter, *jus in bello* and *jus ad bellum* address different legal issues and should not be conflated.[83]

Conflating *jus in bello* and *jus ad bellum* risks misunderstanding and misapplying these concepts. For example, in *jus ad bellum*, proportionality refers to the principle that the overall goal of the State in resorting to war should not be outweighed by the harm that the war is expected to produce.[84] However, the principle of proportionality in *jus in bello* generally refers to the obligations to take feasible precautions in planning and conducting attacks and to refrain from attacks in which the expected loss of civilian life, injury to civilians, and damage to civilian objects incidental to the attack would be excessive.[85] Therefore, although a *jus ad bellum* proportionality analysis might consider the harm suffered by enemy military forces in the fighting, a *jus in bello* proportionality analysis would not.[86]

3.5.2 *Jus in Bello* and *Jus ad Bellum* Generally Operate Independently of One Another. One important attribute of rules for conduct during war (*jus in bello*) is that, in general, they operate independently from rules regarding the resort to force (*jus ad bellum*).

3.5.2.1 *Compliance With* Jus in Bello *Is Required Regardless of Compliance With* Jus ad Bellum. States fighting against one another must adhere to rules relating to the conduct of hostilities (*jus in bello*), regardless of whether a State may be considered the aggressor or whether the initial resort to force was lawful under *jus ad bellum*. For example, the 1949 Geneva Conventions require States to undertake to respect and to ensure respect for the conventions in all circumstances.[87] The phrase "in all circumstances" has been interpreted to mean that a Party's obligations to respect and to ensure respect for the 1949 Geneva Conventions applies regardless of whether a Party to the Convention is the aggressor or lawfully using force in selfdefense.[88] Similarly, once an occupation exists in fact, the law of occupation applies, regardless of whether the invasion was lawful under *jus ad bellum*.[89]

3.5.2.2 *Compliance With* Jus ad Bellum *Is Required Regardless of Compliance With* Jus in Bello. Compliance with *jus ad bellum* is required regardless of compliance with *jus in bello*.[90] For example, a State that complies with *jus in bello* rules may nonetheless commit aggression under *jus ad bellum*.[91] In addition, violations of law of war treaties applicable to noninternational armed conflict generally have not been understood to provide a basis in international law for a non-belligerent State to intervene against the State in that conflict.[92]

3.5.2.3 *Rationales for the Independent Operation of* Jus in Bello *and* Jus ad Bellum. The principle that *jus in bello* rules operate independently of *jus ad bellum* rules is based on principles of sovereignty and humanity, as well as practical considerations. If the law of war only protected parties justly resorting to force, then both sides, believing their opponent's cause to be unjust, could consider themselves free to depart from *jus in bello* rules.[93] As a consequence, both sides could deny protections to their opponent, and no one would benefit from the law of war's humanitarian protections.[94] Moreover, there might not be a competent procedure for deciding which, if any, State has unlawfully resorted to force.[95]

The principle that *jus in bello* rules operate independently of *jus ad bellum* rules is also influenced by the fact that it would be unjust to punish individual military members based on *jus ad bellum* considerations when they have no influence on whether their State has resorted to force lawfully under applicable international law.[96]

3.5.3 *Jus in Bello* and *Jus ad Bellum* Are Sometimes Related. Although as a general matter parties must comport with *jus in bello* rules, regardless of *jus ad bellum* considerations, sometimes *jus ad bellum* issues can affect how *jus in bello* rules operate. For example, the *jus ad bellum* principle of *competent authority* (also called *right authority*) acknowledges that the resort to military force is a prerogative of the State.[97]

Competent authority is reflected in the distinction between international armed conflict and noninternational armed conflict; military operations against another State are fundamentally different as a matter of law than military operations against a non-State armed group.[98] *Competent authority* also is reflected in *jus in bello* rules relating to who is entitled to receive the privileges of combatant status. Private persons captured after engaging in hostilities generally are not entitled to receive the privileges of POW status under the law of war because they lack *competent authority*.[99]

As another example, the *jus ad bellum* issue of whether a disputed territory belongs to a State affects whether the law of belligerent occupation applies to that territory because the law of belligerent occupation only applies to territory that belongs to an enemy State.[100]

3.5.4 U.N. Security Council Decisions and *Jus in Bello*. The Charter of the United Nations provides the modern treaty framework for *jus ad bellum*, and assigns important responsibilities to the U.N. Security Council.[101]

In theory, decisions by the Security Council, acting under Chapter VII of the Charter of the United Nations, could create obligations that conflict with, and prevail over, obligations in law of war treaties or customary international law.[102] In practice, however, the U.N. Security Council frequently has affirmed the obligations of States and parties to conflicts to comply with the law of war, including military forces operating pursuant to U.N. Security Council decisions.[103]

The U.N. Security Council also has certain authorities to respond to situations involving violations of the law of war, including establishing commissions of inquiry or authorizing the use of force.[104]

Decisions of the Security Council, however, may alter the obligations of member States of the United Nations under the law of neutrality.[105] In addition, a U.N. Security Council authorization may provide additional authority for an Occupying Power to govern occupied territory.[106]

3.6 RECIPROCITY AND LAW OF WAR RULES

"Reciprocity" sometimes refers to the idea that whether a law of war rule applies to a party to a conflict depends on whether its opponent has accepted and complies with that same rule or a corresponding rule. The requirement to comply with many law of war rules (such as the obligation to treat detainees humanely) does not depend on whether the enemy complies with that rule. Nonetheless, in the law of war, reciprocity may play a role in: (1) whether a rule applies; (2) enforcing a rule; or (3) how a rule operates.

3.6.1 Treaty Provisions That Provide for Reciprocity in the Scope of Application of the Treaty. Considerations of reciprocity – *i.e.*, the degree of confidence as to whether an adversary will, in fact, abide by a certain rule may be a critical factor in the willingness of States to enter into treaty obligations.

Similarly, various treaty provisions also reflect, to varying degrees, the principle that whether a rule is legally binding on a party depends on whether its opponent has accepted and applied that same rule. For example, some law of war treaties have a "general participation clause" – *i.e.*, a clause specifying that the treaty only applies to an armed conflict if all the parties to the armed conflict are also Parties to the treaty.[107] A number of treaties on the law of neutrality have such a clause.[108] Other treaties specify that if both Parties and non-Parties to a treaty are in an armed conflict, then Parties to the treaty remain bound by the treaty in their mutual relations, but not in relation to States that are not Parties to the treaty.[109] Treaties have also provided that if a State in an armed conflict is not a Party to the treaty, but it accepts and applies the treaty's provisions, then the Parties to the treaty are bound by the treaty in relation to that State.[110]

These provisions, however, only determine the application of the treaty as matter of treaty law and not as customary international law. For example, although not all parties to World War II were Parties to Hague IV (thus failing to satisfy the requirements of Hague IV's general participation clause), Hague IV's humanitarian protections were deemed applicable as a matter of customary international law.[111]

3.6.2 Reciprocity in the Enforcement of the Law of War. Reciprocity may be reflected in the enforcement of the law of war. For example, the principle of reciprocity is reflected in the concept of reprisal, which under very limited circumstances permits a belligerent to take action that would otherwise be unlawful in order to remedy an enemy's breach of the law.[112] However, the prohibitions on reprisal in the law of war also may be understood to reflect important limitations on the principle of reciprocity in enforcing the law of war.[113]

Reciprocity is also reflected in the principle of *tu quoque*, which may limit a State's ability to deem unlawful and punish certain conduct by its adversary when that State has chosen

to allow its forces to engage in that same conduct.[114]

Even if the application of a law of war rule does not depend on reciprocity as a matter of law, reciprocity may be important as a practical way of encouraging compliance by the adversary

with the law of war.[115]

3.6.3 Law of War Rules May Incorporate Reciprocity. Apart from affecting whether rules apply, reciprocity may be incorporated into the operation of particular law of war rules. In other words, a law of war rule may operate differently depending upon an opponent's behavior.

3.6.3.1 *Reciprocity – "Golden Rule"*. A principle of reciprocity may be understood to be reflected in law of war rules that reflect the golden rule.[116] For example, the treatment of POWs has been based on the principle that POWs should be treated as the Detaining Power would want its forces held by the enemy to be treated.[117] Similarly, during the process of releasing and repatriating POWs, it is proper to expect that each Party's conduct with respect to the repatriation of POWs will be reasonable and broadly commensurate with the conduct of the other.[118]

3.6.3.2 *Benefits-Burdens Principle in Law of War Rules*. In some cases, the law of war requires that those seeking to obtain certain benefits under the law of war also accept certain burdens as a condition for receiving those benefits.[119] For example, militia and volunteer corps that seek the privileges of combatant status during international armed conflict must meet certain conditions (*e.g.*, conducting their operations in accordance with the laws and customs of war) before receiving such privileges.[120] In addition, hospital ships and coastal rescue craft must not be used for military purposes in order to receive their protection from capture and from being made the object of attack.[121] Similarly, cultural property must not be used for military purposes in order to receive special protection.[122]

As another example, the rights and duties of belligerents and neutrals under the law of neutrality may be understood as correlative or reciprocal. For example, a neutral's valid assertion of its rights may depend on whether it has fulfilled its corresponding neutral duties.[123]

3.6.3.3 *Law of War Duties That Are Reinforced by Corresponding Duties for the Enemy*. Similarly, the ability of a party to comply with a particular duty may be affected by whether its opponent has complied with a corresponding duty. For example, the ability of a party to discriminate in conducting attacks may be affected by whether its adversary has properly distinguished its military objectives from the civilian population and other protected persons and objects.[124]

3.7 APPLYING RULES BY ANALOGY

In some cases, a rule developed specifically for one situation may be a useful and appropriate standard to apply in a different situation. This is sometimes called an application of a rule by analogy. In some cases, there is a treaty requirement to apply rules by analogy; in other cases, it may be appropriate to apply law of war rules by analogy without a treaty-based requirement to do so.

3.7.1 Treaty Requirement to Apply Rules by Analogy. The application of law of war rules by analogy is sometimes required by a treaty provision. For example, under the 1949 Geneva Conventions, neutral or non-belligerent States must apply by analogy the rules relating to the treatment of the wounded and sick and of POWs when interning such persons under their duties of neutrality.[125] Similarly, the GC rules for penal procedures for protected persons in occupied territory apply by analogy to proceedings against internees who are in the national territory of the Detaining Power.[126]

3.7.2 Examples of Law of War Rules and Areas in Which Such Rules Have Been Applied by Analogy. There are other situations in which it may be appropriate to apply law of war rules by analogy, even though there is no treaty requirement to do so. For example, it may be appropriate to apply the GWS and GWS-Sea rules for the respectful treatment and handling of enemy military dead to all persons.[127] Other examples of law of war rules and situations in which it may be appropriate to apply those rules by analogy include: (1) applying *jus in bello* rules in certain situations involving neutral or non-belligerent States; (2) applying law of war rules with a humanitarian purpose in situations outside the context of armed conflict; (3) applying occupation law provisions for the protection of the civilian population in situations not constituting belligerent occupation; and (4) applying certain international armed conflict rules in situations of non-international armed conflict.

3.7.2.1 Jus in Bello *Rules and Situations Involving Neutral or Non-Belligerent States*. Although States developed *jus in bello* rules to address relations between enemies, some *jus in bello* rules may be applied by analogy to other situations, such as relations between a belligerent and a neutral or between co-belligerents. For example, a belligerent might take feasible precautions to protect the civilian population of a neutral or co-belligerent State from its military operations, even though such actions might not be required by the law of war.[128] In addition, although the GC excludes certain persons from the definition of protected person based on their nationality, it may be appropriate to afford such persons the standards of treatment for protected persons.[129]

3.7.2.2 *Law of War Rules and Military Operations Outside the Context of Armed Conflict*. Because law of war rules often reflect elementary considerations of humanity, it may be appropriate to apply such standards to military operations occurring outside the context of armed conflict.[130]

3.7.2.3 *Occupation Law and Situations Not Constituting Belligerent Occupation*. Occupation law may also provide a useful framework for certain situations to which it may not be strictly applicable.[131] For example, it may be appropriate for a State that liberates its ally's territory from enemy control to apply by analogy rules from the law of belligerent occupation to the administration of such territory, pending an agreement with the allied government.[132]

3.7.2.4 *International Armed Conflict Rules and Situations of Non-International Armed Conflict.* Legal rules applicable to international armed conflict may sometimes be applied by analogy to non-international armed conflict.[133]

3.8 END OF HOSTILITIES AND THE APPLICATION OF THE LAW OF WAR

In general, law of war rules for the conduct of hostilities cease to apply when hostilities have ended. However, certain duties that have arisen during hostilities may continue after hostilities have ended, and certain new duties arise at the end of hostilities.

3.8.1 General Cessation of the Application of the Law of War at the End of Hostilities. Hostilities end when opposing parties decide to end hostilities and actually do so, *i.e.*, when neither the intent-based nor act-based tests for when hostilities exist are met.[134] Of course, if the test for the existence of hostilities continues to be met, then hostilities cannot be deemed to have ceased.[135] For example, hostilities may be terminated by:

• an agreement to end hostilities, normally in the form of a treaty of peace;

• unilateral declaration of one of the parties to end the war, provided the other party does not continue hostilities or otherwise decline to recognize the act of its enemy;

• the complete subjugation of an enemy State and its allies; or

• a simple cessation of hostilities.[136]

3.8.1.1 *Agreements to End Hostilities.* Parties to a conflict often have negotiated peace treaties to end hostilities.[137] Armistice agreements, *i.e.*, temporary cease-fires, are negotiated to suspend hostilities.[138] In addition, the U.N. Security Council may require certain steps leading to the end of hostilities.[139]

In drafting and interpreting agreements for the cessation of hostilities, it is important to understand the rules normally applicable to the cessation of hostilities. These agreements may refer to provisions in the Geneva Conventions or other law of war instruments. These agreements may modify or supplement the rules normally applicable to the cessation of hostilities, *e.g.*, by specifying precisely when a legal obligation is triggered or satisfied.

3.8.1.2 *End of Hostilities Absent Written Agreement.* Although States often have concluded agreements to end hostilities, it is possible for hostilities to cease absent a written or formal agreement. For example, an armed conflict may end when a party is fully subjugated.

It may be difficult to determine when an armed conflict has ceased, as opposed, for example, to a lull in hostilities during which opposing forces may simply be reconstituting themselves.[140] Hostilities generally would not be deemed to have ceased without an agreement, unless the conditions clearly indicate that they are not be resumed or there has been a lapse of time indicating the improbability of resumption.[141]

3.8.2 Duties Continuing After Hostilities. Under the law of war, certain duties that have arisen during hostilities may continue after hostilities have ended. For example, POWs are protected by the GPW from the moment they fall into the power of the enemy until their final release and repatriation.[142] Similarly, protected persons whose release, repatriation, or reestablishment may take place after the general close of military operations continue to benefit from the protection of the GC.[143] In addition, duties under occupation law may continue after hostilities have ended.[144]

3.8.3 Duties Arising at the End of Hostilities. Certain obligations are triggered by the end of hostilities. For example, the end of hostilities triggers obligations regarding the marking of minefields, demining, and clearance of unexploded ordnance.[145] In addition, POWs and protected persons, in general, must be released and returned to the party to the conflict to which they belong.[146]

3.9 LAW OF WAR DUTIES ALSO APPLICABLE IN PEACE

Some law of war obligations also apply in peace, *i.e.*, even when a State is not engaged in an armed conflict. For example, States must:

• disseminate information regarding the law of war;[147]

• train their armed forces in accordance with the law of war;[148]

• issue instructions and regulations for their armed forces in conformity with the law of war;[149]

• review the legality of new weapons;[150]

• take appropriate measures to prepare for the safeguarding of cultural property;[151] and

• take other appropriate measures to ensure the implementation and enforcement of law of war treaties.[152]

States that are at peace have obligations under the law of neutrality in relation to States that are at war.[153]

IV – Classes of Persons

Chapter Contents

4.1 INTRODUCTION

This Chapter addresses different classes of persons under the law of war. The law of war has created a framework of classes of persons to help confine the fighting between opposing military forces and thereby to help protect the civilian population from the effects of war.[1] This Chapter addresses issues relating to various classes of people under the law of war including: (1) who is included in the various classes, such as "combatant" and "civilian"; (2) the rights, duties, and liabilities of the persons in each class; and (3) how certain factual categories of persons, such as journalists, police officers, or child soldiers, fall within various classes and are treated under the law of war.

This Chapter briefly addresses specific rules that relate to the rights, duties, and liabilities of persons in the various classes to illustrate and provide an overview of that class. For more information about specific rules, practitioners should refer to the cross-referenced sections that addresses those specific rules.[2]

4.1.1 General Notes on Terminology for Persons in the Law of War. The terms in the law of war that describe different classes of people are often used in confusing and contradictory ways. Although striving to use terms consistently within DoD reduces confusion, understanding the substantive standards that apply to a person in the applicable circumstances is more important than using a particular label or a particular system of classification.

4.1.1.1 *The Same Term Used With Different Meanings*. Sometimes different meanings are given to the same term. For example, someone might be considered a "combatant" in the sense that the person may be made the object of attack, but the person would not necessarily be a "combatant" in the sense that the person is privileged to engage in hostilities.[3] Similarly, one source might use the term "noncombatant" to mean all persons who are not combatants, including persons placed *hors de combat* and civilians.[4] Alternatively, another source might use the term "noncombatant" to refer specifically to persons who are members of the armed forces, but who are not combatants.[5] In the past, some commentators have used "noncombatants" of the armed forces to refer to all members of the armed forces serving in combat service support or sustainment roles.[6] In contemporary parlance, however, the term "noncombatant" should generally be used to mean military medical and religious personnel,[7] but also can include those combatants placed *hors de combat*.[8]

4.1.1.2 *Different Terms Used to Describe the Same Concept*. Different legal sources may use different terms to refer to the same class of people under the law of war. For example, one source might use "belligerent," whereas another source might use "combatant" to refer to the same class of people under the law of war.[9]

4.1.2 Classes and Categories Are Only the Starting Point for Legal Analysis. When analyzing a person's rights, duties, and liabilities under the law of war, it is important to analyze the specific question in light of the applicable facts. Determining what class a person falls into, such as "combatant," "civilian," or "unprivileged belligerent," may be only the first step in a legal analysis. For example, whether a person may be the object of attack, may be detained, is entitled to POW status, or may be punished for their actions are all different questions. Although these questions are often related to one another and associated with the general classes of "combatant" and "civilian," each question requires its own specific analysis. This specific analysis should be done in each case, applying the legal rules to the facts, rather than deriving an answer based on a conclusory labeling of a person as, for example, an "enemy combatant."[10] Indeed, some persons might, for some purposes, be treated like "combatants," but for other purposes be treated like "civilians."[11]

4.2 THE ARMED FORCES AND THE CIVILIAN POPULATION

The law of war has recognized that the population of an enemy State is generally divided into two classes: the armed forces and the civilian population, also sometimes called, respectively, "combatants" and "civilians." This division results from the principle of distinction.[12]

4.2.1 Development of the Distinction Between the Armed Forces and the Civilian Population. A citizen or national of a State that is a party to a conflict, as one of the constituents of a State that is engaged in hostilities, may be subjected to the hardships of war by an enemy State.[13] However, because the ordinary members of the civilian population make no resistance, it has long been recognized that there is no right to make them the object of attack.[14] Thus, States have departed from ancient and medieval practices of war between entire peoples, and instead, as much as possible, have treated war as a contention between the professional military forces of warring States.[15] This separation of the armed forces and the civilian population has greatly mitigated the evils of war.[16]

4.2.2 No Person May Claim the Distinct Rights Afforded to Both Combatants and Civilians at the Same Time. The classes of combatants and civilians have distinct rights, duties, and liabilities; no person may claim the distinct rights afforded both classes at the same time.[17] For example, a person may not claim the combatant's right to attack enemy forces while also claiming the civilian's right not to be made the object of attack.[18]

4.2.3 Mixed Cases. Certain classes of persons do not fit neatly within the dichotomy of the armed forces and the civilian population, *i.e.*, combatants and civilians. Each of these particular classes has some attributes of combatant status and some attributes of civilian status; in certain respects persons in these classes are treated like combatants, but in other respects they are treated like civilians. These classes may be classified into three groups: (1) certain personnel engaged in humanitarian duties; (2) certain authorized supporters of the armed forces; and (3) unprivileged belligerents.

4.2.3.1 *Certain Personnel Engaged Humanitarian Duties*. Certain categories of personnel have humanitarian duties that involve them in hostilities but also entitle them to special protections:

- military medical and religious personnel;[19]
- authorized staff of voluntary aid societies;[20]
- staff of a recognized aid society of a neutral country;[21]
- auxiliary medical personnel;[22] and
- personnel engaged in the protection of cultural property.[23]

4.2.3.2 *Certain Authorized Supporters of the Armed Forces*. Certain categories of persons are not members of the armed forces, but are nonetheless authorized to support the armed forces in the fighting:

• persons authorized to accompany the armed forces, but who are not members thereof;[24]

and

• members of the crews of merchant marine vessels or civil aircraft of a belligerent.[25]

4.2.3.3 *Unprivileged Belligerents*. Unprivileged belligerents generally are subject to the liabilities of both combatant and civilian status, and include:[26]

• persons engaging in spying, sabotage, and similar acts behind enemy lines;[27] and

• private persons engaging in hostilities.[28]

4.3 LAWFUL COMBATANTS AND UNPRIVILEGED BELLIGERENTS

In addition to distinguishing between the armed forces and the civilian population, the law of war also distinguishes between "privileged" and "unprivileged," or "lawful" and "unlawful" combatants. As the Supreme Court has explained: Lawful combatants are subject to capture and detention as prisoners of war by opposing military forces. Unlawful combatants are likewise subject to capture and detention, but in addition they are subject to trial and punishment by military tribunals for acts which render their belligerency unlawful.[29]

"Unlawful combatants" or "unprivileged belligerents" are persons who, by engaging in hostilities, have incurred one or more of the corresponding liabilities of combatant status (*e.g.*, being made the object of attack and subject to detention), but who are not entitled to any of the distinct privileges of combatant status (*e.g.*, combatant immunity and POW status).[30]

4.3.1 "Unprivileged Belligerents" as a Category in Treaty Law. States have, in a few cases, explicitly recognized in treaties certain categories of unprivileged belligerents, such as spies and saboteurs.[31] However, States have generally refrained from explicitly recognizing unprivileged belligerents as a class in treaties in the way that classes of lawful combatants have been defined.[32]

Law of war treaties have been understood to reflect restrictions on the conduct of hostilities by States,[33] and States have been reluctant to conclude treaties to afford unprivileged enemy belligerents the distinct privileges of POW status or the full protections afforded civilians.[34]

Although seldom explicitly recognized as a class in law of war treaties, the category of unprivileged belligerent may be understood as an implicit consequence of creating the classes of lawful combatants and peaceful civilians.[35] The concept of unprivileged belligerency, *i.e.*, the set of legal liabilities associated with unprivileged belligerents, may be understood in opposition to the rights, duties, and liabilities of lawful combatants and peaceful civilians. Unprivileged belligerents include lawful combatants who have forfeited the privileges of combatant status by engaging in spying or sabotage, and private persons who have forfeited one or more of the protections of civilian status by engaging in hostilities.[36]

4.3.2 Combatant – Notes on Terminology.

4.3.2.1 *"Combatant" and "Belligerent"*. "Combatant" and "belligerent" have sometimes been used interchangeably and, in this usage, they generally describe individuals who are not "civilians." "Belligerent," however, has also sometimes used to describe States and to contrast such States with "neutral" or "non-belligerent" States.[37] "Belligerent" has also been used to contrast armed groups that have "belligerent rights" with armed groups that lack such rights, such as "insurgents."[38]

4.3.2.2 *"Lawful," "Privileged," and "Qualified"*. The distinction between "lawful" and "unlawful" combatants has sometimes been called a distinction between "privileged" and "unprivileged" belligerents, *i.e.*, distinguishing between persons who are entitled to the privileges of combatant or belligerent status, and those who are not.[39] This distinction has also sometimes been called a distinction between "qualified" and "unqualified" belligerents, *i.e.*, distinguishing between persons who have met the qualifications to receive the privileges of combatant status and those who have not.[40]

4.3.2.3 *"Combatant" Used Without Modification*. "Combatant" and "belligerent," when used without modification (such as "lawful" or "unlawful," or "privileged" or "unprivileged"), have often referred implicitly to lawful or privileged combatants.[41]

However, in some cases, "combatant" or "belligerent" has been used to refer to all persons who engage in hostilities, without taking a position as to whether they are entitled to receive the privileges of combatant status.

4.3.2.4 *General Usage of "Combatant" in This Manual*. This manual generally uses "combatant" to refer implicitly to lawful or privileged combatants.

This manual generally uses the term "unprivileged belligerent" (instead of, *e.g.*, "unlawful combatant," "unlawful belligerent," "unprivileged combatant," etc.) to refer to persons who are subject to one or more of the liabilities of combatant status, but are not entitled to receive its distinct privileges.

4.3.3 Types of Lawful Combatants. Three classes of persons qualify as "lawful" or "privileged" combatants:

• members of the armed forces of a State that is a party to a conflict, aside from certain categories of medical and religious personnel;[42]

• under certain conditions, members of militia or volunteer corps that are not part of the armed forces of a State, but belong to a State;[43] and

• inhabitants of an area who participate in a kind of popular uprising to defend against foreign invaders, known as a *levée en masse*.[44]

4.3.4 Types of Unprivileged Belligerents. Unprivileged belligerents may generally be classified into two categories that may be distinguished from one another by the presence or absence of State authorization: • persons who have initially qualified as combatants (*i.e.*, by falling into one of the three categories mentioned above), but who have acted so as to forfeit the privileges of combatant status by engaging in spying or sabotage;[45] and

• persons who never meet the qualifications to be entitled to the privileges of combatant status, but who have, by engaging in hostilities, incurred the corresponding liabilities of combatant status (*i.e.*, forfeited one or more of the protections of civilian status).[46]

These two categories of unprivileged belligerents generally receive the same treatment.[47] However, the distinction that the first category has State authorization, while the second category does not, may be important and create different legal results. For example, the combatant who spies regains the entitlement to the privileges of combatant status upon returning to friendly lines, but the private person who spies cannot regain a status to which the person was never entitled.[48] Similarly, acts of unprivileged belligerency on the high seas may constitute piracy, a crime under international law, although similar acts by persons acting under State authority, even if they were not members of the armed forces, could not constitute piracy.[49]

4.4 RIGHTS, DUTIES, AND LIABILITIES OF COMBATANTS

Combatants have a special legal status, *i.e.,* certain rights, duties, and liabilities. As discussed below, combatants may engage in hostilities and are liable to being made the object of attack by enemy combatants. Combatants must conduct their operations in accordance with the law of war. They have the right to POW status if they fall into the power of the enemy during international armed conflict. Combatants have legal immunity from domestic law for acts done under military authority and in accordance with the law of war.

4.4.1 Combatants – Conduct of Hostilities. In general, combatants may engage in hostilities and may be made the object of attack by enemy combatants.[50] However, combatants placed *hors de combat* must not be made the object of attack.[51]

Combatants must conduct their operations in accordance with the law of war. For example, combatants must take certain measures to distinguish themselves from the civilian population.[52] Combatants also may not kill or wound the enemy by resort to perfidy.[53]

Combatants may make enemy combatants and other military objectives the object of attack, but may not make the civilian population and other protected persons and objects the object of attack.[54] Combatants must exercise due regard to reduce the risk of incidental harm to the civilian population and other persons and objects that may not be made the object of attack.[55] Chapter V addresses in detail the rules that combatants must follow in the conduct of hostilities.

4.4.2 Combatants – POW Status During Detention. Combatants are liable to capture and detention by enemy combatants, but are entitled to POW status when they fall into the power of the enemy during international armed conflict. POWs, like all detained individuals, must be treated humanely.[56] In addition, POWs are afforded a variety of privileges in detention in accordance with the GPW, such as camp canteens, advances of pay, and permission to wear their badges of rank, nationality, or decorations.[57] POWs also have duties in detention, such as identifying themselves to their captors,[58] and they are subject to the laws, regulations, and orders of the Detaining Power.[59] Chapter IX addresses in detail the treatment of POWs and their duties.

In general, POWs shall be released and repatriated without delay after the cessation of active hostilities.[60] However, seriously wounded, injured, or sick POWs should be returned before the end of hostilities.[61] In addition, after the hostilities have ended, certain POWs may be held in connection with criminal proceedings.[62]

In general, combatants retain their right to POW status and treatment, even if they are alleged to have committed crimes before capture.[63] For example, POWs are entitled to a variety of rights in relation to judicial proceedings against them.[64] In addition, POWs serving disciplinary punishment shall continue to receive the benefits of the GPW, except insofar as these benefits are necessarily rendered inapplicable by the mere fact that the POW is confined.[65] Combatants captured while engaged in spying or sabotage forfeit their entitlement to POW status.[66] In cases of doubt as to whether a detainee is entitled to POW status, that person should be afforded the protections of POW status until their status has been determined by a competent tribunal.[67]

4.4.3 Combatants - Legal Immunity From a Foreign State's Domestic Law. International law affords combatants a special legal immunity from the domestic law of the enemy State for their actions done in accordance with the law of war.[68] This legal immunity is sometimes called the "combatant's privilege" or "combatant immunity." This means that a combatant's "killing, wounding, or other warlike acts are not individual crimes or offenses,"[69] if they are done under military authority and are not prohibited by the law of war.[70] Similarly, a combatant's warlike acts done under military authority and in accordance with the law of war also do not create civil liability.[71]

Combatants lack legal immunity from an enemy State's domestic law for acts that are prohibited by the law of war.[72] Also, combatants lack legal immunity from an enemy State's domestic law while engaging in spying or sabotage.[73] Combatants, however, must receive a fair and regular trial before any punishment.[74]

4.4.3.1 *Combatants - Legal Immunity and POW Status*. The "combatant's privilege" from liability under domestic law has been associated with POW status.[75] In that vein, U.S. courts have inferred from provisions of the GPW the combatant's privilege against being prosecuted by capturing States.[76] However, the legal immunity that combatants may be afforded is not the same as POW status. For example, a combatant's conduct may be protected by legal immunities even when that person is not in the power of the enemy and thus is not a POW. As another example of how POW status and legal immunity may differ, the GPW generally affords the same treatment to all classes of POWs identified in Article 4. However, not all the categories of POWs identified in Article 4 of the GPW, such as persons authorized to accompany the armed forces, receive the general license to commit belligerent acts that is afforded members of the armed forces.[77]

4.4.3.2 *Combatants – Legal Immunity and Sovereignty*. In addition to being

associated with humanitarian principles governing the treatment of POWs, the combatant's privilege has also been viewed as an application of the immunity that international law affords States from each other's jurisdiction.[78] In this view, "the act of the soldier who conforms to the law of war and does not engage in private acts of warfare is an act of state depriving the enemy state of jurisdiction."[79] This view of the combatant's privilege requires that combatants act under the commission of a belligerent State.[80] This view also reflects the principle that only States may authorize the resort to force.[81]

4.4.4 Nationality and Combatant Status.

4.4.4.1 *Nationals of Neutral States in Enemy Forces*. Members of enemy armed forces may include nationals of neutral or non-belligerent States. For example, the U.S. armed forces include many foreign nationals, and the United States could be engaged in hostilities when those foreign nationals' home States are not. Nationals of a neutral or non-belligerent State who are members of the armed forces of a belligerent State should be treated like other members of that State's armed forces.[82] For example, such nationals are entitled to POW status if they fall into the power of the enemy during international armed conflict.[83]

4.4.4.2 *Nationals of a State Who Join Enemy Forces*. The special privileges that international law affords combatants generally do not apply between a national and his or her State of nationality.[84] For example, provisions of the GPW assume that POWs are not nationals of the Detaining Power.[85] Thus, international law does not prevent a State from punishing its nationals whom it may capture among the ranks of enemy forces.[86] This rule is significant in non-international armed conflicts in which a State is fighting a rebel group composed of its own citizens.[87]

Although, as a matter of international law, nationals may not assert the privileges of combatant status against their own State, they may be subject to the liabilities of combatant status in relation to their own State under that

State's domestic law. For example, under U.S. law, U.S. nationals who join enemy forces have been subject to the liabilities of combatant status, such as potentially being made the object of attack or detained.[88]

4.4.4.3 *Nationals of Allied or Co-Belligerent States.* Nationals of an allied or cobelligerent State who are serving with enemy forces are in a position that is similar to the position of nationals of a State who are serving with enemy forces. If the nationals of an allied or co-belligerent State who are serving with enemy forces are captured by a State, they may be transferred to their State of nationality (*i.e.*, the co-belligerent or allied State), which is not required to afford them POW status.[89] However, U.S. practice as the Detaining Power in this situation has been to afford POW treatment to such individuals if they claim such protection.[90]

4.5 ARMED FORCES OF A STATE

Members of the armed forces of a State, including members of all groups that are part of the armed forces of a State, but excluding certain medical and religious personnel,[91] receive combatant status (*i.e.*, its rights, duties, and liabilities) by virtue of that membership.[92] This section addresses various classes of persons within the armed forces of a State.

4.5.1 Components of Armed Forces.

The armed forces of a State may include a variety of components, such as militia or volunteer corps that form part of those armed forces.[93]

The U.S. armed forces include members of the active duty military, the reserve forces, and the National Guard. U.S. armed forces also include the Coast Guard, which normally operates under the Department of Homeland Security.[94]

The U.S. armed forces may also include the Commissioned Corps of the U.S. Public Health Service, which normally operates under the Department of Health and Human Services.[95] Similarly, members of the Commissioned Corps of the National Oceanic and Atmospheric Administration, which normally operates under the Department of Commerce, may also become part of the U.S. armed forces.[96] Members of these and other organizations assigned to, and serving with, the U.S. armed forces may be subject to the Uniform Code of Military Justice.[97] The U.S. armed forces may also include the volunteer auxiliary of the Air Force when the services of the Civil Air Patrol are used in certain missions.[98]

4.5.1.1 *Reserve Armed Forces.* Although domestic law sometimes differentiates between the reserve and active components of the armed forces for the purpose of entitlement to benefits and other matters, international law treats members of the reserve forces that are part of the armed forces of a State the same as other members of the armed forces. In the United States, reserve armed forces include the reserve components of the Army, Navy, Air Force, Marine Corps, and Coast Guard, as well as the Army National Guard of the United States and the Air National Guard of the United States.[99]

4.5.2 Classes of Persons Within the Armed Forces.

4.5.2.1 *Special Operations Forces.* Special operations forces may be described as military forces specially organized, trained, and equipped to achieve military, political, economic, and psychological objectives by unconventional military means in hostile, denied, or politically sensitive areas.[100] As members of the armed forces, special operations forces have the same rights, duties, and liabilities as other members of the armed forces.[101] Nonetheless, in the past, some States have illegitimately questioned whether special operations forces are entitled to the privileges of combatant status. For example, during World War II, Hitler directed that German forces summarily execute captured Allied special operations forces.[102] Post-World War II war crimes tribunals found that this order was not a legitimate reprisal, violated the prohibition against executions without a fair trial, and improperly denied POW status to soldiers wearing a uniform behind enemy lines.[103]

Special operations forces personnel, like other members of the armed forces, remain entitled to the privileges of combatant status, unless they temporarily forfeit such privileges by engaging in spying or sabotage.[104] In some cases, military personnel who do not wear the standard uniform of their armed forces may nonetheless remain entitled to the privileges of combatant status because the wearing of such uniforms does not constitute the element of "acting clandestinely or under false pretenses."[105] For example, special operations forces have sometimes dressed like friendly forces.[106] Special operations forces personnel remain entitled to the privileges of combatant status even when operating detached from the main body of forces behind enemy lines.[107]

4.5.2.2 *Members Trained as Medical Personnel, but Not Attached to the Medical Service.* Members of the armed forces might have medical training but not be designated as military medical personnel. For example, before joining the armed forces, a person might have been trained as a nurse or physician, and after joining the armed forces might not be designated as part of the medical corps. Because such personnel have not been designated as military medical personnel, they are combatants, like other members of the armed forces.[108] However, if they fall into the power of the enemy during international armed conflict, such personnel may be required to tend to fellow POWs, in light of their previous training. In particular, POWs who, though not attached to the medical service of their armed forces, are physicians, surgeons, dentists, nurses, or medical orderlies may be required by the Detaining Power to exercise their medical functions in the interests of POWs dependent on the same Power.[109] In that case, they shall continue to be POWs, but shall receive the same treatment as corresponding medical personnel retained by the Detaining Power.[110] They shall be exempted from any other work under Article 49 of the GPW.[111]

4.5.2.3 *Members Who Are Ministers of Religion Without Having Officiated as Chaplains to Their Own Forces*. Members of the armed forces might be ministers of religion, but might not be designated as military religious personnel.[112] Because such personnel have not been designated as military religious personnel, they are combatants, like other members of the armed forces.[113] However, if they fall into the power of the enemy during international armed conflict, such personnel may minister to fellow POWs. In particular, POWs who are ministers of religion, without having officiated as chaplains to their own forces, shall be at liberty, whatever their denomination, to minister freely to the members of their community.[114] For this purpose, they shall receive the same treatment as chaplains retained by the Detaining Power.[115] They shall not be obliged to do any other work.[116]

4.5.2.4 *Draftees*. Some States require military service for categories of their nationals. The United States employs all-volunteer armed forces, although it has drafted its nationals into military service in prior conflicts. Under international law, a draftee, *i.e.*, a person who has been compelled to join a State's armed forces, is to be treated the same as other members of the armed forces.[117]

4.5.2.5 *Deserters*. A deserter from the armed forces of a belligerent who falls into the power of the enemy in international armed conflict is a POW.[118] Similarly, a deserter who is interned by a neutral State would also be treated as a POW.[119] The deserter's relationship with his or her armed forces is a question of that State's domestic law and not international law. States generally forbid members of their armed forces from desertion and generally regard members of the armed forces who desert as continuing to be members of their armed forces. Deserters who are subsequently captured by their own armed forces are not POWs because they are not in the power of the enemy and because the privileges of combatant status are generally understood not to apply, as a matter of international law, between nationals and their State of nationality.[120]

4.5.2.6 *Defectors*. Defectors are persons from one side's armed forces who voluntarily join the armed forces of the opposing side. They are generally not regarded as POWs while serving in their new armed force.[121]

Defectors serving in the forces of the enemy who are captured by the State to which they originally owed an allegiance generally would not be entitled to POW status because the privileges of combatant status are generally understood not to apply, as a matter of international law, between nationals and their State of nationality.[122]

States may not compel POWs, retained personnel, or protected persons in their power to defect and serve in their armed forces.[123]

4.5.3 Regular Armed Forces Who Profess Allegiance to a Government or an Authority Not Recognized by the Detaining Power. During international armed conflict, members of regular armed forces who profess allegiance to a Government or an authority not recognized by the Detaining Power are treated as members of the armed forces of a State.[124] Under Article 4A(3) of the GPW, they receive POW status, and they should also receive the rights, duties, and liabilities of combatants.[125]

Article 4A(3) of the GPW was developed to address situations like those that had occurred during World War II, when members of a military force continued fighting after their State had been occupied.[126] For example, military forces might continue to fight for a government-in-exile or for a government that had ceased to exist.[127] Such a government would provide the *right authority* for its regular armed forces to participate in the ongoing war and to receive POW status upon capture by the enemy.[128] Members of those forces were sometimes denied POW status by an enemy State, even though other States recognized the group to which they belonged as a co-belligerent force.[129]

4.5.4 Persons Belonging, or Having Belonged, to the Armed Forces of an Occupied State. Under Article 4B(1) of the GPW, persons belonging, or having belonged, to the armed forces of an occupied State should be treated as POWs if, while hostilities are continuing outside occupied territory, the Occupying Power considers it necessary, by reason of their allegiance to the armed forces, to intern them.[130]

Article 4B(1) of the GPW seeks to address the proper status of an army demobilized by the Occupying Power while a portion of those same armed forces continue the struggle. When the forces are demobilized, they are treated as civilians, but when recalled for internment based on their prior service, they are treated as POWs.[131] In particular, States developed this provision to address Germany's practice during World War II of arresting demobilized military personnel from occupied States.[132] These personnel were often interned and sought to escape to join the ongoing fighting. This provision was promulgated to ensure that individuals in similar circumstances would receive POW treatment if they were interned. For example, the rules for the parole of POWs would apply to them.[133] Similarly, the rules relating to POW escape would also apply to them.[134] So, a demobilized person who disobeyed an internment order and attempted to escape to rejoin his or her armed force would, like a POW, be subject, at most, to disciplinary punishment in respect of the act of escape.[135]

Persons belonging, or having belonged, to the armed forces of an occupied State would only be entitled to receive POW treatment while an international armed conflict continues.[136] For example, this provision would not apply to a situation like the occupation of Japan after World War II because all hostilities had ended.[137]

4.6 OTHER MILITIA AND VOLUNTEER CORPS

Under certain conditions, members of militia and volunteer corps that are not part of the armed forces of a State qualify as combatants and receive the rights, duties, and liabilities of combatant status.[138] More specifically, Article 4(A)(2) of the GPW defines prisoners of war to include: Members of other militias and members of other volunteer corps, including those of organized resistance movements, belonging to a Party to the conflict and operating in or outside their own territory, even if this territory is occupied, provided that such militias or volunteer corps, including such organized resistance movements, fulfil the following conditions:

(a) that of being commanded by a person responsible for his subordinates;

(b) that of having a fixed distinctive sign recognizable at a distance;

(c) that of carrying arms openly;

(d) that of conducting their operations in accordance with the laws and customs of war.[139]

Under these conditions, which are discussed below, members of these armed groups may operate as combatants in or outside their own territory, even if this territory is occupied. By contrast, a *levée en masse* may only be formed on the approach of the enemy to non-occupied territory.[140]

4.6.1 GPW 4A(2) Conditions in General. The conditions set forth in Article 4A(2) of the GPW were derived from conditions found in the Regulations annexed to the 1899 Hague II and the 1907 Hague IV.[141] These conditions reflect the attributes common to regular armed forces of a State.[142] By seeking to ensure that participants in hostilities are sufficiently disciplined, law abiding, and distinguishable from the civilian population, these conditions help protect the civilian population from the hardships of war. In addition, these conditions contribute to the military effectiveness of the force that satisfies the conditions.[143]

These conditions may be understood to reflect a burdens-benefits principle, *i.e.*, the receipt of certain benefits in the law of war (*e.g.*, privileges of combatant status) requires the assumption of certain obligations.[144]

4.6.1.1 *GPW 4A(2) Conditions Required on a Group Basis*. The armed group, as a whole, must fulfill these conditions for its members to be entitled to the privileges of combatant status. For example, if a member of an armed group met these requirements, but the armed group did not, that member would not be entitled to the privileges of combatant status because the armed group failed to satisfy the conditions.[145] Similarly, isolated departures from a condition by a member of the armed group (*e.g.*, a failure to comply with the conditions by a member of the armed group that was not directed by the armed group's leader) would not prevent the armed group from satisfying these conditions.

4.6.1.2 *AP I and the GPW 4A(2) Conditions*. AP I changed, for its Parties, the conditions under which armed groups that are not part of a State's armed forces may qualify for combatant status.[146] The United States has objected to the way these changes relaxed the requirements for obtaining the privileges of combatant status, and did not ratify AP I, in large part, because of them.[147] A chief concern has been the extent to which these changes would undermine the protection of the civilian population.[148] The United States has expressed the view that it would not be appropriate to treat this provision of AP I as customary international law.[149]

4.6.1.3 *Application of GPW 4A(2) Conditions to the Armed Forces of a State.* The text of the GPW does not expressly apply the conditions in Article 4A(2) of the GPW to the armed forces of a State.[150] Thus, under the GPW, members of the armed forces of a State receive combatant status (including its privileges and liabilities) by virtue of their membership in the armed forces of a State.[151] Nonetheless, the GPW 4A(2) conditions were intended to reflect attributes of States' armed forces.[152] If an armed force of a State systematically failed to distinguish itself from the civilian population and to conduct its operations in accordance with the law of war, its members should not expect to receive the privileges afforded lawful combatants.[153] Similarly, members of the armed forces engaged in spying or sabotage forfeit their entitlement to the privileges of combatant status if captured while engaged in those activities.[154]

4.6.2 Belonging to a Party to the Conflict. The armed group must belong to a party to the conflict.[155] The requirement of "belonging" to a party establishes that the armed group fulfills a *jus ad bellum* requirement of *right authority*, *i.e.*, it is acting on the authority of a State.[156] This requirement recognizes that members of a non-State armed group are not entitled to the privileges of combatant status even if that armed group satisfies the other conditions.[157] The State's authority may be granted by its representatives orally; it need not be granted in writing, such as through a commission or warrant. In all cases, however, opposing parties to the conflict must be able to discern that the armed group enjoys such authority. A State's formal acknowledgement that an armed group "belongs" to it is sufficient. On the other hand, the bare claim by the armed group that it acts on behalf of a State would be insufficient. In some cases, State support and direction to the armed group may establish that it belongs to the State.[158]

4.6.3 Being Commanded by a Person Responsible for His or Her Subordinates. The armed group must be commanded by a person responsible for his or her subordinates; the armed group must have a commander with effective authority over the armed group.[159] This requirement helps ensure that the armed group has sufficient discipline and organization to conduct its operations in accordance with the law of war.[160] The commander may derive his or her authority over the armed group by a regular or temporary commission from a State. However, a commander may derive his or her command from another position or authority. For example, the armed group may be formed informally and may have elected the commander as its leader. In practice, a State may provide members of the armed group with certificates or distinctive badges to show that they are officers, or military personnel responsible to higher authority, and not private persons acting on individual initiative.[161]

The authority of the commander over his or her subordinates gives rise to a corresponding duty to ensure that the armed group's members conduct their operations in accordance with the law of war.[162]

4.6.4 Having a Fixed Distinctive Sign Recognizable at a Distance. Members of the armed group should display a fixed distinctive sign or other device recognizable at a distance. The essence of this requirement is that members of the armed group are distinguishable from the civilian population.[163] By helping to ensure that members of the armed group can be visually distinguished from civilians, this requirement helps protect the civilian population from being erroneously made the object of attack.[164]

4.6.4.1 *Distinctive Sign.* The requirement does not specify a particular sign or emblem that persons must wear.[165] Wearing a military uniform satisfies this condition. However, a full uniform is not required.[166] The sign suffices if it enables the person to be distinguished from the civilian population. For example, a helmet or headdress that makes the silhouette of the individual readily distinguishable from that of a civilian can meet this requirement. Similarly, a partial uniform (such as a uniform jacket or trousers), load bearing vest, armband, or other device could suffice, so long as it served to distinguish the members from the civilian population. Formally notifying enemy forces of the distinctive sign is not required, and the proposal to add such a requirement was not accepted by States.[167] Such notification, however, may avoid misunderstanding and facilitate claims of POW status for captured members of the armed group.

4.6.4.2 *Fixed.* The text of the GPW indicates that the sign should be "fixed." This requirement has been interpreted to mean that the sign must be such that it cannot be easily removed and disposed of at the first sign of danger.[168] In practice, however, it would be important to assess whether members of the armed group are functionally distinguishable from the civilian population, even if the distinctive sign that they wear is not permanent and could be removed.[169]

4.6.4.3 *Visible at a Distance.* "Distance" has not been defined, but may be interpreted as requiring that the sign be easily distinguishable by the naked eye of ordinary people at a distance at which the form of the individual can be determined.[170]

4.6.5 Carrying Arms Openly. Members of the armed group must carry their arms openly.[171] This requirement is not satisfied if the armed group makes a practice of carrying only concealed weapons or of hiding weapons on the approach of enemy forces to avoid identification as fighters.[172]

4.6.6 Conducting Their Operations in Accordance With the Laws and Customs of War. The armed group, as a whole, must conduct its operations in accordance with the law of war. Evidence that an armed group enforced the law of war (such as by promulgating instructions regarding law of war requirements and punishing violations by its members) would help establish that an armed group meets this condition.

4.7 *LEVÉE EN MASSE*

A *levée en masse* is a spontaneous uprising of the inhabitants of non-occupied territory who, on the approach of the enemy in an international armed conflict, take up arms to resist the invading forces, without having time to form themselves into regular armed units.[173] Participants in a *levée en masse* are entitled to the privileges of combatant status, provided that they carry their arms openly and respect the laws and customs of war.[174]

4.7.1 Conditions for a *Levée en Masse*. The following discussion elaborates upon some of the conditions for a *levée en masse*.

4.7.1.1 *Spontaneous*. A *levée en masse* is a spontaneous uprising in which members have not had time to form into regular armed units.[175] Thus, unlike other categories of lawful combatants, persons who join a *levée en masse* need not wear a distinctive sign nor be organized under a responsible command. The spontaneity of their response generally precludes their being able to take such measures.

4.7.1.2 *Inhabitants*. A *levée en masse* is understood to reflect the right of inhabitants to resist approaching enemy forces. Non-inhabitants who travel to a territory to resist invading forces would not be entitled to participate in a *levée en masse*.[176]

4.7.1.3 *Approach of the Enemy to Non-Occupied Territory*. A *levée en masse* may only be formed on the approach of the enemy to non-occupied territory. Once the belligerent has established an occupation, the local population should comply with the orders of the occupation force, and a *levée en masse* may not be formed.[177] Similarly, even before the establishment of an occupation, inhabitants of areas that have already been invaded may not form a *levée en masse*.[178]

By contrast, members of an organized resistance movement under Article 4A(2) of the GPW may operate as combatants in or outside their own territory, even if this territory is occupied.[179]

4.7.2 *Levée en Masse* – Discerning Participants. Should some inhabitants form a *levée en masse* to defend an area, it may be justifiable for the invading force to detain all persons of military age in that area and treat them as POWs.[180] Even if an inhabitant who joined or participated in the *levée en masse* lays down arms, if he or she is later captured, he or she may be detained as a POW.[181] If a person joins a *levée en masse*, he or she may be held as a POW even if he or she actually took no part in fighting.[182] If detained, a person believed to have participated in a *levée en masse* should be treated as a POW until a determination is otherwise made by a competent tribunal.[183] If a competent tribunal determines that a person did not join the *levée en masse* but instead committed private acts of hostility against enemy military forces,[184] the person may be treated as an unprivileged belligerent.[185]

4.8 RIGHTS, DUTIES, AND LIABILITIES OF CIVILIANS

Like combatants, members of the civilian population also have certain rights, duties, and liabilities under the law of war. Civilians may not be made the object of attack. If detained, civilians are entitled to humane treatment and a variety of additional protections. Civilians lack the combatant's privilege, and may be punished, after a fair trial, by an enemy State for engaging in hostilities against it. Civilians – Notes on Terminology. Like other terms, "4.8.1 civilian" is used in a variety of different ways in the law of war.[186]

4.8.1.1 *"Civilian" Versus "Military"*. Sometimes, "civilian" is used to describe persons who are not military personnel. For example, persons authorized to accompany the armed forces are often called "civilians" in this sense, even though they are POWs if they fall into the power of the enemy during international armed conflict.[187]

4.8.1.2 *"Civilian" Versus "Combatant"*. "Civilian" is also often used to refer to persons who are not "combatants." Since "combatant" is often used in different ways, "civilian," correspondingly, is also used in different ways. For example, sometimes "civilians" is used to refer to persons who lack the right to participate in hostilities. Other times "civilian" is used to refer to persons who neither have the right to participate in hostilities nor have in fact participated in hostilities.[188] For example, "civilian" casualty reports generally exclude insurgents or terrorists, even though some might call such persons "civilians" because they are not entitled to participate in hostilities.

4.8.1.3 *"Civilian" in the GC.* The GC does not define "civilian," although it uses the word.[189] The GC uses the term "protected person" to refer to persons protected by the Convention.[190] The GC excludes from the definition of "protected person" those persons who are protected under the other 1949 Geneva Conventions, *e.g.*, POWs and retained personnel.[191]

In some cases, "protected persons" can include a person "detained as a spy or saboteur, or as a person under definite suspicion of activity hostile to the security of the Occupying Power," although such persons are not entitled to all the protections applicable to protected persons.[192]

4.8.1.4 *"Civilian" in AP I and AP II.* AP I defines "civilian" in opposition to "combatant"; under AP I, anyone who is not a "combatant" is, by definition, a "civilian."[193] The United States has objected to AP I's definition of combatant.[194] AP II uses the term "civilian" without providing a definition.[195]

4.8.1.5 *General Usage of "Civilian" in This Manual.* This manual generally uses "civilian" to mean a member of the civilian population, *i.e.*, a person who is neither part of nor associated with an armed force or group, nor otherwise engaging in hostilities.

4.8.2 Civilians – Conduct of Hostilities. Civilians may not be made the object of attack, unless they take direct part in hostilities.[196] The wounded and sick, as well as the infirm, and expectant mothers, shall be the object of particular protection and respect.[197] Civilians may be killed incidentally in military operations; however, the expected incidental harm to civilians may not be excessive in relation to the anticipated military advantage from an attack,[198] and feasible precautions must be taken to reduce the risk of harm to civilians during military operations.[199]

Civilians who engage in hostilities thereby forfeit the corresponding protections of civilian status and may be liable to treatment in one or more respects as unprivileged belligerents.[200]

4.8.3 Civilians – Detention. In general, civilians may be subject to non-violent measures that are justified by military necessity, such as searches, or temporary detention.[201] Belligerents or Occupying Powers may take necessary security measures in relation to civilians, including internment or assigned residence for imperative reasons of security.[202]

Enemy civilians who are interned during international armed conflict or occupation generally are classified as "protected persons" under the GC and receive a variety of protections.[203] Chapter X addresses in detail the required treatment of enemy civilian internees during international armed conflict and occupation. In all circumstances, detained civilians must be treated humanely.[204] Special categories of civilians, such as children, may require additional consideration during detention.[205]

4.8.4 Civilians – Legal Liability Under an Enemy State's Domestic Law. Unlike combatants, civilians lack the combatant's privilege excepting them from the domestic law of the enemy State. Civilians who engage in hostilities may, after a fair trial, be punished by an opposing State.[206]

A State that is an Occupying Power has additional authorities over enemy civilians that extend beyond the ability to punish their unauthorized participation in hostilities.[207]

Note, however, the special cases of persons authorized to accompany the armed forces,[208] members of the merchant marine and civil aircraft,[209] and participants in a *levée en masse*.[210]

4.9 MILITARY MEDICAL AND RELIGIOUS PERSONNEL

"Medical personnel exclusively engaged in the search for, or the collection, transport or treatment of the wounded or sick, or in the prevention of disease, staff exclusively engaged in the administration of medical units and establishments, as well as chaplains attached to the armed forces" are treated as a special category under the law of war.[211] This manual generally refers to them as military medical and religious personnel. Under certain circumstances, (1) the authorized staff of voluntary aid societies and (2) the staff of a recognized aid society of a neutral country are treated like military medical and religious personnel.[212]

4.9.1 Types of Military Medical and Religious Personnel. Military medical and religious personnel include: (1) medical personnel exclusively engaged in the search for, or the collection, transport, or treatment of the wounded or sick, or in the prevention of disease; (2) staff exclusively engaged in the administration of medical units and establishments; and (3) chaplains attached to the armed forces.[213]

4.9.1.1 *Medical Personnel Exclusively Engaged in Medical Duties*. Medical personnel exclusively engaged in the search for, or the collection, transport, or treatment of the wounded or sick, or in the prevention of disease include military physicians, dentists, nurses, orderlies, stretcher-bearers, and other persons belonging to the armed forces who give direct care to the wounded and sick.[214]

Persons who are exclusively engaged in the prevention of disease also qualify as military medical personnel. For example, veterinary personnel qualify for military medical status, if they are exclusively engaged in providing health services for military personnel (*e.g.*, performing food safety inspections and ensuring that animal illnesses do not spread to humans). However, as a general matter, veterinary personnel would not qualify for military medical status based on being part of the veterinary service.[215]

4.9.1.2 *Staff Exclusively Engaged in Support to Medical Units and Establishments*. In addition to persons directly engaged in medical duties, medical personnel also include staff exclusively engaged in the administration of medical units and establishments. These persons also receive status as medical personnel provided that they are exclusively assigned to the Medical Service.[216] For example, Medical Service Corps personnel, and individuals with a non-medical Military Occupation Specialty permanently assigned to a medical unit or facility (such as its cooks, clerks and supply personnel, or crews operating permanent medical aircraft), would qualify for military medical status.

4.9.1.3 *Chaplains Attached to the Armed Forces*. Chaplains attached to the armed forces include any cleric, regardless of faith, who is attached to the armed forces of a belligerent and assigned duties exclusively of a religious or spiritual nature.[217]

4.9.2 Requirements for Military Medical and Religious Status. To acquire and retain military medical and religious status, members of the armed forces must (1) belong to a force whose members qualify for POW status; (2) be designated as exclusive medical or religious personnel by their armed forces; and (3) serve exclusively in a medical or religious capacity. Underlying military medical and religious status is the principle that the armed forces have committed to use these personnel exclusively in a humanitarian role; thus, these personnel have a "supra-national" and "quasi-neutral character" because their humanitarian duties place them "above the conflict."[218]

4.9.2.1 *Belong to a Force Whose Members Qualify for POW Status*. To acquire military medical and religious status, a person must belong to an armed force whose members qualify for POW status, *i.e.*, the armed forces of a State, regular armed forces who profess allegiance to a government or an authority not recognized by the Detaining Power, or other militia and volunteer corps meeting the conditions in the GPW.[219] For example, medical personnel belonging to non-State armed groups would not be entitled to retained personnel status under the GWS, GWS-Sea, and GPW because their members do not qualify for POW status.

4.9.2.2 *Designated by Their Armed Forces*. To acquire military medical and religious status, members of the armed forces must be designated as such by their armed forces, usually by being part of the official medical or religious service.[220] A member of the armed forces cannot designate himself or herself as military medical or religious personnel. Thus, members of the armed forces do not acquire military medical and religious status merely by performing medical or religious functions or by having medical or religious training. For example, a member of a special operations unit might be trained as a medical specialist, but not be designated as military medical personnel, because that person is expected to perform both combatant and medical duties. That person, therefore, would not receive the rights, duties, and liabilities related to military medical status. He or she, however, may be treated like military medical personnel while detained by the enemy by being required to perform medical duties.[221]

4.9.2.3 *Exclusively Engaged in Humanitarian Duties*. In order to establish and maintain their status as military medical or religious personnel, these personnel must serve exclusively in a humanitarian capacity. This assignment must generally be permanent.[222] Military physicians or other medical specialist personnel who are not exclusively engaged in humanitarian duties, such as duties that involve committing acts harmful to the enemy, would not be entitled to military medical status.[223] For example, an Army Medical Corps or Medical Service Corps officer serving as the commander of a tactical convoy would not be entitled to military medical status.[224] Similarly, in general, persons who engage in combat search and rescue missions would not be exclusively engaged in humanitarian duties, since preventing the capture of combatants by the adversary constitutes an act harmful to the enemy.[225] However, these individuals may be treated like auxiliary medical personnel or members of the armed forces who are trained in medical care, but who are not attached to the medical service.[226]

The requirement that military medical and religious personnel serve exclusively in a humanitarian capacity not only requires that they refrain from acts harmful to the enemy, but also has been interpreted as an affirmative obligation to serve in a humanitarian capacity. For example, captured military medical personnel who refuse to perform their medical duties to care for fellow prisoners would not be entitled to retained personnel status.[227]

4.10 RIGHTS, DUTIES, AND LIABILITIES OF MILITARY MEDICAL AND RELIGIOUS PERSONNEL

Although military medical and religious personnel are members of, or attached to, the armed forces and are in many respects treated like combatants, they are afforded special privileges so that they may fulfill their humanitarian duties. They must be respected and protected in all circumstances. In turn, they generally may not commit acts harmful to the enemy. If they fall into the power of the enemy during international armed conflict, they may be retained to care for, or minister to, POWs.

4.10.1 Military Medical and Religious Personnel - Conduct of Hostilities. Military medical and religious personnel may not be made the object of attack.[228] However, military medical and religious personnel must accept the risks of incidental harm from military operations.[229]

Military medical and religious personnel generally may not commit acts harmful to the enemy (*e.g.*, resisting lawful capture by the enemy military forces).[230] Military medical and religious personnel, however, may employ arms in self-defense or in defense of their patients against unlawful attacks.[231]

Military medical and religious personnel who take actions outside their role as military medical and religious personnel forfeit the corresponding protections of their special status and may be treated as combatants or auxiliary medical personnel, as appropriate.[232]

4.10.2 Military Medical and Religious Personnel - Detention. If military medical and religious personnel fall into the power of the enemy during international armed conflict, they are not held as POWs, but instead are held as retained personnel.[233] They should present their identity cards to demonstrate their status as retained personnel.[234] They are retained so that they may fulfill their humanitarian duties to care for, or minister to, POWs.[235]

Although they are not held as POWs, military medical and religious personnel receive, at a minimum, the protections of POW status.[236] In addition, retained personnel shall be granted all facilities necessary to provide for the medical care of, and religious ministration to, POWs.[237]

For example, retained personnel may not be compelled to carry out any work other than that concerned with their medical or religious duties.[238] Retained personnel shall be authorized to visit periodically POWs situated in working detachments or in hospitals outside the camp.[239] Retained personnel, through their senior officer in each camp, have the right to deal with the competent authorities of the camp on all questions relating to their duties.[240] If they are not needed to care for, or minister to, POWs, and if military requirements permit, retained personnel should be returned to the forces to which they belong so that they may continue to care for, or minister to, members of their armed forces.[241] The parties to the conflict would establish special agreements to establish the procedures for repatriation.[242]

4.11 AUTHORIZED STAFF OF VOLUNTARY AID SOCIETIES

States may choose to employ the staff of National Red Cross Societies and that of other Voluntary Aid Societies, like military medical and religious personnel. If States subject such staff to military laws and regulations, then such personnel are to be treated like military medical and religious personnel.[243] States must notify other Parties to the 1949 Geneva Conventions before actually employing such personnel.[244]

4.11.1 American National Red Cross. Under U.S. law, the American National Red Cross is a voluntary aid society authorized to support the U.S. armed forces in time of war.[245] American National Red Cross personnel who support the U.S. armed forces in military operations in this way are subject to the Uniform Code of Military Justice.[246]

4.12 STAFF OF A RECOGNIZED AID SOCIETY OF A NEUTRAL COUNTRY

The 1949 Geneva Conventions recognize that neutral States may lend their recognized aid societies to a party to the conflict by placing those personnel and units under the control of that party to the conflict.[247] The neutral Government shall notify this consent to the adversary of the State that accepts such assistance, and the party to a conflict that accepts such assistance must notify enemy States before using it.[248] This assistance is not considered as interference in the conflict by the neutral State.[249]

The staff of a recognized aid society of a neutral country who have been lent to a party to the conflict must be furnished with an identity card similar to that provided to retained personnel before leaving their neutral State.[250] Such personnel who have fallen into the hands of the adverse party may not be detained.[251] Unless otherwise agreed, such personnel shall have permission to return to their country, or if this is not possible, to the territory of the party to the conflict in whose service they were, as soon as a route for their return is open and military considerations permit.[252] Pending their release, such personnel shall continue their work under the direction of the adverse party; they shall preferably be engaged in the care of the wounded and sick of the party to the conflict in whose service they were.[253] On their departure, they shall take with them their effects, personal articles and valuables and the instruments, arms, and if possible, the means of transport belonging to them.[254]

Such personnel, while in the power of a party to the conflict, should be treated on the same basis as corresponding personnel of the armed forces of that party to the conflict; food in particular must be sufficient as regards quantity, quality, and variety to keep them in a normal state of health.[255] They may also fly their national flag.[256]

4.13 AUXILIARY MEDICAL PERSONNEL

Auxiliary medical personnel are members of the armed forces specially trained for employment, should the need arise, as hospital orderlies, nurses, or auxiliary stretcher-bearers, in the search for or the collection, transport, or treatment of the wounded and sick in international armed conflict.[257] In general, auxiliary medical personnel are treated like combatants; however, while carrying out medical duties, they must distinguish themselves by wearing a white armlet bearing the distinctive sign (*e.g.*, the red cross), and they may not be made the object of attack. Auxiliary Medical Personnel – U.S. Armed Forces.

4.13.1 The recent practice of the United States has not been to employ persons as auxiliary medical personnel. Rather, the U.S. armed forces have employed military medical and religious personnel.[258] In addition, the U.S. armed forces have given members of the armed forces additional training in combat medicine but have not designated them as military medical personnel or as auxiliary medical personnel.[259]

Thus, such personnel have not worn the distinctive emblem while engaging in medical duties, and they may be made the object of attack by the enemy. Although the recent practice of the United States has not been to employ persons as auxiliary medical personnel, the United States may later decide to employ auxiliary medical personnel, and enemy military forces may employ such personnel. Moreover, certain members of the medical corps, who do not qualify as military medical personnel because they perform duties inconsistent with exclusive medical status, may be treated like auxiliary medical personnel.[260]

4.13.2 Acquiring Auxiliary Medical Status. As with military medical and religious status, members of the armed forces do not acquire auxiliary medical status simply by performing medical duties.[261] For example, a combatant who treats fellow combatants on the battlefield does not automatically acquire auxiliary medical status. Similarly, persons do not acquire auxiliary medical status only because they happen to have medical training. In order to acquire auxiliary medical status, a person must receive appropriate training and be designated as such by his or her armed forces.[262] Those armed forces must provide proper identification to such persons, including an armband and a special identity document.[263]

4.13.3 Auxiliary Medical Personnel – Conduct of Hostilities. Auxiliary medical personnel may not be made the object of attack when carrying out their medical duties.[264] Auxiliary medical personnel shall wear, but only while carrying out medical duties, a white armlet bearing in its center the distinctive sign in miniature; the armlet shall be issued and stamped by the military authority.[265] In addition, auxiliary medical personnel must abstain from acts harmful to the enemy while carrying out their medical duties.[266] When the above conditions are not present, auxiliary medical personnel may be made the object of attack on the same basis as other combatants.

4.13.4 Auxiliary Medical Personnel – Detention. Auxiliary medical personnel are POWs when detained by the enemy during international armed conflict, but may be required to perform their medical duties, as needed.[267] Auxiliary medical personnel are not subject to the repatriation provisions that apply specifically to retained personnel.[268]

4.14 PERSONNEL ENGAGED IN DUTIES RELATED TO THE PROTECTION OF CULTURAL PROPERTY

During armed conflict, different classes of persons may be engaged in duties related to the protection of cultural property. These classes of persons may include: specialist personnel in the armed forces, armed custodians specially empowered to guard cultural property, as well as persons who are engaged in duties of control in accordance with the Regulations for the Execution of the 1954 Hague Cultural Property Convention. So far as consistent with the interests of security, such personnel should be respected and permitted to carry out their duties for the protection of cultural property.

4.14.1 Personnel Engaged in the Protection of Cultural Property. As far as is consistent with the interests of security, personnel engaged in the protection of cultural property shall, in the interests of such property, be respected and, if they fall into the hands of the opposing party, shall be allowed to continue to carry out duties whenever the cultural property for which they are responsible has also fallen into the hands of the opposing party.[269] Such personnel are analogous to military medical and religious personnel who also shall continue to carry out their medical and spiritual duties when they have fallen into the hands of the enemy.[270]

No special agreement may be concluded that would diminish the protection afforded by the 1954 Hague Cultural Property Convention to the personnel engaged in the protection of cultural property.[271]

4.14.1.1 *Specialist Personnel in the Armed Forces*. States, especially Parties to the 1954 Hague Cultural Property Convention, may have within their armed forces, services or specialist personnel whose purpose is to secure respect for cultural property and to cooperate with the civilian authorities responsible for safeguarding it.[272] The United States has long had such personnel in its armed forces.[273]

For example, during World War II, Allied forces dedicated a specific group of personnel who were tasked to save as much of the culture of Europe as they could during combat.[274] These personnel worked to mitigate combat damage to churches and museums and to locate moveable works of art that were stolen or missing.[275]

4.14.1.2 *Armed Custodians Specially Empowered to Protect Cultural Property*.

States may use armed custodians who are specially empowered to protect cultural property. The guarding of cultural property under special protection by armed custodians specially empowered to do so, however, shall not be deemed to be a use for military purposes that would deprive cultural property of special protection.[276]

4.14.2 Persons Responsible for the Duties of Control in Accordance With the Regulations for the Execution of the 1954 Hague Cultural Property Convention. A number of persons are responsible for the duties of control in accordance with the Regulations for the Execution of the 1954 Hague Cultural Property Convention. These individuals may include: (1) the Director- General of the United Nations Educational, Scientific and Cultural Organization (UNESCO); (2) a State-appointed representative for cultural property; (3) delegates of the Protecting Powers; (4) a Commissioner-General for Cultural Property; and (5) inspectors and experts proposed by the Commissioner-General for Cultural Property. The Commissioners-General for Cultural Property, delegates of the Protecting Powers, inspectors, and experts shall in no case exceed their mandates. In particular, they shall take account of the security needs of the Party to the 1954 Hague Cultural Property Convention to which they are accredited and shall in all circumstances act in accordance with the requirements of the military situation as communicated to them by that State.[277]

4.14.2.1 *Director-General of UNESCO*. Under the 1954 Hague Cultural Property Convention, the Director-General of the United Nations Educational, Scientific, and Cultural Organization (UNESCO) plays an important role. For example, among other duties, the Director-General compiles and periodically revises an international list consisting of all persons nominated by participating States as qualified to carry out the functions of Commissioner General for Cultural Property.[278]

4.14.2.2 *State-Appointed Representative for Cultural Property*. As soon as any Party to the 1954 Hague Cultural Property Convention is engaged in an international armed conflict, that Party shall appoint a representative for cultural property situated in its territory; if that Party is in occupation of another territory, it shall appoint a special representative for cultural property situated in that territory.[279]

4.14.2.3 *Delegates of the Protecting Powers*. A Protecting Power shall appoint delegates to perform certain functions in the protection of cultural property.[280] The delegates of the Protecting Powers shall take note of violations of the 1954 Hague Cultural Property Convention, investigate, with the approval of the Party to which they are accredited, the circumstances in which they have occurred, make representations locally to secure their cessation, and, if necessary, notify the Commissioner-General of such violations. They shall keep the Commissioner-General informed of their activities.[281]

4.14.2.4 *Commissioner-General for Cultural Property*. The Regulations for the Execution of the 1954 Hague Cultural Property Convention provide for the appointment of a Commissioner-General for cultural property, but always with the approval of the Party to which he or she will be accredited.[282]

The Commissioner-General exercises a number of functions, including proposing the appointment of inspectors and experts, ordering an investigation with the agreement of the Party to which he or she is accredited, and drawing up reports on the application of the Convention.[283]

4.14.2.5 *Inspectors and Experts Proposed by the Commissioner-General for Cultural Property.* The Commissioner-General for Cultural Property may propose, for the approval of the Party to which the Commissioner-General is accredited, inspectors of cultural property to be charged with specific missions.[284] The Commissioner-General, delegates, and inspectors may have recourse to the services of experts, who will also be proposed for the approval of the Party to which the Commissioner-General is accredited.[285]

In some cases, inspectors may be entrusted with the functions of delegates of the Protecting Powers.[286]

4.14.3 Identifying Personnel Engaged in Duties for the Protection of Cultural Property. Persons responsible for duties of control in accordance with the Regulations for the Execution of the 1954 Hague Cultural Property Convention, and persons engaged in duties for the protection of cultural property, are to carry a special identity card bearing the distinctive emblem.[287] In addition, such persons may wear an armlet bearing the distinctive emblem issued and stamped by the competent authorities.[288] Such persons may not, without legitimate reason, be deprived of their identity card or of the right to wear the armlet.[289] The distinctive emblem displayed on an armlet or special identity card is to be displayed once (as opposed to repeated three times).[290]

4.15 PERSONS AUTHORIZED TO ACCOMPANY THE ARMED FORCES

Under the law of war, persons who are not members of the armed forces, but are authorized to accompany them, fall into a special category. Although they are often referred to as "civilians" because they are not military personnel, they differ materially from the civilian population because these persons are authorized – and in some cases, are ordered – to accompany military forces into a theater of operations to support the force.[291] DoD policies have often addressed the use of non-military personnel to support military operations.[292]

Persons authorized to accompany the armed forces may not be made the object of attack unless they take direct part in hostilities. They may, however, be detained by enemy military forces, and are entitled to POW status if they fall into the power of the enemy during international armed conflict. They have legal immunity from the enemy's domestic law for providing authorized support services to the armed forces.

4.15.1 Persons Authorized to Accompany the Armed Forces – Notes on Terminology. In the past, the category of persons who "accompany" or "follow" the armed forces has referred to journalists, private clergy, and sutlers,[293] who traditionally were subject to the jurisdiction of the army that they accompanied. [294]These persons, to the extent that they were simply following the armed forces for their own purposes and not to provide support to military operations, were subject to detention, not as POWs, but like civilians, only if specifically necessary.[295] On the other hand, non-military persons who were serving the armed forces, such as civilian employees of the war department, were detained as POWs.[296]

Article 4A(4) of the GPW reflects the modern practice and rule.[297] Persons authorized to accompany the armed forces under Article 4A(4) include employees of the Department of Defense, employees of other government agencies sent to support the armed forces, and other authorized persons working on government contracts to support the armed forces. DoD practice has been to permit a broad range of civilians to be authorized to accompany U.S. forces.[298]

Persons authorized to accompany the armed forces are often referred to, and treated as, "civilians," since they are not members of the armed forces. However, as discussed below, persons authorized to accompany the armed forces are also treated like "combatants" in some respects.

4.15.2 Persons Authorized to Accompany the Armed Forces – Conduct of Hostilities.

The rules relating to the conduct of hostilities for persons authorized to accompany the armed forces are similar in some respects to the rules relating to military medical and religious personnel. For example, persons in both classes: (1) generally may not be made the object of attack; (2) accept certain risks of incidental harm from enemy military operations; and (3) have the right of self-defense against unlawful attacks.[299]

4.15.2.1 *Liability to Being Made the Object of Attack.* For the purposes of determining whether they may be made the object of attack, persons authorized to accompany the armed forces are treated as civilians. They may not be made the object of attack unless they take direct part in hostilities.[300]

4.15.2.2 *Employment in Hostilities.* The law of war does not prohibit persons authorized to accompany the armed forces from providing authorized support that constitutes taking direct part in hostilities. Even if the authorized support that they provide constitutes taking a direct part in hostilities, such persons retain their entitlement to POW status under Article 4A(4) of the GPW.[301] Although international law does not prohibit States from using persons authorized to accompany the armed forces to provide support that constitutes direct participation in hostilities, commanders should exercise care in placing such personnel in situations in which an attacking enemy may consider their activities to constitute taking a direct part in hostilities, as there may be legal and policy considerations against such use. Commanders may not employ persons authorized to accompany the armed forces to perform duties and functions traditionally performed by a military person if such action is taken for the purpose of shielding a military objective from attack.[302]

4.15.2.3 *Increased Risk of Incidental Harm*. Persons authorized to accompany the armed forces should expect that they have an increased risk of death or injury incidental to an enemy attack because of their deliberate proximity to military operations and military objectives.[303] For example, civilian members of military aircrews should expect that they might be incidentally injured in an attack against the military aircraft in which they are working.[304]

Similarly, in some cases, the location at which persons authorized to accompany the armed forces are serving may itself be a military objective.[305] Furthermore, in some cases, an enemy could reasonably conclude that the mere presence of a person at a location (*e.g.*, a remote military base) indicates that they are a combatant or are directly participating in hostilities, and therefore could lawfully be made the object of attack.

4.15.2.4 *Self-Defense and Arming*. Persons authorized to accompany the armed forces have a right of self-defense against unlawful attacks, such as attacks by bandits. Persons authorized to accompany the armed forces should not resist capture by enemy military forces whom they expect to respect their status under the law of war. If persons authorized to accompany the armed forces make any resistance to enemy military forces, then they may be regarded as taking a direct part in hostilities, and may be made the object of attack.[306] Their resistance to capture by enemy military forces whom they expect to respect their status under the law of war would be incompatible with the principle that a person may not claim the distinct rights of both combatants and civilians at the same time.[307]

The arming of persons authorized to accompany the armed forces is analogous to the arming of military medical and religious personnel.[308] DoD practice has been to permit commanders to authorize persons authorized to accompany the armed forces to carry defensive weapons if necessary.[309]

A State may decide to arm persons authorized to accompany the armed forces for personal security and self-defense purposes without incorporating them into the armed forces or otherwise authorizing them to act as a combatant

4.15.2.5 *Wearing of Uniform*. The law of war does not prohibit persons authorized to accompany the armed forces from wearing a uniform, including the uniform of armed forces that they accompany. Recent U.S. practice, however, has been to ensure that persons accompanying U.S. forces wear clothing that distinguishes them from members of the armed forces in order to prevent confusion about their status.[310]

Persons who wear a military uniform accept the risk of being made the object of attack by enemy forces, as the enemy would reasonably believe them to be lawful objects of attack. However, the mere wearing of a uniform or being authorized by a State to wear a uniform does not necessarily authorize that person to act as a combatant.

4.15.2.6 *Provision of Security Services*. Persons authorized to accompany the armed forces who provide security against criminal elements generally would not be viewed as taking a direct part in hostilities (and do not forfeit their protection from being made the object of attack).[311] However, providing such services to defend against enemy armed forces of a State would be regarded as taking a direct part in hostilities (and would forfeit their protection from being made the object of attack).[312]

DoD policies have addressed the use of non-military personnel to provide security services for DoD components.[313] Where there has been a significant risk of attack by enemy armed forces of a State, DoD practice generally has been to use military personnel to provide security.

4.15.3 Persons Authorized to Accompany the Armed Forces – Detention. For the purposes of detention, persons authorized to accompany the armed forces are treated like combatants. These persons may be detained by the enemy and are entitled to POW status during international armed conflict.[314]

When held as POWs, persons authorized to accompany the armed forces are to receive advance of pay from the Detaining Power, and they should be assigned equivalent ranks to those of members of the armed forces for this purpose.[315]

Persons who are authorized to accompany the armed forces must be issued an identity card to confirm their status.[316] Presenting the identification card is not a prerequisite for POW status, but it helps captured persons establish to enemy forces that they are entitled to POW status.[317]

4.15.4 Persons Authorized to Accompany the Armed Forces – Liability Under Domestic Law for Participation in Hostilities. Unlike combatants, persons authorized to accompany the armed forces receive no general license to participate in hostilities. However, international law contemplates that persons authorized to accompany the armed forces may lawfully support armed forces in the conduct of hostilities. Such persons should not be liable under an enemy State's domestic law for providing authorized support services.[318] For example, they should not be prosecuted for offenses of aiding the enemy. Persons authorized to accompany the armed forces may not be punished by an enemy State for authorized support activities or for defending themselves against unlawful attacks. This protection would not apply with respect to acts by persons authorized to accompany the armed forces that are prohibited by the law of war. Persons authorized to accompany the armed forces should not engage in unauthorized participation in hostilities. Such activity would be treated like engagement in private acts of hostility,[319] and such persons would be in the position of unprivileged belligerents in relation to those activities.[320] There may be additional considerations in determining which prosecution forum is appropriate because persons authorized to accompany the armed forces are not themselves members of the armed forces.[321]

Commanders who use persons authorized to accompany the armed forces could, under certain circumstances, be prosecuted for war crimes committed by such personnel under theories of command responsibility or other theories of individual liability.[322]

4.16 CREWS OF MERCHANT MARINE VESSELS OR CIVIL AIRCRAFT

Crews of merchant marine vessels or civil aircraft of a belligerent fall into a special category, and are in many respects treated like persons authorized to accompany the armed forces.[323] Under certain circumstances, crews of merchant marine vessels or civil aircraft of a neutral that engage in hostilities may be treated like the crews of belligerent vessels or aircraft.[324]

4.16.1 Merchant or Civil Crews - Conduct of Hostilities. Merchant vessels or civil aircraft of a belligerent may be used to support military operations, such as by conveying goods and military personnel to theaters of active military operations. Enemy merchant vessels or civil aircraft may be captured.[325] Belligerent merchant vessels or civil aircraft may resist attacks by enemy forces, including by eventually seizing the attacking vessels or aircraft.[326] However, merchant vessels or civil aircraft should not commit hostile acts in offensive combat operations.[327]

4.16.2 Merchant or Civil Crews – Detention. Members of the crews of merchant marine vessels or civil aircraft of a belligerent are entitled to POW status, if they fall into the power of the enemy during international armed conflict.[328]

The GPW contemplates that certain members of the crews of merchant marine vessels or civil aircraft of a belligerent may benefit from more favorable treatment under international law; *i.e.*, they would not be detained as POWs. During wartime, enemy merchant seamen have customarily been subject to capture and detention.[329] However, the 1907 Hague XI provides that the crews of enemy merchant ships that did not take part in hostilities were not to be held as POWs provided "that they make a formal promise in writing, not to undertake, while hostilities last, any service connected with the operations of the war."[330] Although these provisions proved ineffective during World War I and World War II,[331] the GPW allows for the possibility that they might apply.[332]

4.17 SPIES, SABOTEURS, AND OTHER PERSONS ENGAGING IN SIMILAR ACTS BEHIND ENEMY LINES

Spying, sabotage, and similar acts behind enemy lines have a dual character under the law of war; States are permitted to employ persons who engage in these activities, but these activities are punishable by the enemy State. Belligerents may employ spies and saboteurs consistent with the law of war. However, any person (including individuals who would otherwise receive the privileges of lawful combatants) engaging in spying, sabotage, or similar acts behind enemy lines, is regarded as an unprivileged belligerent while doing so. These persons forfeit entitlement to the privileges of combatant status and may be punished after a fair trial if captured.

4.17.1 Spies, Saboteurs, and Other Persons Engaging in Similar Acts Behind Enemy Lines – Notes on Terminology. Spying has been given a technical definition in the Hague IV Regulations.[333] However, as discussed below, certain conduct, most notably sabotage, has been treated as having the same legal consequences as spying and has even been called spying, even though such acts do not meet the definition of spying in the Hague IV Regulations.[334] In some cases, "saboteurs" has been used in a purely factual way to refer to those persons engaged in damaging or destroying enemy materiel.[335] In other cases, "saboteurs" has been used as a legal term of art to refer to those persons who are engaged in damaging or destroying enemy material and who, by definition, would not be entitled to receive the privileges of combatant status.[336]

4.17.1.1 *General Usage of "Spies and Saboteurs" in This Manual*. This manual generally uses the term "spying and sabotage" to refer to that general category of secretive, hostile activities that, when performed behind enemy lines, deprives that person of entitlement to the privileges of combatant status. In some places, the phrase "spying, sabotage, and similar acts behind enemy lines" is used to make more clear that this general category of acts is broader than only spying and sabotage.

4.17.2 Spies. A person may only be considered a spy when, (1) acting clandestinely or under false pretenses, (2) in the zone of operations of a belligerent, (3) he or she obtains, or endeavors to obtain, information, (4) with the intention of communicating it to the hostile party.[337] During war, any person—military or civilian—whose actions meet all of these elements may be considered a spy under the law of war. The following discussion elaborates upon the elements of spying.

4.17.2.1 *Acting Clandestinely or Under False Pretenses*. Acting "clandestinely or under false pretenses" means deliberately concealing or misrepresenting one's identity and conduct.[338] For example, a member of the armed forces fulfills this element when the person wears a disguise, such as civilian clothes or the uniform of the enemy, so that the enemy will fail to identify the person as a member of the opposing armed force. Persons who act openly, such as by wearing the uniform of the armed forces to which they belong, do not meet this element.[339] For example, members of a ground reconnaissance team or couriers who wear their normal uniforms would not meet this element.[340] In addition, observers on military reconnaissance aircraft have not been regarded as acting clandestinely or under false pretenses.[341]

4.17.2.2 *In the Zone of Operations of a Belligerent*. A person must engage in acts of espionage in the zone of operations of a belligerent to be considered a spy.[342] "Zone of operations" has been construed broadly to include areas supporting the war effort.[343] However, a person who engages in surveillance or information gathering from outside territory controlled by a hostile party would not meet this element, and would not be considered a spy.

4.17.2.3 *Obtains, or Endeavors to Obtain, Information*. A person may be punished as a spy regardless of whether that person succeeds in obtaining information or in transmitting it to the enemy.[344] A person, however, must obtain or attempt to obtain information to be considered a spy.

4.17.2.4 *With the Intention of Communicating It to the Hostile Party*. A person must act with the intention of communicating the sought-after information to a hostile party to the conflict to be considered a spy within the meaning of this rule. However, a person who seeks to send information to a State not involved in the conflict may still commit acts punishable by the offended State, and that person's conduct may fall within the broader category of secretive, hostile acts behind enemy lines.[345]

4.17.3 Saboteurs and Other Persons Engaging in Secretive, Hostile Acts Behind Enemy Lines. In addition to spies, other persons acting clandestinely or under false pretenses with a hostile purpose behind enemy lines have also been treated like spies under the law of war. Like spies, persons engaged in these secretive, hostile activities behind enemy lines have also been deprived of the privileges of combatant status and often been punished. For example, saboteurs[346] acting clandestinely or under false pretenses in the zone of operations of a belligerent are treated as spies.[347] However, activities besides sabotage that are helpful to one side's war effort that are done behind the other side's lines have been punishable as well, often under the rubric of "war treason,"[348] "secretly entering the lines,"[349] or "activities hostile to the security of the State."[350]

These kinds of activities almost necessarily take on the character of spying.[351] In many cases, these actions have been reported as spying.[352] However, the actual purpose of these activities may not be to gain or transmit intelligence information, but to take other sorts of actions that would further the war effort. Thus, a belligerent's presence on the territory controlled by an opposing State with a hostile purpose while clandestinely or under false pretenses, suffices to make that person liable to treatment as a spy under the law of war.[353]

4.17.4 Spying and Sabotage Permissible Under the Law of War. Under the law of war, belligerents may employ spies and saboteurs.[354]

Spying and sabotage are not prohibited by any law of war treaty to which the United States is a Party. For example, spying and sabotage are not prohibited by the 1949 Geneva Conventions, nor defined as a "grave breach" of those conventions.[355] Similarly, spying and sabotage also have not been listed as war crimes punishable under the statutes of international criminal tribunals.[356] In addition, law of war treaties that regulate, but do not prohibit, spying, recognize implicitly that belligerents may use this method of warfare.[357]

Although spying and sabotage are not prohibited by the law of war, acting clandestinely or under false pretenses could, in some circumstances, constitute "feigning a protected status," one of the elements of perfidy.[358] Persons engaged in these activities and commanders who employ them should take special care not to kill or wound by resort to perfidy.

4.17.5 Spying and Sabotage – Forfeiture of the Privileges of Combatant Status. Although the law of war allows belligerents to employ spies, saboteurs, and other persons engaged in secretive hostile activities behind enemy lines, the law of war also permits belligerents to take additional measures to defend against these persons. These individuals, by acting clandestinely or under false pretenses, fail to distinguish themselves as combatants generally must do.[359] Thus, persons otherwise entitled to privileges of combatant status, including POW status, forfeit their entitlement to those privileges while engaged in spying, sabotage, or other hostile, secretive activities behind enemy lines.[360] Although not explicitly reflected in the GPW, this understanding was the general understanding at the 1949 Diplomatic Conference[361] and is reflected in other treaties,[362] judicial decisions,[363] military manuals,[364] and scholarly works.[365]

4.17.5.1 *Liability of Persons Not Captured While Spying for Previous Acts of Espionage*. Persons who qualify for the privileges of combatant status who engage in spying, and then return to friendly lines, incur no responsibility or liability for previous acts of espionage.[366] Persons who have never qualified for the privileges of combatant status would not benefit from this rule because they cannot regain a status that they did not receive in the first place.[367]

4.17.5.2 *Cases of Doubt*. During international armed conflict, should there be any doubt as to whether persons suspected of committing a belligerent act and having fallen into the hands of the enemy are entitled to POW status, such persons are entitled to have their status determined by a competent tribunal and should be treated as POWs pending that determination.[368]

4.18 PRIVATE PERSONS WHO ENGAGE IN HOSTILITIES

In general, private persons who engage in hostilities forfeit many of the protections to which members of the civilian population are entitled, and are liable to treatment in one or more respects as unprivileged belligerents.[369]

4.18.1 Private Persons Who Engage in Hostilities – Notes on Terminology. This section refers to "private persons" who engage in hostilities rather than "civilians" who engage in hostilities for three reasons. First, an emphasis on "private" persons is consistent with longstanding explanations of the principle of distinction.[370] Second, it may be analytically unhelpful to focus on "civilians" because private persons who engage in hostilities are liable to treatment in one or more respects as combatants. Third, non-military personnel belonging to a State (*e.g.*, persons authorized to accompany the armed forces), who are often called "civilians," raise a different set of issues that merit special consideration as opposed to the general case of a private person who decides to engage in hostilities.

For the purpose of applying different law of war rules, different formulations have been used to describe when a person has engaged in hostilities. Although these phrases often refer to the same conduct, the context in which each term is applied is important; whether a phrase includes a particular type of conduct may depend on the particular legal rule in question. Whether a particular person is liable as a consequence of his or her conduct to treatment as a combatant (*e.g.*, being made the object of attack, internment, or prosecution for acts of unprivileged belligerency) must be assessed with reference to the specific legal rule at issue rather than based on the use of a conclusory label, such as "enemy combatant."[371]

This manual generally uses the phrase "engaging in hostilities" in a broad sense to refer to any of those actions that could cause a person to forfeit one or more protections under the law of war. When discussing specific legal rules, on the other hand, this manual uses the particular language of the rule at issue, rather than the phrase "engaging in hostilities." For example, this manual generally reserves the use of the phrase "taking a direct part in hostilities" to address the rule applicable when a civilian forfeits his or her protection from being made the object of attack.[372]

4.18.2 Private Persons Who Engage in Hostilities – Denial of the Distinct Protections Afforded Peaceful Civilians. Private persons who engage in hostilities forfeit many of the protections afforded civilians under the law of war.[373] The principle of military necessity can justify taking military action (such as detention) to address the threat posed by such persons.[374] The forfeiture of many of the protections of civilian status is also reflected in the principle that a person may not claim the distinct rights of both combatants and civilians at the same time.[375]

4.18.3 Private Persons Who Engage in Hostilities – Lack of the Privileges of Combatant Status. Private persons who engage in hostilities are not entitled to the privileges of combatant status and may be punished, after a fair trial.[376] The law of war does not condone the "farmer by day and guerilla by night."[377]

The denial of the privileges of combatant status to private persons engaging in hostilities may be justified on the basis that such persons act inconsistently with the *jus ad bellum* principle of competent authority, under which the resort to armed force is a prerogative of the State.[378] These individuals lack the principal qualification for entitlement to the privileges of combatant status—State authorization.[379] This requirement for State authorization has been traced to medieval law of arms.[380]

Denying private persons who engage in hostilities the privileges of combatant status has also been viewed as justified on humanitarian grounds. Private persons who engage in hostilities risk undermining the protections afforded the civilian population.[381] And, private persons who engage in hostilities generally have not been trained in the law of war and are not subject to the same disciplinary regime as members of the armed forces.[382] Thus, their participation in hostilities has been associated with the commission of war crimes.[383]

4.18.4 Activities That Constitute "Engaging in Hostilities" by Private Persons. Certain activities constitute "engaging in hostilities," *i.e.*, when done by private persons, make them liable to treatment in one or more respects by the enemy State as unprivileged belligerents.

4.18.4.1 *Being Part of a Hostile, Non-State Armed Group.* Being part of a non- State armed group that is engaged in hostilities against a State is a form of engaging in hostilities that makes private persons liable to treatment in one or more respects as unprivileged belligerents by that State.[384] Being part of a non-State armed group may involve formally joining the group or simply participating sufficiently in its activities to be deemed part of it. Whether a person has joined a non-State armed group may be a difficult factual question. Non-State armed groups may not use formal indicia of membership (*e.g.*, uniforms or identity cards), or members of these groups may seek to conceal their association with the group.[385] It may be appropriate to use circumstantial or functional information to assess whether a person is part of a non-State armed group.[386]

In addition, non-State armed groups may also rely on individuals who are not members of the groups, but who are functionally part of those organizations. Their support might be particularly important to enable the non-State armed group to conduct and sustain its operations.[387] For example, these individuals might participate sufficiently in the activities of the group or support its operations substantially in a way that is analogous to the support that persons authorized to accompany the armed forces provide to the armed forces.[388] These individuals may be regarded as constructively part of the group, even if they are, in fact, not formal members of the group.[389]

4.18.4.2 *Other Hostile Acts.* In addition to being part of or substantially supporting a hostile, non-State armed group, other hostile acts can make a person liable to treatment in one or more respects as an unprivileged belligerent under the law of war. For example, private persons who bear arms against enemy personnel or who attempt to kill or injure enemy personnel would become liable to being made the object of attack.[390] Performing spying, sabotage, and other hostile activities behind enemy lines would render a person liable to prosecution for such conduct.[391]

4.18.5 Private Persons Who Engage in Hostilities and the Law of War. As in the cases of spying and sabotage, under international law private enemy nationals are not generally regarded as being prohibited from engaging in hostilities during international armed conflict. International law does recognize, however, that opposing States may punish such persons.[392] Nonetheless, like other forms of unprivileged belligerency, private engagement in hostilities has been described in some contexts as a "war crime" or as a violation of international law or the law of war.[393]

In discussions of the status of private acts of hostility under international law, the point is often made that international law does not require States to prevent what they may regard as acts of patriotism and heroism.[394] However, outside the context of patriotic acts of resistance by persons in international armed conflict, private acts of hostility often carry additional sanction under international law. For example, persons who set off in private military expeditions against a foreign State from a State that has peaceful relations with the foreign State, have been subject to punishment under the law of nations.[395] Similarly, private acts of hostility, under certain conditions, may be regarded as piracy.[396] In contemporary parlance, private acts of hostility are often punished as "terrorism."[397] The unauthorized use of violence by private persons to achieve political ends has been viewed as contrary to the principles of democratic States.[398] Moreover, States have obligations under international law to repress terrorism, especially when conducted on their territory against other States.[399]

4.19 RIGHTS, DUTIES, AND LIABILITIES OF UNPRIVILEGED BELLIGERENTS

Unprivileged belligerents have certain rights, duties, and liabilities. In general, unprivileged belligerents lack the distinct privileges afforded to combatants and civilians, and are subject to the liabilities of both classes. Unprivileged belligerents generally may be made the object of attack by enemy combatants. They, however, must be afforded fundamental guarantees of humane treatment if *hors de combat*. Unprivileged belligerents may be punished by enemy States for their engagement in hostilities if they are convicted after a fair trial.

4.19.1 Rules Applicable to the Treatment of Unprivileged Belligerents. Although unprivileged belligerents have not been recognized and protected in treaty law to the same extent as peaceful civilians and lawful combatants,[400] basic guarantees of humane treatment in customary international law (*i.e.*, elementary considerations of humanity) protect unprivileged belligerents.[401] Moreover, some treaty protections apply to certain unprivileged belligerents.[402]

In some cases, U.S. practice has, as a matter of domestic law or policy, afforded unprivileged belligerents more favorable treatment than they would be entitled to receive under international law.[403] Nonetheless, U.S. practice has also recognized that unprivileged belligerents should not be afforded the distinct privileges afforded lawful combatants and the protections afforded peaceful civilians under the law of war.[404]

4.19.2 Unprivileged Belligerents – Conduct of Hostilities. Although unprivileged belligerents lack the right to engage in hostilities, international law nevertheless requires that they observe the same duties as lawful combatants during their conduct of hostilities.[405] In addition, unprivileged belligerents are generally subject to the same liabilities to which combatants are subject in the conduct of hostilities; *i.e.*, they may be made the object of attack by enemy combatants. Unprivileged belligerents placed *hors de combat*, however, may not be made the object of attack, and must be treated humanely.

4.19.3 Unprivileged Belligerents – Detention. Unprivileged belligerents are liable to capture and detention, like lawful combatants.

4.19.3.1 *Humane Treatment*. Although unprivileged belligerents are not entitled to the privileges of POW status, unprivileged belligerents, like all other detained persons, must be treated humanely. In particular, they, like all other detainees, must receive, at a minimum, the fundamental guarantees of humane treatment described in Common Article 3 of the 1949 Geneva Conventions.[406] In addition, the United States has explicitly supported, out of a sense of legal obligation, the fundamental guarantees reflected in Article 75 of AP I as minimum standards for the humane treatment of all persons detained during international armed conflict.[407]

4.19.3.2 *Cases of Doubt*. During international armed conflict, in cases of doubt as to whether a person held as an unprivileged belligerent is, in fact, a POW or entitled to POW treatment, that person should enjoy POW protections until their status is assessed by a competent tribunal.[408]

4.19.3.3 *Necessary Security Measures*. A State may take necessary security measures with regard to unprivileged belligerents. Since they have engaged in hostilities, unprivileged belligerents, or persons suspected of being unprivileged belligerents, may be denied certain privileges to which they might otherwise be entitled under the law of war. For example, the rights of unprivileged belligerents who are entitled to "protected person" status under the GC would be subject to derogation for security reasons.[409] However, unprivileged belligerents who are protected by the GC should be afforded its full protections when feasible.[410]

4.19.3.4 *Duration of Detention*. Unprivileged belligerents who are detained in order to prevent their further participation in hostilities generally must be released when hostilities have ended, unless there is another legal basis for their detention.[411] DoD practice has been to review periodically the detention of all persons not afforded POW status or treatment.[412]

4.19.4 Unprivileged Belligerents – Liability for Participation in Hostilities. Although international law affords lawful combatants a privilege or immunity from prosecution,[413] unprivileged belligerents lack such protection.[414] A State may punish unprivileged enemy belligerents,[415] subject to applicable requirements, such as a fair trial.[416]

4.19.4.1 *Unprivileged Belligerency and the Law of War – Notes on Terminology*. In contemporary parlance, spying and other forms of unprivileged belligerency generally have not been referred to as "violations of the law of war" or "war crimes."[417] For example, spying is permissible under the law of war in the sense that belligerents are not prohibited from employing spies; these activities are punishable but not prohibited under international law.[418] However, in some cases, offenses related to unprivileged belligerency have been characterized as violations of the law of war.[419] For example, spying has been called a violation of the law of war or "war crime."[420] Spying and other acts of unprivileged belligerency have been called offenses against the law of nations or law of war – the punishment of these offenders has been viewed as a recognized incident or exercise of a belligerent's war powers under the law of war.[421]

The difference in these characterizations may be traced to different definitions of "war crime" that have been used over time. The difference in these characterizations may also be traced to different definitions of the "law of war." If one views the law of war as only containing prohibitions, the punishment of unprivileged belligerents, like all exercises of the war powers, emanates from the domestic law of the belligerent State.[422] On the other hand, if one views the law of war as also including sources of authority, the punishment of unprivileged belligerents is also grounded in the international law of war.[423] Although the relationship between unprivileged belligerency and the law of war has been characterized in different ways, it is well-accepted that States may punish unprivileged enemy belligerents after a fair trial.

4.20 CHILDREN

The GC provides special protection for children in order to protect them against the dangers of war. In addition, certain provisions of treaties and U.S. law seek to restrict the use or recruitment of children in armed conflict.

4.20.1 Specific Protections for All Children During International Armed Conflict.

4.20.1.1 *Children Under Fifteen Who Are Orphaned or Separated*. The parties to the conflict shall take the necessary measures to ensure that children under fifteen, who are orphaned or are separated from their families as a result of the war, are not left to their own resources, and that their maintenance, the exercise of their religion, and their education are facilitated in all circumstances.[424] The maintenance of the children concerned means their feeding, clothing, and accommodation, care for their health, and, where necessary, medical and hospital treatment.[425]

Their education shall, as far as possible, be entrusted to persons of a similar cultural tradition.[426] This provision is intended to exclude inappropriate religious or political propaganda designed to influence them.[427] The parties to the conflict shall facilitate the reception of such children in a neutral country for the duration of the conflict with the consent of the Protecting Power, if any, and under due safeguards for the observance of the principles stated in the first paragraph of Article 24 of the GC.[428]

4.20.1.2 *Identification for Children Under Twelve*. The parties to the conflict shall endeavor to arrange for all children under twelve to be identified by the wearing of identity discs, or by some other means.[429] The age of twelve was chosen because it was considered that children over twelve are generally capable of stating their own identity.[430]

4.20.2 Protection for Alien Children in a Belligerent's Home Territory. In a belligerent's home territory, the GC provides that certain aliens (*i.e.*, those qualifying for protected person status under the GC) who are children under fifteen years, pregnant women, and mothers of children under seven years shall benefit from any preferential treatment to the same extent as the nationals of the State concerned.[431]

4.20.3 Protection for Children in Occupied Territory. Provisions of the GC address the protection of children in occupied territory.[432]

4.20.4 Protection for Children in the Context of Internment Under the GC. The GC provides certain protections for children in the context of internment. Provisions of the GC address:

• the internment of children with their parents;[433]

• special treatment for children during internment, including

o ensuring the education of children and young people, including school attendance,[434] o additional food for children under fifteen in proportion to their physiological needs,[435] and

o special playgrounds reserved for children and young people;[436] and

• special agreements for the release, the repatriation, the return to places of residence, or the accommodation in a neutral country of, among others, children, pregnant women, and mothers with infants and young children.[437]

4.20.5 Child Soldiers. Certain provisions of treaties and U.S. law seek to restrict the use or recruitment of children in armed conflict. If children are nonetheless employed in armed conflict, they generally are treated on the same basis as adults, although children may be subject to special treatment in detention because of their age. Prohibitions on the use or recruitment of children also apply in non-international armed conflict.[438]

4.20.5.1 *U.S. Offense of Recruiting or Using Child Soldiers*. U.S. law makes it a crime, under certain circumstances, to recruit, enlist, or conscript a person to serve in an armed force or group, while such person is under 15 years of age.[439] U.S. law also makes it a crime to use a person under 15 years of age to participate actively in hostilities.[440] These restrictions in U.S. law are similar to provisions in treaties to which the United States is not a Party.[441]

4.20.5.2 *Child Soldiers Protocol*. As a Party to the Optional Protocol to the Convention on the Rights of the Child on the involvement of children in armed conflict,[442] the United States must "take all feasible measures to ensure that members of [its] armed forces who have not attained the age of 18 years do not take a direct part in hostilities."[443] The United States made a statement of understanding regarding the meaning of the phrase "direct part in hostilities" in the Child Soldiers Protocol.[444]

As a Party to the Child Soldiers Protocol, the United States also has an obligation to ensure that persons who have not attained the age of 18 years are not compulsorily recruited into its armed forces.[445] In a declaration deposited pursuant to Article 3(2) of the Child Soldiers Protocol,[446] the United States described its measures in place to comply with this obligation and stated that the minimum age at which it permitted persons to be voluntarily recruited in the U.S. Armed Forces is 17 years of age.[447]

4.20.5.3 *Treatment of Child Soldiers*. In general, children receive the rights, duties, and liabilities of combatant status on the same basis as other persons. For example, there is no age requirement for someone to receive POW status. Similarly, whether a civilian is considered to be taking a direct part in hostilities does not depend on that person's age. Children who have participated in hostilities or been associated with an armed force who are detained might require additional consideration because of their age. For example, rules for the additional provision for their education might be applicable.[448] Similarly, it might be appropriate to take into account the age of a defendant in determining liability or punishment for violations of the law of war.

4.21 MERCENARIES

The act of being a mercenary is not a crime in customary international law nor in any treaty to which the United States is a Party.[449] Under the customary law of war and the GPW, "mercenaries" receive the rights, duties, and liabilities of combatant status on the same basis as other persons. For example, being a "mercenary" or paid for participating in hostilities does not deprive a person of POW status, if that person otherwise meets the requirements for POW status, such as by being a member of a militia that is part of the armed forces of a State.[450] Mercenaries must comply with the law of war and may be tried and punished for violations of the law of war on the same basis as other persons. States that employ mercenaries are responsible for their conduct, including their compliance with the law of war. Mercenaries are often nationals of States that are not parties to a conflict. In general, a national of a neutral State who, during an international armed conflict, commits hostile acts against a State or who voluntarily enlists in the armed forces of one of the parties should not be more severely treated by the State against whom he or she has abandoned his or her neutrality than a national of the other belligerent State could be for the same act.[451]

4.21.1 Treaties on Mercenaries. A number of treaty provisions are intended to repress mercenary activities. The United States has not accepted any such provision because these efforts are not consistent with fundamental principles of the law of war.[452] AP I denies "mercenaries" the right to be lawful combatants or POWs.[453] This provision in AP I was adopted because certain States wanted to condemn mercenary groups that had played a role in armed conflicts that had taken place in Sub-saharan Africa since 1960.[454] However, AP I defines "mercenary" very narrowly.[455] For example, any member of the armed forces of a party to the conflict, or any member of the armed forces of any other State, who is sent on official duty as a member of its armed forces, is not a mercenary as defined by AP I.[456]

In addition, although, under AP I, mercenaries lack the right to be a combatant or a POW, Parties to AP I may nonetheless decide as a matter of policy to treat enemy mercenaries as lawful combatants or POWs.

Shortly after the adoption of AP I, the Convention for the Elimination of Mercenarism in Africa was adopted, which uses the same definition of mercenary as AP I,[457] but creates greater obligations for Parties to punish and repress mercenarism.[458] The United States is not a Party to this treaty.

The International Convention against the Recruitment, Use, Financing and Training of Mercenaries of December 4, 1989, defines "mercenary" slightly more broadly than AP I, applies to both international and non-international armed conflicts, and defines various offenses related to the recruiting, use, financing, or training of mercenaries.[459] The United States is not a Party to this treaty.

4.22 AP I PROVISIONS ON CIVIL DEFENSE PERSONNEL

Articles 61-67 of AP I address "civil defence", *i.e.*, the performance of certain humanitarian tasks intended to benefit the civilian population.[460] The United States has supported the principle that civilian civil defense organizations and their personnel be respected and protected as civilians and be permitted to perform their civil defense tasks except in cases of imperative military necessity.[461] However, a number of military operational problems have been identified with respect to the system of protection for civil defense established by AP I, and these provisions of AP I may be understood not to preclude an attack on an otherwise lawful military objective.[462]

4.23 LAW ENFORCEMENT OFFICERS

In general, police officers receive the rights, duties, and liabilities of civilian status.[463] However, law enforcement agencies are in many respects similar to military forces. They are authorized to use necessary force on behalf of the State, are generally distinguishable from private citizens, and are often organized like military forces. In cases where States choose to use police officers as part of the armed forces, they receive the rights, duties, and liabilities of combatant status.[464]

4.23.1 Police as Civilians. In general, members of law enforcement agencies have civilian status.[465] Furthermore, routine domestic law enforcement is part of the general protection of the civilian population and does not constitute "taking a direct part in hostilities" that would deprive police officers of their protection from being made the object of attack.[466]

4.23.2 Police With a Military Role. Some States use police forces in a paramilitary capacity or use military forces in a police role. Members of the armed forces engaged in police roles are combatants.

The extent to which police officers are treated as combatants largely depends on whether the State decides to use them in that capacity. States may decide to make law enforcement agencies part of their armed forces.[467] Members of these law enforcement agencies, like other members of those armed forces, receive combatant status by virtue of their membership in the armed forces.[468] In addition, States may authorize members of the law enforcement agencies to accompany their armed forces without incorporating them into their armed forces. These persons have the legal status of persons authorized to accompany the armed forces.[469]

4.23.3 Police in Non-International Armed Conflict. Police officers may play a larger role in armed conflicts between States and insurgent or terrorist groups because in such conflicts the State may treat all enemy persons' participation in hostilities as criminal.[470]

4.24 JOURNALISTS

In general, journalists are civilians and are protected as such under the law of war. Journalists play a vital role in free societies and the rule of law and in providing information about armed conflict.[471] Moreover, the proactive release of accurate information to domestic and international audiences has been viewed as consistent with the objectives of U.S. military operations.[472] DoD operates under the policy that open and independent reporting is the principal means of coverage of U.S. military operations.[473] In addition to responding to press inquiries and providing briefings to members of the press on U.S. military operations, DoD practice has also been to embed journalists with units during military operations.[474] Embedded journalists are assigned to a unit, and they eat, sleep, and move with the unit.[475]

4.24.1 Status of Journalists – Generally Civilian. In general, journalists are protected as civilians;[476] *i.e.*, engaging in journalism does not constitute taking a direct part in hostilities such that such a person would be deprived of protection from being made the object of attack.[477] Journalists do not form a distinct class of persons under the law of war, but instead receive protection through the general protections afforded civilians.[478] Thus, in general, the rights, duties, and liabilities applicable to civilians also apply to journalists.[479] Although journalism is regarded as a civilian activity, the fact that a person performs such work does not preclude that person from otherwise acquiring a different status under the law of war, such as the status of persons authorized to accompany the armed forces or of combatants.

4.24.1.1 *War Correspondents and Other Journalists Authorized to Accompany the Armed Forces*. Journalists authorized to accompany the armed forces have the status of such persons.[480] For example, if they fall into the power of the enemy during international armed conflict, they are entitled to POW status.[481] Such journalists are sometimes called "war correspondents," which is a term used in Geneva Conventions of 1949.[482]

Since journalists authorized to accompany the armed forces are liable to become POWs, they must be issued identification cards so that they may establish their entitlement to POW status upon falling into the power of the enemy during international armed conflict.[483]

4.24.1.2 *Combatants Performing Journalistic Work*. Although generally journalists are civilians, journalists are not precluded from being considered combatants, whether privileged or unprivileged, if they otherwise acquire such status. For example, members of the armed forces sometimes serve as journalists or in some other public affairs capacity, and these persons have the same status as other privileged combatants.[484] Non-State armed groups sometimes use their members[485] for propaganda or other media activities, and such personnel are not precluded from being considered unprivileged belligerents.[486] In addition, although engaging in journalism would not be a basis to consider a person an unprivileged belligerent, an unprivileged belligerent would not be precluded from being considered as such because he or she works as a journalist.

4.24.2 General Rules for the Treatment of Civilian Journalists and Journalists Authorized to Accompany the Armed Forces. Although journalists generally are protected as civilians as a legal matter, the factual circumstances of armed conflict often expose journalists exercising their profession in such a situation to dangers not normally encountered by other civilians.[487]

4.24.2.1 *Journalists – Risks in Areas of Military Operations*. Journalists who enter areas of military operations assume a significant risk that they could be injured or killed incidental to an enemy attack or from other dangers.[488] Unfortunately, journalists have sometimes been killed while reporting in close proximity to combat operations.[489] To minimize the risk of mistakenly being made the object of attack, journalists often seek to distinguish themselves from military forces.[490] Where possible, efforts should be made to distinguish between the activities of journalists and the activities of enemy forces, so that journalists' activities (*e.g.*, meetings or other contacts with enemy personnel for journalistic purposes) do not result in a mistaken conclusion that a journalist is part of enemy forces. In addition, civilian journalists and journalists authorized to accompany the armed forces should not take any action adversely affecting their status as civilians if they wish to retain protection as a civilian.[491] For example, relaying target coordinates with the specific purpose of directing an artillery strike against opposing forces would constitute taking a direct part in hostilities that would forfeit protection from being made the object of attack.[492]

4.24.2.2 *Journalists – Reporting on Military Operations*. As noted above, DoD operates under the policy that open and independent reporting is the principal means of coverage of U.S. military operations. Under the law of war, there is no special right for journalists to access areas of military operations without the consent of the State conducting those operations.[493] The law of war does not prohibit States from taking security measures to reduce the risk of disclosure of sensitive military information, such as numbers of military personnel, types of on-hand equipment, unit locations, and plans for future operations.[494] For example, military security ground rules may be developed for media personnel.[495] The particular restrictions imposed by a State are likely to depend on its domestic law and policy. Access to military operations that is afforded to media personnel must be consistent with law of war obligations,[496] including obligations to protect prisoners of war and other detainees against public curiosity.[497]

4.25 DELEGATES OR REPRESENTATIVES OF THE PROTECTING POWERS

4.25.1 Appointment of Delegates of the Protecting Powers. The Protecting Powers may appoint, apart from their diplomatic or consular staff, delegates from among their own nationals or the nationals of other neutral Powers to carry out its duties under the GPW and the GC.[498] These delegates shall be subject to the approval of the Power with which they are to carry out their duties.[499]

4.25.2 Duties of the Representatives or Delegates of the Protecting Power. The parties to the conflict shall facilitate to the greatest extent possible the task of the representatives or delegates of the Protecting Powers.[500]

The representatives or delegates of the Protecting Powers shall not in any case exceed their mission under the 1949 Geneva Conventions.[501] They shall, in particular, take account of the imperative necessities of security of the State wherein they carry out their duties.[502]

4.25.3 Restrictions on Representatives of the Protecting Powers. A belligerent may impose legitimate security restrictions on the activities of the delegates or representatives of the Protecting Powers working in its territory or in its facilities. However, belligerents should only restrict the activities of the representatives or delegates of the Protecting Powers "as an exceptional and temporary measure when this is rendered necessary by imperative military necessities."[503] For example, a belligerent may postpone a visit by Protecting Power representatives to a POW camp for security or humanitarian reasons, such as tactical movement of its own forces or to protect Protecting Power personnel from explosive remnants of war being cleared from recent military operations along the route to the POW camp.

4.26 ICRC AND OTHER IMPARTIAL HUMANITARIAN ORGANIZATIONS

The 1949 Geneva Conventions contemplate that the ICRC and other impartial humanitarian organizations may, subject to the consent of the parties to the conflict concerned, provide humanitarian aid and seek to ensure the protection of war victims in armed conflict.[504] In some cases, the ICRC or another impartial humanitarian organization may assume the humanitarian functions performed by the Protecting Powers.[505]

4.26.1 "Impartial". To be an "impartial humanitarian organization" under the 1949 Geneva Conventions, a humanitarian organization must be impartial. The requirement of impartiality distinguishes these humanitarian organizations from those with an allegiance to a party to the conflict, such as national red cross societies.[506]

4.26.2 Consent of the Parties to the Conflict Concerned. The activities of the ICRC or other impartial humanitarian organizations in a State's sovereign territory, or in the area of a State's military operations, are subject to the consent of that State.[507] The requirement of State consent is based on the State's sovereign right to control access to its territory and a belligerent's right to control access to its military operations or territory it occupies.[508] For example, members of impartial humanitarian organizations, like other civilians, may be removed from the vicinity of military objectives for their protection.[509]

States may grant access on a case-by-case basis; access granted to one impartial humanitarian organization does not constitute entitlement of access for other humanitarian organizations. Impartial humanitarian organizations that have been granted access must also act within the terms of this consent.[510] For comparison, the activities of the Protecting Power are also subject to the consent of the affected States.[511]

States should not arbitrarily withhold their consent to the activities of humanitarian organizations, even in non-international armed conflict.[512] Where a State has accepted the services of an impartial humanitarian organization, the State must not regard such service, such as the provision of medical care, as unlawful and subject to punishment.[513]

4.26.2.1 *Impartial Humanitarian Organizations – Conditions on Access*. States may attach conditions to their consent, including necessary security measures. For example, in the past, armed groups have sometimes attempted to use humanitarian organizations as cover for participation in hostilities.[514] In addition to legitimate military considerations, other considerations may also limit access by impartial humanitarian organizations to military operations. For example, the GPW obligates a Detaining Power to protect POWs from "public curiosity," which may entail limiting access to POWs by private organizations.[515]

4.26.2.2 *Impartial Humanitarian Organizations – CCW Amended Mines Protocol Obligation*. In certain cases, however, there may be an obligation to take measures to help protect the personnel of certain humanitarian organizations that are performing functions with the consent of the Party to the CCW Amended Mines Protocol on whose territory the functions are performed from the effects of mines, booby-traps, and other devices.[516]

4.26.3 Special Status of the ICRC. The 1949 Geneva Conventions explicitly recognize the special position of the ICRC among impartial humanitarian organizations.[517] Similarly, Congress has specifically authorized – and the President has designated – the ICRC to be extended the same privileges and immunities that are afforded to public international organizations in which the United States participates.[518] The President has also recognized the role of the ICRC in visiting individuals detained in armed conflict.[519] The United States has relied on the ICRC's capacity, particularly in conflict situations, and has contributed substantially to the ICRC's work.[520] The United States has maintained a very constructive dialogue with the ICRC.

The ICRC does important work in visiting detainees, facilitating communication between detainees and their families, organizing relief operations, and undertaking similar humanitarian activities during armed conflicts.

For example, the ICRC has performed the functions of a Protecting Power during armed conflict.[521] In addition, the ICRC has performed the functions of the Central Information Agency for POWs and protected persons during international armed conflict.[522] In order to facilitate access, the ICRC conducts its visits to detention facilities on a confidential basis, a practice that DoD has sought to respect.[523] The ICRC has issued policy proposals or interpretative guidance on a variety of international law issues. Although the ICRC's proposals and interpretations do not have binding legal effect, they have often been helpful to States. In some cases, the United States and other States have not accepted the ICRC's proposals or interpretations and instead expressed opposing views. For example, the United States has not accepted the ICRC's study on customary international humanitarian law nor its "interpretive guidance" on direct participation in hostilities.[524]

4.27 DETERMINING THE STATUS OF DETAINEES IN CASES OF DOUBT

4.27.1 Identification Cards Used to Help Clarify Status. The 1949 Geneva Conventions contemplate that identification cards will be used to help clarify the status of detainees in international armed conflict.[525] Parties to the GPW must provide identity cards to persons under their jurisdiction who are liable to become POWs.[526] Similarly, Parties to the GWS must provide retained personnel with a special identity card that denotes their status.[527] Capturing units should not take these identity cards from POWs or retained personnel.[528] In addition, States should retain duplicate copies of identification cards that they issue.[529] In contemporary practice, modern storage systems, such as computer databases, are used rather than storing duplicates of the issued identification cards.[530]

Producing an identification card to capturing forces is not necessarily a prerequisite for a person to be entitled to a particular status. Identification cards may become lost, damaged, or stolen during military operations, so a failure to produce an identity card does not necessarily mean that person lacks a particular status.[531]

4.27.2 POW Protections for Certain Persons Until Status Has Been Determined. Capturing personnel may be unable to establish a detainee's status, including whether that person is entitled to POW status under the GPW. For example, a detainee might have lost his or her identity card or the detainee might be a deserter who does not wish to admit that he or she is a member of enemy armed forces. During international armed conflict, should any doubt arise as to whether persons, having committed a belligerent act and having fallen into the hands of the enemy, belong to any of the categories enumerated in Article 4 of the GPW, such persons shall enjoy the protection of the GPW until such time as their status has been determined by a competent tribunal.[532]

4.27.3 Competent Tribunal to Assess Entitlement to POW Status or Treatment. The "competent tribunal" in Article 5 of the GPW is often called an "Article 5 tribunal." In some cases, courts have undertaken to assess whether a detainee is entitled to POW status,[533] but a "competent tribunal" generally entails the Detaining Power convening an administrative board.[534] The GPW affords the Detaining Power substantial discretion regarding the composition and procedures of an Article 5 tribunal.[535]

Article 5 only requires a tribunal if there is "any doubt" regarding a person's entitlement to POW status or treatment.[536] For example, if there was no doubt that the armed group to which a person belongs fails to qualify for POW status, then the GPW would not require a tribunal to adjudicate the person's claim to POW status by virtue of membership in that group.[537]

4.27.4 Tribunals to Assess Other Detainee Issues. By its terms, Article 5 of the GPW only addresses a person's entitlement to POW status or treatment. However, an administrative process may be appropriate to address status questions besides entitlement to POW status or treatment, such as whether detainees are retained personnel or civilians. DoD practice has been to use Article 5 tribunals or similar administrative tribunals to address those issues.[538] For example, DoD used administrative tribunals to address a variety of detainee issues during detention operations in Viet Nam,[539] Panama,[540] the Persian Gulf in 1991,[541] Iraq in 2003,[542] Afghanistan,[543] and Guantanamo.[544]

V – The Conduct of Hostilities

Chapter Contents

5.1 INTRODUCTION

The right of States engaged in armed conflict to adopt means and methods of warfare is not unlimited.[1] This Chapter addresses the law of war rules on the conduct of hostilities during international armed conflict, such as the rules applicable to conducting attacks, the seizure and destruction of enemy property, and deception.

This Chapter, however, does not address all rules related to the conduct of hostilities. For example, rules on the use of certain types of weapons are addressed in Chapter VI.[2] In addition, later chapters address conduct of hostilities rules and issues specific to the Naval (Chapter XIII), Air (Chapter XIV), Space (Chapter XIV), and Cyber (Chapter XVI) domains, and Chapter XVIII addresses non-international armed conflict.[3]

Violations of the rules in the conduct of hostilities may be violations of criminal law, although it is not a purpose of this Chapter to address liability under criminal law.

5.1.1 Notes on Terminology – "Means and Methods of Warfare". In general, *method of warfare* has referred to how warfare is conducted, while *means of warfare* has referred to weapons or devices used to conduct warfare.[4]

For example, an analysis of a *method of warfare* might consider the way in which an artillery projectile may be employed, particularly where employment could have an adverse effect on the civilian population. On the other hand, an analysis of the legality of the *means of warfare* might consider the legality of the way in which the artillery projectile is designed to kill or injure enemy combatants.[5]

Although the terms *means of warfare* and *methods of warfare* lack an established, specific meaning, in some cases, the phrase "method of warfare" may be a term of art in a legal instrument that has been specifically interpreted by the United States. For example, the Chemical Weapons Convention obligates Parties to undertake not to use riot control agents as a "method of warfare," and the United States has interpreted that prohibition not to include certain uses of riot control agents.[6]

5.1.2 Implementation of Law of War Obligations in the Conduct of Hostilities During Military Operations. The law of war rules for the conduct of hostilities have been implemented during military operations through rules of engagement and other military orders.[7]

5.1.2.1 *Applying Standards in the Conduct of Hostilities That Pose More Restrictions on Military Operations and That Are More Protective of Civilians Than Required By the Law of War*. U.S. military operations are routinely subject to more restrictions and apply standards that are more protective of civilians than required by the law of war.[8] For example, military commanders often seek to reduce the risk of civilian casualties by taking additional precautions even when such measures are not required by the law of war.[9] Similarly, there are cases in which, for military or policy reasons, an attack is not conducted, even though the attack would be legally permissible.[10]

As with other cases in which a higher standard is applied as a matter of policy, violations of such standards are not necessarily violations of the law of war.[11]

5.1.2.2 *Applicability of International Obligations to Persons With the Domestic Authority to Implement the Obligations*. As with other aspects of the law of war, the persons within a party to a conflict who are responsible for implementing an obligation, are those persons with the authority to make the decisions and judgments required by the law of war.[12] For example, the decisions and judgments required by the principle of proportionality normally require authority over military operations, and ordinarily would not require military personnel to take actions, such as providing warnings, when a superior commander has already determined that such actions are not feasible.[13] As another example, a pilot generally would be entitled to rely on the determination by headquarters that a given target is, in fact, a military objective.[14]

5.1.3 Principles for the Conduct of Hostilities. The principles that provide the foundation for the law of war are discussed in Chapter II, and these principles apply to the conduct of hostilities. As noted in Chapter II, specific rules on the conduct of hostilities are derived from, and must be interpreted consistent with, the basic principles that underlie the law of war.[15] Moreover, where no specific rule applies, law of war principles provide a general guide for conduct during war, including the conduct of hostilities.[16]

5.2 OVERVIEW OF RULES FOR THE PROTECTION OF CIVILIANS

The protection of civilians[17] against the harmful effects of hostilities is one of the main purposes of the law of war.[18] Many of the rules for the protection of civilians are derived from the principles of distinction and proportionality.[19] Specific rules for the protection of civilians may be grouped into two categories: (1) essentially negative duties to respect civilians and to refrain from directing military operations against them; and (2) affirmative duties to take feasible precautions to protect civilians and other protected persons and objects.

5.2.1 Responsibility of the Party Controlling Civilian Persons and Objects. The party controlling civilians and civilian objects has the primary responsibility for the protection of civilians and civilian objects.[20] The party controlling the civilian population generally has the greater opportunity to minimize risk to civilians.[21] Civilians also may share in the responsibility to take precautions for their own protection.[22] For example, civilians may decide to take measures for their own protection upon receiving a warning to stay away from military objectives.[23]

5.2.2 Essentially Negative Duties to Respect Civilians and to Refrain From Directing Military Operations Against Them. In general, military operations must not be directed against enemy civilians.[24] In particular:

- Civilians must not be made the object of attack;[25]
- Military objectives may not be attacked when the expected incidental loss of civilian life, injury to civilians, and damage to civilian objects would be excessive in relation to the concrete and direct military advantage expected to be gained;[26]
- Civilians must not be used as shields or as hostages;[27] and

• Measures of intimidation or terrorism against the civilian population are prohibited, including acts or threats of violence, the primary purpose of which is to spread terror among the civilian population.[28]

5.2.2.1 *Non-Violent Measures That Are Militarily Necessary*. The principle that military operations must not be directed against civilians does not prohibit military operations short of violence that are militarily necessary. For example, such operations may include:

• stopping and searching civilians for weapons and to verify that they are civilians;[29]

• temporarily detaining civilians for reasons of mission accomplishment, self-defense, or for their own safety;[30]

• collecting intelligence from civilians, including interrogating civilians;[31]

• restricting the movement of civilians or directing their movement away from military operations for their own protection;[32] or

• seeking to influence enemy civilians with propaganda.[33]

5.2.2.2 *Military Operations Intended to Benefit Civilians*. The principle that military operations not be directed against civilians does not preclude military operations intended to benefit civilians. Such operations may include humanitarian assistance operations, noncombatant evacuation operations, civil affairs operations, or civil-military operations. During counter-insurgency operations, military operations to protect civilians and to help obtain their support may be particularly important.[34]

5.2.3 Affirmative Duties to Take Feasible Precautions for the Protection of Civilians and Other Protected Persons and Objects. Parties to a conflict must take feasible precautions to reduce the risk of harm to the civilian population and other protected persons and objects.[35] Feasible precautions to reduce the risk of harm to civilians and civilian objects must be taken when planning and conducting attacks.[36] Feasible precautions should be taken to mitigate the burden on civilians when seizing or destroying enemy property.[37] It is specifically provided that feasible precautions must be taken in connection with certain types of weapons.[38] Feasible precautions to reduce the risk of harm to civilians must also be taken by the party subject to attack.[39]

5.2.3.1 *Feasible Precautions – Notes on Terminology*. Although this manual primarily uses "feasible," other adjectives have been used to describe the obligation to take feasible precautions during armed conflict. The words "practicable,"[40] "reasonable,"[41] "due,"[42] and "necessary"[43] have been used to describe this obligation. Some sources may distinguish between whether a precaution is "feasible" or "reasonable" or "practical."[44] It is also possible to use these words interchangeably with one another, *e.g.*, such that feasible precautions are reasonable precautions.[45]

Insofar as these words are used to describe the legal obligation to take precautions, these words are used as terms of art that do not necessarily take their ordinary meaning and are subject to a substantial degree of interpretation and discretion in their application to a given circumstance.

Thus, this manual does not distinguish between, for example, the obligation to take "reasonable" precautions versus the obligation to take "feasible" precautions versus the obligation to take "practical" precautions. This manual uses "feasible" to mean the standard of precautions that are required by the law of war, and that standard is explained in more detail in this section.

Some sources describe the obligation to take precautions as one in which "all feasible" precautions must be taken or that "everything feasible" must be done. Stating the requirement in terms of "all feasible" precautions could entail articulating a narrower interpretation of what is considered "feasible" rather than articulating a difference in what the law ultimately requires. In any event, as discussed below, the obligation to take feasible precautions does not require everything that is capable of being done – a standard that would be almost impossible to meet if interpreted literally and that is inconsistent with the definition of feasible precautions that provides for taking into account humanitarian and military considerations. In addition, the fact that a precaution was taken does not necessarily mean that taking the precaution was required as a matter of law.

5.2.3.2 *What Precautions Are Feasible*. The standard for what precautions must be taken is one of due regard or diligence, not an absolute requirement to do everything possible.[46] A wanton disregard for civilian casualties or harm to other protected persons and objects is clearly prohibited.[47]

Feasible precautions are those that are practicable or practically possible, taking into account all circumstances ruling at the time, including humanitarian and military considerations.[48] These circumstances may include:

- the effect of taking the precaution on mission accomplishment;[49]
- whether taking the precaution poses a risk to one's own forces or presents other security risks;
- the likelihood and degree of humanitarian benefit from taking the precaution;
- the cost of taking the precaution, in terms of time, money, or other resources; or
- whether taking the precaution forecloses alternative courses of action.

For example, a commander may determine that a precaution would not be feasible because it would result in increased operational risk (*i.e.*, a risk of failing to accomplish the mission) or an increased risk of harm to his or her forces.[50]

Similarly, taking a precaution would not be required if it is assessed not to yield a humanitarian benefit. For example, issuing a warning before an attack would not be required when only military objectives are expected to be harmed and civilians are not present.[51] Moreover, it would not be required to use distinctive emblems to identify civilians and civilian objects as such, if enemy forces are likely to use that information to direct attacks against those civilians and civilian objects.[52]

Since what precautions are feasible depends greatly on the context and other military considerations,[53] it would be inaccurate to conclude that specific precautions are required as a general rule. For example, there is not a general requirement to use precision-guided munitions.[54] Nonetheless, military commanders must make reasonable efforts to reduce the risk of harm to civilians and civilian objects.

5.2.3.3 *The Taking of Precautions and Law, Policy, and Practice*. There may be questions about the extent to which precautions are legally required or are instead policy decisions. The obligation to take feasible precautions is a legal requirement. However, the determination of whether a precaution is feasible involves significant policy, practical, and military judgments, which are committed to the responsible commander to make in good faith based on the available information. In assessing whether the obligation to take feasible precautions has been satisfied after the fact, it will be important to assess the situation that the commander confronted at the time of the decision and not to rely on hindsight.

As discussed above, it is not the case that the legal requirement to take feasible precautions requires whatever may be done. Rather, it is possible for precautions to be taken, as a matter of practice or policy, that are not required as a matter of law, and the U.S. military frequently has done so.

In this connection, it should also be noted that mere poor military judgment (such as mistakes or accidents in conducting attacks that result in civilian casualties) is not by itself a violation of the obligation to take precautions.

5.2.3.4 *Requirements to Take Precautions Regarding Specific Weapons*. In addition to the general obligation to take feasible precautions to reduce the risk of harm to the civilian population and other protected persons and objects, it is specifically provided that precautions be taken in connection with certain types of weapons:

- mines, booby-traps, or other devices;[55]
- incendiary weapons;[56]
- laser systems;[57] and
- explosive ordnance.[58]

5.2.3.5 *AP I Obligation to Take Constant Care to Spare Civilians and Civilian Objects*. Parties to AP I have agreed that "[i]n the conduct of military operations, constant care shall be taken to spare the civilian population, civilians and civilian objects."[59] Although this obligation is susceptible to a range of interpretations, Parties to AP I may also interpret it in a manner that is consistent with the discussion in this section.

5.3 ASSESSING INFORMATION UNDER THE LAW OF WAR

Commanders and other decision-makers must make decisions in good faith and based on the information available to them. Even when information is imperfect or lacking (as will frequently be the case during armed conflict), commanders and other decision-makers may direct and conduct military operations, so long as they make a good faith assessment of the information that is available to them at that time.[60]

5.3.1 Law of War Rules Often Depend on Difficult Factual Assessments. Many of the rules for the conduct of hostilities require determinations of fact that may be difficult to make. The special circumstances of armed conflict often make an accurate determination of facts very difficult.[61] For example, combatants must make decisions while enemy forces are attempting to attack them[62] and while enemy forces are seeking to deceive them.[63] In addition, the importance of prevailing during armed conflict[64] often justifies taking actions based upon limited information that would be considered unreasonable outside armed conflict.[65] Thus, for example, it may be difficult to discern whether a person is a combatant, a civilian, or a civilian taking a direct part in hostilities. Similarly, it may be difficult to assess the degree to which an object makes an effective contribution to the adversary's military action or to assess the concrete and direct military advantage anticipated from an attack.

5.3.2 Decisions Must Be Made in Good Faith and Based on Information Available at the Time. Decisions by military commanders or other persons responsible for planning, authorizing, or executing military action must be made in good faith and based on their assessment of the information available to them at the time.[66] A large number of States have recognized this principle.[67] This principle has also been reflected in the decisions of courts assessing individual responsibility under the law of war, which have declined to second-guess military decisions with the benefit of hindsight.[68]

The requirement that military commanders and other decision-makers make decisions in good faith based on the information available to them recognizes that decisions may be made when information is imperfect or lacking, which will often be the case during armed conflict.

5.4 RULES ON CONDUCTING ASSAULTS, BOMBARDMENTS, AND OTHER ATTACKS

Combatants may conduct assaults, bombardments, and other attacks, but a number of rules apply to these operations.

5.4.1 Notes on Terminology – Protection From "Attack As Such," "Being Made the Object of Attack," "Direct Attack," and "Intentional Attack" Versus Protection From "Incidental" or "Collateral" Harm. A variety of formulations have been commonly used to distinguish between: (1) the protection from being made the object of attack (*i.e.*, the attack is purposefully directed against that person or object) and (2) the protection from the incidental effects of an attack (*i.e.*, the object or person is not the object of the attack, but is collaterally harmed by the attack).

These situations are treated quite differently under the law of war.[69] In the former case, it is often said that protected persons and objects are protected "as such,"[70] from "direct attack,"[71] from "intentional attack,"[72] "from attack directed exclusively against them,"[73] or from being made the "object of attack."[74] Sometimes a combination of these formulations has been used. In some cases, a text may not use any qualification (*e.g.*, "direct attack" or "as such,"), but is understood to refer only to the first category of protection. For example, Article 52 of AP I provides that "[a]ttacks shall be limited strictly to military objectives."[75] However, this article has been understood to comprise only an obligation not to direct attacks against civilian objects and not to address the question of incidental harm resulting from attacks directed against military objectives.[76]

This manual generally uses the phrase "object of attack" to convey the idea that the attack is being purposefully directed against that person or object.

This manual generally uses the term "incidental harm" to refer to the death or injury to civilians, or the destruction of or damage to civilian objects, that is incidentally (but potentially knowingly) caused as a collateral consequence of an attack directed against a military objective.[77]

5.4.2 Overview of Rules in Conducting Attacks. Parties to a conflict must conduct attacks in accordance with the principles of distinction and proportionality. In particular, the following rules must be observed:

• Combatants may make military objectives the object of attack, but may not direct attacks against civilians, civilian objects, or other protected persons and objects.[78]

• Combatants must refrain from attacks in which the expected loss of life or injury to civilians, and damage to civilian objects incidental to the attack, would be excessive in relation to the concrete and direct military advantage expected to be gained.[79]

• Combatants must take feasible precautions in planning and conducting attacks to reduce the risk of harm to civilians and other persons and objects protected from being made the object of attack.[80]

• In conducting attacks, combatants must assess in good faith the information that is available to them.[81]

- Combatants may not kill or wound the enemy by resort to perfidy.[82]
- Specific rules apply to the use of certain types of weapons.[83]

5.4.3 Assessing Information in Conducting Attacks. Persons who plan, authorize, or make other decisions in conducting attacks must make the judgments required by the law of war in good faith and on the basis of information available to them at the time.[84] For example, a commander must, on the basis of available information, determine in good faith that a target is a military objective before authorizing an attack against that target. Similarly, the expected incidental damage to civilians or civilian objects must be assessed in good faith, given the information available to the commander at the time.

In making the judgments that are required by the law of war rules governing attacks, persons may rely on information obtained from other sources, including human intelligence or other sources of information.[85] For example, in a long-distance attack, a commander may rely on information obtained from aerial reconnaissance and intelligence units in determining whether to conduct an attack.[86]

5.4.3.1 *Heightened Identification Requirements in Conducting Attacks*. Although

doing so would impose greater restrictions than the requirements of the law of war, applying heightened standards of identification may be a policy choice to reduce the risk of incidental harm in conducting an attack.[87]

5.4.3.2 *AP I Presumptions in Favor of Civilian Status in Conducting Attacks*. In the context of conducting attacks, certain provisions of AP I reflect a presumption in favor of civilian status in cases of doubt. Article 52(3) of AP I provides that "[i]n case of doubt whether an object which is normally dedicated to civilian purposes, such as a place of worship, a house or other dwelling or a school, is being used to make an effective contribution to military actions, it shall be presumed not to be so used."[88] Article 50(1) of AP I provides that "[i]n case of doubt whether a person is a civilian, that person shall be considered to be a civilian." Under customary international law, no legal presumption of civilian status exists for persons or objects, nor is there any rule inhibiting commanders or other military personnel from acting based on the information available to him or her in doubtful cases.[89] Attacks, however, may not be directed against civilians or civilian objects based on merely hypothetical or speculative considerations regarding their possible current status as a military objective. In assessing whether a person or object that normally does not have any military purpose or use is a military objective, commanders and other decision-makers must make the decision in good faith based on the information available to them in light of the circumstances ruling at the time. A legal presumption of civilian status in cases of doubt may demand a degree of certainty that would not account for the realities of war.[90] Affording such a presumption could also encourage a defender to ignore its obligation to separate military objectives from civilians and civilian objects.[91] For example, unprivileged belligerents may seek to take advantage of a legal presumption of civilian status.[92] Thus, there is concern that affording such a presumption likely would increase the risk of harm to the civilian population and tend to undermine respect for the law of war.

In applying AP I rules on "doubt," some Parties to AP I have interpreted these rules in a more limited way (*e.g.*, applying a "substantial doubt" standard) than AP I's text would suggest.[93]

5.4.4 Failure by the Defender to Separate or Distinguish Does Not Relieve the Attacker of the Duty to Discriminate in Conducting Attacks. A party that is subject to attack might fail to take feasible precautions to reduce the risk of harm to civilians, such as by separating the civilian population from military objectives.[94] Moreover, in some cases, a party to a conflict might attempt to use the presence or movement of the civilian population or individual civilians in order to shield military objectives from seizure or attack.[95]

When enemy persons engage in such behavior, commanders should continue to seek to discriminate in conducting attacks and to take feasible precautions to reduce the risk of harm to the civilian population and civilian objects.[96] However, the ability to discriminate and to reduce the risk of harm to the civilian population likely will be diminished by such enemy conduct. In addition, such conduct by the adversary does not increase the legal obligations of the attacking party to discriminate in conducting attacks against the enemy.[97] Violations by the defending party are not a basis for that party to assert additional legal rights against the attacking party.[98]

5.4.5 Permissible Location of Attacks. In general, attacks may be conducted against military objectives wherever located, outside neutral territory.[99] Attacks, however, may not be conducted in special zones established by agreement between the belligerents, such as hospital, safety, or neutralized zones.[100]

Attacks on military objectives in the enemy rear or diversionary attacks away from the current theaters or zones of active military operations are lawful.[101] The law of war does not require that attacks on enemy military personnel or objectives be conducted near ongoing fighting, in a theater of active military operations, or in a theater of active armed conflict.[102] There are many examples of lawful attacks taking place far from where the fighting was previously taking place.[103]

For policy or operational reasons, military orders, such as applicable rules of engagement, may limit the locations where attacks on otherwise lawful military objectives may be conducted.

5.4.6 Force That May Be Applied Against Military Objectives. In the absence of expected harm to civilians and civilian objects or of wanton destruction that is not justified by military necessity, the law of war imposes no limit on the degree of force that may be directed against enemy military objectives, including enemy military personnel (but not including enemy personnel who are placed *hors de combat*).[104] For example, the principle of military necessity does not require that only the minimum force that is actually necessary in a specific situation may be used against military objectives.[105] Instead, the broader imperatives of winning the war may be considered, and overwhelming force may be used against enemy military objectives.[106] In addition, combatants need not offer opponents an opportunity to surrender before carrying out an attack.[107]

In particular, the following practices are not prohibited: (1) surprise attacks; (2) attacks on retreating forces; (3) harassing fires; and (4) attacks on specific individuals.

5.4.6.1 *Surprise Attacks*. The law of war does not prohibit the use of surprise to conduct attacks, such as the use of surprise in ambushes, sniper attacks, air raids, and attacks by special operations forces carried out behind enemy lines.[108] There is no requirement that an enemy combatant must be warned before being attacked.[109] Rather, warning requirements apply only with respect to the civilian population and certain other protected units, vessels, and facilities, such as military medical units and facilities.[110]

5.4.6.2 *Attacks on Retreating Forces*. Enemy combatants remain liable to attack when retreating.[111] Retreat is not the same as surrender. Retreating forces remain dangerous as the enemy force may recover to counterattack, consolidate a new defensive position, or assist the war effort in other ways.[112] Retreat may also be a ruse.[113] Retreating enemy combatants may have the same amount of force brought to bear upon them as an attacking military force, and a military commander is under no obligation to limit force directed against enemy combatants because they are, or appear to be, in retreat.

5.4.6.3 *Harassing Fires*. Harassing fires against enemy combatants are not prohibited. (Such action is clearly distinguishable from attacks to terrorize or otherwise harm the civilian population, which are, of course, prohibited.) Harassing fires are delivered on enemy locations for the purpose of disturbing enemy forces' rest, curtailing their movement, or lowering their morale.[114]

5.4.6.4 *Attacks on Specific Individuals*. Military operations may be directed against specific enemy combatants.[115] U.S. forces have often conducted such operations.[116]

Prohibition Against Declaring That No Quarter Be Given.

5.4.7 It is forbidden to declare that no quarter will be given.[117] This means that it is prohibited to order that legitimate offers of surrender will be refused or that detainees, such as unprivileged belligerents, will be summarily executed.[118] Moreover, it is also prohibited to conduct hostilities on the basis that there shall be no survivors, or to threaten the adversary with the denial of quarter.[119] This rule is based on both humanitarian and military considerations.[120] This rule also applies during non-international armed conflict.[121]

5.4.8 Obligation of Combatants to Distinguish Themselves When Conducting Attacks. Combatants have certain obligations to distinguish themselves that include, but are not limited to, those times when they conduct attacks.[122] For example, militia and volunteer corps must wear fixed, distinctive insignia, including when they are conducting attacks.[123] In addition, combatants may not kill or wound by resort to perfidy.[124] Combatants may not fight in the enemy's uniform.[125] Lastly, persons engaging in spying or sabotage risk additional penalties under the domestic law of enemy States.[126]

5.4.8.1 *Fighting Out of Uniform*. Although military operations generally are conducted while wearing a uniform or other distinctive emblems, there may be occasions, such as a surprise attack by enemy forces, when military personnel are unable to dress in their uniforms before resisting the enemy's assault.[127]

Military personnel not in uniform may resist an attack, so long as they are not wearing the enemy's uniform and do not kill or wound treacherously.[128] For example, military personnel not in uniform who resist an attack, and who do not purposefully seek to conceal their status as combatants, commit no violation of the law of war and remain entitled to the privileges of combatant status. The normal wearing of uniforms or other distinctive emblems, however, should resume as soon as practicable because such wear helps protect the civilian population from erroneous attack by helping to distinguish military forces from the civilian population.[129]

5.4.8.2 *AP I Obligation for Combatants to Distinguish Themselves During Attacks or Military Operations Preparatory to an Attack.* AP I provides that: "combatants are obliged to distinguish themselves from the civilian population while they are engaged in an attack or in a military operation preparatory to an attack."[130]

The AP I provision only partially describes the obligation under customary international law of combatants to distinguish themselves from the civilian population. Under customary international law, the obligation of combatants to distinguish themselves is a general obligation that the armed forces have as a group and is not limited to times when they are engaged in an attack or in a military operation preparatory to an attack.[131] Moreover, measures such as wearing insignia or other distinctive emblems may be of less practical significance during an attack.

During an attack, combatants are likely to be distinguishable based on their activities more than any insignia or devices they are wearing.

5.5 DISCRIMINATION IN CONDUCTING ATTACKS

Under the principle of distinction, combatants may make enemy combatants and other military objectives the object of attack, but may not make the civilian population and other protected persons and objects the object of attack.[132]

Persons, Objects, and Locations That Are Not Protected

5.5.1 From Being Made the Object of Attack. Combatants may make enemy combatants and other military objectives the object of attack.[133]

5.5.2 Persons, Objects, and Locations That Are Protected From Being Made the Object of Attack. Persons, objects, and locations that are not military objectives may not be made the object of attack. In particular, the following persons and objects may not be made the object of attack, unless that protection is specifically forfeited under the circumstances:

• Persons, such as

o individual civilians and the civilian population;[134]

o military medical and religious personnel, including military medical units;[135]

o combatants placed *hors de combat*;[136] and

o *parlementaires*;[137] and

• Objects and locations, provided they are not military objectives, such as

o military medical transport, facilities, and equipment;[138]

o cultural property;[139]

o historic monuments, museums, scientific, artistic, educational and cultural institutions;[140]

o places of worship;

o undefended villages, towns, or cities;[141]

o hospital and safety zones, and persons and objects within these zones;[142] and

o neutralized zones, and persons and objects within neutralized zones.[143]

5.6 MILITARY OBJECTIVES

Military objectives refers to persons and objects that may be made the object of attack. Certain classes of persons and objects are categorically recognized as military objectives. Apart from these classes that are categorically military objectives, other objects are assessed as to whether they meet the definition of "military objective."

5.6.1 Military Objective – Notes on Terminology. The term *military objective* has been used in various treaties as a term of art to mean a person or object that may lawfully be made the object of attack.[144]

5.6.1.1 *Persons and Objects as Military Objectives.* Although enemy combatants may be made the object of attack, some sources do not classify persons as *military objectives*, and limit the term *military objective* to objects.

Definitions of *military objectives* in treaties have defined the term *military objective* insofar as objects (rather than persons) are concerned.[145] In addition, the treaty definitions have been written with the purpose of explaining when objects that normally are civilian objects have become military objectives under the circumstances.

This manual uses the term *military objective* to include persons who may be made the object of attack.[146]

5.6.1.2 *Dual-Use Objects*. Sometimes, "dual-use" is used to describe objects that are used by both the armed forces and the civilian population, such as power stations or communications facilities. However, from the legal perspective, such objects are either military objectives or they are not; there is no intermediate legal category.[147] If an object is a military objective, it is not a civilian object and may be made the object of attack.[148] However, it will be appropriate to consider in applying the principle of proportionality the harm to the civilian population that is expected to result from the attack on such a military objective.[149]

5.6.2 Persons Who Are Military Objectives. Certain classes of persons are military objectives and may be made the object of attack.[150] These classes of persons include:

• combatants, such as military ground, air, and naval units, or unprivileged belligerents;[151]

and

• civilians taking a direct part in hostilities.[152]

However, the following classes of persons are not military objectives:

• military medical and religious personnel, unless they commit acts harmful to the enemy;[153]

• military medical units, unless they have forfeited their protected status,[154]

• combatants placed *hors de combat*;[155] and

• *parlementaires*.[156]

5.6.3 Objects That Are Military Objectives. Military objectives, insofar as objects are concerned, include "any object which by its nature, location, purpose or use makes an effective contribution to military action and whose total or partial destruction, capture or neutralization, in the circumstances ruling at the time, offers a definite military advantage."[157]

This definition of military objective may be viewed as a way of evaluating whether military necessity exists to attack an object.[158] It may also be applied outside the context of conducting attacks to assess whether the seizure or destruction of an object is justified by military necessity.[159]

5.6.4 Objects Categorically Recognized as Military Objectives. Two types of objects are categorically recognized as military objectives. In other words, the definition of military objective is always considered to be met as a matter of law with regard to these objects. Thus, these objects may be made the object of attack without specifically applying the analysis discussed in § 5.6.5 (Definition of Military Objective for Objects: A Two-Part Test).

5.6.4.1 *Military Equipment and Bases*. First, certain objects belonging to the armed forces and used in military operations are recognized as military objectives.[160] This includes:

• Military ground, air, and naval equipment (other than identifiable medical equipment or transport),[161] including vehicles, ships, weapons, munitions, and supplies, such as

o strategic and tactical integrated air defense systems;

o missile launching equipment and positions; and

o command and control equipment.

• Military bases, such as army, air, and naval bases (other than military medical facilities, POW camps, and civilian internee camps), whether used for training, billeting, or staging, or offensive or defensive purposes, such as

o headquarters or command and control facilities;

o defense ministries; and

o intelligence facilities.

5.6.4.2 *Objects Containing Military Objectives*. Second, objects that contain military objectives are military objectives.[162] For example,

• Storage and production sites for military equipment, including

o missile production and storage facilities; and

o nuclear, biological, and chemical weapons research and production facilities; and

• Facilities in which combatants are sheltering or billeting.[163]

Definition of Military Objective for Objects: A Two-5.6.5 Part Test. The definition of military objective insofar as objects are concerned may be divided into two parts, both of which must be met for the object to be considered a military objective: (1) that the object somehow makes an effective contribution to military action; and (2) attacking, capturing, or neutralizing the object, in the circumstances, offers a definite military advantage.[164]

Generally, the reason why the object meets the first part of the definition also satisfies the second part of the definition. In other words, attacking the object in the circumstances will offer a definite military advantage because it seeks to preclude the object from effectively contributing to the enemy's military action. Moreover, the two parts are not necessarily connected because the concept of definite military advantage is broader than simply denying the adversary the benefit of an object's effective contribution to its military operations.[165] These broader aspects of "military advantage" may also be relevant in evaluating whether an attack is expected to be excessive under the principle of proportionality.[166]

The following discussion elaborates upon the definition of *military objective*.

5.6.6 By Its Nature, Location, Purpose, or Use Makes an Effective Contribution to Military Action. The first part of the test is whether the object, by its nature, location, purpose, or use makes an effective contribution to the enemy's military action.

5.6.6.1 *Nature, Location, Purpose, or Use*. The nature, location, purpose, or use of the object may contribute to the object making an effective contribution to the enemy's military action. The issue is whether, in total, an effective contribution is made; one factor alone need not provide the effective contribution. In addition, nature, location, purpose, or use need not be viewed as mutually exclusive concepts; rather, these concepts may be understood to overlap.

"Nature" refers to the type of object and may be understood to refer to objects that are *per se* military objectives. For example, military equipment and facilities, by their nature, make an effective contribution to military action.[167] On the other hand, "nature" can also be understood to refer to objects that may be used for military purposes as discussed below. The location of an object may provide an effective contribution to military action. For example, during military operations in urban areas, a house or other structure that would ordinarily be a civilian object may be located such that it provides cover to enemy forces or would provide a vantage point from which attacks could be launched or directed. The word "location" also helps clarify that an area of land can be militarily important and therefore a military objective.[168]

"Use" refers to the object's present function. For example, using an otherwise civilian building to billet combatant forces makes the building a military objective.[169] Similarly, using equipment and facilities for military purposes, such as using them as a command and control center or a communications station, would result in such objects providing an effective contribution to the enemy's military action. "Purpose" means the intended or possible use in the future.[170] For example, runways at a civilian airport could qualify as military objectives because they may be subject to immediate military use in the event that runways at military air bases have been rendered unserviceable or inoperable.[171] Similarly, the possibility that bridges or tunnels would be used to assist in the adversary's military operations in the future could result in such objects providing an effective contribution to the enemy's military action, even though they are not being used at that moment for such purposes.

5.6.6.2 *Make an Effective Contribution to Military Action*. The object must make or be intended to make an effective contribution to military action; however, this contribution need not be "direct" or "proximate."[172] For example, an object might make an effective, but remote, contribution to the enemy's military action and nonetheless meet this aspect of the definition. Similarly, an object might be geographically distant from most of the fighting and nonetheless satisfy this element. Military action has a broad meaning and is understood to mean the general prosecution of the war.[173] It is not necessary that the object provide immediate tactical or operational gains or that the object make an effective contribution to a specific military operation. Rather, the object's effective contribution to the war-fighting or war-sustaining capability of an opposing force is sufficient.[174] Although terms such as "war-fighting," "war-supporting," and "war sustaining" are not explicitly reflected in the treaty definitions of military objective, the United States has interpreted the military objective definition to include these concepts.[175]

5.6.7 Whose Total or Partial Destruction, Capture, or Neutralization, in the Circumstances Ruling at the Time, Offers a Definite Military Advantage. In addition to making an effective contribution to the adversary's military action, the attack of the object must also, in the circumstances ruling at the time, offer a definite military advantage for the object to be considered a "military objective."

5.6.7.1 *Capture or Neutralization*. The definition of military objective incorporates considerations beyond whether an object's destruction is justified. It also incorporates considerations of whether the object's capture and neutralization would offer a military advantage.

Capture refers to the possibility of seizure (rather than destruction), which would confer a military advantage. For example, the seizure of a city may be a military objective because of its strategic location.[176]

Neutralization refers to a military action that denies an object to the enemy without capturing or destroying it. For example, a specific area of land may be neutralized by planting landmines on or around it, and thus denying it to the enemy.[177]

5.6.7.2 *In the Circumstances Ruling at the Time*. The attack of the object must, "in the circumstances ruling at the time," offer a definite military advantage for the object to be considered a military objective.

Nonetheless, the purpose (*i.e.*, future use) of the object can be considered in whether an object provides an effective contribution to the adversary's military action.[178] In addition, the definite military advantage offered by the attack need not be immediate, but may be assessed in the full context of the war strategy.[179]

5.6.7.3 *Definite Military Advantage*. "Definite" means a concrete and perceptible military advantage, rather than one that is merely hypothetical or speculative.[180] The advantage need not be immediate.[181] For example, the military advantage in the attack of an individual bridge may not be seen immediately (particularly if, at the time of the attack, there is no military traffic in the area), but can be established by the overall effort to isolate enemy military forces on the battlefield through the destruction of bridges. "Military advantage" refers to the advantage anticipated from an attack when considered as a whole, and not only from its isolated or particular parts.[182] Similarly, "military advantage" is not restricted to immediate tactical gains, but may be assessed in the full context of the war strategy.[183]

The definite military advantage offered by damaging, destroying, or neutralizing the object may result from denying the enemy the ability to use this object in its military operations (*i.e.*, to benefit from the object's effective contribution to the military action). For example, the attack or seizure of objects with a common military purpose, such as bridges used, or potentially available to be used, in lines of communication would offer a definite military advantage. The military advantage from an attack is not limited to denying the enemy the benefit of that object's contribution to its military action. For example, in a diversionary attack, the military advantage to be gained from attacking an object would result from diverting enemy forces' resources and attention.[184] The military advantage from an attack may involve a variety of other considerations, including improving the security of the attacking force.

The military advantage from an attack may result from harm to the morale of enemy forces.[185] Diminishing the morale of the civilian population and their support for the war effort does not provide a definite military advantage.[186] However, attacks that are otherwise lawful are not rendered unlawful if they happen to result in diminished civilian morale.[187]

The military advantage expected to be gained from an attack might not be readily apparent to the enemy or to outside observers because, for example, the expected military advantage might depend on the commander's strategy or assessments of classified information.

5.6.8 Examples of Objects Often Regarded as Military Objectives. The following types of objects generally have met the definition of "military objective" in past conflicts, but may not be military objectives in all circumstances: (1) leadership facilities; (2) communications objects; (3) transportation objects; (4) places of military significance; and (5) economic objects associated with military operations or with war-supporting or war-sustaining industries. This list of examples is not exclusive (*i.e.*, an object could fall in more than one category in this list), and this list is not exhaustive (*i.e.*, an object outside these categories may nonetheless meet the definition of military objective). Lastly, this list is not conclusive, *i.e.*, whether an example is, in fact, a military objective, must be assessed according to the definition of military objective.

5.6.8.1 *Examples of Military Objectives – Leadership Facilities*. Facilities used by enemy leaders as headquarters for military operations or otherwise to command military operations have often been regarded as military objectives.[188] In some cases, enemy leaders themselves may be made the object of attack.[189]

5.6.8.2 *Examples of Military Objectives – Communications Objects*. Communications objects, such as facilities, networks, and equipment that could be used for command and control of military operations or intelligence gathering, have often been regarded as military objectives.[190]

5.6.8.3 *Examples of Military Objectives – Transportation Objects*. Transportation objects, including facilities (*e.g.*, port facilities and airfields) and equipment that could be part of lines of communication (*e.g.*, highways, railroads, waterways, and bridges connecting military forces with logistics depots and storage areas), have often been regarded as military objectives.[191]

5.6.8.4 *Examples of Military Objectives – Places of Military Significance*. Areas of land that are militarily significant may constitute military objectives.[192] For example, anti- tank mines may be laid on such areas in order to block enemy forces' tanks.[193] Areas of land that have been regarded as military objectives have included, for example:

• road networks;[194]

• known or suspected enemy avenues of approach or withdrawal;

• mountain passes, hills, defiles, and bridgeheads;[195] and

• villages, towns, or cities whose seizure is militarily important.[196]

5.6.8.5 *Examples of Military Objectives – Economic Objects Associated With Military Operations*. Economic objects associated with military operations or with war supporting or war-sustaining industries have been regarded as military objectives. Electric power stations are generally recognized to be of sufficient importance to a State's capacity to meet its wartime needs of communication, transport, and industry so as usually to qualify as military objectives during armed conflicts.[197]

Oil refining and distribution facilities and objects associated with petroleum, oil, and lubricant products (including production, transportation, storage, and distribution facilities) have also been regarded as military objectives.[198]

5.7 COMBATANTS

In general, combatants, whether privileged or unprivileged, may be made the object of attack, provided they have not been placed *hors de combat*.

5.7.1 Armed Forces and Groups and Liability to Being Made the Object of Attack. Membership in the armed forces or belonging to an armed group makes a person liable to being made the object of attack regardless of whether he or she is taking a direct part in hostilities.[199]

This is because the organization's hostile intent may be imputed to an individual through his or her association with the organization. Moreover, the individual, as an agent of the group, can be assigned a combat role at any time, even if the individual normally performs other functions for the group.[200]

Thus, combatants may be made the object of attack at all times, regardless of the

activities in which they are engaged at the time of attack.[201] For example, combatants who are standing in a mess line, engaging in recreational activities, or sleeping remain the lawful object of attack, provided they are not placed *hors de combat*.[202]

5.7.1.1 *U.S. Practice in Declaring Forces Hostile*. In DoD practice, an armed force or group may be designated as hostile (also known as declaring the force hostile) in rules of engagement. This means that personnel to whom such rules of engagement apply are authorized to attack the members of the group. In DoD practice, the authority to designate a group as hostile has been limited to only certain officials.[203]

5.7.2 Categories of Persons Who Are Combatants for the Purpose of Assessing Their Liability to Attack. The following categories of persons are combatants who may be made the object of attack because they are sufficiently associated with armed forces or armed groups:

• members of the armed forces of a State;[204]

• members of militia and volunteer corps;[205]

• participants in a *levée en masse*;[206]

• persons belonging to non-State armed groups;[207] and

• leaders whose responsibilities include the operational command and control of the armed forces or of a non-State armed group.[208]

5.7.3 Persons Belonging to Non-State Armed Groups. Like members of an enemy State's armed forces, individuals who are formally or functionally part of a non-State armed group that is engaged in hostilities may be made the object of attack because they likewise share in their group's hostile intent.[209]

5.7.3.1 *Formal Membership*. Formal membership in an armed group might be indicated by formal or direct information or by other types of information. In some cases, there might be formal or direct information indicating membership in the group. This might include:

• using a rank, title, or style of communication;

• taking an oath of loyalty to the group or the group's leader;

• wearing a uniform or other clothing, adornments, or body markings that identify members of the group; or

• documents issued or belonging to the group that identify the person as a member, such as membership lists, identity cards, or membership applications.[210]

Although in some cases this type of formal or direct information might be available, in many cases it will not be available because members of these groups seek to conceal their association with that group. In such cases, the following types of information might indicate that a person is a member of a non-State armed group:

• acting at the direction of the group or within its command structure;

• performing a function for the group that is analogous to a function normally performed by a member of a State's armed forces;

• taking a direct part in hostilities, including consideration of the frequency, intensity, and duration of such participation;[211]

• accessing facilities, such as safe houses, training camps, or bases used by the group that outsiders would not be permitted to access;[212]

• traveling along specific clandestine routes used by those groups;[213] or

• traveling with members of the group in remote locations or while the group conducts operations.[214]

5.7.3.2 *Functional Membership*. Some non-State armed groups might not be organized in a formal command structure, as generally is required for POW status during international armed conflict.[215] Such groups might lack a formal distinction between those members and non-members who nonetheless participate in the hostile activities of the group.[216] An individual who is integrated into the group such that the group's hostile intent may be imputed to him or her may be deemed to be functionally (*i.e.*, constructively) part of the group, even if not formally a member of the group. The integration of the person into the non-State armed group and the inference that the individual shares the group's intention to commit hostile acts distinguish such an individual from persons who are merely sympathetic to the group's goals.[217]

The following may indicate that a person is functionally a member of a non-State armed group:

• following directions issued by the group or its leaders;[218]

• taking a direct part in hostilities on behalf of the group on a sufficiently frequent or intensive basis;[219] or

• performing tasks on behalf of the group similar to those provided in a combat, combat support, or combat service support role in the armed forces of a State.

5.7.3.3 *Dissociation or Renunciation*. A person may not be made the object of attack based on his or her association with a non-State armed group if that association has clearly been severed. Relevant factors in determining when an individual has unambiguously ceased to be a member of a non-State armed group may include:

• whether the individual has formally ceased to be a member of the group, such as by filing

relevant paperwork or by otherwise formally renouncing any allegiance to the group; • whether there are concrete and verifiable facts or persuasive indicia that he or she has affirmatively returned to peaceful pursuits, such as by participating in a reconciliation program and swearing an oath of loyalty to the government; and

• the amount of time that has passed since the person participated in the activities of the group in question, if coupled with other indicia of dissociation or renunciation.[220] The onus is on the person having belonged to the armed group to demonstrate clearly and affirmatively to the opposing forces that he or she will no longer participate in the activities of the group.[221] Moreover, if persons who have dissociated from an armed group rejoin the group or fail to cease permanently their participation in hostilities, they may be made the object of attack.[222]

5.7.4 Leaders. Military leaders are subject to attack on the same basis as other members of the armed forces. Similarly, leaders of non-State armed groups are also subject to attack on the same basis as other members of the group. There is no objection to making a specific enemy leader who is a combatant the object of attack.[223]

Leaders who are not members of an armed force or armed group (including heads of State, civilian officials, and political leaders) may be made the object of attack if their responsibilities include the operational command or control of the armed forces. For example, as the commander-in-chief of the U.S. armed forces, the President would be a legitimate target in wartime, as would, for example, the Prime Minister of a constitutional monarchy. In contrast, the reigning monarch of a constitutional monarchy with an essentially ceremonial role in State affairs may not be made the object of attack.

In addition to leaders who have a role in the operational chain of command, leaders taking a direct part in hostilities may also be made the object of attack.[224] Planning or authorizing a combat operation is an example of taking a direct part in hostilities.[225]

As a matter of practice, attacks on the national leadership of an enemy State have often been avoided on the basis of comity and to help ensure that authorities exist with whom peace agreements may be concluded.

5.8 CIVILIANS TAKING A DIRECT PART IN HOSTILITIES

Civilians who take a direct part in hostilities forfeit protection from being made the object of attack.

5.8.1 Civilians Taking a Direct Part in Hostilities – Notes on Terminology. This manual uses the phrase "direct part in hostilities" to indicate what activities cause a civilian to forfeit his or her protection from being made the object of attack. This usage does not mean that the United States has adopted the direct participation in hostilities rule that is expressed in Article 51 of AP I.

5.8.1.1 *"Active" Versus "Direct"*. The phrases "*active* part in hostilities" and "*direct* part in hostilities" have been used to describe when civilians forfeit their protection from being made the object of attack. As noted above, this manual uses "direct" rather than "active" in this context, although as discussed below, this usage should not be regarded as indicating a substantive difference between "active" and "direct."

Common Article 3 of the 1949 Geneva Conventions refers to "[p]ersons taking no active part in the hostilities." AP I and AP II use the phrase "direct part in hostilities."[226] In addition, AP I uses the phrase "direct part in hostilities" to address other situations apart from the protection of civilians.[227]

Although the words *active* and *direct* can mean different things in the English language, the terms have sometimes been treated as the same for the purpose of applying the direct participation in hostilities rule.[228] One of the reasons for treating the terms the same is that although the English language version of the 1949 Geneva Conventions uses "active," and the English language versions of AP I and AP II use "direct," the French language versions of these treaties use the same word, "*directement*."[229] Because the English and French language versions of the 1949 Geneva Conventions, AP I, and AP II are equally authentic, States negotiating these treaties may not have intended a difference between "active" and "direct."[230]

Another reason for treating the terms "active" and "direct" the same in this context is that they are understood to be terms of art addressing a particular legal standard, and there are a range of views as to what that legal standard means. Thus, there may be different views about what the underlying standard means, even when there is agreement on the appropriate term to describe that standard. Accordingly, there seems to be little value in distinguishing between the two terms for the purposes of applying this legal rule.

5.8.1.2 *AP I, Article 51(3) Provision on Direct Participation in Hostilities*. Although, as drafted, Article 51(3) of AP I does not reflect customary international law, the United States supports the customary principle on which Article 51(3) is based.[231] Similarly, although parts of the ICRC's interpretive guidance on the meaning of direct participation in hostilities are consistent with customary international law, the United States has not accepted significant parts of the ICRC's interpretive guidance as accurately reflecting customary international law.[232] But some States that are Parties to AP I may interpret and apply Article 51(3) of AP I consistent with the customary international law standard.

5.8.2 Persons to Whom This Rule Applies. For the purpose of applying the rule discussed in this section, "civilians" are persons who do not fall within the categories of combatants listed in § 5.7.2 (Categories of Persons Who Are Combatants for the Purpose of Assessing Their Liability to Attack). Accordingly, for the purposes of this section, "civilians" include:

• members of the civilian population;[233]

• persons authorized to accompany the armed forces;[234] and

• members of the merchant marine and civil aircraft of parties to a conflict.[235]

5.8.2.1 *Persons Belonging to Hostile, Non-State Armed Groups*. Some States may choose to characterize persons who belong to hostile, non-State armed groups that do not qualify for status as lawful combatants as "civilians" who may not be attacked unless they are taking a direct part in hostilities. However, these States may also characterize the act of joining and remaining a member of an armed group that is engaged in hostilities as a form of taking a direct part in hostilities that continuously deprives these individuals of their protection from being made the object of attack.[236]

The U.S. approach has generally been to refrain from classifying those belonging to non- State armed groups as "civilians" to whom this rule would apply. The U.S. approach has been to treat the status of belonging to a hostile, non-State armed group as a separate basis upon which a person is liable to attack, apart from whether he or she has taken a direct part in hostilities.[237] Either approach may yield the same result: members of hostile, non-State armed groups may be made the object of attack unless they are placed *hors de combat*. However, practitioners, especially when working with coalition partners, should understand that different legal reasoning is sometimes applied in reaching that result.

5.8.3 "Taking a Direct Part in in Hostilities". Unlike the treaty definition of "military objective" for objects,[238] the United States is not a Party to a treaty with a comparable provision defining taking a direct part in hostilities for the purpose of assessing what conduct renders civilians liable to being made the object of attack.

At a minimum, taking a direct part in hostilities includes actions that are, by their nature and purpose, intended to cause actual harm to the enemy.[239] Taking a direct part in hostilities extends beyond merely engaging in combat and also includes certain acts that are an integral part of combat operations or that effectively and substantially contribute to an adversary's ability to conduct or sustain combat operations.[240] However, taking a direct part in hostilities does not encompass the general support that members of the civilian population provide to their State's war effort, such as by buying war bonds.[241]

Whether an act by a civilian constitutes taking a direct part in hostilities is likely to depend highly on the context, such as the weapon systems or methods of warfare employed by the civilian's side in the conflict.[242] For example, in some contexts, training and logistical support may be viewed as taking a direct part in hostilities, while in other contexts it might not.[243] The following considerations may be relevant:[244]

• the degree to which the act causes harm to the opposing party's persons or objects, such as

o whether the act is the proximate or "but for" cause of death, injury, or damage to persons or objects belonging to the opposing party; or

o the degree to which the act is likely to affect adversely the military operations or military capacity of the opposing party;

• the degree to which the act is connected to the hostilities, such as

o the degree to which the act is temporally or geographically near the fighting; or

o the degree to which the act is connected to military operations;

• the specific purpose underlying the act, such as

o whether the activity is intended to advance the war aims of one party to the conflict to the detriment of the opposing party;

• the military significance of the activity to the party's war effort, such as

o the degree to which the act contributes to a party's military action against the

opposing party;

o whether the act is of comparable or greater value to a party's war effort than acts that are commonly regarded as taking a direct part in hostilities;[245]

o whether the act poses a significant threat to the opposing party;

• the degree to which the activity is viewed inherently or traditionally as a military one, such as

o whether the act is traditionally performed by military forces in conducting military operations against the enemy (including combat, combat support, and

combat service support functions);[246] or

o whether the activity involves making decisions on the conduct of hostilities, such as determining the use or application of combat power.

5.8.3.1 *Examples of Taking a Direct Part in Hostilities*. The following acts are generally considered taking a direct part in hostilities that would deprive civilians who perform them of protection from being made the object of attack. These examples are illustrative and not exhaustive:

• taking up or bearing arms against the opposing party, or otherwise personally trying to kill, injure, or capture personnel or damage material belonging to the opposing party,[247] such as

o defending military objectives against enemy attack (*e.g.,* manning an antiaircraft gun, acting as a bodyguard for an enemy combatant);[248]

o acting as a member of a weapons crew;

o engaging in an act of sabotage; or

o emplacing mines or improvised explosive devices;

• preparing for combat and returning from combat;

• planning, authorizing, or implementing a combat operation against the opposing party, even if that person does not personally use weapons or otherwise employ destructive force in connection with the operation;[249]

• providing or relaying information of immediate use in combat operations, such as o acting as an artillery spotter or member of a ground observer corps or otherwise relaying information to be used to direct an airstrike, mortar attack, or ambush;[250]

and

o acting as a guide or lookout for combatants conducting military operations;[251]

• supplying weapons and ammunition, whether to conventional armed forces or non-state armed groups, or assembling weapons (such as improvised explosive devices) in close geographic or temporal proximity to their use,[252] such as

o delivering ammunition to the front lines; or

o outfitting and preparing a suicide bomber to conduct an attack.

5.8.3.2 *Examples of Acts Not Considered Taking a Direct Part in Hostilities*. The following acts are generally not considered taking a direct part in hostilities that would deprive civilians who perform them of protection from being made the object of attack. These examples are illustrative and not exhaustive:

• mere sympathy or moral support for a party's cause;

• general contributions made by citizens to their State's war effort (*e.g.*, buying war bonds or paying taxes to the government that will ultimately be used to fund the armed forces);

• police services (*e.g.*, police officers who maintain public order against common criminals during armed conflict);[253]

• independent journalism or public advocacy (*e.g.*, opinion journalists who write columns supporting or criticizing a State's war effort);[254]

• working in a munitions factory or other factory that is not in geographic or temporal proximity to military operations but that is supplying weapons, materiel, and other goods useful to the armed forces of a State;[255] or

• providing medical care or impartial humanitarian assistance.[256]

Although performing these activities does not make a person liable to being made the object of attack, performing these activities also does not immunize a person from attack if that person takes a direct part in hostilities or is otherwise lawfully made the object of attack.

5.8.3.3 *Taking a Direct Part in Hostilities and Standards for the Use of Force in Self-Defense.* In the practice of the United States, the U.S. armed forces have been authorized to use necessary and proportional force in self-defense in response to hostile acts or demonstrated hostile intent.[257]

In some cases, hostile acts or demonstrated hostile intent may also constitute taking a direct part in hostilities. However, hostile acts and demonstrated hostile intent in some respects may be narrower than the concept of taking a direct part in hostilities. For example, although supplying weapons and ammunition in close geographic or temporal proximity to their use is a common example of taking a direct part in hostilities, it would not necessarily constitute a hostile act or demonstrated hostile intent.

On the other hand, hostile acts and demonstrated hostile intent in some respects may be broader than the concept of taking a direct part in hostilities. For example, the use of force in response to hostile acts and demonstrated hostile intent applies outside hostilities, but taking a direct part in hostilities is limited to acts that occur during hostilities. Thus, the concept of taking a direct part hostilities must not be understood to limit the use of force in response to hostile acts or demonstrated hostile intent.

In the practice of the United States, offensive combat operations against people who are taking a direct part in hostilities have been authorized through specific rules of engagement.

5.8.4 Duration of Liability to Attack. There has been a range of views about the duration for which civilians who have taken a direct part in hostilities forfeit protection from being made the object of attack.[258]

In the U.S. approach, civilians who have taken a direct part in hostilities must not be made the object of attack after they have permanently ceased their participation because there would be no military necessity for attacking them. Persons who take a direct part in hostilities, however, do not benefit from a "revolving door" of protection. There may be difficult cases not clearly falling into either of these categories, and in such situations a case-by-case analysis of the specific facts would be needed.[259]

5.8.4.1 *Permanently Ceased Participation in Hostilities.* If a civilian has permanently ceased participation in hostilities, then that person must not be made the object of attack because there is no military necessity for doing so.[260] The assessment of whether a person has permanently ceased participation in hostilities must be based on a good faith assessment of the available information.[261]

For example, a civilian might have engaged in an isolated instance of taking a direct part in hostilities. This isolated instance is likely to have involved multiple acts, because taking a direct part in hostilities normally includes deploying or moving to a position of attack and exfiltrating from an attack.[262] However, if this participation was an isolated instance that will not be repeated, then no military necessity for attacking that person exists after he or she has ceased taking a direct part in hostilities. Accordingly, the civilian must not be made the object of attack after he or she has ceased taking a direct part in hostilities. However, there may be other legal consequences from this isolated instance of participation. For example, such civilians may, depending on the circumstances, be detained, interned, or prosecuted because of these actions.[263]

5.8.4.2 *No "Revolving Door" Protection.* The law of war, as applied by the United States, gives no "revolving door" protection; that is, the off-and-on protection in a case where a civilian repeatedly forfeits and regains his or her protection from being made the object of attack depending on whether or not the person is taking a direct part in hostilities at that exact time.[264] Thus, for example, persons who are assessed to be engaged in a pattern of taking a direct part in hostilities do not regain protection from being made the object of attack in the time period between instances of taking a direct part in hostilities.[265] A "revolving door" of protection would place these civilians who take a direct part in hostilities on a better footing than lawful combatants, who may be made the object of attack even when not taking a direct part in hostilities.[266] The United States has strongly disagreed with posited rules of international law that, if accepted, would operate to give the so-called "farmer by day, guerilla by night" greater protections than lawful combatants.[267] Adoption of such a rule would risk diminishing the protection of the civilian population.

5.8.5 Civilians Who Take a Direct Part in Hostilities and the Law of War. Although the concept of direct participation in hostilities may be discussed in contexts besides targeting, such as in the context of criminal liability or detention, there are often significant differences between "taking a direct part in hostilities" for targeting purposes and the standards used for assessing whether a civilian may be detained or prosecuted.

For example, whether someone may be made the object of attack for taking a direct part in hostilities is different from whether he or she may be prosecuted for his or her actions. In some cases, domestic criminal liability for support to enemy armed groups is much broader than what acts constitute "taking a direct part in hostilities."

Similarly, the authority to detain enemy persons during wartime extends beyond detaining those who have taken a direct part in hostilities.[268]

5.9 PERSONS PLACED HORS DE COMBAT

Persons, including combatants, placed *hors de combat* may not be made the object of attack. Persons placed *hors de combat* include the following categories of persons, provided they abstain from any hostile act and do not attempt to escape:

• persons in the power of an adverse party;

• persons not yet in custody, who have surrendered;

• persons who have been rendered unconscious or otherwise incapacitated by wounds, sickness, or shipwreck; and

• persons parachuting from aircraft in distress.

5.9.1 *Hors de Combat* – Notes on Terminology. *Hors de combat* is a French phrase that means "out of the battle." It is generally used as a term of art to mean persons who may not be made the object of attack because they are out of the fighting and who therefore must be treated humanely.

Hors de combat is used in Common Article 3 of the 1949 Geneva Conventions[269] and has been defined in Article 41 of AP I.[270]

5.9.2 Persons in the Power of an Adverse Party. Persons in the power of an adverse party include all persons detained by an adverse party, such as POWs, unprivileged belligerents, retained personnel, and civilian internees. As with other categories of persons *hors de combat*, detainees must refrain from hostile acts or attempts to escape in order to be considered *hors de combat*.[271]

5.9.3 Persons Who Have Surrendered. Persons who are not in custody but who have surrendered are *hors de combat* and may not be made the object of attack.[272] In order to make a person *hors de combat*, the surrender must be (1) genuine; (2) clear and unconditional; and (3) under circumstances where it is feasible for the opposing party to accept the surrender.[273]

5.9.3.1 *Genuine*. The offer to surrender must be genuine. In addition to being legally ineffective, feigning the intent to surrender can constitute perfidy.[274]

5.9.3.2 *Clear and Unconditional*. The offer to surrender must be clear and unconditional. Any arms being carried should be laid down. All hostile acts or resistance, or manifestations of hostile intent, including efforts to escape or to destroy items, documents, or equipment to prevent their capture by the enemy, would need to cease immediately for the offer to be clear and unconditional. Raising one's hands above one's head to show that one is not preparing to fire a weapon or engage in combat is often a sign of surrender. Waving a white flag technically is not a sign of surrender, but signals a desire to negotiate.[275]

The surrender must be "at discretion," *i.e.*, unconditional.[276] A person who offers to surrender only if certain demands are met would not be *hors de combat* until that offer has been accepted.

5.9.3.3 *Under Circumstances in Which It Is Feasible to Accept*. For an offer of surrender to render a person *hors de combat*, it must be feasible for the opposing party to accept the offer.[277] By way of comparison, a city may not be declared "undefended" (and thus essentially surrendered) if it is not open for immediate physical occupation by opposing military forces.[278]

The feasibility of accepting the surrender refers to whether it is practical and safe for the opposing force to take custody of the surrendering persons in the circumstances. For example, consider the situation of enemy soldiers who man an antiaircraft gun and shoot at an enemy aircraft, and then who raise their hands as if to surrender seconds before a second aircraft attacks their position. In the circumstances, it would not be feasible for the crew of the attacking aircraft to land and accept their surrender.[279] Similarly, a soldier fifty meters from an enemy defensive position in the midst of an infantry assault by his unit could not throw down his weapon and raise his arms (as if to indicate his desire to surrender) and reasonably expect that the defending unit will be able to accept and accomplish his surrender while resisting the ongoing assault by his unit.[280]

Although the feasibility of accepting surrender includes consideration of whether it is feasible to take custody of the persons offering to surrender, this does not include consideration of whether it is feasible to care for detainees after taking custody. Offers to surrender may not be refused because it would be militarily inconvenient or impractical to guard or care for detainees.[281]

5.9.4 Persons Rendered Unconscious or Otherwise Incapacitated by Wounds, Sickness, or Shipwreck. Persons who have been rendered unconscious or otherwise incapacitated by wounds, sickness, or shipwreck, such that they are no longer capable of fighting, are *hors de combat*.[282]

Those "rendered unconscious" does not include persons who simply fall asleep. Sleeping combatants generally may be made the object of attack.[283]

Shipwrecked combatants include those who have been shipwrecked from any cause and includes forced landings at sea by or from aircraft.[284]

Persons who have been incapacitated by wounds, sickness, or shipwreck are in a helpless state, and it would be dishonorable and inhumane to make them the object of attack.[285] In order to receive protection as *hors de combat*, the person must be wholly disabled from fighting.[286] On the other hand, many combatants suffer from wounds and sickness, but nonetheless continue to fight and would not be protected.[287]

In many cases, the circumstances of combat may make it difficult to distinguish between persons who have been incapacitated by wounds, sickness, or shipwreck and those who continue to fight.[288] If possible, those seeking protection as wounded, sick, or shipwrecked, should make their condition clear.[289]

5.9.5 Persons Parachuting From an Aircraft in Distress. In general, persons, such as aircrew or embarked passengers, parachuting from an aircraft in distress are treated as though they are *hors de combat*, *i.e.*, they must not be made the object of attack.[290] This protection is provided because a person descending by parachute is temporarily *hors de combat* just like someone who is shipwrecked[291] or unconscious.[292]

5.9.5.1 *No Hostile Acts or Attempts to Evade Capture*. As with other categories of persons *hors de combat*, the protection from being made the object of attack is forfeited if the persons engage in hostile acts or attempt to evade capture.

Routine "slipping" to steer a parachute or similar actions to facilitate a safe parachute landing do not constitute acts of evasion.

5.9.5.2 *Persons Deploying Into Combat by Parachute*. Persons deploying into combat by parachute may be attacked throughout their descent, and upon landing. Persons deploying into combat by parachute may include special operations or reconnaissance personnel, combat control teams, or airborne forces (*i.e.*, specialized combat forces trained to arrive at military objectives by parachute drops).

Persons deploying into combat by parachute may be attacked even if they deploy from an aircraft in distress (*e.g.*, the enemy has attacked the aircraft to resist the assault). It may be the case, however, that airborne forces are parachuting from an aircraft in distress outside the context of an airborne assault. Since they would not be "deploying into combat," they would be *hors de combat* while descending by parachute.

5.10 PROPORTIONALITY IN CONDUCTING ATTACKS

In accordance with the principle of proportionality, combatants must not exercise the right to engage in attacks against military objectives in an unreasonable or excessive way.[293] Therefore, when prosecuting attacks against military objectives (*i.e.*, the persons and objects that may be made the object of attack), combatants must exercise due regard to reduce the risk of incidental harm to the civilian population and other persons and objects that may not be made the object of attack. In particular, the following rules apply:

• Combatants must take feasible precautions in planning and conducting attacks to reduce the risk of harm to civilians and other persons and objects protected from being made the object of attack;[294] and

• Combatants must refrain from attacks in which the expected loss of civilian life, injury to civilians, and damage to civilian objects incidental to the attack would be excessive in relation to the concrete and direct military advantage expected to be gained.[295]

5.10.1 Scope of Application of the Principle of Proportionality in Conducting Attacks. The principle of proportionality in conducting attacks imposes duties that apply to the protection of persons and objects that may not be made the object of attack. This principle does not impose obligations intended to reduce the risk of harm to military objectives (*i.e.*, persons and objects that may be made the object of attack).

Although the prohibition on attacks expected to cause excessive harm to civilians and civilian objects generally does not require consideration of military personnel and objects, feasible precautions must be taken to reduce the risk of harm to military personnel and objects that are protected from being made the object of attack, such as military personnel placed *hors de combat*.

5.10.1.1 *Proportionality in Conducting Attacks and Military Objectives*. The principle of proportionality in conducting attacks does not impose obligations intended to reduce the risk of harm to military objectives (*i.e.*, persons and objects that may be made the object of attack).[296] For example, an attack against an enemy combatant might also injure other enemy combatants who were not the specific targets of the attack. The principle of proportionality in conducting attacks would not require that efforts be made to reduce the likelihood of harm to other enemy combatants or damage to other military objectives, even if this harm were an unintended result of the attack.

5.10.1.2 *Proportionality in Conducting Attacks and Protected Military Personnel and Facilities*. The prohibition on attacks expected to cause excessive incidental harm requires consideration of civilians and civilian objects, but this prohibition generally does not require consideration of military personnel and objects, even if they may not be made the object of attack, such as military medical personnel, the military wounded and sick, and military medical facilities. For example, treaty provisions articulating a prohibition on attacks expected to cause excessive incidental harm do not reflect protections for military personnel who are protected from being made the object of attack.[297] Those planning or conducting attacks may consider such military personnel as a matter of practice or policy in applying the prohibition on attacks expected to cause excessive incidental harm.

The exclusion of protected military personnel and military medical facilities from this prohibition reflects such factors as, among others, the general impracticality of prohibiting attacks on this basis during combat operations.[298] For example, the expected incidental harm to a sick-bay on a warship would not serve to exempt that warship from being made the object of attack.

Nonetheless, feasible precautions must be taken to reduce the risk of harm to military personnel and objects that are protected from being made the object of attack.[299] For example, in the context of a deliberate, planned bombardment of a military objective near an identifiable military hospital, it may be feasible to take precautions to reduce the risk of harming the military hospital, and such precautions must be taken.

5.10.2 Responsibility of Commanders for Implementing the Principle of Proportionality in Conducting Attacks. Commanders, at all levels, have a great responsibility to exercise the leadership necessary to reduce the risk of harm to civilians and civilian objects.[300] Although the specific responsibilities that individual combatants have to implement the principle of proportionality will depend on their role and assigned military duties in planning or conducting an attack, commanders are responsible for making the decisions and judgments required by the principle of proportionality in conducting attacks.[301]

5.10.2.1 *Need for Competent Authority to Make the Decisions and Judgments Required by the Principle of Proportionality*. As with other aspects of the law of war, the persons within a party to a conflict who are responsible for making the decisions and judgments required by the principle of proportionality are those who have the authority to make these decisions and judgments.[302] The specific responsibilities that individual combatants have to implement the principle of proportionality will depend on their role and assigned military duties in planning or conducting an attack.[303]

The decisions and judgments required by the principle of proportionality normally require authority over military operations. For example, assessing the military advantage expected to be gained from an attack may require knowing the broader strategy being employed by the attacking party or knowing intelligence information about the strategic and operational context in which the attack takes place.[304] Lower-level personnel may not be privy to the strategic or operational significance of a specific attack, and thus may not be competent to evaluate the expected military advantage of the attack against the expected harm to civilians and civilian objects.[305] Similarly, determinations about the feasibility of taking a precaution may require the authority to direct and manage resources (*e.g.*, allocating weapons systems and intelligence assets) or judgments about the acceptable degree of risk (*e.g.*, to the lives of friendly forces and to mission accomplishment).[306] For example, the principle of proportionality ordinarily does not require military personnel to take actions, such as providing warnings, when a superior commander has already determined that such actions are not feasible because, for instance, such actions could endanger mission accomplishment and the safety of friendly forces.

In the absence of specific direction to the contrary, subordinate commanders or engagement authorities have the authority to make the corresponding decisions required by the law of war, such as the decision to cancel or suspend an attack in light of new information, in order to effectuate the commander's intent.[307]

5.10.2.2 *"Reasonable Military Commander" Standard for Proportionality Decisions and Judgments*. The commander's decisions on proportionality must be reasonable.[308] For example, the commander must be able to explain the expected military importance of the target and why the anticipated civilian collateral injury or damage is not expected to be excessive.[309]

5.10.2.3 *Subjective Aspects of the Decisions and Judgments Required by the Principle of Proportionality*. The decisions and judgments required by the principle of proportionality in conducting attacks have subjective aspects.[310]

Although the decisions and judgments required by the principle of proportionality must be applied with common sense and good faith,[311] there nonetheless might be variation in how reasonable persons would apply the principle of proportionality in a given circumstance.[312] The principle of proportionality typically involves the comparison of "unlike quantities and values."[313] Moreover, there may be disagreements among experts as to the precise requirements of the law.

For example, it could often be the case that reasonable persons might disagree as to whether the expected civilian casualties from an attack would be excessive.[314] Similarly, reasonable commanders might make different decisions in applying the principle of proportionality.[315]

A degree of deference should be given to commanders when assessing after the fact whether a decision-maker has complied with the principle of proportionality.[316] In addition, an assessment of whether a decision-maker has complied with the legal requirements must be based on the information available to that person at the time.[317]

5.10.2.4 *Duty Not to Comply With Clearly Illegal Orders and the Principle of Proportionality*. In the context of the principle of distinction, it would often be clear whether a given situation implicates the duty not to comply with clearly illegal orders to commit law of war violations such as the duty of a subordinate to refrain from complying with an order to attack the civilian population. However, the nuances involved in applying the principle of proportionality could make it more difficult to know whether an order given is clearly illegal. The duty not to comply with orders that are clearly illegal also applies to violations of the principle of proportionality, in particular, the case of a commander who orders subordinates to conduct an attack that is expected to result in civilian casualties that the commander himself or herself acknowledges would be excessive.

Nonetheless, this duty must be understood in light of application of the principle that law of war obligations are implemented by those with responsibility to make the decisions and judgments required by the law of war, and in particular the point that subordinates might not be competent to evaluate whether the requirements of proportionality had been met.[318] Subordinates ordinarily do not have an obligation to second-guess the judgments of their superiors in relation to the principle of proportionality. As with other aspects of the law of war, subordinates are not required to screen the orders of superiors for questionable points of legality, and may, absent specific knowledge to the contrary, presume that orders have been lawfully issued.[319] If questions arise, service members should apply their training, refer to applicable policy and regulations, and direct questions through appropriate channels.[320]

5.10.2.5 *Discretion of Subordinates in the Conduct of Attacks*. Commanders often provide subordinates substantial discretion in the tactics of an attack, such as in the selection of individual targets and in the timing of the attack. In the absence of specific direction to the contrary, subordinate commanders or engagement authorities have the authority to make the corresponding decisions required by the law of war, such as the decision to cancel or suspend an attack in light of new information, in order to effectuate the commander's intent.[321] This is especially important in situations where the subordinate is not in direct communication with the commander and the commander does not have situational awareness. For example, a commander may provide subordinates broad discretion to conduct an attack in which subordinates discover an unanticipated risk of civilian casualties. If receiving updated instructions in light of this new information was not possible, the subordinates would be understood to have the authority to take additional precautions to mitigate this unanticipated risk or even to refrain from conducting the attack if the subordinates believe that such measures would best achieve the commander's intent if the commander had known such information.

5.10.3 Implementation of Proportionality Through Military Procedures. As with other aspects of the law of war, during U.S. military operations, commanders have implemented the requirements of the principle of proportionality through military procedures, such as rules of engagement, doctrine, standard operating procedures, and special instructions.[322] In U.S. practice, implementing procedures sometimes place more restrictions on military operations and are more protective of civilians than what would be required by the law of war.[323] Military authorities have developed procedures to assess before an attack the expected collateral damage or effects of an attack.[324] For example, lists of sensitive objects for which responsible commanders have imposed requirements for additional review or higher-level approval before the objects may be attacked (sometimes referred to as "no-strike lists") have been developed,[325] as have procedures for presenting to more senior commanders for decision potential attacks on targets that involve higher risks of incidental harm.[326]

These procedures help ensure the sound implementation of the requirements of the principle of proportionality. As a case in point, procedures for requesting higher-level approval of targets that involve high risks of incidental harm allow for a better evaluation of the expected military advantage from the attack (as it is likely that more senior commanders have a more comprehensive understanding of the strategic and operational context), and allow more senior commanders to implement additional precautions with the greater resources under their control.[327] Similarly, rules of engagement and military orders can be used to delineate the circumstances and limitations under which subordinates will initiate and/or continue an attack.[328]

5.10.4 Time Constraints and the Application of Proportionality. The time available to make decisions and to take precautions is a major factor affecting the implementation of the requirements of the principle of proportionality. For example, the available time affects what precautions are feasible, and the available time is a constraint in assessing whether an attack is expected to cause excessive incidental harm. Thus, although the legal standard of proportionality does not distinguish as such between, *e.g.*, dynamic and deliberate targeting, the availability of more time in the context of deliberate targeting generally allows for more actions to be taken to reduce the risk of civilian casualties.

Training on, and application of, targeting procedures can help forces implement the requirements of the principle of proportionality even when there is little time to make targeting decisions or to take precautions. Moreover, as with other law of war requirements, the principle of proportionality does not preclude commanders from taking decisive action in doubtful cases, but commanders must make decisions in good faith on the basis of the information available to them.[329]

5.10.5 Relationship Between the Requirement to Take Feasible Precautions in Planning and Conducting Attacks and the Prohibition on Attacks Expected to Cause Excessive Incidental Harm. This manual describes the requirement to take feasible precautions in planning and conducting attacks and the prohibition on attacks expected to cause excessive incidental harm as rules that are based on the principle of proportionality. Other sources may describe the relationships between the rules differently.[330] But this manual adopts this approach because the requirement to take feasible precautions in planning and conducting attacks and the prohibition on attacks expected to cause excessive incidental harm are fundamentally connected and mutually reinforcing obligations. U.S. military manuals reflect the fundamental connection between the requirement to take feasible precautions in planning and conducting attacks and the prohibition on attacks expected to cause excessive incidental harm.[331] This connection is also reflected in the negotiating history of the treaty provisions that sought to codify the principle of proportionality.[332] For example, the obligation to take precautions has been viewed as following from the prohibition on attacks expected to result in excessive harm.[333] In addition, applying the prohibition on attacks expected to result in excessive harm has been viewed as a precaution.[334]

The requirement to take feasible precautions in planning and conducting attacks and the prohibition on attacks expected to cause excessive incidental harm are mutually reinforcing obligations. Taking precautions helps ensure that the prohibition is not violated, and ensuring that the precautionary obligation has been complied with may help with the assessment of whether a strike would be consistent with the prohibition.[335] However, feasible precautions must be taken even when the attack is not expected to cause excessive incidental harm, and, even when feasible precautions have been taken, an attack could still be prohibited as one expected to cause excessive incidental harm.

5.11 PROPORTIONALITY – FEASIBLE PRECAUTIONS IN PLANNING AND CONDUCTING ATTACKS TO REDUCE THE RISK OF HARM TO PROTECTED PERSONS AND OBJECTS

Combatants must take feasible precautions in planning and conducting attacks to reduce the risk of harm to civilians and other persons and objects protected from being made the object of attack.[336]

As discussed in § 5.2.3.2 (What Precautions Are Feasible), what precautions are feasible depends greatly on the context.[337] Moreover, the requirement to take feasible precautions should be applied as part of the principle of proportionality in conducting attacks, and the discussion in this section should be considered in light of the discussion in § 5.10 (Proportionality in Conducting Attacks).[338]

Feasible precautions in planning and conducting attacks may include, but are not limited to, the following:

- assessing the risks to civilians;[339]
- identifying zones in which military objectives are more likely to be present or civilians are more likely to be absent;[340]
- providing effective advance warning before an attack that may affect the civilian population;[341]
- adjusting the timing of an attack;[342]
- cancelling or suspending attacks based on new information raising concerns of expected civilian casualties;[343]
- weaponeering (*e.g.*, selecting appropriate weapons, aim points);[344] and
- selecting military objectives.[345]

5.11.1 Assessing the Risks to Civilians. Assessments of the risks to civilians can include: 1) general assessments of the risks to civilians and the effectiveness of efforts to reduce the risk to civilians; 2) pre-strike assessments of expected civilian casualties; and 3) after-action assessments and investigations.

5.11.1.1 *General Assessments of the Risks to Civilians and the Effectiveness of Efforts to Reduce the Risks to Civilians*. In the planning of military operations, general assessments (*i.e.*, assessments not necessarily done in relation to a particular attack) of the risks to civilians and efforts to reduce the risks to civilians can assist in efforts to reduce the risk of incidental harm. For example, military authorities could identify places or objects posing significant risks of incidental harm if struck and could compile them into a list. They could then use such information in planning efforts and when conducting specific attacks to reduce the risk of incidental harm.

5.11.1.2 *Pre-Strike Assessments of Expected Civilian Casualties*. Pre-strike assessments of expected civilian casualties can assist in determining what other precautions are feasible, as well as in determining whether a strike is expected to result in excessive incidental harm.[346]

5.11.1.3 *After-Action Assessments and Investigations*. Battle damage assessments, as well as other reviews or investigations of incidents involving civilian casualties, may be appropriate in order to identify measures to mitigate the likelihood of future incidents of civilian casualties.[347] Similarly, review or investigation may be appropriate after "close call" incidents that do not cause civilian casualties but in which post-incident reporting reveals that a high risk of causing such casualties was not fully understood when relevant decisions were made.[348]

After such reviews, appropriate actions should be taken to reduce the risk of such incidents in the future.[349] For example, such actions could include repairing or inspecting equipment, retraining personnel, or changing applicable rules of engagement or operational procedures.

Although battle damage assessments or other after-action reviews or investigations of whether and how many civilians were harmed by an attack may serve a number of useful purposes, the actual results of an attack do not always provide useful information in assessing whether the requirements of the principle of proportionality were met in conducting the attack. The requirements of the principle of proportionality in conducting attacks are to be met based on the information available at the time, and the responsibility of commanders is not to be assessed on the basis of information known only after the fact.

5.11.2 Identifying Zones in Which Military Objectives Are More Likely to Be Present or Civilians Are More Likely to Be Absent. Identifying and designating zones in which military objectives are more likely to be present or in which civilians are more likely to be absent may also reduce the risk of harm to civilians or other protected persons and objects. For example, a commander could decide to impose greater restrictions on attacks in areas in which civilians are known to be present.[350] Similarly, it may be possible to identify areas in which objects of the greatest military importance are located and to place additional restrictions on attacks conducted outside these areas in order to limit unnecessary destruction.[351]

5.11.3 Adjusting the Timing of the Attack. Adjusting the timing of an attack may reduce the risk of incidental harm. For example, attacking a military objective when civilians are less likely to be present may be appropriate.[352] Similarly, it may be appropriate to wait until enemy forces have departed from populated areas before attacking such forces in order to reduce the risk of civilian casualties.[353]

5.11.4 Canceling or Suspending Attacks Based on New Information Raising Concerns of Expected Civilian Casualties. Commanders may decide to cancel or suspend attacks based on new information of expected civilian casualties or authorize subordinates to do so, in order to reduce the risk of harm to civilians and civilian objects.[354]

Subordinates generally would not be authorized to suspend or cancel attacks contrary to specific military orders because they disagree with the commander's decisions and judgments in relation to the principle of proportionality.[355] However, subordinate commanders or engagement authorities would be required to report promptly new information of expected civilian casualties or to cancel or suspend attacks in appropriate circumstances. For example, aircrews have refrained from conducting an attack or have redirected a munition toward a different impact location when concerns about civilian harm have become unexpectedly apparent.[356]

Subordinates should also generally be understood to have the authority to make decisions required by the law of war, such as the decision to cancel or suspend an attack in light of new information, in order to effectuate the commander's intent.[357] This is especially true when those commanders will not be able to maintain situational awareness of the risks of civilian casualties. The decision to cancel or suspend an attack based on new information raising concerns of expected civilian casualties does not necessarily mean that the attack would have been unlawful, such as by violating the prohibition against attacks expected to cause excessive incidental harm.[358]

5.11.5 Providing Effective Advance Warning Before an Attack That May Affect the Civilian Population. Unless circumstances do not permit, effective advance warning must be given of an attack that may affect the civilian population.[359]

5.11.5.1 *Other Warning Requirements.* In addition, warning requirements exist before certain medical units, vessels, or facilities forfeit their protection from being made the object of attack:

- military medical units and facilities;[360]
- ground medical transports;[361]
- hospital ships and sick-bays in warships;[362]
- civilian hospitals;[363] and
- civilian hospital convoys.[364]

5.11.5.2 *Effective Advance Warning.* The purpose of a warning is to facilitate the protection of the civilian population so that civilians and the authorities in control of the civilian population can take measures to reduce the risk that civilians will be harmed by military operations. Although there is no set form for warnings, a warning should be designed to accomplish this purpose to the extent feasible.

Warnings may be communicated to the authorities in control of the civilian population, such as the national leadership of the enemy State.[365] Warnings may also be delivered directly to the civilian population through military information support operations (*e.g.*, broadcasts,[366] leaflets[367]) advising the civilian population of risk of harm if they are near military objectives. Low passes of aircraft or warning shots may also be appropriate in certain circumstances.[368] The warnings may be general in nature; giving the specific time and place of an attack is not required.[369]

5.11.5.3 *That May Affect the Civilian Population.* The purpose of a warning to facilitate the protection of the civilian population is so that civilians and the authorities in control of the civilian population can take measures to reduce the risk that civilians will be harmed by military operations. If the civilian population will not be affected by an attack, then there is no obligation to provide a warning to facilitate its protection.[370]

5.11.5.4 *Unless Circumstances Do Not Permit.* Circumstances not permitting the giving of advance warning include where giving a warning would be incompatible with legitimate military requirements, such as exploiting the element of surprise in order to provide for mission accomplishment and preserving the security of the attacking force.[371]

5.11.6 Weaponeering (E.g., Selecting Appropriate Weapons, Aim Points). Depending on the circumstances, the process of "weaponeering" in a strike against a target can be used to reduce the risk of harm to civilians and civilian objects by selecting munitions of appropriate size and type, as well as appropriate aim points, while offering the same or superior military advantage in neutralizing or destroying a military objective.[372] For example, it may be advantageous to employ incendiary weapons in attacking an adversary's repository of biological weapons so as to prevent the biological agents from adversely affecting the civilian population.[373]

As with other precautions, the decision of which weapon to use will be subject to many practical considerations.[374] For example, a commander may decide not to use precision-guided munitions or cyber tools[375] because he or she determines it is necessary to preserve those capabilities for other engagements.[376]

5.11.7 Selecting Military Objectives. In planning or conducting an attack, the selection of a military objective may reduce the risk of incidental harm. In particular, when attempting to achieve a particular military advantage through an attack, a commander may confront a choice among several military objectives for achieving that advantage. When facing such a choice, provided that all other factors are equal, the object to be selected for attack shall be the object the attack on which may be expected to cause the least danger to civilian lives and to civilian objects. For example, in seeking to deny an adversary the ability to use a railroad network, it may be possible to disable the railroad network just as effectively by striking the railroad lines away from inhabited areas as by striking the railroad station located near civilians.[377] This rule does not dictate what choices a commander should consider, but presupposes that the commander is confronted with a choice of potential military objectives to attack to achieve the desired military advantage. In other words, this rule does not prevent the commander from attacking multiple military objectives or pursuing every military advantage he or she believes warrants pursuit.

When the choice of military objectives involves different risks and benefits potentially yielding different military advantages, this rule does not require that the object that may be expected to cause the least danger to civilian lives and to civilian objects be chosen for attack. For example, a commander could decide to attack a military objective involving higher risks of civilian casualties because the attack on that objective affords a greater likelihood of achieving the military advantage.

5.11.7.1 *AP I Provision on Choice Possible Between Several Military Objectives*. AP I provides that "[w]hen a choice is possible between several military objectives for obtaining a similar military advantage, the objective to be selected shall be that the attack on which may be expected to cause the least danger to civilian lives and to civilian objects."[378] The United States has expressed the view that this language is not a part of customary international law.[379]

Whether this AP I provision is consistent with customary international law would depend on how AP I Parties interpret it. If this AP I provision is interpreted as described in § 5.11.7 (Selecting Military Objectives), then this provision could be understood to be consistent with customary international law.

5.12 PROPORTIONALITY – PROHIBITION ON ATTACKS EXPECTED TO CAUSE EXCESSIVE INCIDENTAL HARM

Combatants must refrain from attacks in which the expected loss of civilian life, injury to civilians, and damage to civilian objects incidental to the attack would be excessive in relation to the concrete and direct military advantage expected to be gained.[380]

This prohibition should be applied as part of the principle of proportionality in conducting attacks, and the discussion in this section should be considered in light of the discussion in § 5.10 (Proportionality in Conducting Attacks).[381]

5.12.1Expected Loss of Civilian Life, Injury to Civilians, and Damage to Civilian Objects Incidental to the Attack. The totality of the expected loss of civilian life, injury to civilians, and damage to civilian objects incidental to the attack must be considered. AP I uses the phrase "a combination thereof" and an "or" to convey that the totality of the expected loss of civilian life, injury to civilians, and damage to civilian objects incidental to the attack is to be considered.[382]

5.12.1.1 *Priority in Considering the Loss of Civilian Life and Damage to Civilian Objects*. In light of the humanitarian objectives of the law of war, the expected loss of civilian life and injury to civilians should be given greater consideration than the expected damage to civilian objects.[383] Similarly, the expected damage to civilian objects (such as schools, hospitals, and religious facilities) should be given greater consideration when such damage is expected to involve the risk of harming civilians present inside such objects. In light of the great importance of cultural property, expected damage to cultural property should be afforded greater consideration than expected damage to ordinary property.[384]

5.12.1.2 *Loss of Life, Injury, and Damage Versus Mere Inconveniences or Temporary Disruptions*. Expected loss of civilian life, injury to civilians, and damage to civilian objects must be considered. Mere inconveniences or temporary disruptions to civilian life need not be considered in applying this rule.[385]

For example, although the actual damage to a civilian marketplace from an attack on a nearby military objective would be considered, the temporary disruption to commerce from the closure of the marketplace due to the nearby attack would not need to be considered.

5.12.1.3 *Foreseeable Harms Versus Remote Harms*. The expected loss of civilian life, injury to civilians, and damage to civilian objects is generally understood to mean such immediate or direct harms foreseeably resulting from the attack. Remote harms that could result from the attack do not need to be considered in applying this prohibition. The exclusion of remote harms is based on the difficulty in accurately predicting the myriad of remote harms from the attack (including the possibility of unrelated or intervening actions that might prevent or exacerbate such harms) as well as the primary responsibility of the party controlling the civilian population to take measures to ensure that population's protection.[386]

For example, if the destruction of a power plant would be expected to cause the loss of civilian life or injury to civilians very soon after the attack due to the loss of power at a connected hospital, then such harm should be considered in assessing whether an attack is expected to cause excessive harm. On the other hand, the attacker would not be required to consider the economic harm that the death of an enemy combatant would cause to his or her family, or the loss of jobs due to the destruction of a tank factory.[387] Similarly, in determining the expected loss of civilian life, injury to civilians, and damage to civilian objects, the attacker would not be required to consider the possibility that a munition might not detonate as intended and might injure civilians much later after the attack. This is due to the difficulty in assessing such risks and the responsibility of the party controlling the territory and the civilian population to take steps with regard to the protection of the civilian population from unexploded ordnance.[388]

5.12.1.4 *Harms Caused by the Attacking Force Versus Harms Caused by Enemy Action or Beyond the Control of Either Party*. The harm caused by the attacking forces' actions in conducting the attack must be considered. Harm caused by enemy action, or beyond the control of either party, need not be considered. For example, a commander directing an air operation would not need to consider civilians injured or killed by counter-attacks from enemy air defense measures, such as spent surface-to- air measures or antiaircraft projectiles.[389] Similarly, the risk to the civilian population from the legitimate deception activities of the defending force, such as jamming, smoke, or chaff, would not need to be considered by the attacking force, although the defending force should consider such risks as part of its obligations to take feasible precautions in defense.[390]

When the attacking force causes harms that are the responsibility of the defending force due to its use of voluntary human shields or due to the employment of civilian personnel in or on military objectives, the responsibility of the defending force is a factor that may be considered in determining whether such harm is excessive.[391]

5.12.2 Concrete and Direct Military Advantage Expected to Be Gained. The "concrete and direct military advantage expected to be gained" must be considered. The considerations in assessing a "definite military advantage" in the definition of "military objective" are also relevant in assessing the "concrete and direct military advantage expected to be gained."[392] For example, the military advantage may not be merely hypothetical or speculative, although there is no requirement that the military advantage be "immediate." There must be a good faith expectation that the attack will make a relevant and proportional contribution to the goal of the military attack involved.[393] Such goals may include: (1) denying the enemy the ability to benefit from the object's effective contribution to its military action (*e.g.*, using this object in its military operations); (2) improving the security of the attacking force;[394] and (3) diverting enemy forces' resources and attention.

5.12.2.1 *Considering the Advantage From the Attack as a Whole – In the Operational and Strategic Context*. The military advantage expected from an attack is intended to refer to the advantage gained from the attack considered as a whole, rather than only from isolated or particular parts of an attack.[395] Similarly, "military advantage" is not restricted to immediate tactical gains, but may be assessed in the full context of the war strategy.[396] The military importance of a target often turns on its relationship to other targets within an operational system, and the effect that disabling the target will have on the functions that comprise the adversary's ability to wage war.[397] For example, an attack against a communications relay that was a military objective might be expected to yield a much greater military advantage when the attack was part of a strategy of coordinated attacks to disable the adversary's command and control network than as an isolated attack.[398]

Consideration of the military advantage of the entire attack in its operational and strategic context ensures the proper consideration and allocation of responsibility to the appropriate levels within the military hierarchy.[399]

Lower-level personnel may not be competent to evaluate the broader strategic and operational implications of the entire attack.[400]

5.12.3 Determining Whether the Expected Incidental Harm Is Excessive. A potential attack against a military objective is prohibited by this rule only when the expected incidental harm is excessive in relation to the concrete and direct military advantage expected to be gained. Determining whether the expected incidental harm is excessive does not necessarily lend itself to quantitative analysis because the comparison is often between unlike quantities and values.[401] The evaluation of expected incidental harm in relation to expected military advantage intrinsically involves both professional military judgments as well as moral and ethical judgments evaluating the risks to human life (*e.g.*, civilians at risk from the attack, friendly forces or civilians at risk if the attack is not taken). On the one hand, striking an ammunition depot or a terrorist training camp would not be prohibited because a farmer is plowing a field in the area.[402] On the other hand, an extraordinary military advantage would be necessary to justify an operation posing risks of collateral death or injury to thousands of civilians.[403] In less clear-cut cases, the legal question of whether the expected incidental harm is excessive may be a "highly open-ended" inquiry, and the answer may be "subjective and imprecise."[404] For this reason, States have chosen to apply a "clearly excessive" standard for determining whether a criminal violation has occurred.[405] In determining whether the expected incidental harm would be excessive, the totality of the circumstances must be considered. This holistic judgment should consider any relevant moral, legal, and military factors.

5.12.3.1 *The Military Concept of Economy of Force*. As with the obligation to take feasible precautions, the prohibition on attacks expected to cause excessive incidental harm involves military judgments.[406] Use of excessive and indiscriminate force is costly, highly inefficient, and wasteful of scarce resources.

The principle of proportionality has been viewed as a legal restatement of the military concept of economy of force.[407] Economy of force seeks to allocate the maximum possible combat power on primary efforts through the judicious employment and distribution of forces and the measured allocation of available combat power.[408]

Although mere poor military judgment or substandard operational practice is not by itself unlawful, if a strike would be so wasteful of resources as to be inconsistent with the principle of economy of force, this is a factor counseling against taking the contemplated strike.

5.12.3.2 *Feasibility of Taking Additional Precautions to Reduce the Risk of Incidental Harm*. Incidental harm is more likely to be regarded as excessive in attacks in which it would be feasible to take additional precautions and those precautions are not taken.[409] The obligation to take feasible precautions is fundamentally connected to the prohibition on attacks expected to cause excessive incidental harm, and the two obligations are mutually reinforcing.[410] Thus, although the obligation to take feasible precautions applies even if an attack is expected not to cause excessive incidental harm, incidental harm expected to result from strikes in which additional precautions are feasible, but have not been taken, would be more likely to be considered excessive.

5.12.3.3 *Civilian Workers Who Support Military Operations In or On Military Objectives*. In general, reasonable steps must be taken to separate the civilian population from military objectives, such as by removing members of the civilian population from the vicinity of military objectives and combat operations.[411] However, sometimes civilian personnel work in or on military objectives in order to support military operations. For example, civilian workers sometimes serve as members of military aircrews, as technical advisers on warships, and as workers in munitions factories. Such persons assume a certain risk of injury.[412] Provided such workers are not taking a direct part in hostilities, those determining whether a planned attack would be excessive must consider such workers, and feasible precautions must be taken to reduce the risk of harm to them. Those making such determinations may consider all relevant facts and circumstances. The expected harm to civilian members of military aircrews on military aircraft or civilian technical advisers on warships does not render military aircraft or warships immune from attack.[413] Similarly, munitions factories have not been deemed to be immune from attack just because of the presence of civilian workers inside the factory.[414]

5.12.3.4 *Enemy Use of Human Shields*. Adversary use of human shields can present complex moral, ethical, legal, and policy considerations.[415] Use of civilians as human shields violates the rule that protected persons may not be used to shield, favor, or impede military operations.[416] The party that employs human shields in an attempt to shield military objectives from attack assumes responsibility for their injury, although the attacker may share this responsibility if it fails to take feasible precautions.[417]

If civilians are being used as human shields, provided they are not taking a direct part in hostilities, they must be considered as civilians in determining whether a planned attack would be excessive, and feasible precautions must be taken to reduce the risk of harm to them.[418]

However, the enemy use of voluntary human shields may be considered as a factor in assessing the legality of an attack. Based on the facts and circumstances of a particular case, the commander may determine that persons characterized as voluntary human shields are taking a direct part in hostilities.[419]

5.13 ATTACKS ON FACILITIES,WORKS, OR INSTALLATIONS CONTAINING DANGEROUS FORCES

Certain facilities containing dangerous forces, such as dams, nuclear power plants, or facilities producing weapons of mass destruction, may constitute military objectives. There may be a number of reasons for their attack, such as denial of electric power to military sources, use of a dangerous facility (*e.g.*, by causing release from a dam) to damage or destroy other military objectives, or to pre-empt enemy release of the dangerous forces to hamper the movement or advance of U.S. or allied forces.[420]

Attack of facilities, works, or installations containing dangerous forces, such as dams, nuclear power plants, or facilities producing weapons of mass destruction, is permissible so long as it is conducted in accordance with other applicable rules, including the rules of discrimination and proportionality.[421] In light of the increased potential magnitude of incidental harm, additional precautions, such as weaponeering or timing the attack such that weather conditions would minimize dispersion of dangerous materials, may be appropriate to reduce the risk that the release of these dangerous forces may pose to the civilian population.[422]

5.13.1 AP I Provisions on Works and Installations Containing Dangerous Forces. Article 56 of AP I provides special rules for works and installations containing dangerous forces. For example, "[w]orks or installations containing dangerous forces, namely dams, dykes and nuclear electrical generating stations, shall not be made the object of attack, even where these objects are military objectives, if such attack may cause the release of dangerous forces and consequent severe losses among the civilian population." In addition, Article 56 of AP I provides immunity from attack to combatants and military equipment stationed or placed around works and installations containing dangerous forces "for the sole purpose of defending the protected works."

The United States has objected to this article of AP I.[423] In ratifying AP I, other States have taken reservations from this article.[424] Insofar as Article 56 of AP I deviates from the regular application of the principles of distinction and proportionality, the U.S. view has been that it does not reflect customary international law applicable in international and noninternational armed conflicts.

5.14 FEASIBLE PRECAUTIONS TO REDUCE THE RISK OF HARM TO PROTECTED PERSONS AND OBJECTS BY THE PARTY SUBJECT TO ATTACK

Outside the context of conducting attacks (such as when conducting defense planning or other military operations), parties to a conflict should also take feasible precautions to reduce the risk of harm to protected persons and objects from the effects of enemy attacks.[425] In particular, military commanders and other officials responsible for the safety of the civilian population must take reasonable steps to separate the civilian population from military objectives and to protect the civilian population from the effects of combat. As discussed in § 5.2.3.2 (What Precautions Are Feasible), what precautions are feasible depends greatly on the context.[426] Feasible precautions by the party subject to attack may include the following:

5.14.1 Refraining From Placing Military Objectives in Densely Populated Areas. It may be appropriate to avoid placing military objectives, such as the armed forces, in urban or other densely populated areas, in order to reduce the risk of incidental harm to the civilian population.[427]

However, it often may not be feasible to refrain from placing military objectives in densely populated areas. Legitimate military reasons often require locating or billeting military forces in urban areas or other areas where civilians are present. For example, forces may be housed in populated areas to take advantage of existing facilities, such as facilities for shelter, health and sanitation, communications, or power. In some cases, especially during counterinsurgency operations or in non-international armed conflict generally, the protection of the civilian population may be increased by placing military forces in densely populated areas to protect the civilian population from enemy attack and influence.[428]

5.14.2 Removing Civilians and Civilian Objects From the Vicinity of Military Objectives. It may be appropriate to remove civilians and civilian objects from the vicinity of military objectives.[429] In particular, civilian hospitals should be situated as far as possible from military objectives.[430]

The evacuation of civilians from areas likely to be attacked is advisable when there is immediate danger and where it would be likely to involve less hardship and danger to civilians than leaving them in place.[431]

Voluntary removal of civilians may be accomplished through the use of warnings.[432] When removing civilians from the vicinity of military objectives, it may be appropriate to establish safety, hospital, or neutralized zones so that civilians have safe places to move toward.[433] It also may be appropriate to conclude local agreements for the removal of civilians from besieged or encircled areas.[434]

During international armed conflict, the forcible transfer or deportation of protected persons from occupied territory is prohibited.[435]

5.14.3 Establishing Areas Where Civilians or the Wounded and Sick Are Protected. Certain zones or localities may be established through the agreement of parties to a conflict to shelter civilians or wounded and sick combatants from the effects of attacks.[436] In this context, "locality" means a specific place of limited area, generally containing buildings, while "zone" means a relatively large area of land and may include one or more localities.[437]

5.14.3.1 *Civilian Hospital and Safety Zones and Localities*. During international armed conflict, States may establish hospital and safety zones and localities to protect certain persons from the effects of war, namely, wounded, sick, and aged persons, children under fifteen, expectant mothers, and mothers of children under seven.[438] Parties to a conflict may conclude agreements on the mutual recognition of the hospital zones and localities they have created, drawing upon model agreements that are annexed to the 1949 Geneva Conventions.[439] The establishment of a zone only binds an adverse party when it agrees to recognize the zone.[440]

These zones and localities may be combined with hospital zones established for the wounded and sick of the armed forces.[441]

Protecting Powers and the ICRC may seek to facilitate the institution and recognition of civilian hospital and safety zones and localities.[442]

5.14.3.2 *Military Hospital Zones and Localities*. States may establish in their own territory, and if the need arises, in occupied areas, hospital zones and localities so organized as to protect the wounded and sick (as defined in the GWS[443]) from the effects of war, as well as the personnel entrusted with the organization and administration of these zones and localities and with the care of the persons assembled within them.[444]

5.14.3.3 *Neutralized Zones*. It may be appropriate to conclude agreements with opposing forces to establish neutralized zones in regions where fighting is taking place.[445] Neutralized zones differ from civilian hospital and safety zones in that they are intended to protect a broader group of persons and in that they are generally established on a temporary basis in regions where fighting is taking place.[446]

These neutralized zones are to shelter: (a) wounded and sick combatants or noncombatants; and (b) civilian persons who take no part in hostilities, and who, while they reside in the zones, perform no work of a military character.[447]

Agreements should be made in writing, signed by representatives of the parties to the conflict, and should establish: (a) the location of the zone; (b) the administration of the zone; (c) the food supply of the zone; (d) the supervision of the zone; and (e) the beginning and duration of the neutralization of the zone.[448]

5.14.4 Using Distinctive and Visible Signs to Identify Protected Persons and Objects as Such. It may be appropriate to identify protected persons and objects, as such, through the use of distinctive and visible signs. For example, it may be appropriate to identify civilian hospitals[449] or civilian air raid shelters in order to facilitate their protection during enemy bombardment.[450]

Signs indicating civilian objects, as such, should be notified to the opposing party so that the opposing party knows to refrain from bombarding places or buildings bearing these signs. Where enemy forces seek to make peaceful civilians the object of attack, then it would not be appropriate to distinguish civilian objects. Such precautions would not be feasible because they would not be expected to yield a humanitarian benefit.[451]

5.14.5 Carrying Arms Openly and Wearing of Distinctive Emblems by the Armed Forces to Distinguish Themselves From the Civilian Population. Although the practices of carrying arms openly and wearing uniforms and distinctive emblems by armed forces predates contemporary doctrines of feasible precautions, these practices may be viewed as feasible precautions that help protect the civilian population from being erroneously made the object of attack.[452]

This requirement of armed forces to wear distinctive emblems has been understood as a requirement for the armed forces as a group.[453] However, it applies, to some extent, on an individual basis in the context of liability for spying or sabotage.[454] The obligation of armed forces to distinguish themselves includes, but is not limited to, those times when they conduct attacks.[455] However, the requirement to wear distinctive emblems is not an absolute one, even during combat.[456]

5.15 UNDEFENDED CITIES, TOWNS, AND VILLAGES

Attack, by whatever means, of a village, town, or city that is undefended is prohibited.[457] Undefended villages, towns, or cities may, however, be captured.

5.15.1 "Undefended" – Notes on Terminology. The term *undefended city* (or *town* or *village*, or any other populated area) is a term of art in the law of war and should not be confused with a place that simply lacks defensive capabilities. Historically, open or undefended status for a town, village, or city would be sought as opposing military forces approached and the military forces previously controlling the city abandoned it.[458] Undefended or open status would essentially surrender the city to the opposing force; this minimized injury to the inhabitants and damage to civilian objects within the city because the city could be occupied without resistance or bypassed.[459]

5.15.1.1 *Non-defended Versus Undefended.* Article 59 of AP I uses the term "non-defended" rather than "undefended," but uses essentially the same criteria as FM 27-10's discussion of "undefended" places within the meaning of Article 25 of the Hague IV Regulations. The change from "undefended" in the Hague IV Regulations and Hague IX, to "non-defended" in AP I, appears to reflect a difference in translation from French to English rather than a substantive difference.

The original text of the Hague IV Regulations and Hague IX is French.[460] In the French versions of the text, both the Hague IV Regulations and Hague IX use a dependent clause "that are not defended" in order to modify the list of places.[461] Rather than a dependent clause, AP I uses an adjective -- "non-defended" modifying the word "localities."[462] The difference in English terms seems to reflect this underlying change in the French phrasing. Moreover, the title of the chapter in Hague IX—"Du bombardement des ports, villes, villages, habitations ou bâtiments non défendus" (emphasis added)—uses the same French phrase as is used in AP I, yet the last two words of the title of the chapter in Hague IX have typically been translated as "undefended" rather than "non-defended." Thus, the consistency in terminology across the French texts suggests that even though the English terminology has varied slightly, no substantive change has been intended.

5.15.1.2 *Open Versus Undefended.* If "undefended" is simply understood in a factual sense to mean a lack of defenses and "open" is simply understood in a factual sense to mean "open" for entry by the enemy, then distinctions could be drawn between "open" cities and "undefended" cities, because not all cities lacking defenses would be open for entry by enemy forces. For example, a city that did not itself have defenses, but was located deep behind the State's defensive lines would not be "open" for entry by enemy forces. Most sources, however, use "open" and "undefended" interchangeably.[463] It appears that these terms tend to be used interchangeably because both terms generally refer to a specific legal category of populated areas that are protected by the law of war from attack.[464]

5.15.2 "By Whatever Means". The phrase "by whatever means" was inserted in the Hague IV Regulations to clarify that bombing attacks by air were included in the rule.[465]

5.15.3 Declaration of a City as "Undefended". An undefended city may be established through negotiations with opposing forces,[466] or unilaterally by the party to the conflict in control of it. If the latter, the intent and actions of that party should be communicated to opposing military forces through a declaration. A town, village, or city may be declared "undefended" when it is near, or in, a zone where opposing armed forces are in contact with one another and it is open for immediate physical occupation by an adverse party without resistance.[467]

5.15.3.1 *Open for Immediate Physical Occupation*. The area in question must be open for immediate physical occupation by opposing military ground forces.[468] Thus, a city in rear areas behind enemy lines cannot be "undefended." For example, a party to a conflict cannot declare a city or other population center to be undefended if it is several hundred miles behind the area of ground conflict where it would not be feasible for opposing ground forces to occupy it. By way of comparison, it would not be effective for a person to seek to surrender to an opposing force's aircraft that could not physically take custody of that person.[469]

5.15.3.2 *Necessity for a Declaration of Undefended Status*. Although there is no explicit treaty requirement that a city be declared undefended before achieving that status, the practice has been to make such declarations to the opposing party. Without them, a belligerent would have no way of knowing whether a city was truly open for occupation without resistance. For example, a city may not satisfy the conditions described below for undefended status, or it may contain defenses that are concealed or that are operated from other areas.[470]

5.15.3.3 *Refusal to Recognize a Declaration.* Belligerents may refuse to recognize a declaration that a city is undefended if they assess that it does not satisfy all of the necessary conditions, although they should notify the opposing belligerent of that decision.[471] Absent or until recognition, military objectives in a city unilaterally designated as undefended remain subject to attack.[472]

5.15.3.4 *Authority to Make a Declaration or to Recognize a Declaration.* Declaration of a city being open or recognition of an opposing party's declaration may only be accomplished by those with authority to enforce the corresponding obligations, which may be the immediate military commander, theater commander, or national command authorities.[473] These obligations may extend more broadly than control over the city itself as, for example, when the city's defenses are operated from a different location.

5.15.4 Conditions Necessary for a Place to Maintain Undefended Status. Once a party to a conflict has validly declared a city to be undefended, the city must also satisfy certain conditions. If the city fails to satisfy these conditions, it would not be entitled to undefended status, or, if previously granted, it would lose that status.[474] Persons and objects within that city, however, may still receive other protections for civilians and civilian objects.[475]

5.15.4.1 *Combatants and Mobile Military Equipment Have Been Evacuated.* All combatants, as well as their mobile weapons and mobile military equipment, must have been evacuated.[476] However, the presence of military medical personnel, the wounded and sick, and civilian police forces for the purpose of maintaining local law and order would not cause a city designated as undefended to lose that status.[477]

5.15.4.2 *No Hostile Use of Fixed Military Facilities.* No hostile use shall be made of fixed military installations or establishments within the city.[478]

5.15.4.3 *No Hostile Acts Against the Occupying Force.* Hostile acts may not be committed by the local civilian authorities or the civilian population against the occupying military force.[479]

5.15.4.4 *No Activities In Support of Military Operations.* No activities in support of military operations may be undertaken.[480] For example, factories in the city should not be used to manufacture munitions, and ports or railroads should not be used to transport military supplies.[481]

5.15.5 Agreements to Establish Undefended Cities. Even if the conditions described above are not met, parties to a conflict may agree between themselves to treat an area as an undefended city.[482] The agreement should be in writing and should define and describe, as precisely as possible, the limits of the undefended city (such as the exact geographic limits of the locality, when the location is to begin to receive protection as undefended, and the duration of undefended status). If necessary, methods of supervision to ensure that the city continues to fulfill the conditions should be specified.[483] It may also be appropriate for agreements to specify: (1) rules on marking the city and agreed signs;[484] (2) persons authorized to enter the city; and (3) whether and under what conditions the city may be occupied by enemy forces.

5.15.6 Actions of the Occupying Force. The occupying force may use the city, town, or village for military operations, including by establishing defenses to prevent its recapture by the enemy.[485] In such cases, the city would lose its undefended status, and, for example, military objectives within the city may be attacked.[486] However, if the occupying force acts in a manner consistent with the conditions necessary for the inhabited place to maintain undefended status,[487] the place would remain undefended for the purpose of protecting it against military operations by all parties to the conflict.

5.16 PROHIBITION ON USING PROTECTED PERSONS AND OBJECTS TO SHIELD, FAVOR, OR IMPEDE MILITARY OPERATIONS

Parties to a conflict may not use the presence or movement of protected persons[488] or objects:[489] (1) to attempt to make certain points or areas immune from seizure or attack; (2) to shield military objectives from attack; or (3) otherwise to shield or favor one's own military operations or to impede the adversary's military operations.

5.16.1 Protected Persons and Objects. In particular, the civilian population,[490] protected persons under the GC,[491] POWs,[492] fixed medical establishments and medical units, *parlementaires* and other persons protected by a flag of truce,[493] and cultural property are protected persons and objects that may not be used for the above purposes. Although persons and objects that are protected by the law of war may not be used in these ways, this rule does not prohibit a party from using what would otherwise be a civilian object for military purposes and thereby converting it to a military objective that is not protected by the law of war. For example, a building that previously was a civilian object could be used for military purposes (including as cover) and would not implicate this rule because it would no longer be a protected object. Similarly, this rule does not prohibit persons who would otherwise be civilians from participating in hostilities or otherwise assuming the risks inherent in supporting military operations.[494]

5.16.2 Intent to Endanger Protected Persons or Objects to Deter Enemy Military Operations. The essence of this rule is to refrain from deliberately endangering protected persons or objects for the purpose of deterring enemy military operations.[495] This absolute duty to refrain from purposeful misconduct may be contrasted with the affirmative obligation that parties have to take feasible precautions to separate the civilian population,[496] civilian internees,[497] POWs,[498] fixed medical establishments and medical units,[499] and cultural property,[500] from the dangers of military operations.

An adversary's intention to violate this rule is likely to be clear because that adversary normally would make it apparent to the opposing party that attacks against the military objective being shielded would risk harming protected persons or objects.[501]

On the other hand, in the absence of purposeful action to put protected persons and objects at risk of harm from enemy military operations, there would be no violation of this rule. For example, this rule would not prohibit restricting the movement of civilians in order to conduct military operations without their interference. In addition, it would also not prohibit the evacuation of civilians for their own security or for imperative military reasons.[502] In addition, this rule must not be understood to alter the rules on spying and sabotage. In other words, this rule does not prohibit a spy from hiding among civilians in order to gather intelligence. However, the spy runs the risk of prosecution by the enemy State if he or she is discovered and captured while spying.

5.16.3 Prohibition on Taking Hostages. The taking of hostages is prohibited.[503] This prohibition is understood to include a prohibition against threatening to harm detainees whose lives are valued by the adversary in order to induce the adversary not to attack, but is also understood to include the prohibition against using hostages as human shields.[504]

The taking of hostages is also prohibited in non-international armed conflict.[505]

5.16.4 Use of Human Shields by Defenders Does Not Relieve Attackers of Duties in Conducting Attacks. Violations by the adversary of the prohibition on using protected persons and objects to shield, favor, or impede military operations do not relieve those conducting attacks from their obligation to seek to discriminate between lawful and unlawful objects of attack.[506] However, such violations by the adversary may impair the attacking force's ability to discriminate and increase the risk of harm to protected persons and objects.[507]

Although human shields who are civilians are to be considered as civilians in determining whether a planned attack would be excessive, enemy use of voluntary human shields may be considered as a factor in assessing the legality of an attack.[508] Feasible precautions must be taken to reduce the risk of harm to human shields.

5.16.5 Responsibility for Harm to Human Shields. The party that employs human shields in an attempt to shield military objectives from attack assumes responsibility for their injury, although the attacker may share this responsibility if it fails to take feasible precautions.[509]

5.17 SEIZURE AND DESTRUCTION OF ENEMY PROPERTY

Outside the context of attacks, certain rules apply to the seizure and destruction of enemy property:

• Enemy property may not be seized or destroyed unless imperatively demanded by the necessities of war;

• Public movable property and certain types of private movable property may customarily be captured as war booty;

• Enemy private movable property that is not susceptible of direct military use may be appropriated only to the extent that such taking is permissible in occupied areas;

• Pillage is prohibited; and

• Feasible precautions should be taken to mitigate the burden on civilians, but there is no obligation to compensate the owners of enemy property that is lawfully damaged.

During occupation, other rules relating to the treatment of enemy property apply.[510]

5.17.1 Definition of Enemy Property. All property located in enemy territory is regarded as enemy property regardless of its ownership.[511]

5.17.2 Enemy Property – Military Necessity Standard. Enemy property may not be seized or destroyed unless imperatively demanded by the necessities of war.[512] In particular, devastation or destruction may not be pursued as an end in itself.[513] The measure of permissible seizure or destruction of enemy property is found in the strict necessities of war.[514] There must be some reasonable connection between the seizure or destruction of the enemy property and the overcoming of enemy forces.[515]

Extensive destruction and appropriation of property protected by the GC, not justified by military necessity and carried out unlawfully and wantonly, constitutes a grave breach of the GC.[516]

5.17.2.1 *Using the Military Objective Definition to Assess Whether the Seizure or Destruction of Enemy Property Is Justified by Military Necessity*. The definition of military objective (that has been developed to assess whether objects may be made the object of attack) may be applied outside the context of attacks to assess whether an object's seizure or destruction is justified by imperative military necessity.[517] After all, if sufficient military necessity exists to justify attacking an object as reflected by that object meeting the definition of military objective, then imperative military necessity would also exist to justify seizing or destroying that object by measures short of attack. On the other hand, if imperative military necessity justifies the seizure or destruction of an object, that object does not necessarily satisfy the definition of a military objective. Rather, the object must be assessed as to whether it meets the definition of military objective and may be made the object of attack.

5.17.2.2 *Seizure or Destruction of Enemy Property to Support Military Operations*. It may be imperatively demanded by the necessities of war to seize or destroy enemy property in order to support military operations. These examples are illustrative and not exhaustive:

• using enemy land for the construction of military bases, air fields, and other facilities to support military operations;[518]

• using enemy buildings to billet military personnel and others supporting them, to house the wounded and sick, for observation and reconnaissance, for concealment and cover, and for defensive purposes;[519]

• demolishing, cutting down, or removing enemy walls, forests, and buildings to clear a field of fire or to provide construction material;[520] or

• seizing means of transportation for use in military operations.[521]

5.17.2.3 *Destruction of Enemy Property to Diminish the Enemy's Ability to Conduct or Sustain Operations*. It may be imperatively demanded by the necessities of war to seize or destroy enemy property in order to diminish the enemy's ability to conduct or sustain operations, such as railways, lines of communication, and other war fighting and war sustaining infrastructure.[522] For instance, Union forces destroyed the Confederacy's cotton during the Civil War in order to deprive the Confederacy of the ability to fund its military operations.[523]

Similarly, coalition forces during Operation ENDURING FREEDOM destroyed narcotics in order to weaken the Taliban and al Qaeda's ability to finance their operations.[524]

5.17.2.4 *Incidental Damage to Enemy Property*. Military necessity also justifies damage to property incidental to combat operations that is reasonably related to overcoming enemy forces.[525] For example, the movement of armed forces and equipment may damage roads or fields.[526]

Enemy Movable Property on the Battlefield (5.17.3 War Booty). Any enemy public movable property on the battlefield may be seized.[527] Property is public if it belongs to the enemy State or an agency of that State.[528]

In general, enemy private movable property on the battlefield may be seized if the property is susceptible to direct military use, *i.e.*, it is necessary and indispensable for the conduct of war.[529] This includes arms, ammunition, military papers, or property that can be used as military equipment (*e.g.*, as a means of transportation or communication).[530] However, certain types of military equipment used for clothing, feeding, or personal protection (*e.g.*, helmets and gas masks) may not be seized from POWs, and should remain in their possession.[531]

5.17.3.1 *Private Military Property Not Susceptible to Direct Military Use*. Private enemy property that is not susceptible to direct military use (*e.g.*, arms, ammunition, military papers, or military equipment) may be appropriated only to the extent such taking is permissible in occupied areas.[532] In particular, receipts should be given and compensation paid, when feasible.[533]

5.17.3.2 *Ownership of Captured or Found Property on the Battlefield.* All enemy property that is captured or found on the battlefield becomes the property of the capturing State.[534] During international armed conflict, personal property recovered from enemy dead becomes the property of the capturing State for the purpose of returning it to the next-of-kin of the deceased.[535] The individual military member or person accompanying the armed forces who captures or finds enemy property acquires no title or claim.[536]

Failure to turn over such property to the proper authorities, or disposal of such property for personal profit, is punishable under Article 103 of the Uniform Code of Military Justice.[537]

5.17.4 Pillage Prohibited. Pillage is prohibited, both in general[538] and specifically with respect to:

- the wounded, sick, shipwrecked, and dead;[539]
- POWs;[540]
- protected persons under the GC;[541]
- persons in occupied territory[542] and in areas of non-international armed conflict;[543] and
- cultural property.[544]

In certain cases, there is an affirmative obligation to prohibit and prevent pillage.[545]

5.17.4.1 *Pillage – Definition and Notes on Terminology. Pillage* is the taking of private or public movable property (including enemy military equipment) for private or personal use.[546] It does not include an appropriation of property justified by military necessity.[547] For example, if ordinary requisitions are unavailable, the taking of food does not constitute pillage.[548] Pillage also has been referred to as "looting"[549] and "plunder."[550]

5.17.4.2 *History of Pillage.* In the medieval era, pillage served as a form of compensation for private armies, but it ceased to be regarded as lawful with the widespread adoption of standing armies at the end of the Eighteenth Century. The prohibition on pillage has, in part, been intended to maintain discipline among the armed forces.[551] For example, a member of the armed forces who before or in the presence of the enemy quits his or her place of duty to plunder or pillage is guilty of the offense of misbehavior before the enemy.[552]

5.17.5 Feasible Precautions Should Be Taken to Mitigate the Burden on Civilians. In seizing or destroying enemy property, feasible precautions should be taken to mitigate the burdens imposed on civilians.

For example, in the practice of the United States, religious buildings, shrines, and consecrated places employed for worship are used only for aid stations, medical installations, or the housing of wounded personnel awaiting evacuation, provided in each case that a situation of emergency requires such use.[553]

Similarly, if armed forces use a private residence, they must treat its inhabitants and owners humanely and with as much consideration as the circumstances permit.[554] In particular, the armed forces should generally allow the inhabitants to continue to live in part of their own homes, and should not expel them in the absence of alternate shelter.[555] But if imperative military necessity requires the removal of the inhabitants, then efforts should be made to give them notice and to aid them in taking their essential possessions.[556] If the armed forces take anything, they should leave a note to this effect.[557]

However, there is no obligation to protect abandoned property in the area of active operations.[558]

5.17.5.1 *Compensation for Property Seizure or Damage.* Although reasonable efforts should be made to spare civilians from unnecessary harm when seizing or destroying enemy property, the law of war imposes no obligation to compensate for loss of, or damage to, private property imperatively demanded by the necessities of war, including damage incidental to combat operations.[559] However, if time allows, a record of the use or damage should be kept, or given to the owner, so that in the event of funds being provided by either belligerent at the close of hostilities to compensate the owners, there may be evidence to assist the assessors.[560] As a matter of practice, during counter-insurgency operations, U.S. forces have often made payments to, or taken other actions on behalf of, civilians suffering loss.[561]

5.18 PROTECTION OF CULTURAL PROPERTY DURING HOSTILITIES

Certain types of property receive additional protection as cultural property. Cultural property, the areas immediately surrounding it, and appliances in use for its protection should be safeguarded and respected.

Some obligations with respect to cultural property apply during non-international armed conflict.[562] There are also obligations with respect to cultural property during occupation[563] and peacetime.[564] Certain treaty obligations with respect to cultural property may only apply on the territory of Parties to the 1954 Hague

Cultural Property Convention,[565] but the United States has previously identified some of these obligations as customary international law.[566] DoD personnel, therefore, in the absence of contrary guidance by competent authority, should act as if they were legally bound by the rules for the protection of cultural property in the 1954 Hague Cultural Property Convention during hostilities even when conducting operations in the territory of a State that is not a Party to the 1954 Hague Cultural Property Convention.[567]

5.18.1 Definition of Cultural Property. For the purpose of the 1954 Hague Cultural Property Convention and this manual, cultural property includes, irrespective of origin or ownership:[568]

- movable or immovable property of great importance to the cultural heritage of every people;[569]
- buildings intended to shelter cultural property;[570] and
- centers containing monuments.[571]

5.18.1.1 *Definition of Cultural Property – Notes on Terminology*. "Cultural property" is a term of art that is defined in the 1954 Hague Cultural Property Convention. The definition in the 1954 Hague Cultural Property Convention may be more limited than cultural property described and protected by other instruments.

For example, the Lieber Code contemplates protection for property belonging to "establishments of an exclusively charitable character, to establishments of education, or foundations for the promotion of knowledge, whether public schools, universities, academies of learning or observatories, museums of the fine arts, or of a scientific character," as well as "[c]lassical works of art, libraries, scientific collections, or precious instruments, such as astronomical telescopes."[572] The Hague IV Regulations seek to protect "buildings dedicated to religion, art, science, or charitable purposes, historic monuments."[573] Hague IX seeks to protect "sacred edifices, buildings used for artistic, scientific, or charitable purposes, [and] historic monuments."[574] The Roerich Pact seeks to protect "historic monuments, museums, scientific, artistic, educational and cultural institutions."[575] AP I seeks to protect "historic monuments, works of art or places of worship which constitute the cultural or spiritual heritage of peoples."[576]

Because the definition of cultural property in the 1954 Hague Cultural Property Convention is different from the categories of property protected by other instruments, the scope of objects being protected by each instrument is not the same (although there may be overlap). Nevertheless, the protections afforded cultural property by the 1954 Hague Cultural Property Convention are supplementary to those afforded by earlier treaties, although the distinctive emblem for cultural property established in the 1954 Hague Cultural Property Convention is to be used where appropriate instead of emblems established in earlier treaties.[577]

5.18.1.2 *Movable or Immovable Property of Great Importance to the Cultural Heritage of Every People*. Property must be "of great importance to the cultural heritage of every people" to qualify as cultural property. Ordinary property (such as churches or works of art) that are not of great importance to the cultural heritage of every people would not qualify as cultural property, although such property may benefit from other protections, such as those afforded civilian objects or enemy property.[578]

The question of whether cultural property is "of great importance to the cultural heritage of every people" may involve delicate and somewhat subjective judgments.[579] Items that can easily be replaced would not qualify as being of great importance.[580] On the other hand, irreplaceable items may be of great cultural importance, even if they have little monetary value.[581]

Cultural property may include, but is not limited to, the following types of property (provided the property is of great importance to the cultural heritage of every people):

- monuments of architecture, art, or history, whether religious or secular;[582]
- archaeological sites;
- groups of buildings that, as a whole, are of historical or artistic interest;[583]
- works of art;
- manuscripts, books, and other objects of artistic, historical, or archaeological interest;
- scientific collections;[584] and
- important collections of books or archives or of reproductions of the property defined above.[585]

Natural sites of great beauty are not included within the definition of cultural property.[586]

5.18.1.3 *Buildings Intended to Preserve Cultural Property*. Cultural property is also understood to include buildings whose main and effective purpose is to preserve or exhibit movable cultural property, such as museums, large libraries, and depositories of archives, and refuges intended to shelter cultural property in the event of armed conflict.[587]

For these types of buildings, protection is gained from the structure's purpose and contents as opposed to the physical structure itself constituting immovable property of great importance to the cultural heritage of every people. The building must be intended to contain and, in fact, contain, conserve, or exhibit movable cultural property as its primary purpose.[588] Alternatively, the structure may be intended to store movable cultural property for protection in the event of armed conflict. Some of these refuges may be subject to special protection.[589]

5.18.1.4 *Centers Containing Monuments*. Cultural property also includes centers containing monuments, *i.e.*, centers containing a large amount of cultural property or buildings intended to preserve cultural property.[590] This includes monument complexes or major groups of buildings containing movable and immovable cultural property.

5.18.2 Respect and Safeguarding of Cultural Property. The protection of cultural property comprises the safeguarding of and respect for such property.[591] Safeguarding consists of the affirmative acts that are to be taken to ensure the protection of cultural property.[592] On the other hand, respect essentially requires parties to refrain from certain acts, such as placing cultural property in peril or causing damage to it.[593]

Safeguarding and respect are mutually reinforcing obligations that help ensure the protection of cultural property.

5.18.2.1 *Peacetime Obligations to Prepare for the Safeguarding of Cultural Property*. Parties to 1954 Hague Cultural Property Convention are obliged to undertake to prepare in time of peace for the safeguarding of cultural property situated within their own territory against the foreseeable effects of an armed conflict, by taking such measures as they consider appropriate.[594] This obligation reflects the opinion that measures for the protection of cultural property during armed conflict, in order to be effective, must begin in peacetime.[595]

The failure of a State to fulfill its obligations in peacetime to safeguard cultural property during armed conflict does not relieve other States of their obligation to respect cultural property.[596]

5.18.3 Refraining From Any Use for Purposes That Are Likely to Expose It to Destruction or Damage. In general, no use should be made of cultural property, its immediate surroundings, or appliances in use for its protection, for purposes that are likely to expose it to destruction or damage in the event of armed conflict.[597] However, such use is permissible when military necessity imperatively requires such use.[598]

Uses that would be likely to expose cultural property to destruction or damage in the event of armed conflict would include: (1) using the cultural property for military purposes;[599] (2) placing military objectives near cultural property; or (3) using the cultural property in such a way that an adversary would likely regard it as a military objective.[600] For example, such uses would include billeting military personnel in buildings that constitute cultural property, or emplacing artillery, mortars, or anti-air systems on the grounds of cultural property. In addition, it is prohibited to use deliberately the threat of potential harm to cultural property to shield military objectives from attack, or otherwise to shield, favor, or impede military operations.[601] There is no waiver of this obligation in cases of imperative military necessity.

5.18.3.1 *Imperative Military Necessity Waiver*. Cultural property, its immediate surroundings, and appliances in use for its protection may be used for purposes that are likely to expose it to destruction or damage if military necessity imperatively requires such use. The requirement that military necessity imperatively require such acts should not be confused with convenience or be used to cloak slackness or indifference to the preservation of cultural property.[602] This waiver of obligations with respect to cultural property is analogous to the requirement that enemy property may only be seized or destroyed if imperatively required by the necessities of war.[603]

5.18.4 Other Feasible Precautions to Reduce the Risk of Harm to Cultural Property. Other feasible precautions should be taken to reduce the risk of harm to cultural property.[604]

Such precautions may include:

- determining the location of cultural property and disseminating that information among the armed forces;[605]
- compiling and promulgating lists of cultural property and areas that are not to be attacked;[606]
- establishing civil authorities to assume responsibility for planning and acting to ensure respect for cultural property under its control;[607]
- physically shielding cultural property from harm; and

• establishing refuges and evacuating movable cultural property to them.[608]

5.18.5 Refraining From Any Act of Hostility. In general, acts of hostility may not be directed against cultural property, its immediate surroundings, or appliances in use for its protection.[609] Acts of hostility may, however, be directed against cultural property, its immediate surroundings, or appliances in use for its protection, when military necessity imperatively requires such acts.[610]

5.18.5.1 *Imperative Military Necessity Waiver*. Acts of hostility may be directed against cultural property, its immediate surroundings, or appliances in use for its protection when military necessity imperatively requires such acts. The requirement that military necessity imperatively require such acts should not be confused with convenience or be used to cloak slackness or indifference to the preservation of cultural property.[611] This waiver of obligations with respect to cultural property is analogous to the requirement that enemy property may only be seized if imperatively required by the necessities of war.[612]

For example, if cultural property is being used by an opposing force for military purposes, then military necessity generally would imperatively require its seizure or destruction.[613]

Similarly, if an opposing force uses cultural property and its immediate surroundings to protect military objectives, then the attack of those military objectives may be imperatively required by military necessity.[614] Or, if a military objective was located near cultural property, the protection afforded the area surrounding the cultural property could be subject to waiver for reasons of imperative military necessity such that the attack of the military objective would be permissible, despite its proximity to cultural property.[615]

Even where the waiver of the protection afforded cultural property, its immediate surroundings, or appliances in use for its protection may be warranted for reasons of imperative military necessity, the risk of harm to the cultural property must be considered in a proportionality analysis[616] and feasible precautions should be taken to reduce the risk of harm to the cultural property.[617] And, even where waiver of the protection of cultural property may be warranted as a matter of law, as a matter of policy, decision-makers may still seek to refrain from harming cultural property.[618]

5.18.6 Prohibition Against and Prevention of Theft, Pillage, Misappropriation, or Acts of Vandalism, and Prohibition Against Requisition of Foreign Cultural Property. Any form of theft, pillage, or misappropriation of, and any acts of vandalism directed against, cultural property are prohibited.[619] In addition, Parties to the 1954 Hague Cultural Property Convention shall refrain from requisitioning movable cultural property situated in the territory of another Party to the 1954 Hague Cultural Property Convention.

There are no "imperative military necessity" waivers from these obligations. Pillage is prohibited, even if the property does not constitute cultural property.[620]

5.18.6.1 *Obligation to Stop or Prevent Theft, Pillage, or Misappropriation of, and Acts of Vandalism Against, Cultural Property*. Military commanders have an obligation to take reasonable measures to prevent or stop any form of theft, pillage, or misappropriation of, and any acts of vandalism directed against, cultural property. This obligation may also be viewed as part of the obligation to take feasible precautions to reduce the risk of harm to cultural property.[621] What efforts would be reasonable would depend on a variety of factors, such as availability of forces, the commander's mission, and enemy threats. What efforts would be reasonable would also depend on the information available to the commander at the time.[622] Some measures to ensure adherence to this obligation may include training members of the armed forces and persons authorized to accompany the armed forces on the protection of cultural property in general (and the prohibition of theft, pillage, or vandalism in particular) as part of general procedures for the dissemination and implementation of the law of war.[623] Other measures may include: (1) issuing and enforcing orders to the same effect;[624] (2) investigating alleged or suspected offenses and taking disciplinary or punitive action, as appropriate under U.S. law;[625] and (3) taking measures to facilitate the recovery and return of stolen or misappropriated cultural property.

Although it may not always be feasible to prevent or stop theft, pillage, or misappropriation of, or acts of vandalism against, cultural property, efforts to identify cultural property within an area of operations and to secure it from theft or pillage are a prudent part of the planning process of military operations. For example, such efforts may deny opposing forces the opportunity to exploit harm to cultural property for propaganda purposes.

5.18.6.2 *Prohibition Against Requisition of Movable Cultural Property Situated in the Territory of Another Party to the 1954 Hague Cultural Property Convention*. Parties to the 1954 Hague Cultural Property

Convention shall refrain from requisitioning movable cultural property situated in the territory of another Party to the 1954 Hague Cultural Property Convention.[626]

This rule makes no distinction between movable property of nationals of the country in which it is located and property of nationals of another High Contracting Party. As such, it would also protect from requisition any movable property from a third State temporarily on loan for an exhibition within the territory of a State participating in an armed conflict.[627]

5.18.7 Marking of Cultural Property With the Distinctive Emblem. Cultural property may bear a distinctive emblem so as to facilitate its recognition.[628] This distinctive emblem is described in Article 16 of the 1954 Hague Cultural Property Convention.[629] There is a degree of discretion as to the placing of the distinctive emblem and its degree of visibility.[630] The emblem may be displayed on flags or armlets, and it may be painted on an object or represented in any other appropriate form.[631]

5.18.7.1 *Distinctive Emblem Facilitates Identification Rather Than Confers Status as Such*. As is the case with distinctive emblems for medical personnel, units, transport, and facilities, the distinctive emblem for cultural property facilitates identification, rather than confers cultural property status as such.[632] For example, cultural property may be protected, even though it is not marked with the distinctive emblem. On the other hand, a distinctive emblem placed on property does not make that property cultural property (and grant corresponding protections) if that property does not meet the criteria for cultural property (or the corresponding criteria for protection).[633]

5.18.7.2 *Display of the Distinctive Emblem for Cultural Property – Once Versus Three Times*. In some cases, the distinctive emblem for cultural property is to be used once, while in other cases, the distinctive emblem for cultural property is to be displayed three times in a triangular formation (one shield below the other two shields).[634]

The distinctive emblem for cultural property used once may only be used as a means of identification of:

• cultural property not under special protection;

• the persons responsible for the duties of control in accordance with the Regulations for the Execution of the 1954 Hague Cultural Property Convention;[635]

• the personnel engaged in the protection of cultural property;[636] and

• the identity cards mentioned in the Regulations for the Execution of the 1954 Hague Cultural Property Convention.[637]

The distinctive emblem, repeated three times, may be used only as a means of identification of:[638]

• immovable cultural property under special protection;[639]

• the transport of cultural property under the special protection or in urgent cases;[640] and

• improvised refuges, under the conditions provided for in the Regulations for the Execution of the 1954 Hague Cultural Property Convention.[641]

5.18.7.3 *Display of the Distinctive Emblem on Immovable Cultural Property*. The distinctive emblem for cultural property may not be placed on any immovable cultural property unless at the same time there is displayed an authorization duly dated and signed by the competent authority of the relevant State.[642]

5.18.7.4 *Other Uses of the Distinctive Emblem for Cultural Property Prohibited During Armed Conflict*. During armed conflict, the use of the distinctive emblem for cultural property in any other cases than those mentioned in the 1954 Hague Cultural Property Convention, and the use for any purpose whatsoever of a sign resembling the distinctive emblem, is forbidden.[643]

5.18.8 Special Protection for Certain Cultural Property. A highly limited category of cultural property receives special protection by its entry in an International Register.[644]

The standards governing the protection of such cultural property are essentially the same as the standards governing the protection of cultural property under customary law[645] and the protection afforded all cultural property under the 1954 Hague Cultural Property Convention.[646]

However, the special procedures applicable to cultural property under special protection facilitate its protection beyond that afforded cultural property that is not under special protection. Special protection can apply to:[647]

• a limited number of refuges intended to shelter movable cultural property in the event of armed conflict;

• a limited number of centers containing monuments and other immovable property of very great importance; and

• improvised refuges under the procedures specified in the Regulations for the Execution of the 1954 Hague Cultural Property Convention.[648]

Property granted special protection must be marked with the distinctive emblem for special protection, and opened to international control as provided for in the Regulations for the Execution of the 1954 Hague Cultural Property Convention.[649]

5.18.8.1 *Conditions for the Granting of Special Protection – Adequate Distance From Military Objectives.* In order to qualify for special protection, these categories of property must: (1) be situated an adequate distance from any large industrial center, or from any important military objective constituting a vulnerable point; and (2) not be used for military purposes.

The property must be situated at an adequate distance from any large industrial center or from any important military objective, such as an airport, railway station, or marshalling yards, port facilities, key lines of communication, command and control activity, or other defense facility.[650] The term "adequate distance" is not defined within the 1954 Hague Cultural Property Convention, but relies upon determinations and actions by the State Party requesting "special protection" status, and by States Parties in determining whether the distance set forth by the requesting nation is sufficient to support approval of the request.

Immovable cultural property of very great importance adjacent to a military objective may retain its special protection status provided, in the event of armed conflict, all use of the military objective ceases.[651] For example, if the military objective in question is an airport, port facility, or railway station or marshaling yards, all activity must cease, and military traffic must be diverted, in order for the site to maintain its protected status. A refuge for storing and protecting movable cultural property may also be placed under special protection, whatever its location, if it is so constructed that, in all probability, it will not be damaged by bombs.[652]

5.18.8.2 *Conditions for the Granting of Special Protection – No Use for Military Purposes.* A "center containing monuments"[653] shall be deemed to be used for military purposes whenever it is used for the movement of military personnel or material, even in transit. The same shall apply whenever activities directly connected with military operations, the stationing of military personnel, or the production of war material are carried on within the center.[654] The guarding of cultural property under special protection by armed custodians specially empowered to do so, or the presence, in the vicinity of such cultural property, of police forces normally responsible for the maintenance of public order, however, shall not be deemed to be use for military purposes.[655] Such activities may fulfill the obligation to safeguard cultural property, especially against looters.[656] Employment of law enforcement personnel to protect cultural property is analogous to activities of armed military medical personnel deployed in and around military medical units and facilities in order to provide security from criminal acts.[657]

5.18.8.3 *Marking of Cultural Property Under Special Protection.* The distinctive emblem for cultural property is to be displayed three times in a triangular formation to indicate cultural property under special protection.[658]

The emblem shall be visible from the ground: (a) at regular intervals sufficient to indicate clearly the perimeter of a center containing monuments under special protection; and (b) at the entrance to other immovable cultural property under special protection.[659]

5.18.8.4 *Withdrawal of Special Protection.* If a Party to 1954 Hague Cultural Property Convention commits, in respect of any item of cultural property under special protection, a violation of its obligations to refrain from any act of hostility directed against such property and from use of such property or its surroundings for military purposes, the opposing Party shall, so long as this violation persists, be released from the obligation to ensure the immunity of the property concerned.[660] Nevertheless, whenever possible, the latter Party shall first request the cessation of such violation within a reasonable time.[661]

Apart from violations by the opposing Party, immunity shall be withdrawn from cultural property under special protection only in exceptional cases of unavoidable military necessity, and only for such time as that necessity continues.[662] Such necessity can be established only by an officer commanding a force the equivalent of a division in size or larger.[663] However, such necessity should not be construed to impose an unreasonable and disproportionate responsibility on the attacker to avoid damage to cultural property.[664] Moreover, all property may be attacked using any lawful and proportionate means if required by military necessity and notwithstanding possible collateral damage to such property.[665]

Whenever circumstances permit, the opposing Party shall be notified, a reasonable time in advance, of the decision to withdraw immunity.[666] The Party withdrawing immunity is also to notify the Commissioner-General for Cultural Property as soon as possible as provided for in the Regulations for the Execution of the 1954 Hague Cultural Property Convention, in writing, stating the reasons for withdrawing immunity.[667]

5.18.9 Transport of Cultural Property. The transport of cultural property may take place: (1) under special protection through procedures provided in the Regulations for the Execution of the 1954 Hague Cultural Property Convention; and (2) in urgent cases without such procedures. Additional rules apply to the transport of cultural property in occupied territory.[668]

The distinctive emblem for cultural property should be placed on vehicles of transport so as to be clearly visible in daylight from the air as well as from the ground.[669]

5.18.9.1 *Transport Under Special Protection*. Transport exclusively engaged in the transfer of cultural property, whether within a territory or to another territory, may, at the request of the State concerned, take place under special protection in accordance with the conditions specified in the Regulations for the Execution of the 1954 Hague Cultural Property Convention.[670] No acts of hostility may be directed against transport under special protection.[671] A request for such transport should be addressed to the Commissioner-General for Cultural Property and should contain all relevant information, including the reason for the transfer, the approximate number and importance of objects to be transferred, their present location, the intended destination, the means of transport to be used, the route to be followed, and the date proposed for transfer.[672] The Commissioner-General for Cultural Property determines whether such transfer is justified and notifies parties to the conflict concerned of the transfer.[673]

Transport under special protection shall take place under the international supervision provided for in the Regulations for the Execution of the 1954 Hague Cultural Property Convention and shall display the distinctive emblem described in Article 16.[674] The distinctive emblem for cultural property is to be displayed repeated three times in a triangular formation to indicate that the transport occurs under special protection.[675]

The transport of cultural property to the territory of another State is subject to additional rules.[676]

5.18.9.2 *Transport in Urgent Cases*. If a State that is a Party to the 1954 Hague Cultural Property Convention considers that the safety of certain cultural property requires transfer, and that the matter is of such urgency that the request procedure to the Commissioner- General for Cultural Property cannot be followed, the transport may display the distinctive emblem for cultural property, provided that a request for transport under special protection has not already been made and refused.[677] As far as possible, notification of transfer should be made to the opposing Parties.[678] Nevertheless, transport conveying cultural property to the territory of another State may not display the distinctive emblem unless immunity has been expressly granted to it.[679]

As far as possible, necessary precautions shall be taken to avoid acts of hostility directed against transport of cultural property under the protection provided in urgent cases and while displaying the distinctive emblem.[680]

5.18.9.3 *Immunity From Seizure, Capture, and Prize*. Transport of cultural property under special protection or under the protection provided in urgent cases is immune from seizure, placing in prize, or capture.[681] Means of transport exclusively engaged in the transport of such property is similarly immune.[682] However, such immunity does not limit the right of visit and search of such transportation or property.[683]

5.18.10 AP I Provision on the Protection of Objects Which Constitute the Cultural or Spiritual Heritage of Peoples. Article 53 of AP I provides certain protections to "historic monuments, works of art or places of worship which constitute the cultural or spiritual heritage of peoples."[684] This provision has been interpreted in a more limited fashion than its text would suggest. For example, based on the statements of national delegations, including the U.S. delegation, during the negotiations of this provision, it appears that objects that qualify for special protection under Article 53 of AP I are substantially those that qualify for special protection under the 1954 Hague Cultural Property Convention.[685]

In addition, protection is waived when an object is used in support of the military effort or for military purposes.[686] In addition, Canada has interpreted Article 53 to permit waiver of its obligations when military necessity imperatively requires such waiver.[687]

5.19 SIEGES AND ENCIRCLED AREAS

It is lawful to besiege enemy forces. Commanders must seek to make arrangements to permit the passage of certain consignments and should seek to make arrangements for the passage of certain categories of civilians, and of religious and medical personnel. Different rules apply to blockade.[688]

5.19.1 Siege and Encirclement Permissible. It is lawful to besiege enemy forces, *i.e.*, to encircle them with a view towards inducing their surrender by cutting them off from reinforcements, supplies, and communications with the outside world.[689] In particular, it is permissible to seek to starve enemy forces into submission.[690]

5.19.1.1 *Belligerent Authority to Exercise Control in the Immediate Vicinity of Military Operations*. The conduct of a siege or encirclement may require the imposition of measures of control to ensure that outsiders may not deliver supplies to enemy forces. Thus, the right to conduct a siege or encirclement impliedly recognizes the authority of the military commander to exercise control (*e.g.*, stopping, searching, and diverting traffic) over civilians and other persons in the immediate vicinity of military operations. For example, commanders may also impose certain restrictions on neutral vessels or aircraft (such as restricting communications) within the immediate vicinity of the belligerent's operations.[691]

5.19.2 Removal and Passage of Certain Personnel – Vulnerable Civilians, Diplomatic and Consular Personnel, the Wounded and Sick, and Medical Personnel. Although the commander of the force laying siege has the right to forbid all communications and access between the besieged place and the outside, the parties to the conflict should attempt to conclude local agreements for the removal of wounded, sick, infirm and aged persons, children, and maternity cases, or for the passage of ministers of all religions, medical personnel, and medical equipment on their way to such areas.[692] Concluding such agreements is not compulsory. A commander of an encircling force is not required to agree to the passage of medical or religious personnel, supplies, and equipment if he or she has legitimate military reasons denying such requests (*e.g.*, if denying passage may increase the likelihood of surrender of enemy forces in the encircled area). Nonetheless, commanders should make reasonable, good-faith efforts to do so when possible.[693]

Diplomatic and consular personnel of a neutral State should not be prevented from leaving a besieged place before hostilities commence, but this privilege cannot be claimed while hostilities are in progress. If diplomatic and consular personnel of a neutral State voluntarily decide to remain, they are subject to the same risks as other inhabitants.[694]

5.19.3 Passage of Relief Consignments. Commanders should make arrangements to permit the free passage of certain consignments:

- all consignments of medical and hospital stores and objects necessary for religious worship intended only for civilians; and
- all consignments of essential foodstuffs, clothing, and tonics (*i.e.*, medicine) intended for children under fifteen, expectant mothers, and maternity cases.[695]

However, allowing passage of these items is not required by the party controlling the area unless that party is satisfied that there are no serious reasons for fearing that:

- the consignments may be diverted from their destination;
- the control may not be effective; or
- a definite advantage may accrue to the military efforts or economy of the enemy.[696]

5.19.3.1 *Technical Arrangements for the Passage and Distribution of Relief Consignments*. Commanders may require, as a condition for allowing the passage of consignments listed above, that the consignments be distributed under the local supervision of the Protecting Powers.[697] Commanders may prescribe other rules and regulations for how consignments are to be distributed, although consignments should be forwarded as rapidly as possible.[698]

5.19.4 Attacks by Forces Laying Siege. The general rules that apply to conducting attacks also apply to attacks by forces laying siege.[699]

5.19.4.1 *Use of Force to Drive Fleeing Civilians Back to Besieged Areas*. In the past, it was permissible, but an extreme measure, to refuse to allow civilians to leave a besieged locality and to use force to drive any who attempted to flee back to the besieged locality.[700] However, such actions would now be prohibited because they would not be consistent with the duty to take feasible precautions for the protection of civilians.[701] Permitting civilians to flee may also be militarily advantageous.[702]

5.19.5 Duties of Forces Under Siege to Take Feasible Precautions to Reduce the Risk of Incidental Harm. Besieged forces should take feasible precautions to reduce the risk of incidental harm.[703] For example, it may be appropriate to mark protected buildings to indicate their protected status to enemy forces.[704] Similarly, it may be appropriate to concentrate the wounded and sick and civilians in areas remote from military objectives[705] or in areas neutralized by arrangement with the enemy.[706]

5.20 STARVATION

Starvation is a legitimate method of warfare, but it must be conducted in accordance with the principles of distinction and proportionality, as well as other law of war rules. Starvation of civilians as a method of combat is also prohibited in non-international armed conflict.[707]

5.20.1 Starvation – Distinction. It is a legitimate method of war to starve enemy forces.[708] For example, it is permitted to destroy food intended as sustenance for enemy forces with a view towards weakening them and diverting their resources.[709] Enemy forces, for the purpose of this rule, means those persons constituting military objectives.[710]

Starvation specifically directed against the enemy civilian population, however, is prohibited.[711] For example, it would be prohibited to destroy food or water supplies for the purpose of denying sustenance to the civilian population.

5.20.2 Starvation – Proportionality. Military action intended to starve enemy forces, however, must not be taken where it is expected to result in incidental harm to the civilian population that is excessive in relation to the military advantage anticipated to be gained.[712]

Feasible precautions to reduce the risk of harm to the civilian population or other reasonable measures to mitigate the burden to the civilian population may also be warranted when seeking to starve enemy forces.[713] For example, it may be appropriate to seek to compensate civilians whose food has been inadvertently destroyed.[714] Moreover, an Occupying Power would have additional duties to ensure food and water for the civilian population.[715]

5.20.3 Starvation and Other Law of War Rules. Starvation as a method of warfare must comply with other applicable law of war rules. For example, it would be unlawful to poison food or water.[716] Additionally, starvation, for example, may involve sieges or encirclement, blockades, attacks, or the seizure and destruction of enemy property.[717] In each case, the rules applicable to those situations must be followed.

5.20.4 AP I Provision on Objects Indispensable to the Survival of the Civilian Population. Article 54(2) of AP I makes it prohibited for Parties to AP I "to attack, destroy, remove or render useless objects indispensable to the survival of the enemy civilian population, such as foodstuffs, agricultural areas for the production of foodstuffs, crops, livestock, drinking water installations, and supplies for irrigation works, for the specific purpose of denying them for their sustenance value to the civilian population or to the adverse party, whatever the motive, whether in order to starve out civilians, cause them to move away, or for any other motive."[718] This rule would not apply to attacks that are carried out for specific purposes other than to deny sustenance.[719] For example, this rule would not prohibit destroying a field of crops to prevent it from being used as concealment by the enemy or destroying a supply route that is used to move military supplies but is also used to supply the civilian population with food.[720]

Similarly, this AP I prohibition does not apply to objects that would otherwise be covered by it if those objects are used by an adverse party "as sustenance solely for the members of its armed forces" or "if not as sustenance, then in direct support of military action."[721] Actions against this latter category of objects forfeiting protection, however, may not be taken if they "may be expected to leave the civilian population with such inadequate food or water as to cause its starvation or force its movement."[722] Further exception is made for a State to engage in a "scorched earth" defense of a party's own territory.[723]

When adopted, this AP I prohibition was novel and the product of extensive diplomatic negotiation. Given the intricacy of this provision of AP I, it would be difficult to conclude that all of its particulars reflect customary international law. Nonetheless, the United States has supported the underlying principle that starvation of civilians may not be used as a method of warfare, as discussed in the sub-sections above.[724]

5.21 OVERVIEW OF GOOD FAITH, PERFIDY, AND RUSES

Absolute good faith with the enemy must be observed as a rule of conduct.[725] The line between those deceptions that good faith permits and those that good faith prohibits may appear indistinct and has varied according to State practice.[726] Good faith prohibits:

- killing or wounding enemy persons by resort to perfidy;[727]
- misusing certain signs;[728]
- fighting in the enemy's uniform;[729]
- feigning non-hostile relations in order to seek a military advantage;[730] and
- compelling nationals of a hostile party to take part in the operations of war directed against their own country.[731]

However, good faith permits:

- ruses of war or other lawful deceptive activities;[732]
- intelligence collection;[733]
- military information support operations, including propaganda;[734] and
- certain types of bribery.[735]

5.21.1 Harms From Breaches of Good Faith. Breaches of good faith may: (1) undermine the protections afforded by the law of war to classes of persons and objects; (2) impair nonhostile relations between opposing belligerents; and (3) damage the basis for the restoration of peace short of complete annihilation of one belligerent by another.

Breaches of good faith, principally perfidious conduct, may undermine the protections afforded by the law of war to civilians, persons who are *hors de combat*, or certain other classes of persons and objects.[736]

Breaches of good faith also impair other non-hostile relations between opposing belligerents during the war. For example, such breaches may diminish the legitimate activities that depend upon a degree of mutual respect and trust between hostile forces, such as effecting surrender or collection of the dead, wounded, or sick (enemy or friendly force) on the battlefield. It also may make it more difficult for military commanders to ensure that their forces comply with the law of war where treacherous acts by the enemy have resulted in casualties among their own forces.

Breaches of good faith also damage, and may destroy, the basis for restoration of peace short of the complete annihilation of one belligerent by another.[737] A degree of mutual respect and trust is essential for the negotiation of cease-fires, truces, surrenders, and other agreements necessary to bring an end to hostilities.

5.22 TREACHERY OR PERFIDY USED TO KILL OR WOUND

During international armed conflict, it is prohibited to kill or wound the enemy by resort to perfidy.[738]

5.22.1 Definition of Perfidy. Acts of *perfidy* are acts that invite the confidence of enemy persons to lead them to believe that they are entitled to, or are obliged to accord, protection under the law of war, with intent to betray that confidence.[739]

The key element in perfidy is the false claim to protections under the law of war in order to secure a military advantage over the opponent.[740] The claim must be to legal protections.[741]

5.22.1.1 *Perfidy and Treachery – Notes on Terminology*. "Treachery" and "perfidy" have been used interchangeably.[742] Article 23(b) of the Hague IV Regulations uses the word "treacherously," which was also used in Article 13(b) of the 1874 Brussels Declaration.[743] Sometimes, "perfidy" is used to refer to conduct that is, by definition, prohibited.[744]

Other times, as, for example, in AP I, "perfidy" is used to refer to a certain type of deception, which might not, by itself, be prohibited (*e.g.*, feigning death would not be prohibited in order to facilitate escape).[745] It will be important to examine closely the context to determine which usage is intended. This manual generally uses "perfidy" in the latter sense.

5.22.2 Perfidy - "Kill or Wound". It is prohibited to use perfidy to kill or wound the enemy. It may not be prohibited to invite the confidence of the adversary that he or she is obligated to accord protection under the law of war, for certain purposes (*e.g.*, to facilitate spying, sabotage, capturing enemy personnel, or evading enemy forces). However, such deception may not rely on certain signs and symbols.[746] Moreover, persons who use perfidy to engage in spying and sabotage may forfeit POW status or be liable to certain penalties under the domestic law of enemy States.[747]

5.22.2.1 *AP I Prohibition on "Capture" by Resort to Perfidy*. In addition to killing or wounding, Article 37 of AP I prohibits "capture" by resort to perfidy. The Department of Defense has not interpreted customary international law to prohibit U.S. forces from seeking to capture by resort to perfidy.

5.22.3 Examples of Killing or Wounding by Resort to Perfidy. Examples of killing or wounding by resort to perfidy include:

- feigning an intent to negotiate under a flag of truce and then attacking,[748] which takes advantage of the rule that flags of truce may not be used to shield military operations.[749]
- feigning surrender and then attacking,[750] which takes advantage of the rule that the enemy may not attack those who have surrendered;[751]
- feigning of death or incapacitation by wounds or sickness[752] and then attacking, which takes advantage of the respect afforded the dead[753] or the protection afforded those who are *hors de combat*;[754]
- feigning civilian status and then attacking;[755] and
- calling out "do not fire; we are friends" and then attacking.[756]

5.23 USE OF ENEMY FLAGS, INSIGNIA, ANDMILITARY UNIFORMS

During international armed conflict, improper use of enemy flags, military emblems, insignia, or uniforms, is prohibited.[757]

5.23.1 Improper Use of Enemy Flags, Insignia, and Military Uniforms Prohibited. In general, the use of enemy flags, insignia, and military uniforms is prohibited during combat, but is permissible outside of combat.[758]

A similar rule is applied during naval operations.[759]

5.23.1.1 *Improper Use of Enemy Uniforms and "Perfidy"*. The prohibition on the use of enemy uniforms in combat has been described as a prohibition against using enemy uniforms to kill or wound treacherously.[760] However, feigning enemy military status is not technically "perfidy" as the term is used in this manual because perfidy requires the feigning of protected status, and the law of war generally does not protect enemy military personnel from being made the object of attack.[761]

5.23.1.2 *Use of Enemy Uniforms and Other Equipment Without the Intent to Deceive*. Uses of enemy uniforms and other equipment without any intention to deceive the enemy are not prohibited. For example, if there is a shortage of clothing it may be necessary to use enemy uniforms.[762] Similarly, the immediate use of captured enemy weapons or equipment during combat is permissible. When circumstances permit, however, enemy insignia should generally be removed or alternative measures taken to distinguish U.S. forces from the enemy.[763]

5.23.1.3 *Deceptive Use of Enemy Flags, Insignia, and Military Uniforms Outside of Combat*. It is a legitimate ruse to use enemy flags, insignia, and military uniforms outside of combat.[764]

5.23.1.4 *Use of Enemy Uniforms to Evade Capture or Escape*. Military personnel, such as aircrew downed behind enemy lines, may use enemy uniforms to evade capture. Similarly, escaping prisoners of war may use enemy military uniforms to facilitate their escape from a POW camp to return to friendly lines.[765]

However, those using enemy uniforms to evade capture or escape must not engage in combat while in the enemy's uniform, and, if they are not escaping POWs, they may be liable to treatment as spies and saboteurs if caught behind enemy lines.[766]

5.23.1.5 *Use of Enemy Codes, Passwords, and Countersigns Not Restricted*. The prohibition on misuse of enemy flags, insignia, and military uniforms refers only to concrete visual objects, rather than enemy codes, passwords, and countersigns.[767] Enemy codes, passwords, and countersigns may be used as a ruse to aid military operations.[768] Enemy use of these measures is permissible because military forces are expected to take measures to guard against the use of their codes, passwords, and countersigns by the enemy.

5.23.2 Persons Using Enemy Uniforms May Be Liable to Treatment as Spies and Saboteurs. Although the use of enemy uniforms outside of combat generally is neither prohibited by the customary law of war nor by law of war treaties to which the United States is a Party, combatants captured by an opposing party behind the opposing party's lines while wearing the uniform of the opposing party may be liable to treatment as spies and saboteurs.[769]

5.23.3 AP I and the Use of Enemy Flags, Insignia, or Uniforms to Shield, Favor, Protect, or Impede Military Operations. In addition to prohibiting the use of enemy uniforms while engaging in attacks, AP I also prohibits the use of flags, military emblems, insignia, or uniforms of an enemy in order to shield, favor, protect, or impede military operations.[770] Canada has taken a reservation from this rule.[771]

This provision of AP I is unclear as to what uses would be permissible and what uses would be prohibited.[772] However, because the United States is not a Party to AP I and because the rule is not part of customary international law, U.S. military personnel are not subject to this more restrictive rule.

5.24 IMPROPER USE OF CERTAIN SIGNS

Certain signs, symbols, or signals reflect a status that receives special protection under the law of war, and thus these signs may not be improperly used. They may not be used: (1) while engaging in attacks; (2) in order to shield, favor, or protect one's own military operations; or (3) to impede enemy military operations. Thus, their use may be improper even when that use does not involve killing or wounding, and they may not be used to facilitate espionage (except for signs, emblems, or uniforms of a neutral or non-belligerent State). The following types of signs are subject to this broader rule against improper use:

5.24.1 Signs, Emblems, or Uniforms of a Neutral or Non-Belligerent State. During international armed conflict, the use of signs, emblems, or uniforms of a neutral or other nation not a party to the conflict is prohibited.[773]

However, recognized exceptions exist concerning espionage[774] and warfare at sea.[775]

5.24.2 Distinctive Emblems of the Geneva Conventions (*e.g.*, Red Cross). The distinctive emblems of the red cross, red crescent, and red crystal are symbols that identify military medical and religious personnel, medical units, and medical transports, or certain other categories of persons engaged in humanitarian work as personnel and objects entitled to special protection.[776] These emblems may not be used except to identify these protected persons and objects.[777]

5.24.3 Markings for POW or Civilian Internee Camps. Only POW camps under the GPW should be marked using internationally agreed symbols, such as the PW or PG designation.[778]

Only civilian internee camps under the GC should be marked with an IC designation.[779]

5.24.4 Markings for Hospital, Safety, or Neutralized Zones. Markings that distinguish hospital or safety zones or neutralized zones established under the 1949 Geneva Conventions may not be used for other purposes.[780]

5.24.5 Distinctive and Visible Signs to Identify Civilian Objects as Such. Certain distinctive and visible signs that are intended to identify that an object is protected as a civilian object under the law of war must not be used for other purposes.[781]

5.24.6 Markings for Cultural Property. The distinctive emblem for cultural property may not be used for other purposes.[782]

5.24.7 Flags of Truce. The improper use of a flag of truce is strictly prohibited.[783]

5.25 RUSES OF WAR AND OTHER LAWFUL DECEPTIONS

Ruses of war are considered permissible.[784] In general, a belligerent may resort to those measures for mystifying or misleading the enemy against which the enemy ought to take measures to protect itself.[785] Apart from ruses of war, certain other deceptions are not prohibited, but may expose combatants employing them to liability as spies and saboteurs.

5.25.1 Definition of Ruses of War. Ruses of war are acts that are intended to mislead an adversary or to induce him to act recklessly, but that do not infringe upon any rule of international law applicable in armed conflict and that are not perfidious because they do not invite the confidence of an adversary with respect to protection under that law.[786] Ruses of war are methods, resources, and techniques that can be used either to convey false information or deny information to opposing forces. They can include physical, technical, or administrative means, such as electronic warfare measures, flares, smoke, chaff, aerosol material, or dissemination devices.

5.25.1.1 *Acts That Are Intended to Mislead an Adversary or to Induce an Adversary to Act Recklessly*. Ruses may be used for a variety of purposes, such as:

• facilitating surprise attacks or ambushes,[787] such as by

o misleading the enemy as to the planned targets or locations of military

operations;[788]

o baiting the enemy into a trap; or

o distracting or disorienting the enemy;

• inducing enemy forces to waste their resources;

• inducing enemy forces to surrender by falsely alleging military superiority;

• provoking friendly fire among enemy forces; or

• causing confusion among enemy forces.

5.25.1.2 *But That Do Not Infringe Upon Any Rule of International Law Applicable in Armed Conflict.* According to the definition in AP I, *ruses* do not infringe upon any rule of international law applicable in armed conflict. For example, misusing certain signs and symbols would not constitute ruses.[789] Similarly, although fighting in the enemy's uniform would not be perfidy since enemy military personnel are not generally protected by the law of war, fighting in the enemy's uniform also would not be a ruse, since such action would infringe upon the rule against improper use of the enemy's uniform.[790]

5.25.1.3 *And That Are Not Perfidious Because They Do Not Invite the Confidence of an Adversary With Respect to Protection Under That Law.* The ruses described in AP I do not invite the confidence of the enemy with respect to protection under the law of war.[791] For example, pretending to be a civilian would not constitute a ruse, although it may be a permissible deception in certain circumstances.[792]

5.25.1.4 *Ruses – Notes on Terminology.* The term "ruse" is sometimes used as a legal term of art to mean deceptions that are not prohibited by the law of war. In other cases, "ruse" may be used as a factual term to mean a deception that may or may not be prohibited by the law of war, and thus the distinction is made between "lawful" and "unlawful" ruses.[793]

In this manual, the term "ruse" is used to indicate a certain category of permissible deceptions.

5.25.2 Examples of Ruses. Often, ruses of war operate by misleading the enemy as to the identity, strength, position, or disposition of one's own forces. Ruses of war include, but are not limited to:[794]

• decoys or dummy materials, such as dummy weapons, equipment, ships, aircraft, and buildings;[795]

• feigned activity or inactivity, such as

o simulating quiet;

o feigned flights, retreats, attacks, marches,[796] movements (*e.g.*, approaching a destination indirectly[797]), supply movements, operations, withdrawals, or camp;

• mimicking other friendly forces;[798]

• camouflage or deceptive lighting;[799]

• removing or altering identifying information, including moving identifying landmarks;

• planting false information in a manner that allows enemy forces to intercept it,[800] such as through the use of

o false messages among one's own forces;

o intensifying or minimizing message traffic; or

o bogus messages, dispatches, or newspapers;

• lying in the course of a POW interrogation in answering questions where no obligation to answer correctly exists;

• feigning enemy military status by using enemy flags, insignia, or military uniforms outside of combat;[801] or

• using enemy codes, signals, or passwords.[802]

5.25.2.1 *Mimicking Other Friendly Forces.* It is a permissible ruse of war for combatant forces to mimic other friendly forces. For example, small forces may simulate larger units.[803] Units of one type may pretend to be units of another type.[804] Markings that identify equipment or personnel as belonging to a particular unit may be removed. Individuals or units may also dress like friendly forces.[805]

5.25.3 Examples of Other Deceptions That Are Not Prohibited. Certain deceptions may not meet the definition of "ruses" because they may invite the confidence of an adversary with respect to protection under the law of war. Nevertheless, these deceptions are not technically prohibited by the law of war, even though, in some cases, they may expose the persons performing them to liability under an enemy State's domestic law.[806] These include:

• feigning incapacitation or death in order to escape the enemy,[807] such as o a soldier who feigns death in the hope that he or she will be left alone by enemy forces, permitting him or her subsequently to take evasive actions to return to friendly forces without engaging the enemy;[808] and

o an aircraft crew that feigns loss of control and the appearance that the aircraft was about to crash in order to dissuade further enemy attack and to break contact with enemy forces;[809]

• feigning civilian status to facilitate spying or sabotage, such as

o false use of journalist credentials; and

• using an informal ceasefire for the purpose of collecting wounded and dead and to withdraw unseen by the enemy.[810]

5.26 NON-FORCIBLE MEANS AND METHODS OF WARFARE

In general, propaganda, information gathering, and bribery are permissible means and methods of warfare.

5.26.1 Propaganda. In general, the use of propaganda is permissible under the law of war, even when it encourages acts that violate an enemy State's domestic law or is directed towards civilian or neutral audiences. However, certain types of propaganda are prohibited.

5.26.1.1 *Propaganda – Notes on Terminology*. Propaganda has been referred to as psychological warfare, psychological operations (PSYOP), or military information support operations (MISO).[811] The word "propaganda" often has been used with a pejorative connotation to refer to misleading or false statements, but "propaganda" is used in this section without intending any such connotation.

5.26.1.2 *Propaganda Generally Permissible*. In general, propaganda is a permissible means of warfare.[812] Propaganda has been disseminated through a variety of communications media, including printed materials, loudspeakers, radio or television broadcast, aircraft, or the internet. Propaganda is sometimes used with bribery[813] or to support intelligence gathering.[814] Propaganda may be directed at enemy civilians and neutrals.[815] Propaganda may encourage enemy persons to commit acts that would violate the domestic law of the enemy State.[816] For example, it would be permissible to encourage enemy combatants to defect, desert, or surrender.[817] Similarly, it is generally permissible to encourage insurrection among the enemy civilian population.[818]

5.26.1.3 *Prohibited Types of Propaganda*. Propaganda must not: (1) incite violations of the law of war; nor (2) itself violate a law of war rule. Propaganda must not incite acts that are prohibited by the law of war.[819] For example, propaganda intended to incite attacks against civilians is prohibited. In certain cases, individuals may be liable for instigating or inciting violations of the law of war.[820]

Propaganda is also prohibited when it would violate other law of war rules. For example, it is specifically prohibited for an Occupying Power to use propaganda that aims at securing voluntary enlistment of protected persons in its armed or auxiliary forces.[821] Similarly, it is prohibited to declare that no quarter will be given, and propaganda in the form of a declaration to the adversary that no quarter will be given would be prohibited.[822] In addition, propaganda would be prohibited if it constituted a measure of intimidation or terrorism against the civilian population, such as the threats of violence whose primary purpose is to spread terror among the civilian population.[823] Similarly, propaganda may not be used to subject a detainee to public curiosity or other humiliating or degrading treatment.[824] Additionally, the delivery of the propaganda should be consistent with other law of war obligations.[825]

5.26.2 Information Gathering. The employment of measures necessary for obtaining information about the enemy and their country is considered permissible.[826]

Information gathering measures, however, may not violate specific law of war rules.[827] For example, it would be unlawful, of course, to use torture or abuse to interrogate detainees for purposes of gathering information. Similarly, it is prohibited to make improper use of a flag of truce to obtain information.[828] Some forms of deception used to gather information, although not prohibited by the law of war, may put the persons engaging in them at risk of being treated as spies.[829]

5.26.3 Bribery or Offering of Rewards. In general, it is permissible to offer rewards for assistance in the conduct of hostilities, including rewards intended to corrupt enemy combatants or civilians.[830] Rewards, however, may not be offered for the commission of violations of the law of war, and rewards may not be offered for the killing of enemy persons.

5.26.3.1 *Prohibition on Offering Rewards for Enemy Persons Dead or Alive*. It is forbidden to place a price on the head of enemy persons or to offer a reward for enemy persons "dead or alive."[831] Such actions encourage the denial of quarter or encourage private persons to take up arms whose participation in hostilities is often undisciplined and associated with the commission of war crimes.[832] This prohibition extends to offers of rewards for the killing or wounding of all enemies, including specific individuals or a class of enemy persons (*e.g.*, officers).

However, this rule would not prohibit offering rewards for the capture of unharmed enemy personnel generally or of particular enemy personnel.[833] Similarly, this rule does not prohibit offering rewards for information that may be used by combatants to conduct military operations that attack enemy combatants.

5.27 PROHIBITION AGAINST COMPELLING ENEMY NATIONALS TO TAKE PART IN THE OPERATIONS OF WAR DIRECTED AGAINST THEIR OWN COUNTRY

During international armed conflict, it is prohibited to compel the nationals of the hostile party to take part in the operations of war directed against their own country, even if they were in the belligerent's service before the commencement of the war.[834] Underlying this prohibition is the principle that States must not compel foreign nationals to commit treason or otherwise to violate their allegiance to their country.[835]

5.27.1 Additional restrictions are applicable with respect to POWs, retained personnel, protected persons in the home territory of a belligerent, and protected persons in occupied territory.[836] Compel. This prohibition applies to attempts to compel enemy nationals, rather than measures short of compulsion, such as bribing enemy nationals or seeking to influence them through propaganda.[837] However, it is specifically prohibited for an Occupying Power to use propaganda that aims at securing voluntary enlistment of protected persons in its armed or auxiliary forces.[838]

5.27.2 Nationals of a Hostile Party … Against Their Own Country. This rule applies to nationals of a hostile party; States are not prohibited by the law of war from compelling their nationals to serve in the armed forces.[839] Similarly, this rule would not prohibit States from compelling persons to betray an allegiance to a non-State armed group during non-international armed conflict.[840]

VI – Weapons

Chapter Contents

6.1 INTRODUCTION

This Chapter addresses the legal review of new weapons and legal rules specific to certain types of weapons, including whether certain weapons are illegal *per se*, specific rules for the use of certain other weapons, and obligations that may arise after certain weapons are used.

In addition to the law of war, other law, policy, and regulation may apply to certain weapons. Such law, policy, or regulation may include arms control agreements, U.S. domestic law, U.S. policy, applicable rules of engagement, and other military orders. This Chapter notes some provisions of such law, policy, and regulation where relevant, but does not seek to address exhaustively all of them.[1] In advising on the legality of specific weapons, one must take care to identify all applicable restrictions and not rely unduly on the absence of discussion in this Chapter as reflecting the absence of any such restrictions.

6.1.1 Legality of the Weapon Itself (Per Se) Distinguished From the Legality of the Use of the Weapon. A weapon may be illegal *per se* if a treaty to which the United States is a Party or customary international law has prohibited its use under all circumstances.[2] For example, the use of "blinding laser" weapons is prohibited, regardless of how they are used.[3] On the other hand, most weapons are not illegal *per se*. That is, their use may be lawful in some circumstances, although unlawful in others, such as if they are used to attack combatants placed *hors de combat*.[4]

Law of war issues related to targeting (*e.g.*, the requirement that an attack may only be directed against a military objective) generally are not determinative of the lawfulness of a weapon. However, weapons that are inherently indiscriminate are prohibited.[5] In addition, certain weapons, such as mines, are subject to specific rules on their use in order to reduce the risk of harm to the civilian population.[6]

6.2 DOD POLICY OF REVIEWING THE LEGALITY OF WEAPONS

As provided in DoD issuances, DoD policy for many years has required the legal review of the intended acquisition or procurement of weapons or weapon systems; this review includes ensuring that such acquisition or procurement is consistent with the law of war.[7] These DoD policy requirements have been implemented in Military Department regulations.[8]

The fact that the legality of a weapon is discussed in this manual does not obviate the requirement under applicable directives or regulations for a legal review of a weapon or weapon system that is to be acquired or procured.

6.2.1Review of New Types of Weapons. The development of new types of weapons has often resulted in public denunciation of their allegedly cruel effects and in attempts to prohibit their use in armed conflict.[9] This has been true of the crossbow, siege engines for hurling projectiles, firearms, gunpowder, bayonets, and other weapons that have since been widely recognized as not prohibited by the law of war.

Like other aspects of the law of war, the rules relating to weapons are generally characterized as prohibitive law forbidding certain weapons or the use of weapons in certain instances rather than positive law authorizing the weapon or its use.[10] The lawfulness of the use of a type of weapon does not depend on the presence or absence of authorization, but, on the contrary, on whether the weapon is prohibited.[11] Thus, the mere fact that a weapon is novel or employs new technology does not mean that the weapon is illegal.[12] The law of war does not require States to establish a general practice of using a weapon before it is to be regarded as legal. Moreover, it would appear absurd to suggest that a new type of weapon should automatically be prohibited because there is no State practice supporting such use, or to suggest that States must continue using a weapon in each conflict simply to maintain its legality.

6.2.2 Questions Considered in the Legal Review of Weapons for Consistency With U.S. Law of War Obligations. The review of the acquisition or procurement of a weapon for consistency with U.S. law of war obligations should consider three questions to determine whether the weapon's acquisition or procurement is prohibited:

- whether the weapon's intended use is calculated to cause superfluous injury;[13]
- whether the weapon is inherently indiscriminate;[14] and
- whether the weapon falls within a class of weapons that has been specifically prohibited.[15]

If the weapon is not prohibited, the review should also consider whether there are legal restrictions on the weapon's use that are specific to that type of weapon.[16] If any specific restrictions apply, then the intended concept of employment of the weapon should be reviewed for consistency with those restrictions.

Lastly, it may be appropriate to advise whether other measures should be taken that would assist in ensuring compliance with law of war obligations related to the type of weapon being acquired or procured. For example, it may be appropriate to advise on the need for training programs and other practical measures, such as promulgating doctrine and rules of engagement related to that type of weapon.[17]

6.2.3 AP I Requirement for Legal Review of a New Weapon, Means, or Method of Warfare. Article 36 of AP I provides:

In the study, development, acquisition or adoption of a new weapon, means or method of warfare, a High Contracting Party is under an obligation to determine whether its employment would, in some or all circumstances, be prohibited by this Protocol or by any other rule of international law applicable to the High Contracting Party. The DoD policy and practice of conducting legal reviews of weapons preceded this AP I provision.[18]

6.2.4 Policy and Processes for Review of Applicable Arms Control Obligations. DoD has had a separate, but complementary, policy and practice requiring review of its activities (*e.g.*, research, development, and testing of weapons) to ensure that these activities are consistent with the arms control agreements to which the United States is a Party.[19]

6.3 OTHER PRACTICES TO HELP IMPLEMENT LAW OF WAR OBLIGATIONS WITH RESPECT TO WEAPONS

A number of practices contribute to the effective implementation of law of war obligations with respect to weapons (although these practices are not necessarily required by the law of war): (1) using weapons in accordance with their design intent and the doctrine that has been promulgated for their use; (2) refraining from modifying weapons without proper authorization; (3) refraining from using personal firearms during military operations; and (4) refraining from using captured weapons in combat, except on a field-expedient basis.

6.3.1 Using Weapons in Accordance With Their Design Intent and Promulgated Doctrine. The use of weapons in accordance with their design intent and the doctrine that has been promulgated for their use contributes to compliance with law of war obligations with respect to weapons because the legality of the weapon would have been reviewed based on its intended use.[20] However, the use of weapons outside their design intent does not necessarily violate the law of war.

Certain weapon systems may also be subject to law of war restrictions on how they are to be used, such as incendiary weapons or mines.[21] In such cases, the use of the weapon in accordance with doctrine and policy is required by international law because that doctrine and policy incorporates certain international obligations.

6.3.1.1 *Using Weapons for Purposes Other Than Those for Which a Weapon or Tool Is Designed.* Although in some cases military orders may restrict the use of certain weapons to only certain authorized purposes, there is not a law of war rule that requires that weapons or tools only be used for the purposes for which they were designed. For example, an entrenching tool may be designed for digging fighting positions.

However, its use as a weapon is not prohibited by the law of war.[22] Similarly, it is not prohibited to use a laser that is not designed to blind enemy persons as a weapon to blind an attacking enemy.[23] It also generally would not be prohibited to use a weapon that has been designed to destroy enemy material, such as a large-caliber machine gun, against enemy personnel.[24]

6.3.2 Refraining From Modifying Weapons Without Proper Authorization. For a variety of reasons, DoD practice has been not to permit the modification of weapons without proper authorization.

In particular, weapons must not be modified in the field for the purposes of aggravating the harm inflicted on incapacitated persons.[25] For example, substances designed to inflame wounds must not be added to bullets or edged weapons for the purpose of making wounds more painful or difficult to treat.

6.3.3 Refraining From the Use of Personal Firearms and Ammunition. DoD personnel have normally not been permitted to deploy with, or use, personal firearms or ammunition during military operations.[26] This practice is based on a variety of reasons, including to help ensure compliance with the law of war and U.S. domestic law.

6.3.4 Refraining From the Use of Captured Weapons, Except on a Field-Expedient Basis. Under DoD practice, captured or other foreign weapons have been used on a field-expedient basis, but otherwise have not been used in combat unless they have undergone the required legal review and have been duly issued to personnel.

Any weapon, ammunition, or munition captured, found, or recovered on the battlefield by DoD personnel is the property of the United States and not the individual who recovered it, regardless of whether the property is of U.S., enemy, allied, or unknown origin.[27]

6.4 PROHIBITED WEAPONS

Two general prohibitions apply to all types of weapons. In addition, certain types of weapons are specifically prohibited.

6.4.1 General Prohibitions Applicable to All Types of Weapons. Two fundamental prohibitions based in customary international law apply to all weapons. It is prohibited to use:

- weapons calculated to cause superfluous injury;[28] or
- inherently indiscriminate weapons.[29]

6.4.2 Specifically Prohibited Types of Weapons. In addition, the use of the following types of weapons is prohibited by treaty or customary international law:

- poison, poisoned weapons, poisonous gases, and other chemical weapons;[30]
- biological weapons;[31]
- certain environmental modification techniques;[32]
- weapons that injure by fragments that are non-detectable by X-rays;[33]
- certain types of mines, booby-traps, and other devices;[34] and
- blinding lasers.[35]

6.4.3 Relationship Between the Specific Prohibitions and the General Prohibitions. When a specific prohibition on a weapon is viewed as reflecting customary international law, such prohibition may sometimes be viewed as, in effect, a declaration that the weapon is prohibited as one calculated to cause superfluous injury or as one that is inherently indiscriminate.[36]

However, a specific treaty prohibition on a weapon does not necessarily reflect a State Party's view that the weapon is prohibited by customary international law (*e.g.*, that the weapon is calculated to cause superfluous injury or is inherently indiscriminate). For example, the mere prohibition or restriction of a weapon in the CCW or its Protocols does not necessarily reflect the conclusion that the weapon is calculated to cause superfluous injury or is inherently indiscriminate.[37]

6.5 LAWFUL WEAPONS

Apart from the categories of weapons described in § 6.4 (Prohibited Weapons), all other types of weapons are lawful for use by the U.S. armed forces; that is, they are not illegal *per se*. In other words, other types of weapons are prohibited only to the extent that they fall under the categories of weapons described in § 6.4 (Prohibited Weapons). For example, a landmine is not necessarily a legally prohibited weapon; it is only prohibited if it falls under one of the specific classes of prohibited mines listed in § 6.12.4 (Prohibited Classes of Mines, Booby-Traps, and Other Devices). The rules that apply to the use of all weapons are addressed in § 5.4.2 (Overview of Rules in Conducting Attacks).

6.5.1 Certain Types of Weapons With Specific Rules on Use. Certain types of weapons, however, are subject to specific rules that apply to their use by the U.S. armed forces. These rules may reflect U.S. obligations under international law or national policy. These weapons include:

- mines, booby-traps, and other devices (except certain specific classes of prohibited mines, booby-traps, and other devices);[38]
- cluster munitions;[39]
- incendiary weapons;[40]
- laser weapons (except blinding lasers);[41]
- riot control agents;[42]
- herbicides;[43]
- nuclear weapons;[44] and
- explosive ordnance.[45]

6.5.2 Other Examples of Lawful Weapons. In particular, aside from the rules prohibiting weapons calculated to cause superfluous injury and inherently indiscriminate weapons,[46] there are no law of war rules specifically prohibiting or restricting the following types of weapons by the U.S. armed forces:

- edged or pointed weapons, including weapons with serrated edges or entrenching tools used as weapons;[47]
- small arms, cannons, and other guns, including shotguns, exploding bullets, expanding bullets, suppressors, or large-caliber guns;[48]
- blast weapons;[49]
- fragmentation weapons;[50]
- depleted uranium munitions;[51]
- remotely piloted aircraft;[52]
- autonomy in weapons systems;[53] and
- non-lethal weapons.[54]

6.5.3 Edged or Pointed Weapons. Stabbing or cutting weapons (*e.g.*, knives, bayonets, swords, lances, and tomahawks) are not specifically prohibited or restricted by the law of war. Like other weapons, however, they must not be calculated to cause superfluous injury.[55] Thus, it is prohibited to modify the weapon for the purpose of making wounds more difficult to treat, such as by adding a barbed head to a lance or by applying to the blade or point a poison or substance that is designed to inflame wounds.[56]

6.5.3.1 *Serrated Edges*. The law of war does not prohibit the use of serrated edged weapons by military forces, including against enemy personnel. Many bayonets or knives have a serrated edge (a formation resembling the toothed edge

of a saw). Provided that the design intent of the serrated edge is not to aggravate suffering unnecessarily, such as by making the wound more difficult to treat, the serrated edge is not prohibited. For example, a serrated edge may improve the capabilities of the blade as a multipurpose field utility tool, rather than be intended to increase the pain and suffering of enemy personnel injured by the blade.

6.5.3.2 *Use of Tools to Stab or Cut*. Edged tools, such as over-size knives, machetes, and entrenching tools, have often been employed both as tools and as weapons. Such tools are not prohibited by the law of war, and the law of war does not prohibit the use of tools as weapons to bludgeon, stab, or cut.

6.5.4 Small Arms, Cannons, and Other Guns. Small arms, cannons, and other guns are not specifically restricted or prohibited by the law of war. In particular, there are no specific rules against shotguns, exploding bullets, expanding bullets, suppressors, or anti-personnel use of larger caliber guns or cannons.

6.5.4.1 *Small Arms, Cannons, and Other Guns – Other Rules on Weapons, as Applicable.* Although no law of war rule regulates small arms, cannons, and other guns as a specific category, these weapons are subject to other law of war rules on weapons, as applicable. For example, the use of ammunition, the primary effect of which is to injure by fragments that in the human body escape detection by X-rays, is prohibited.[57] In addition, poison must not be added to bullets or other projectiles.[58] As with other weapons, small arms, cannons, and other guns must not be calculated to cause superfluous injury.[59] For example, modifying guns or ammunition for the purpose of aggravating the harm inflicted upon incapacitated persons would be prohibited (*e.g.*, adding a substance to bullets with the intent to inflame wounds).[60]

Guns that fall under the definition of incendiary weapons would be regulated as incendiary weapons.[61] However, tracer ammunition is not an "incendiary weapon" because any incendiary effects are incidental.[62] In addition, even certain projectiles designed to have incendiary effects that are combined with other effects (*e.g.*, penetration, blast, or fragmentation effects) would not be regulated as incendiary weapons.[63]

6.5.4.2 *Shotguns.* There is no law of war prohibition on the use (including antipersonnel use) of the shotgun in combat.[64] The fact that multiple fragments are used in many shotgun rounds to wound does not prohibit their use.[65] Shotguns have been used as anti-personnel weapons for close-combat, for guarding enemy prisoners of war, and for anti-materiel purposes during armed conflict for many years.[66] Shotgun munitions may include, for example, buckshot ammunition, flechette ammunition, nonlethal projectiles, and breaching rounds. Breaching rounds are an anti-materiel munition designed to dislodge door hinges or door locks and are used to enable military forces to enter a room or building. As with other weapons designed for anti-material use, the use of an anti-material round against enemy personnel is not prohibited.[67] For example, a breaching round may be used against an enemy combatant who opens the door or appears in the doorway just as the door is about to be breached.

6.5.4.3 *Exploding Bullets.* The use of exploding bullets, including use against enemy personnel, is not prohibited by the law of war. For example, explosive bullets have been lawfully used during armed conflict, against both enemy material and enemy combatants.[68] Exploding bullets, like all weapons, however, are subject to the superfluous injury rule.[69] For example, without a valid military reason, bullets that are specifically designed to explode within the human body would be prohibited, since a non-exploding bullet would be just as effective and efficient at incapacitating an enemy combatant.[70] On the other hand, an explosive bullet that was designed to explode in order to increase the probability of striking an enemy combatant or to strike an enemy combatant behind cover would not be prohibited.[71] Similarly, bullets that explode in order to neutralize enemy material more effectively are not prohibited, and there is no objection to using these bullets also against enemy combatants.[72]

The United States is not a Party to the 1868 St. Petersburg Declaration on Exploding Bullets, and its prohibition on the use of certain projectiles does not reflect customary international law.[73]

6.5.4.4 *Expanding Bullets.* The law of war does not prohibit the use of bullets that expand or flatten easily in the human body. Like other weapons, such bullets are only prohibited if they are calculated to cause superfluous injury.[74] The U.S. armed forces have used expanding bullets in various counterterrorism and hostage rescue operations, some of which have been conducted in the context of armed conflict.

The 1899 Declaration on Expanding Bullets prohibits the use of expanding bullets in armed conflicts in which all States that are parties to the conflict are also Party to the 1899 Declaration on Expanding Bullets.[75] The United States is not a Party to the 1899 Declaration on Expanding Bullets, in part because evidence was not presented at the diplomatic conference that expanding bullets produced unnecessarily severe or cruel wounds.[76]

In 2013, a review conducted by DoD in coordination with the Department of State reconfirmed that the prohibition in the 1899 Declaration on Expanding Bullets did not reflect customary international law.[77] The findings of this review were consistent with the longstanding position of the United States not to become a Party to the 1899 Declaration and not to apply a distinct prohibition against expanding bullets, but instead to regard expanding bullets as prohibited only to the extent that such bullets are calculated to cause unnecessary suffering.[78] In the past, expanding bullets have been viewed as ineffective for military and technical reasons.[79] Non-expanding bullets that were widely used in armed conflict produced wounds comparable to, or more severe than, wounds produced by expanding bullets.[80] In addition, expanding bullets are widely used by law enforcement agencies today, which also supports the conclusion that States do not regard such bullets are inherently inhumane or needlessly cruel.[81]

6.5.4.5 *Rome Statute of the International Criminal Court War Crime Regarding the Use of Expanding Bullets*. The war crime of using expanding bullets that is reflected in the Rome Statute of the International Criminal Court has been interpreted by States only to criminalize the use of expanding bullets that are also calculated to cause superfluous injury and not to create or reflect a prohibition against expanding bullets as such. The Rome Statute of the International Criminal Court lists as a war crime "[e]mploying bullets which expand or flatten easily in the human body, such as bullets with a hard envelope which does not entirely cover the core or is pierced with incisions."[82]

Elements of Crimes have been adopted by the Assembly of States Parties to the Rome Statute to assist the ICC in its interpretation and application of the Article 8 of the Rome Statute, which includes this war crime.[83] The Elements of Crimes explain that this rule is not violated unless, *inter alia*, "[t]he perpetrator was aware that the nature of the bullets was such that their employment would uselessly aggravate suffering or the wounding effect."[84] When adopting an amendment at the Review Conference in 2010 that would apply this crime to non-international armed conflict, Parties to the Rome Statute reiterated this understanding and explained that "the crime is committed only if the perpetrator employs the bullets to uselessly aggravate suffering or the wounding effect upon the target of such bullets, as reflected in customary international law."[85] Thus, this crime only applies to expanding bullets that are also calculated to cause superfluous injury and does not create or reflect a prohibition against expanding bullets as such.[86]

6.5.4.6 *Suppressors (Silencers)*. A suppressor or silencer is a device that is intended to reduce or disguise the sound of a firearm's discharge without affecting its ability to fire. There is no law of war prohibition on the use of a suppressor, and silenced weapons have been a part of the inventory of military forces and law enforcement agencies in many States, including the United States.[87]

6.5.4.7 *Anti-Personnel Use of Larger (e.g., .50) Caliber Guns or Cannons*. The law of war does not prohibit the use of .50 caliber machineguns or other large-caliber guns against enemy personnel.[88] Such use has been widespread and long-standing.[89]

Similarly, cannons and other artillery (*e.g.*, the M242 25mm Bushmaster in the Bradley Fighting Vehicle or the 30mm GAU-8 in the A-10 aircraft) have often been used during armed conflict against enemy material and may also be used against enemy combatants. Although use against individual enemy combatants may not necessarily reflect the most efficient use of resources, such use is not prohibited by the law of war.

6.5.5 Blast Weapons. Blast weapons rely upon the pressure wave caused by an explosion, rather than fragmentation, to cause damage to objects and injury to enemy military personnel. Inflicting blast injury is not prohibited by the law of war.[90]

6.5.6 Fragmentation Weapons. Some munitions operate by projecting a large number of fragments or projectiles on detonation. Such weapons may include fragmenting artillery shells, the Claymore munition, "grape shot," grenades, flechettes,[91] and shotgun buckshot.[92] The fragments or projectiles in such weapons must be detectable by X-rays.[93] However, the use of multiple fragments or projectiles to injure enemy personnel is not prohibited by the law of war.[94] The fact that enemy military personnel in the path of fragment dispersion are likely to suffer multiple wounds does not render the weapon unlawful. Multiple fragments may be necessary to increase the probability of incapacitating enemy combatants.

6.5.7 Depleted Uranium Munitions. Depleted uranium (DU) is used in some munitions because its density and physical properties create a particularly effective penetrating combination to defeat enemy armored vehicles, including tanks.[95]

States have regarded the use of depleted uranium weapons as consistent with their law of war obligations.[96]

The U.S. armed forces have fielded and used depleted uranium munitions. Extensive efforts have been made to study whether there are harmful health effects from exposure to depleted uranium from weapons that use it, but no such effects have been found.[97]

6.5.8 Remotely Piloted Aircraft. There is no prohibition in the law of war on the use of remotely piloted aircraft (also called "unmanned aerial vehicles").[98] Such weapons may offer certain advantages over other weapons systems.[99]

6.5.9 Autonomy in Weapon Systems. The law of war does not prohibit the use of autonomy in weapon systems. DoD has developed policy on the use of autonomy in weapon systems.[100]

6.5.9.1 *Description and Examples of the Use of Autonomy in Weapon Systems*. Some weapons may have autonomous functions. For example, mines may be regarded as rudimentary autonomous weapons because they are designed to explode by the presence, proximity, or contact of a person or vehicle, rather than by the decision of the operator.[101] Other weapons may have more sophisticated autonomous functions and may be designed such that the weapon is able to select targets or to engage targets automatically after being activated by the user. For example, the United States has used weapon systems for local defense with autonomous capabilities designed to counter time-critical or saturation attacks. These weapon systems have included the Aegis ship defense system and the Counter-Rocket, Artillery, and Mortar (C-RAM) system.

6.5.9.2 *No Law of War Prohibition on the Use of Autonomy in Weapon Systems*. The law of war does not specifically prohibit or restrict the use of autonomy to aid in the operation of weapons. In fact, in many cases, the use of autonomy could enhance the way law of war principles are implemented in military operations. For example, some munitions have homing functions that enable the user to strike military objectives with greater discrimination and less risk of incidental harm. As another example, some munitions have mechanisms to self-deactivate or to self-destruct, which helps reduce the risk they may pose generally to the civilian population or after the munitions have served their military purpose.[102]

Although no law of war rule specifically restricts the use of autonomy in weapon systems, other rules may apply to weapons with autonomous functions. For example, to the extent a weapon system with autonomous functions falls within the definition of a "mine" in the CCW Amended Mines Protocol, it would be regulated as such.[103] In addition, the general rules applicable to all weapons would apply to weapons with autonomous functions. For example, autonomous weapon systems must not be calculated to cause superfluous injury or be inherently indiscriminate.[104]

6.5.9.3 *Law of War Obligations of Distinction and Proportionality Apply to Persons Rather Than the Weapons Themselves*. The law of war rules on conducting attacks (such as the rules relating to discrimination and proportionality) impose obligations on persons. These rules do not impose obligations on the weapons themselves; of course, an inanimate object could not assume an "obligation" in any event. Thus, it is not the case that the law of war requires that a weapon determine whether its target is a military objective. Similarly, the law of war does not require that a weapon make other legal determinations such as whether an attack may be expected to result in incidental harm that is excessive in relation to the concrete and direct military advantage expected to be gained. The law of war does not require weapons to make legal determinations, even if the weapon (*e.g.*, through computers, software, and sensors) may be characterized as capable of making factual determinations, such as whether to fire the weapon or to select and engage a target. Rudimentary autonomous weapons, such as mines, have been employed for many years, and there has never been a requirement that such weapons themselves determine that legal requirements are met. Rather, it is persons who must comply with the law of war. For example, persons may not use inherently indiscriminate weapons.[105] In addition, in the situation in which a person is using a weapon that selects and engages targets autonomously, that person must refrain from using that weapon where it is expected to result in incidental harm that is excessive in relation to the concrete and direct military advantage expected to be gained.[106] In addition, the obligation on the person using the weapon to take feasible precautions in order to reduce the risk of civilian casualties may be more significant when the person uses weapon systems with more sophisticated autonomous functions.[107] For example, such feasible precautions a person is obligated to take may include monitoring the operation of the weapon system or programming or building mechanisms for the weapon to deactivate automatically after a certain period of time.

6.5.9.4 *DoD Policy on Autonomy in Weapon Systems*. DoD policy has addressed the use of autonomy in certain types of weapon systems.[108] Under a 2012 policy, certain types of autonomous weapon systems require an additional review by senior DoD officials before formal development and fielding.[109] In addition, the policy establishes rigorous standards for system design, testing of hardware and software, and training of personnel on the proper use of autonomous and semi-autonomous systems. Among other things, the policy requires that military commanders use autonomous and semi-autonomous weapon systems in a manner consistent with their design, testing, certification, operator training, and doctrine.[110]

U.S. domestic law may also be applicable to the development of autonomous functions in certain weapon systems.[111]

6.5.10 Non-Lethal (Less-Lethal) Weapons. The law of war does not prohibit or specifically address non-lethal weapons as a class of weapons. DoD policies regulate non-lethal weapons and require the legal review of their acquisition, as with other types of weapons. There is no requirement to use non-lethal weapons before using lethal weapons against military objectives.

6.5.10.1 *Non-Lethal Weapons – Notes on Terminology*. The United States is not a Party to any law of war treaty that regulates "non-lethal weapons" as a class or category of weapons. Thus, there is no applicable treaty definition of "non-lethal weapons" for the purpose of applying rules related to such weapons. Accordingly, in practice, the contours of the definition of "non-lethal weapons" have been developed as a matter of policy for the purpose of applying DoD policies with respect to "non-lethal weapons."

6.5.10.2 *Description and Examples of Non-Lethal Weapons*. *Non-lethal weapons*, sometimes called "less-lethal weapons," have been defined as weapons, devices, and munitions that are explicitly designed and primarily employed to incapacitate targeted personnel or materiel immediately, while minimizing fatalities, permanent injury to personnel, and undesired damage to property in the target area or environment.[112]

Non-lethal weapons are less lethal than conventional military weapons, and they use means other than gross physical destruction to defeat targets (whether enemy vehicles or personnel). Non-lethal weapons are intended to have reversible effects on personnel and materiel.

Non-lethal weapons, however, do not have a zero probability of producing fatalities or permanent injuries, which is why these weapons are sometimes referred to as less-lethal weapons.[113] The use of non-lethal weapons may still present risks of serious injury or loss of life.

Non-lethal weapons include, but are not limited to:[114]

- riot control agents;[115]
- blunt impact projectiles, such as stingball grenades, beanbag rounds, and rubber bullets;
- dazzling lasers that cause temporary loss of sight;
- the active denial system, a directed energy weapon that uses millimeter wave energy to penetrate 1/64th of an inch into the surface of the skin, causing a rapid and intense heating sensation;
- electrical stun guns that incapacitate by disrupting the body's electrical signals (also called human electro-muscular incapacitation devices or tasers); and
- counter-material weapons, such as the vehicle lightweight arresting device or the portable vehicle arresting barrier, that are designed to stop vehicles without harming the occupants of the vehicles.

6.5.10.3 *No Law of War Prohibitions or Restrictions on Non-Lethal Weapons as Such*. The law of war does not prohibit or specifically address non-lethal weapons as a class of weapons. Rather, such weapons are governed by the general customary rules applicable to weapons and other specific rules, insofar as those rules are applicable. For example, the use of riot control agents is prohibited as a method of warfare.[116]

Weapons designed to use the minimum amount of force and to preserve life are, consistent with the principle of humanity, often less likely to be prohibited or restricted by the law of war.

6.5.10.4 *DoD Policy Requiring a Legal Review of the Acquisition of Non-Lethal Weapons*. DoD policy has addressed non-lethal weapons.[117] Like lethal weapons, non-lethal weapons have been subject to legal review to ensure their compliance with U.S. law of war and arms control obligations.[118]

If acquisition of a non-lethal weapon is being considered, the command seeking to acquire the weapon should inquire through command or service channels whether the applicable legal reviews have been conducted.

6.5.10.5 *No Requirement to Use Non-Lethal Weapons Before Using Lethal Weapons Where Deadly Force Is Warranted*. Where the use of deadly force is warranted, such as against enemy combatants, the law of war does not create a legal obligation to use non-lethal weapons before using lethal weapons.[119] Moreover, under the law of war, the availability of non-lethal weapons does not create the obligation to use these weapons instead of lethal weapons or create a higher standard for the use of force.[120] In addition, the availability of non-lethal weapons does not limit the authority to use force in self-defense.[121]

6.6 WEAPONS CALCULATED TO CAUSE SUPERFLUOUS INJURY

It is especially forbidden to use weapons that are calculated to cause superfluous injury.[122]

6.6.1Superfluous Injury or Unnecessary Suffering – Notes on Terminology. The prohibition against weapons calculated to cause superfluous injury or of a nature to cause unnecessary suffering has been formulated in a variety of ways both in treaties to which the United States is a Party and in other treaties. The United States is not a Party to a treaty that provides a definition of "superfluous injury." Article 23(e) of the 1899 Hague II Regulations prohibits weapons "of a nature to cause superfluous injury." Article 23(e) of the 1907 Hague IV Regulations prohibits weapons "calculated to cause unnecessary suffering." The official texts of both the 1899 and 1907 treaties are French, and the French text of that paragraph is exactly the same in both treaties, even though English translations of the treaties are different. The title of the CCW refers to "Weapons Which May be Deemed to Be Excessively Injurious," and the CCW Preamble also recognizes "the principle that prohibits the employment in armed conflicts of weapons, projectiles and material and methods of warfare of a nature to cause superfluous injury or unnecessary suffering." Treaties that the United States has not ratified have also included this prohibition. The Preamble to the 1868 Declaration of St. Petersburg noted that "the employment of arms which uselessly aggravate the sufferings of disabled men, or render their death inevitable" would be "be contrary to the laws of humanity."[123] AP I Article 35(2) prohibits the use of "weapons, projectiles and material and methods of warfare of a nature to cause superfluous injury or unnecessary suffering."

Although the various formulations may be regarded as describing the same underlying prohibition, the phrase "calculated to cause superfluous injury" may be regarded as the more accurate translation of the French rule stated in the 1907 Hague IV Regulations and as more precisely conveying the intent of the rule.[124]

This manual generally uses the formulation "weapons calculated to cause superfluous injury" or refers to the "superfluous injury rule."

6.6.2 Superfluous Injury Rule – the Principle of Humanity. The superfluous injury rule is an application of the principle of humanity in the context of weapons.[125] The superfluous injury rule prohibits weapons that are designed to increase the injury or suffering of the persons attacked beyond that justified by military necessity.[126]

Thus, the principle of humanity may assist in understanding and applying the rule.[127] For example, weapons that are regarded as lawful in peacetime or that apply only the minimum force necessary in order to avoid death or injury to civilians would not be prohibited under the superfluous injury rule.[128]

6.6.3 Applying the Superfluous Injury Rule. The test for whether a weapon is prohibited by the superfluous injury rule is whether the suffering caused by the weapon provides no military advantage or is otherwise clearly disproportionate to the military advantage reasonably expected from the use of the weapon.[129] Thus, the suffering must be assessed in relation to the military utility of the weapon.[130] Weapons that may cause great injury or suffering or inevitable death are not prohibited, if the weapon's effects that cause such injury are necessary to enable users to accomplish their military missions.[131]

6.6.3.1 *Military Necessity*. The legitimate military necessity that may be considered in assessing the legality of a weapon includes its capability to disable or incapacitate enemy combatants, including, for example:

- the probability of striking a person against whom the weapon is aimed or directed;
- the probability that a person who is struck by the weapon will be incapacitated;
- the speed at which a person who is struck by the weapon will be incapacitated; and
- the number of persons the weapon could strike.[132]

However, the military utility of the weapon is often broader than its capacity to disable enemy combatants.[133] It may include other factors, such as:[134]

- the destruction or neutralization of military materiel;
- the restriction of military movement;
- the interdiction of military lines of communication;
- the effects on the morale, command and control, stamina, and cohesion of opposing forces;
- the weakening of the enemy's war-making resources and capabilities;
- the enhancement of the security of forces employing the weapon or other friendly forces;
- the effectiveness of the weapon against particular types of targets;

- the availability of alternative weapons and their effects;
- the logistics of providing the weapon and its ammunition where and when it is needed;
- the amount of ammunition required;
- the cost of using the weapon in terms of time, money, and other resources;
- the efficacy of the weapon compared to the adversary's defenses;
- the efficacy of the weapon in light of the weapons and tactics of the adversary; or
- the risk to the civilian population when the weapon is used for its intended purposes.[135]

These other factors can justify weapons that inflict injuries on enemy combatants that are much greater than the minimum needed to render them *hors de combat*. For example, an artillery shell designed to destroy field fortifications or heavy material causes injuries to enemy personnel that are much greater than those necessary to make enemy combatants *hors de combat*. However, the artillery shell is not prohibited because these military advantages are not clearly disproportionate to the injuries it inflicts.[136]

6.6.3.2 *Suffering and Injury Inflicted*. The superfluous injury rule addresses the harm inflicted upon the persons who are struck by the weapon. The types of harm addressed by the superfluous rule include: mortality rates, the painfulness of wounds, the severity of wounds, and the ease with which they may be treated (including the incidence of permanent damage or disfigurement).[137] Weapons that have been modified only for the purpose of increasing these types of harms for persons who have already been incapacitated are prohibited by the superfluous injury rule.[138]

6.6.3.3 *Clearly Disproportionate*. A weapon is only prohibited by the superfluous injury rule if the suffering it inflicts is clearly disproportionate to its military utility. This excessiveness should be assessed in light of the State practice and *opinio juris* and the principle of humanity. Because of the difficulty of comparing the military necessity for the weapon and the suffering it is likely to cause, a weapon is only prohibited if the suffering is clearly or manifestly disproportionate to the military necessity.[139] The suffering likely to result from the use of the weapon and its military effectiveness are likely to be difficult to assess, much less to compare to one another.[140]

Like other determinations of customary international law, whether a weapon is calculated to cause superfluous injury is to be determined in light of the practice of States, and in particular, whether States have declined to use the weapon or similar weapons because they are believed to have this effect.[141] The consideration of the military utility of the weapon and the suffering the weapon inflicts should be compared to other destructive mechanisms that States have used or retained for use and consider to be lawful.

A State's acceptance of a treaty prohibition against a weapon should not be understood as necessarily reflecting that State's conclusion that the weapon is calculated to cause superfluous injury. For example, the CCW and its Protocols generally do not reflect a conclusion that the weapons prohibited or restricted by them are calculated to cause superfluous injury.[142] On the other hand, a prohibition or restriction in a treaty can be relevant in determining that a weapon is not calculated to cause superfluous injury. For example, the CCW Amended Mines Protocol prohibits mines, booby-traps, and other devices that are designed or of a nature to cause unnecessary suffering.[143] Thus, this prohibition implicitly recognizes that not all mines, boobytraps, and other devices are calculated to cause superfluous injury.[144] In addition, even if a treaty that declared that certain weapons were calculated to cause superfluous injury, such treaty language would not necessarily mean that the weapon was prohibited by customary international law. For example, specially affected States might decline to ratify the treaty and decide to retain the right to use the weapon.[145]

6.6.3.4 *Superfluous Injury Rule – the Circumstances to Be Assessed and Design Intent*. In determining a weapon's legality, the weapon's planned uses and the normal circumstances in which it would be used should be considered.[146] It is not necessary to assess all possible uses or all possible circumstances.

6.6.4 Weapons Modified for the Purpose of Aggravating the Harm Inflicted on Incapacitated Persons. Weapons that have been modified for the purpose of aggravating the harm inflicted upon incapacitated persons are weapons that are prohibited by the superfluous injury rule.[147] For example, adding poison to bullets or other substances to weapons for the purpose of increasing the pain caused by the wound, to make wounds more difficult to treat, or to make death inevitable, would be prohibited. Such a substance would not increase the military utility of the weapon to incapacitate the enemy (*i.e.*, it would have no corresponding military advantage); thus, the additional injury would be superfluous when compared to the injury caused by the weapon without such a substance.

Weapons that have been modified are readily assessed under the superfluous injury rule because the unmodified version of the weapon provides a clear point of comparison. It is difficult to compare, in the abstract, the military utility of a weapon against the suffering it causes. But, by comparing an existing weapon to a proposed modification, one can evaluate more easily whether the proposed changes are warranted by legitimate military reasons (*e.g.*, to increase the ability of the weapon to incapacitate the enemy) or illegitimate reasons (*e.g.*, cruelty) because, in considering a modification to the weapon, one would expect that a number of relevant factors would be the same as between the existing weapon and the proposed modification.

6.7 INHERENTLY INDISCRIMINATE WEAPONS

Inherently indiscriminate weapons, *i.e.*, weapons that are incapable of being used in accordance with the principles of distinction and proportionality, are prohibited. Such weapons include weapons that are specifically designed to conduct attacks against the civilian population as well as weapons that, when used, would necessarily cause incidental harm that is excessive compared the military advantage expected to be gained from their use.

6.7.1Inherently Indiscriminate Weapons – Principles of Distinction and Proportionality. The prohibition against inherently indiscriminate weapons results from the principles of distinction and proportionality.[148] Attacks must be conducted in accordance with the principles of distinction and proportionality.[149] Consequently, a weapon that, when used, would necessarily violate these rules, would be prohibited.[150]

Few weapons have been understood to be inherently indiscriminate weapons.[151] This may be the case because weapons that are more accurate and precise are more militarily effective; the military and humanitarian interests coincide.[152] On the other hand, the few weapons that have been understood to violate this rule have been directed at illegitimate objectives, such as attacking and terrorizing the civilian population.[153]

6.7.2 Inherently Indiscriminate Weapons – Circumstances to Be Assessed and Design Intent. The test for whether a weapon is inherently indiscriminate is whether its use necessarily violates the principles of distinction and proportionality, *i.e.*, whether its use is expected to be illegal in all circumstances.

Special consideration should be given to the planned or intended uses of the weapon, *i.e.*, those that are reasonably foreseeable. For example, a practitioner conducting a legal review of the proposed acquisition or procurement of a weapon should consider the uses of the weapon that are planned and reflected in the design documents. Practitioners should advise if the planned uses of the weapon are not consistent with the principles of distinction and proportionality, with a view towards ensuring that either the weapon or the planned uses are modified accordingly. The wide range of circumstances in which weapons can lawfully be used should also be considered before concluding that a weapon is prohibited as inherently indiscriminate. For example, in some circumstances, an area of land can be a military objective.[154] Thus, even if it would not be possible for the weapon to be directed against enemy combatants, if the weapon could be directed at specific areas, it would be unlikely that the weapon would be considered inherently indiscriminate. As another example, in some circumstances, feasible precautions can mitigate the incidental harm expected to be caused so that it is not excessive.[155] Whether such precautions could be taken to mitigate the expected incidental harm caused by the weapon under review should be considered before concluding that a weapon is prohibited as inherently indiscriminate.

6.7.3 Weapons Designed to Conduct Attacks Against the Civilian Population. Inherently indiscriminate weapons include those that are specifically designed to be used to conduct attacks against the civilian population, including attacks to terrorize the civilian population. For example, Japanese bombs attached to free-floating, long-range balloons used during World War II were unlawful for this reason.[156] Also, German long-range rockets without guidance systems used during World War II were similarly illegal.[157]

6.7.4 Weapons That Necessarily Cause Excessive Incidental Harm. Indiscriminate weapons also include weapons whose anticipated incidental effects are necessarily excessive compared to the military advantages expected to be gained from using the weapon.[158] To be clear, the principle of proportionality does not prohibit the use of weapons whose destructive force cannot be limited to a specific military objective.[159] Such weapons may be used when their use is required against a target of sufficient military importance to outweigh the incidental harm that is expected to result.[160]

As with the superfluous injury rule, in assessing whether weapons necessarily cause excessive incidental harm, it will be important to consider whether the practice of States demonstrates that they are prohibited as such.[161] For example, the United States regards nuclear weapons not to be inherently disproportionate weapons.[162]

Weapons that necessarily cause excessive incidental harm include "blind" or essentially random weapons that are incapable of being controlled, and thus, cannot, with any degree of certainty, be directed against a military objective.[163] The expected incidental harm from the use of such weapons outweighs the little, if any, military utility of such weapons.[164]

Weapons that necessarily cause excessive incidental harm also include weapons whose uncontrollable nature is such that, even when directed against military objectives, they otherwise are expected invariably to cause excessive incidental harm. For example, using communicable diseases such as the plague as weapons has been prohibited, in part because such use would almost inevitably cause excessive incidental harm among the civilian population compared to the military advantages from their use.[165]

6.8 POISON, POISONED WEAPONS, POISONOUS GASES, AND OTHER CHEMICAL WEAPONS

The use of poison, poisoned weapons, poison and asphyxiating gases, and other chemical weapons is prohibited.

6.8.1Poison and Poisoned Weapons. It is especially forbidden to use poison or poisoned weapons.[166] For example, poisoning the enemy's food or water supply is prohibited.[167]

Similarly, adding poison to weapons is prohibited. The rule against poison and poisoned weapons reflected in the 1899 Hague II Regulations has been interpreted not to include poison gas weapons that were developed during the modern era, which were subsequently prohibited.[168] Poisons are understood to be substances that cause death or disability with permanent effects when, in even small quantities, they are ingested, enter the lungs or bloodstream, or touch the skin.[169] The longstanding prohibition against poison is based on: (1) their uncontrolled character; (2) the inevitability of death or permanent disability; and (3) the traditional belief that it is treacherous to use poison.[170]

6.8.1.1 *Designed to Injure by Poison.* This prohibition on poison applies to weapons that are designed to injure or kill by poison. It does not apply to weapons that injure or cause destruction by other means that also produce toxic byproducts.[171]

6.8.1.2 *Death or Permanent Disability to Persons.* The prohibition on using poison applies to use against human beings.[172] Thus, the prohibition on the use of poison has been understood not to prohibit the use of chemical herbicides that are harmless to human beings.[173] Other rules, however, may apply to the use of herbicides.[174]

6.8.2 Asphyxiating, Poisonous, or Other Gases, and All Analogous Liquids, Materials, or Devices. It is prohibited to use in war asphyxiating, poisonous, or other gases, and all analogous liquids, materials, or devices.[175] The United States has determined that this rule is part of customary international law.[176]

Although the rule appears quite broad, it has been understood only to apply to the use of gases that are designed to kill or injure human beings.[177] Thus, herbicides and riot control agents are not prohibited by this rule.[178] In addition, the rule is understood only to prohibit weapons whose prime, or even exclusive, effect is to poison or asphyxiate.[179] Thus, for example, smoke, flame, napalm, conventional explosives, and nuclear weapons are not covered by this rule, even though these weapons may produce asphyxiating or poisonous byproducts.[180]

6.8.3 Chemical Weapons. Chemical weapons are subject to a number of prohibitions.

6.8.3.1 *Definition of Chemical Weapons.* Under the Chemical Weapons Convention, chemical weapons mean the following, together or separately:

(a) Toxic chemicals and their precursors, except where intended for purposes not prohibited under this Convention, as long as the types and quantities are consistent with such purposes; (b) Munitions and devices, specifically designed to cause death or other harm through the toxic properties of those toxic chemicals specified in subparagraph (a), which would be released as a result of the employment of such munitions and devices;

(c) Any equipment specifically designed for use directly in connection with the employment of munitions and devices specified in subparagraph (b).[181]

Toxic chemicals refer to any chemical that through its chemical action on life processes can cause death, temporary incapacitation, or permanent harm to humans or animals. This includes all such chemicals, regardless of their origin or of their method of production, and regardless of whether they are produced in facilities, in munitions, or elsewhere.[182] Chemicals that only cause harm to plants, such as herbicides, are not covered.[183] In addition, toxic chemicals intended for purposes not prohibited by the Chemical Weapons Convention are also excluded, so long as they are of a type and quantity consistent with these purposes that are not prohibited.[184]

Precursor means any chemical reactant (including any key component of a binary or multicomponent chemical system) that takes part at any stage in the production by whatever method of a toxic chemical.[185] *Key component of a binary or multicomponent chemical system* means the precursor that plays the most important role in determining the toxic properties of the final product and reacts rapidly with other chemicals in the binary or multicomponent system.[186]

Equipment specifically designed for use directly in connection with the employment of such munitions and devices only applies to equipment designed solely for use with chemical weapons and does not, for example, include equipment that is designed also for purposes that are not prohibited.[187]

6.8.3.2 *Prohibitions With Respect to Chemical Weapons*. Chemical weapons are subject to a number of prohibitions. It is prohibited:

• to use chemical weapons;

• to develop, produce, otherwise acquire, stockpile or retain chemical weapons, or transfer,

directly or indirectly, chemical weapons to anyone;

• to engage in any military preparations to use chemical weapons; and

• to assist, encourage, or induce, in any way, anyone to engage in any activity prohibited to a Party to the Chemical Weapons Convention.[188]

These prohibitions apply in any circumstances. For example, chemical weapons may not be used in international armed conflict and non-international armed conflicts.[189] Similarly, chemical weapons may not be used in retaliation after a State has suffered from a chemical weapons attack, even if that attack has been conducted by a State that is not a Party to the Chemical Weapons Convention.[190]

6.8.3.3 *Obligation to Destroy Certain Chemical Weapons and Chemical Weapons Production Facilities*. In addition, a Party to the Chemical Weapons Convention has an obligation to destroy chemical weapons or chemical weapon production facilities it owns or possesses or that are located in a place under its jurisdiction or control.[191] If U.S. armed forces encounter chemical weapons or chemical weapon production facilities during armed conflict, U.S. national authorities should be notified as soon as practicable. In addition, with due regard for safety and security considerations, reasonable efforts should be made to secure and retain information regarding the chemical weapons.[192]

6.8.3.4 *Certain Uses of Toxic Chemicals Not Prohibited*. The Chemical Weapons Convention does not prohibit the use of toxic chemicals and their precursors for certain purposes.[193] Toxic chemicals and their precursors that are used for these purposes are not considered chemical weapons, so long as they are of a type and quantity consistent with these permitted purposes.[194] These purposes include:

• industrial, agricultural, research, medical, pharmaceutical, or other peaceful purposes;

• protective purposes, namely those purposes directly related to protection against toxic chemicals and to protection against chemical weapons;

• military purposes not connected with the use of chemical weapons and not dependent on the use of the toxic properties of chemicals as a method of warfare; and

• law enforcement, including domestic riot control purposes.[195]

Seeking to develop and use means of protection against chemical weapons is permissible, provided such protection is not intended to facilitate the use of chemical weapons or for other purposes prohibited by the Chemical Weapons Convention.[196]

6.9 BIOLOGICAL WEAPONS

Biological weapons, including bacteriological and toxin weapons, are subject to a number of prohibitions and restrictions.

6.9.1 Biological Weapons – Prohibition on Use as a Method of Warfare. It is prohibited to use bacteriological methods of warfare.[197] This prohibition includes all biological methods of warfare and the use in warfare of toxin weapons.[198] For example, it is prohibited to use plague as a weapon.[199] A prohibition against the use of biological weapons may be understood to result from U.S. obligations in the Biological Weapons Convention to refrain from developing, acquiring, or retaining biological weapons.[200] Bacteriological or biological warfare is prohibited, at least in part, because it can have massive, unpredictable, and potentially uncontrollable consequences.[201]

6.9.1.1 *Toxin Weapons*. The term *toxin* refers to poisonous chemical substances that are naturally produced by living organisms, and that, if present in the body, produce effects similar to disease in the human body.[202] Toxins are not living organisms and thus are not capable of reproducing themselves and transmissible from one person to another.[203] Toxin weapons have been regulated in connection with biological weapons because they have been produced in facilities similar to those used for the production of biological agents.[204] However, even toxins that are produced synthetically, and not through biological processes, fall within these prohibitions. Substances that are classified as "toxins" for the purpose of applying the requirements of the Biological Weapons Convention may also be classified as "chemical weapons" that are subject to the requirements of the Chemical Weapons Convention.[205]

6.9.2 Biological Weapons – Prohibition on Development, Acquisition, or Retention. It is also prohibited to develop, produce, stockpile, or otherwise acquire or retain:

• microbial or other biological agents, or toxins whatever their origin or method of production, of types and in quantities that have no justification for prophylactic, protective, or other peaceful purposes; or

• weapons, equipment or means of delivery designed to use such agents or toxins for hostile purposes or in armed conflict.[206]

6.9.3 Biological Weapons – Prohibition on Transfer or Assisting, Encouraging, or Inducing the Manufacture or Acquisition. It is also prohibited to transfer or to assist, encourage, or induce others to acquire biological weapons.[207]

The exchange of equipment, materials, and scientific and technological information for the use of bacteriological and biological agents and toxins for peaceful purposes, such as the prevention of disease, however, is not restricted.[208]

6.10 CERTAIN ENVIRONMENTAL MODIFICATION TECHNIQUES It is prohibited to use environmental modification techniques having widespread, long lasting, or severe effects as a means of destruction, damage, or injury to another Party to the ENMOD Convention.[209] In addition, it is prohibited to assist, encourage, or induce others to use such environmental modification techniques against a Party to the ENMOD Convention.[210]

Environmental Modification Techniques. "6.10.1 Environmental modification techniques" refers to any technique for changing through the deliberate manipulation of natural processes the dynamics, composition, or structure of the earth, including its biota, lithosphere, hydrosphere and atmosphere, or of outer space.[211]

6.10.2 Having Widespread, Long-Lasting, or Severe Effects. The techniques must have widespread, long-lasting, or severe effects. These terms are interpreted as follows: (a) "widespread": encompassing an area on the scale of several hundred square kilometers;

(b) "long-lasting": lasting for a period of months, or approximately a season; (c) "severe": involving serious or significant disruption or harm to human life, natural and economic resources, or other assets.[212]

For example, earthquakes, tsunamis, and cyclones are environmental effects likely to be widespread, long-lasting, or severe that could be caused by the use of environmental modification techniques.[213] On the other hand, dispelling fog to facilitate military or combat operations may involve the use of environmental modification techniques that would not have widespread, long-lasting, or severe effects.[214]

6.10.3 As a Means of Destruction, Damage, or Injury. In order to fall within the ENMOD Convention's prohibitions, the environmental modification techniques must be used as a means of destruction, damage, or injury to another Party to the ENMOD Convention. The ENMOD Convention does not prohibit damage to the environment, but reflects the idea that the environment itself should not be used as an instrument of war.[215]

Weapons or military operations may incidentally have widespread, long-lasting, or severe effects on the environment. Such weapons and military operations are not prohibited by the ENMOD Convention because the harm to the environment is incidental and not intended to be used as a means of destruction, damage, or injury to another Party to the ENMOD Convention. For example, nuclear weapons are not prohibited by the ENMOD Convention because their effects on the environment are a by-product of their use rather than intended as a means of injuring the enemy.[216]

6.10.3.1 *AP I Provisions on Environmental Protection*. In contrast to the ENMOD Convention's prohibition against using certain environmental modification techniques as a means of destruction, damage, or injury, two provisions of AP I specifically address the protection of the environment. Article 35(3) of AP I provides that "[i]t is prohibited to employ methods or means of warfare which are intended, or may be expected, to cause widespread, long term and severe damage to the natural environment." Article 55 of AP I provides: Protection of the natural environment: (1) Care shall be taken in warfare to protect the natural environment against widespread, long-term and severe damage. This protection includes a prohibition of the use of methods or means of warfare which are intended or may be expected to cause such damage to the natural environment and thereby to prejudice the health or survival of the population. (2) Attacks against the natural environment by way of reprisals are prohibited.

The United States has not accepted these provisions [217] and has repeatedly expressed the view that these provisions are "overly broad and ambiguous and 'not a part of customary law.'"[218] Articles 35(3) and 55 of AP I "fail to acknowledge that use of such weapons is prohibited only if their use is clearly excessive in relation to the concrete and direct overall military advantage anticipated."[219]

6.10.4 Environmental Modification Techniques for Peaceful Purposes. The ENMOD Convention does not hinder the use of environmental modification techniques for peaceful purposes.[220]

6.11 WEAPONS INJURING BY FRAGMENTS NOT DETECTABLE BY X-RAYS

It is prohibited to use any weapon the primary effect of which is to injure by fragments that in the human body escape detection by X-rays.[221]

6.11.1 Primary Effect. The prohibition requires that the fragments that are intended to cause injury be made of materials that are detectable in the human body by X-rays. Portions of many munitions, such as the casing, detonator, or timing mechanism, may be made of nonmetallic parts or other substances that are not detectable by X-rays in order to reduce weight, manufacturing costs, etc. These munitions are not prohibited because their primary effect is not to injure by fragments that in the human body escape detection by X-rays.[222]

6.11.2 Fragments That Injure by Penetrating the Human Body. The prohibition applies only to weapons that are intended to injure by penetrating the human body. Thus, for example, a weapon that employs rubber projectiles that are unlikely to penetrate the skin need not be detectable by X-rays.

6.11.3 Fragments Escaping Detection by X-Rays. The fragments must not be such that they escape detection by X-rays. For example, a member of the armed forces engaged in the authorized construction and/or employment of field-improvised booby-traps may not use glass for fragments for the purpose of wounding enemy personnel.[223]

In some cases, the fragments that otherwise would not be detectable by X-rays may be modified in order to ensure that they would be detectable by X-rays. For example, projectiles may be modified with a substance, such as barium sulfate, so that they are readily detectable by X-rays if they happen to penetrate the skin.[224]

6.12 LANDMINES, BOOBY-TRAPS, AND OTHER DEVICES

As a class of weapons, landmines, booby-traps, and other devices are not specifically prohibited under the law of war. However, certain landmines, booby-traps, and other devices are prohibited. In addition, the use of landmines, booby-traps, and other devices is subject to certain restrictions that the United States has accepted in the CCW Amended Mines Protocol.

6.12.1 Definition of Mine. The CCW Amended Mines Protocol defines a "mine" as "a munition placed under, on, or near the ground or other surface area and designed to be exploded by the presence, proximity or contact of a person or vehicle."[225] The term "mine" thus includes both anti-personnel and anti-vehicle mines, including anti-tank mines. The mines to which the CCW Amended Mines Protocol relate are those used on land, including those laid to interdict beaches, waterway crossings, or river crossings, but do not include the use of anti-ship mines at sea or in inland waterways.[226] The rules on naval mines are addressed in § 13.11 (Naval Mines).

The term "mine" does not include hand grenades.[227] A trip-wired hand grenade is not considered a mine, but is considered a booby-trap under the CCW Amended Mines Protocol.[228]

6.12.1.1 *Notes on Terminology - "Landmines" Versus "Mines"*. The CCW Amended Mines Protocol uses the term "mines," but does not address naval mines. This section uses the term "mines" to refer to landmines, including both anti-personnel and anti-vehicle mines.

6.12.1.2 *Designed to Be Exploded by the Presence, Proximity, or Contact of a Person or Vehicle*. An important characteristic of a mine is its designed function of being exploded by the presence, proximity, or contact of a person or vehicle.[229] Command-detonated munitions (*i.e.*, munitions whose explosion is triggered by a decision of an operator) are not "mines" under the CCW Amended Mines Protocol, but may be regulated under the CCW Amended Mines Protocol as "other devices."[230]

Some munitions, such as the Claymore, may be configured for detonation by command, or by trip-wire. When used in trip-wire mode, they are mines and subject to corresponding restrictions. When used in command-detonated mode, they are subject to the restrictions applicable to "other devices."[231]

The design function of being activated by the target also distinguishes a mine from unexploded ordnance that results from the malfunction of a munition.[232] Other rules address unexploded ordnance.[233]

6.12.1.3 *Anti-Personnel Mines*. Under the CCW Amended Mines Protocol, an "anti-personnel mine" is a mine primarily designed to be exploded by the presence, proximity, or contact of a person and that will incapacitate, injure, or kill one or more persons.[234]

"Primarily" was added to ensure that anti-vehicle mines equipped with anti-handling devices (which often cause the mine to detonate by contact of a person) are not treated as antipersonnel mines under the CCW Amended Mines Protocol.[235]

Another element in the definition of "anti-personnel mine" is its effect of incapacitating, injuring, or killing one or more persons. This description was understood to be broad enough to cover the range of hazards posed by anti-personnel mines, but the term "incapacitating" means permanent incapacity.[236] Thus, the definition of "mine" for the purposes of the CCW Amended Mines Protocol does not include non-lethal weapon technology that is designed temporarily to disable, stun, or signal the presence of a person, but not to cause permanent incapacity.[237]

6.12.1.4 *Mines Other Than Anti-Personnel Mines*. The CCW Amended Mines Protocol uses the term "mines other than anti-personnel mines" to refer to anti-vehicle mines. "Anti-vehicle mines" or "mines other than anti-personnel mines" are also sometimes referred to as anti-tank mines.[238] The CCW Amended Mines Protocol defines "mine" in terms of a munition that is designed to be exploded by a person or vehicle.[239] Thus, "mines other than anti-personnel mines" are anti-vehicle mines.[240]

6.12.1.5 *Remotely Delivered Mines*. "Remotely-delivered mine" means a mine that is not directly emplaced, but is instead delivered by artillery, missile, rocket, mortar, or similar means, or dropped from an aircraft.[241]

Mines delivered from a land-based system from less than 500 meters, however, are not considered to be remotely delivered, provided that they are used in accordance with Article 5 and other relevant articles of the CCW Amended Mines Protocol.[242] These mines were excluded from the definition of remotely delivered mines because, when delivered in the prescribed manner, they can be accurately marked and precautions to protect civilians can be reliably maintained.

6.12.1.6 *Mines With Compliant Self-Destruction and Self-Deactivation (SD/SDA) Mechanisms*. Certain types of mines may be equipped with self-destruction or self-neutralization mechanisms, or be self-deactivating.

"Self-destruction [SD] mechanism" means an incorporated or externally attached automatically functioning mechanism that secures the destruction of the munition into which it is incorporated or to which it is attached.[243]

"Self-neutralization mechanism" means an incorporated automatically functioning mechanism that renders inoperable the munition into which it is incorporated.[244] The term is used in Article 6 of the CCW Amended Mines Protocol in relation to remotely delivered mines other than anti-personnel mines. There are no technical specifications for self-neutralization mechanisms in the Technical Annex to the CCW Amended Mines Protocol.[245] "Self-deactivating" (SDA) means automatically rendering a munition inoperable by means of the irreversible exhaustion of a component; for example, a battery that is essential to the operation of the munition could be exhausted as part of a self-deactivating mine.[246]

6.12.2 Definition of Booby-Trap. Under the CCW Amended Mines Protocol, a "boobytrap" means any device or material that is designed, constructed, or adapted to kill or injure, and that functions when a person disturbs or approaches an apparently harmless object or performs an apparently safe act.[247]

This includes field-expedient devices, *e.g.*, a hand grenade attached to a door and rigged to explode when the door is opened (as opening a door is an apparently safe act), as well as devices designed in advance to function as booby-traps (such as explosive devices designed to resemble civilian objects).[248]

6.12.3 Definition of "Other Devices" Similar to Mines. "Other devices" in the CCW Amended Mines Protocol means manually emplaced munitions and devices, including improvised explosive devices, designed to kill, injure, or damage, and which are actuated manually, by remote control, or automatically after a lapse of time.[249]

"Other devices" do not include munitions and devices that are not manually emplaced, such as munitions that are remotely delivered.

"Other devices" do not include hand grenades.[250] However, trip-wired hand grenades are regulated as booby traps.[251]

6.12.4 Prohibited Classes of Mines, Booby-Traps, and Other Devices. Certain types of mines, booby-traps, and other devices are prohibited. These types include:

- mines, booby-traps, and other devices calculated to cause superfluous injury;[252]
- mines, booby-traps, and other devices specifically designed to detonate during detection operations;[253]
- self-deactivating mines with anti-handling devices designed to function after the mine's operation;[254]
- non-detectable anti-personnel mines;[255]
- remotely delivered mines without compliant self-destruction and self-deactivation mechanisms;[256]
- remotely delivered mines other than anti-personnel mines without a feasible SD/SDA mechanism;[257]
- mines produced after December 3, 1998, without identifying information;[258]
- booby-traps and other devices in the form of apparently harmless portable objects specifically designed to explode;[259] and
- certain types of prohibited booby-traps and other devices.[260]

6.12.4.1 *Mines, Booby-Traps, and Other Devices Calculated to Cause Superfluous Injury*. As with other weapons, it is prohibited in all circumstances to use any mine, booby trap, or other device that is designed or of a nature to cause superfluous injury or unnecessary suffering.[261]

For example, mines and booby-traps filled with shards of glass as their primary wounding mechanism would violate this prohibition. Such a weapon would be regarded as calculated to cause superfluous injury because the shards would be undetectable by X-rays and would hinder the medical treatment of the wound without offering any corresponding military advantage.[262] In addition, such a weapon would be prohibited on the grounds that it used fragments nondetectable by X-rays.[263]

6.12.4.2 *Mines, Booby-Traps, and Other Devices Specifically Designed to Detonate During Detection Operations*. It is prohibited to use mines, booby-traps, and other devices that employ a mechanism or device specifically designed to detonate the munition by the presence of commonly available mine detectors as a result of their magnetic or other non-contact influence during normal use in detection operations.[264]

This rule only applies to mines, booby-traps, and other devices that are specifically designed to detonate through the normal use of non-contact mine detectors during mine detection operations. Thus, mines, booby-traps, and other devices specifically designed to detonate upon actual physical contact are permissible.[265] Similarly, anti-handling devices that are designed to detonate a mine upon attempts to defuse or deactivate the mine are not prohibited by this rule.

6.12.4.3 *Self-Deactivating Mines With Anti-Handling Devices Designed to Function After the Mine's Operation*. It is prohibited to use a self-deactivating mine equipped with an anti-handling device that is designed in such a manner that the anti-handling device is capable of functioning after the mine has ceased to be capable of functioning.[266]

The intent is to avoid situations where a self-deactivating mine, the "life" of which is normally limited by the life of its battery, remains dangerous indefinitely as a result of a longlived anti-handling device. This would defeat the purpose of the self-deactivation function by leaving a hazardous mine in place.[267]

6.12.4.4 *Non-Detectable Anti-Personnel Mines*. It is prohibited to use antipersonnel mines that are not detectable as specified in paragraph 2 of the Technical Annex to the CCW Amended Mines Protocol.[268] Paragraph 2 of the Technical Annex specifies that:

(a) With respect to anti-personnel mines produced after 1 January 1997, such mines shall incorporate in their construction a material or device that enables the mine to be detected by commonly-available technical mine detection equipment and provides a response signal equivalent to a signal from 8 grammes or more of iron in a single coherent mass.

(b) With respect to anti-personnel mines produced before 1 January 1997, such mines shall either incorporate in their construction, or have attached prior to their emplacement, in a manner not easily removable, a material or device that enables the mine to be detected by commonly-available technical mine detection equipment and provides a response signal equivalent to a signal from 8 grammes or more of iron in a single coherent mass.

(c) In the event that a High Contracting Party determines that it cannot immediately comply with sub-paragraph (b), it may declare at the time of its notification of consent to be bound by this Protocol that it will defer compliance with sub-paragraph (b) for a period not to exceed 9 years from the entry into force of this Protocol. In the meantime it shall, to the extent feasible, minimize the use of anti-personnel mines that do not so comply.[269]

6.12.4.5 *Remotely Delivered Anti-Personnel Mines Not in Compliance With Self- Destruction and Self-Deactivation Requirements*. It is prohibited to use remotely delivered antipersonnel mines that are not in compliance with the provisions on self-destruction and selfdeactivation in the Technical Annex to the CCW Amended Mines Protocol.[270]

All remotely delivered anti-personnel mines shall be designed and constructed so that no more than 10% of activated mines will fail to self-destruct within 30 days after emplacement, and each mine shall have a back-up self-deactivation feature designed and constructed so that, in combination with the self-destruction mechanism, no more than one in one thousand activated mines will function as a mine 120 days after emplacement.[271]

6.12.4.6 *Remotely Delivered Mines Other Than Anti-Personnel Mines Without a Feasible SD/SDA Mechanism*. It is prohibited to use remotely delivered mines other than antipersonnel mines, unless, to the extent feasible, they are equipped with an effective self-destruction or self-neutralization mechanism and have a back-up self-deactivation feature, which is designed so that the mine will no longer function as a mine when the mine no longer serves the military purpose for which it was placed in position.[272]

6.12.4.7 *Mines Produced After December 3, 1998, Without Identifying Information*. Subparagraph 1(d) of the Technical Annex to the CCW Amended Mines Protocol provides:

The use of mines produced after the entry into force of this Protocol is prohibited unless they are marked in English or in the respective national language or languages with the following information:

(i) name of the country of origin;

(ii) month and year of production; and

(iii) serial number or lot number.

The marking should be visible, legible, durable and resistant to environmental effects, as far as possible.

The CCW Amended Mines Protocol entered into force on December 3, 1998.[273]

6.12.4.8 *Booby-Traps and Other Devices in the Form of Apparently Harmless Portable Objects Specifically Designed to Explode*. It is prohibited in all circumstances to use booby-traps or other devices in the form of apparently harmless portable objects that are specifically designed and constructed to contain explosive material.[274] This prohibition relates to booby-traps manufactured to resemble items, such as watches, personal audio players, cameras, toys, and the like. This prohibition is intended to prevent the production of large quantities of dangerous objects that can be scattered around and are likely to be attractive to civilians, especially children.[275]

This rule does not prohibit the use of booby-traps in connection with non-portable objects, such as a door or gate. This rule does not prohibit field-expedient adaptation, or adaptation in advance, of objects for use as booby-traps or other devices that are not designed or constructed for such use.[276] For example, it would not be prohibited to improvise a booby-trap using a trip-wired hand grenade in the form of an apparently harmless portable object. Such improvisation of booby-traps, for example, to retard an enemy advance, does not pose the same sort of danger to the civilian population as the mass production of objects specifically designed as booby-traps.[277]

6.12.4.9 *Certain Types of Prohibited Booby-Traps and Other Devices*. It is prohibited to use booby-traps and other devices that are in any way attached to or associated with:[278]

- internationally recognized protective emblems, signs, or signals;

o This includes the distinctive emblem of the red cross and red crescent and the distinctive emblem for cultural property.[279]

• sick, wounded, or dead persons;

• burial or cremation sites or graves;

• medical facilities, equipment, supplies, or transportation;

• children's toys or other portable objects or products specially designed for the feeding, health, hygiene, clothing, or education of children;

• food or drink;

o The United States has reserved the right to use other devices to destroy any stock of food or drink that is judged likely to be used by an enemy military force, if due precautions are taken for the safety of the civilian population.[280]

• kitchen utensils or appliances, except in military establishments, military locations, or military supply depots;

o This does not authorize kitchen utensils or appliances in military hospitals, POW camps, or military chapels to be booby-trapped.[281]

• objects clearly of a religious nature;

• historic monuments, works of art, or places of worship that constitute the cultural or spiritual heritage of peoples;[282] or

• animals or their carcasses.

6.12.5 General Rules for Using Mines, Booby-Traps, and Other Devices. The use of mines, booby traps, and other devices is subject to the same rules and principles that govern the use of other weapons to conduct attacks.[283] However, because mines, booby-traps, and other devices explode only after their emplacement, there have been different views as to when the use of a mine constitutes an "attack."[284] Thus, the CCW Amended Mines Protocol seeks to clarify how these principles apply to the use of mines, booby-traps, and other devices. These rules include:

• a prohibition against directing mines, booby-traps, and other devices against the civilian population;

• a prohibition against the indiscriminate use of mines, booby-traps, and other devices;

• an obligation to take feasible precautions to protect civilians from the effects of mines, booby-traps, and other devices;

• an obligation to give effective advance warning of the emplacement of mines, booby- traps, and other devices, unless circumstances do not permit; and

• obligations with respect to the recording of the placement of mines, booby-traps, and other devices.

6.12.5.1 *Prohibition Against Directing Mines, Booby-Traps, and Other Devices Against Civilians*. As with other weapons, mines, booby-traps, and other devices may not be directed against the civilian population, individual civilians, or civilian objects.[285]

6.12.5.2 *Prohibition Against Indiscriminate Use of Mines, Booby-Traps, or Other Devices*. Indiscriminate use of mines, booby-traps, or other devices is prohibited.[286]

This prohibition on indiscriminate use of mines, booby-traps, or other devices prohibits any placement of mines, booby-traps, or other devices that is not on, or directed against, a military objective.[287] On the other hand, it would be permissible to direct mines against enemy forces that are expected to pass through an area. Similarly, it would be permissible to direct mines against a piece of land that is a military objective.[288] In addition, it would be permissible to place mines on land or objects that belong to friendly forces and that the adversary would regard as a military objective, in order to protect such land or objects from interference or attack. This prohibition on indiscriminate use of mines, booby-traps, or other devices also prohibits placing mines, booby-traps, or other devices using a method or means of delivery that cannot be directed at a specific military objective.[289] For example, it would be prohibited to deliver mines using unguided long-distance balloons.[290]

This prohibition on indiscriminate use of mines, booby-traps, or other devices also prohibits placing mines, booby-traps, or other devices that may be expected to cause incidental loss of civilian life, injury to civilians, damage to civilian objects, or a combination thereof, which would be excessive in relation to the concrete and direct military advantage anticipated.[291]

6.12.5.3 *Obligation to Take Feasible Precautions to Protect Civilians From the Effects of Mines, Booby-Traps, and Other Devices*. All feasible precautions shall be taken to protect civilians from the effects of mines, booby-traps, and other devices.[292] This obligation corresponds to a more general obligation to take feasible precautions.[293] Feasible precautions are those precautions that are practicable or practically possible, taking into account all circumstances ruling at the time, including humanitarian and military considerations.[294] These circumstances include, but are not limited to:

• the short- and long-term effect of mines, booby-traps, and other devices upon the local civilian population for the duration of their use;

• possible measures to protect civilians (*e.g.*, fencing, signs, warning, and monitoring);

• the availability and feasibility of using alternatives; and

• the short- and long-term military requirements for a minefield, or for the use of boobytraps or other devices.[295]

6.12.5.4 *Effective Advance Warning of Any Emplacement*. Effective advanced warning shall be given of any emplacement of mines, booby-traps, and other devices that may affect the civilian population, unless circumstances do not permit doing so.[296]

6.12.5.5 *Recording of Minefields, Mined Areas, Mines, Booby-Traps, and Other Devices*. All information concerning minefields, mined areas, mines, booby-traps, and other devices shall be recorded in accordance with the provisions of the Technical Annex to the CCW Amended Mines Protocol.[297]

"Minefield" is a defined area in which mines have been emplaced (including areas free of mines that simulate a minefield), and a "mined area" is an area that is dangerous due to the presence of mines.[298] "Recording" means a physical, administrative, and technical operation designed to obtain, for the purpose of registration in official records, all available information facilitating the location of minefields, mined areas, booby-traps, and other devices.[299] Subparagraph 1(a) of the Technical Annex to the CCW Amended Mines Protocol specifies that the recording of the location of non-remotely delivered mines, minefields, mined areas, booby-traps, and other devices shall be carried out in accordance with the following provisions:

(i) the location of the minefields, mined areas and areas of booby-traps and other devices shall be specified accurately by relation to the coordinates of at least two reference points and the estimated dimensions of the area containing these weapons in relation to those reference points;

(ii) maps, diagrams or other records shall be made in such a way as to indicate the location of minefields, mined areas, booby-traps and other devices in relation to reference points, and these records shall also indicate their perimeters and extent; (iii) for purposes of detection and clearance of mines, booby-traps and other devices, maps, diagrams or other records shall contain complete information on the type, number, emplacing method, type of fuse and life time, date and time of laying, antihandling devices (if any) and other relevant information on all these weapons laid. Whenever feasible the minefield record shall show the exact location of every mine, except in row minefields where the row location is sufficient. The precise location and operating mechanism of each booby-trap laid shall be individually recorded.[300]

Records shall be retained by the parties to the conflict.[301] Copies of records shall be held at a level of command sufficient to guarantee their safety as far as possible.[302]

6.12.6 Rules for Using Non-Remotely Delivered Anti-Personnel Mines With Compliant SD/SDA Mechanisms. Non-remotely delivered anti-personnel mines that comply with the requirements for self-destruction and self-deactivation in the Technical Annex to the CCW Amended Mines Protocol may be used outside of marked areas.[303]

6.12.7 Rules for Using Non-Remotely Delivered Anti-Personnel Mines Without Compliant SD/SDA Mechanisms. Anti-personnel mines other than remotely delivered mines that are not in compliance with the provisions on self-destruction and self-deactivation in the Technical Annex to the CCW Amended Mines Protocol may be used subject to certain requirements. For a number of years, U.S. forces maintained minefields that met these standards.[304] The United States, however, no longer maintains any minefields.[305] First, such weapons are to be placed within a perimeter-marked area that is monitored by military personnel and protected by fencing or other means, to ensure the effective exclusion of civilians from the area. The marking must be of a distinct and durable character, and must be visible to a person who is about to enter the perimeter-marked area.[306] The State Party controlling a minefield is obligated to take all feasible measures to prevent the unauthorized removal, defacement, destruction, or concealment of any device, system, or material used to establish the perimeter of a permanently marked area.[307]

Second, such weapons are to be cleared before the area is abandoned, unless the area is turned over to the forces of another State that accepts responsibility for the maintenance of the protections stated above, and the subsequent clearance of those weapons.[308] This requirement does not preclude agreement, in connection with peace treaties or similar arrangements, to allocate responsibilities under Article 5 of the CCW Amended Mines Protocol in a manner that respects the essential spirit and purpose of Article 5.[309]

A party to a conflict is relieved from compliance with these obligations if compliance is not feasible due to forcible loss of control of the area as a result of enemy military action, including situations where direct enemy military action makes it impossible to comply. If the party that laid the minefield regains control of the area, it must re-assume its responsibilities.[310]

The forces of a party to the conflict gaining control of an area assume the responsibilities, to the maximum extent feasible, until the mines have been cleared.[311]

6.12.8 Rules for Using Non-Remotely Delivered Anti-Personnel Mines That Propel Fragments in a Horizontal Arc of Less Than 90 Degrees (*e.g.*, Claymores). Anti-personnel mines other than remotely delivered mines that propel fragments in a horizontal arc of less than 90 degrees may be used on or above the ground outside of marked areas for up to 72 hours if:

• they are located in immediate proximity to the military unit that emplaced them; and

• the area is monitored by military personnel to ensure the effective exclusion of civilians.[312]

The maintenance of observation over avenues of approach where such mines are deployed constitutes one acceptable form of monitoring to ensure the effective exclusion of civilians.[313]

A Claymore munition used in trip-wire mode is an example of this type of mine.[314] A Claymore used in command-detonated mode is regulated as an "other device."[315]

6.12.9 Rules for Using Remotely Delivered Mines. Remotely delivered mines are subject to special, additional restrictions because their location cannot be marked as accurately as mines placed by hand or by mechanical mine layers. In addition, because remotely delivered mines are often emplaced by long-range means, it may be difficult to ensure that civilians are absent or excluded from areas containing such mines. Remotely delivered mines may not be used unless they have compliant self-destruction and self-deactivation mechanisms.[316]

6.12.9.1 *Effective Advance Warning of the Delivery of Mines Unless Circumstances Do Not Permit*. As with other weapons, effective advance warning shall be given of any delivery or dropping of remotely delivered mines that may affect the civilian population, unless circumstances do not permit doing so.[317]

6.12.9.2 *Recording of Remotely Delivered Mines*. It is prohibited to use remotely delivered mines unless they are recorded in accordance with subparagraph 1(b) of the Technical Annex to the CCW Amended Mines Protocol.318[318] Subparagraph 1(b) of the Technical Annex requires that the estimated location and area of remotely delivered mines be specified by coordinates of reference points (normally corner points) and shall be ascertained and, when feasible, marked on the ground at the earliest opportunity. In addition, the total number and type of mines laid, the date and time of laying, and the self-destruction time periods shall also be recorded.[319]

6.12.10 Rule for Using Booby-Traps and Other Devices. Booby-traps may not be of the prohibited type discussed in § 6.12.4.8 (Booby-Traps and Other Devices in the Form of Apparently Harmless Portable Objects Specifically Designed to Explode) and § 6.12.4.9 (Certain Types of Prohibited Booby-Traps and Other Devices). In addition, as with mines, there are recording obligations with respect to the use of booby-traps and other devices, including the obligation to record the precise location and operating mechanism of each booby-trap laid.[320]

6.12.10.1 *Use of Booby-Traps and Other Devices in Populated Areas When Combat Between Ground Forces Is Not Taking Place or Does Not Appear Imminent*. In addition, booby-traps and other devices may not be used in any city, town, village, or other area containing a similar concentration of civilians in which combat between ground forces is not taking place or does not appear to be imminent, unless either:

• "they are placed on or in the close vicinity of a military objective;"[321] or

• "measures are taken to protect civilians from their effects, for example, the posting of warning [signs, the posting of] sentries, the issue of warnings or the provision of fences."[322]

6.12.11 Obligation to Seek to Protect Certain Groups From the Effects of Minefields, Mined Areas, Mines, Booby-Traps, and Other Devices. Article 12 of the CCW Amended Mines Protocol creates certain obligations to take measures to help protect certain groups from the effects of minefields, mined areas, mines, booby-traps, and other devices. Such obligations may apply with respect to:

• Any United Nations force or mission performing peace-keeping, observation, or similar functions in any area in accordance with the Charter of the United Nations; or

• Any of the following, if they have the consent of the State in whose territory they are operating:[323]

o missions established pursuant to Chapter VIII of the Charter of the United Nations and performing its functions in the area of a conflict;

o humanitarian or fact-finding missions of the United Nations System;

o missions of the ICRC;

o humanitarian missions of a national Red Cross or Red Crescent Society or of their International Federation; and

o missions of inquiry established pursuant to the provisions of the Geneva Conventions of 12 August 1949 and, where applicable, their Additional Protocols.

The obligations to take affirmative measures under Article 12 are subject to practical considerations.[324] In addition, the forces and missions that are afforded protection under Article 12 have corresponding obligations to keep information confidential, to respect applicable laws and regulations, and to refrain from activities that are incompatible with the impartial and international nature of their duties.[325]

6.12.12 Obligations Arising After the Cessation of Active Hostilities. Certain obligations with respect to mines, booby-traps, and other devices are triggered after the cessation of active hostilities.

6.12.12.1 *Take Measures to Protect Civilians From the Effects of Minefields, Mined Areas, Mines, Booby-Traps, and Other Devices in Areas Under Their Control After Hostilities*. Without delay after the cessation of active hostilities, a party to a conflict is to take all necessary and appropriate measures, including the use of records kept in accordance with the CCW Amended Mines Protocol, to protect civilians from the effects of minefields, mined areas, mines, booby-traps, and other devices in areas under their control.[326] Information is to be shared with the other party or parties to the conflict and with the Secretary-General of the United Nations regarding mines laid by the party's forces in areas no longer under their control, subject to reciprocity and security requirements.[327]

6.12.12.2 *Clearance of Minefields, Mined Areas, Mines, Booby-Traps, and Other Devices After Hostilities*. Each Party to the CCW Amended Mines Protocol or party to a conflict is, in accordance with the provisions of the CCW Amended Mines Protocol, responsible for all mines, booby-traps, and other devices employed by it and undertakes to clear, remove, destroy, or maintain them as specified in Article 10 of the CCW Amended Mines Protocol.[328]

Without delay after the cessation of active hostilities, a Party to the CCW Amended Mines Protocol or a party to a conflict has an obligation to ensure that all minefields, mined areas, mines, booby-traps, and other devices in areas under its control are cleared, removed, destroyed, or maintained in accordance with Article 3 and Paragraph 2 of Article 5 of the CCW Amended Mines Protocol.[329]

With respect to mines laid by a Party to the CCW Amended Mines Protocol or a party to a conflict in areas over which the party no longer exercises control, the party is to provide to the party that is now in control of the area, to the extent permitted by such party, technical and material assistance necessary to fulfill its responsibility to ensure that all minefields are cleared, removed, destroyed, or maintained.[330] For example, such assistance may include maps and other information regarding the mines and minefields.

In addition, the parties have an obligation to endeavor to reach agreement among themselves and, where appropriate, with other States and with international organizations, on the provision of technical and material assistance, including, in appropriate circumstances, the undertaking of joint operations necessary to fulfill such responsibilities.[331] There are additional provisions of the CCW Amended Mines Protocol addressing international exchanges of information and cooperation in this respect.[332] However, no provision of the CCW Amended Mines Protocol is to be construed as affecting the discretion of the United States to refuse assistance or to restrict or deny permission for the export of equipment, material, or scientific or technological information for any reason, and the CCW Amended Mines Protocol may not be used as a pretext for the transfer of weapons technology or the provision of assistance to the military mining or military counter-mining capabilities of a State Party to the CCW Amended Mines Protocol.[333]

6.12.13 U.S. Policy on Landmines. Landmines have been addressed by U.S. national policy. The United States has announced a policy: (1) not to use anti-personnel landmines outside the Korean Peninsula; (2) not to assist, encourage, or induce anyone outside the Korean Peninsula to engage in activity prohibited by the Ottawa Convention; (3) to undertake to destroy anti-personnel landmine stockpiles not required for the defense of the Republic of Korea; and (4) not to produce or otherwise acquire any anti-personnel munitions that are not compliant with the Ottawa Convention.[334] The United States has also previously committed not to emplace new persistent anti-personnel or anti-vehicle landmines (*i.e.*, landmines without self-destruction mechanisms and self-deactivation features).[335] U.S. landmines will continue to meet or exceed international standards for self-destruction and self-deactivation.[336] In addition, the United States no longer has any non-detectable mine of any type in its arsenal.[337]

The United States has a long practice of contributing significantly to humanitarian demining efforts and other conventional weapons destruction programs.[338]

6.12.14 Ottawa Convention on Anti-Personnel Landmines. The United States is not a Party to the Convention on the Prohibition of the Use, Stockpiling, Production and Transfer of Anti-Personnel Mines and on Their Destruction.[339] This Convention is commonly called the Ottawa Convention. The United States has announced policy changes that align U.S. anti-personnel landmine policy outside the Korean Peninsula with the key requirements of the Ottawa Convention.[340]

Among other reasons, because the practice of the United States and other States remains inconsistent with the prohibitions of the Ottawa Convention, its prohibitions do not reflect customary international law.[341]

Many U.S. allies and coalition partners, including Australia, Canada, and the United Kingdom, are parties to the Ottawa Convention. Article 1 of the Ottawa Convention prohibits States Parties from using, developing, producing, otherwise acquiring, stockpiling, retaining, or transferring to anyone, directly or indirectly, anti-personnel landmines, or to assist, encourage, or induce, in any way, anyone to engage in any activity prohibited to a State Party under the Convention. Australia, Canada, and the United Kingdom have taken the position that its armed forces would not violate the Ottawa Convention merely by reason of taking part in joint operations with forces of an ally that is not bound by the Ottawa Convention and that uses antipersonnel mines.[342]

6.13 CLUSTER MUNITIONS

Cluster munitions are not specifically prohibited or restricted by the law of war. DoD has policies on cluster munitions. The use of cluster munitions, in certain circumstances, is likely to reduce the risk of incidental harm as compared to other weapons.

6.13.1 Description of Cluster Munitions. Cluster munitions may be described as munitions composed of a non-reusable canister or delivery body containing multiple, conventional explosive submunitions.[343] Some munitions that may contain submunitions are not considered cluster munitions. For example, nuclear, chemical, and biological weapons, as well as obscurants, pyrotechnics, non-lethal systems (*e.g.*, leaflets), non-explosive kinetic effect submunitions (*e.g.*, flechettes or rods), and landmines, generally are not considered cluster munitions.

6.13.2 Use of Cluster Munitions to Reduce the Risk of Incidental Harm. As with incendiary weapons, the use of cluster munitions, in certain circumstances, is likely to result in less incidental harm than the use of other weapons.[344] For example, cluster munitions have been used against military objectives containing dangerous forces, such as dams, in order to reduce the risk that bombardment of these objectives would release such forces and cause incidental harm to the civilian population.[345]

6.13.3 DoD Policy on Cluster Munitions. DoD has developed policies to minimize the negative humanitarian consequences that can sometimes be associated with cluster munitions.[346]

Under a 2008 policy, the U.S. armed forces will, after 2018, only employ cluster munitions containing submunitions that, after arming, do not result in more than 1% unexploded ordnance (UXO) across the range of intended operational environments.[347] Until the end of 2018, cluster munitions that exceed the 1% UXO rate may not be used unless approved by the Combatant Commander.[348]

6.13.4 Convention on Cluster Munitions. The United States is not a Party to the Convention on Cluster Munitions.[349] The United States has determined that its national security interests cannot be fully ensured consistent with the terms of the Convention on Cluster Munitions.[350] The United States has expressed the view that the Convention on Cluster Munitions does not represent an emerging norm or reflect customary international law that would prohibit the use of cluster munitions in armed conflict.[351]

The Convention on Cluster Munitions establishes for its States Parties a number of prohibitions and restrictions on certain types of cluster munitions.[352] These rules include prohibitions on assisting, encouraging, or inducing anyone to engage in any activity prohibited to a State Party.[353] Despite these restrictions, States Parties and their military personnel and nationals may engage in military cooperation and operations with States that are not Parties to the Convention on Cluster Munitions that might engage in activities prohibited to a State Party.[354] Yet, even during such military operations, the Convention on Cluster Munitions does not authorize a State Party:

(a) To develop, produce or otherwise acquire cluster munitions;

(b) To itself stockpile or transfer cluster munitions;

(c) To itself use cluster munitions; or

(d) To expressly request the use of cluster munitions in cases where the choice of munitions used is within its exclusive control.[355]

6.14 INCENDIARY WEAPONS

The use of incendiary weapons is permissible, but subject to certain restrictions in order to reduce the risk of incidental harm to civilians.

6.14.1 Definition of an Incendiary Weapon. An “incendiary weapon” is any weapon or munition that is primarily designed to set fire to objects or to cause burn injury to persons through the action of flame, heat, or combination thereof, produced by a chemical reaction of a substance on the target.[356] Only “pure” incendiaries, such as napalm or the type of incendiary bombs used in World War II and Korea, are regulated as incendiary weapons.[357]

6.14.1.1 *Fire or Burn Injury Produced by Chemical Reaction of a Substance on the Target*. Incendiary weapons use a chemical reaction of a substance on the target to create flame or heat that destroys or injures. On the other hand, laser weapons, even if their primary effect is to set fire to objects or cause burn injuries, do not constitute “incendiary weapons” under CCW Protocol III on Incendiary Weapons because the fire or burn injuries are not produced by a chemical reaction of a substance that is delivered to the target.[358] Similarly, chemical weapons cause death or other harm through the toxic properties of chemicals themselves, rather than the chemicals causing a fire or burn injury.[359]

6.14.1.2 *Examples of Incendiary Weapons*. Examples of incendiary weapons include:[360]

- flame throwers;[361]
- fougasse;[362] and
- shells, rockets, grenades, mines, bombs, and other containers of incendiary substances, such as napalm and thermite.[363]

6.14.1.3 *Excluded From the Definition of Incendiary Weapons – Weapons With Incidental Incendiary Effects*. Incendiary weapons do not include munitions that may have incidental incendiary effects, such as illuminants, tracers, or smoke or signaling systems.[364] For example, white phosphorous is a munition that contains fragments of white phosphorous. It is intended primarily for marking or illuminating a target or masking friendly force movement by creating smoke. Similarly, tracer rounds are not incendiary weapons as they are designed to enable a gunner to direct his or her rounds onto a target rather than to set fire to objects.

6.14.1.4 *Excluded From the Definition of Incendiary Weapons – Certain Combined-Effects Munitions*. Certain combined-effects munitions are excluded from the definition of incendiary weapons. These munitions are designed to combine penetration, blast, or fragmentation effects with an additional incendiary effect, in which the incendiary effect is not specifically designed to cause burn injury to persons, but to be used against objects that are military objectives, such as armored vehicles, aircraft, and installations or facilities. Such weapons may include armor-piercing projectiles, fragmentation shells, explosive bombs, and similar combined-effects munitions.[365]

6.14.2 Use of Incendiary Weapons Is Permissible. Although subject to certain specific restrictions described in § 6.14.3 (Restrictions on the Use of Incendiary Weapons), the use of incendiary weapons, including anti-personnel use, is not prohibited.

6.14.2.1 *Use of Non-Incendiary Weapons to Set Fire to Objects or Cause Burn Injury to Persons*. The use of non-incendiary weapons to set fire to objects or to cause burn injury is technically not subject to the additional restrictions described in § 6.14.3 (Restrictions on the Use of Incendiary Weapons). For example, white phosphorus may be used as an antipersonnel weapon.[366] However, such use must comply with the general rules for the conduct of hostilities, including the principles of discrimination and proportionality.[367] In addition, feasible precautions to reduce the risk of harm to civilians must be taken.[368]

As a practical matter, however, it may be more effective to attack enemy combatants through other means, rather than seeking to cause them burn injury.[369]

6.14.3 Restrictions on the Use of Incendiary Weapons. As with other weapons, it is prohibited to make the civilian population as such, individual civilians, or civilian objects the object of attack by incendiary weapons.[370] In addition, it is specifically prohibited to make forests or other kinds of plant cover the object of attack by incendiary weapons, except when such natural elements are used to cover, conceal, or camouflage combatants or other military objectives, or are themselves military objectives.[371]

Similarly, the risks that the use of an incendiary weapon may pose to the civilian population should be considered in a proportionality analysis.[372]

As with the use of other weapons, commanders must take feasible precautions to reduce the risk of incidental harm to civilians when using incendiary weapons.[373]

6.14.3.1 *Use Against Military Objectives Located in Concentrations of Civilians*. Subject to a reservation by the United States discussed below, two additional rules apply to the use of incendiary weapons against military objectives that are located within a concentration of civilians. A "concentration of civilians" means any concentration of civilians, be it permanent or temporary, such as in inhabited parts of cities, in inhabited towns or villages, or in camps or columns of refugees or evacuees, or groups of nomads.[374]

First, subject to the U.S. reservation discussed below, it is prohibited to use air-delivered incendiary weapons to make any military objective located within a concentration of civilians the object of attack.[375]

Second, subject to the U.S. reservation discussed below, it is also prohibited to use incendiary weapons other than air-delivered incendiary weapons (*e.g.*, ground artillery) to make any military objective located within a concentration of civilians the object of attack. However, it is not prohibited to use such incendiary weapons against a military objective located within a concentration of civilians, when such military objective is clearly separated from the concentration of civilians and all feasible precautions are taken with a view to limiting the incendiary effects to the military objective and to avoiding, and in any event to minimizing, incidental loss of civilian life, injury to civilians, and damage to civilian objects.[376]

6.14.3.2 *U.S. Reservation to CCW Protocol III on Incendiary Weapons*. The United States has taken a reservation to CCW Protocol III on Incendiary Weapons for the purpose of improving the protection of the civilian population.[377] Thus, for the U.S. armed forces, incendiary weapons may be used against military objectives located in concentrations of civilians where it is judged that such use would cause fewer casualties and/or less collateral damage than alternative weapons. However, in such cases, all feasible precautions should be taken with a view to limiting the incendiary effects to the military objective and to avoiding, and in any event minimizing, incidental loss of civilian life, injury to civilians, and damage to civilian objects.[378]

For example, the use of incendiary weapons might be the means of attacking enemy biological weapons facilities that poses the least risk of incidental harm to the civilian population.[379] Similarly, the use of incendiary weapons might reduce the number of sorties or attacks, and thereby further reduce the risk of incidental harm to the civilian population.[380]

6.15 LASER WEAPONS

In general, laser weapons are permissible. They may be used against enemy combatants, including to kill and injure them. However, the use of "blinding laser" weapons is prohibited. In addition, feasible precautions shall be taken in the employment of laser systems to avoid causing permanent blindness.

6.15.1 Prohibition on "Blinding Laser" Weapons. It is prohibited to employ laser weapons specifically designed, as their sole combat function or as one of their combat functions, to cause permanent blindness to unenhanced vision, that is, to the naked eye or to the eye with corrective eyesight devices.[381]

Any use of such weapons is prohibited, including uses that are not intended to blind. Although the U.S. armed forces and other militaries use lasers in a variety of roles, such as range-finding or causing temporary loss of sight, the United States has not developed or fielded a "blinding laser" weapon.

6.15.1.1 *Laser Weapon Specifically Designed, as Its Sole Combat Function or One of Its Combat Functions.* To fall under the above prohibition, the laser device must be specifically designed, as a combat function, to cause permanent blindness to unenhanced vision. On the other hand, blinding as an incidental or collateral effect of the legitimate military employment of laser systems, including laser systems used against optical equipment, is not prohibited.[382] Thus, laser systems and devices used for rangefinding, target designation, jamming, dazzling, communications, and weapons guidance do not fall under this prohibition, even if serious eye injury or permanent blindness may incidentally occur through their use.[383]

Similarly, incidental blinding from the use of lasers that are designed to attack or destroy material or injure persons as a combat function (but are not specifically designed to blind) would be permissible.

6.15.1.2 *Permanent Blindness to Unenhanced Vision.* "Permanent blindness" is defined as irreversible and uncorrectable loss of vision that is seriously disabling with no prospect of recovery. Serious disability is equivalent to visual acuity of less than 20/200 Snellen measured using both eyes.[384]

Lasers that are designed to cause a temporary loss of sight, such as laser "dazzlers," would not fall under this prohibition.[385]

Unenhanced vision means the naked eye or the eye with corrective eyesight devices, such as ordinary glasses or contact lenses.[386] For example, unenhanced vision does not mean vision through direct-view optics, such as binoculars, a telescopic sight, night-vision goggles, or similar devices used to increase vision beyond that of an ordinary person without such enhancement. Persons using direct-view optics may have an increased risk of eye injury from lasers.

6.15.2 Feasible Precautions in the Employment of Laser Systems to Avoid the Incident of Permanent Blindness. In the employment of laser systems, all feasible precautions shall be taken to avoid the incidence of permanent blindness to unenhanced vision.[387] Feasible precautions may be understood to include those that are practicable or practically possible taking into account all circumstances ruling at the time, including humanitarian and military considerations.[388] Such precautions include training of the armed forces and other practical measures, such as promulgating doctrine and rules of engagement.[389]

6.15.2.1 *Use of Non-Prohibited Laser Systems to Blind Is Not Absolutely Prohibited.* Although feasible precautions should be taken to avoid blinding, the use of laser systems (that are not prohibited as blinding laser weapons) to blind enemy combatants is not prohibited if avoiding such harm is not feasible, *i.e.*, if such harm is militarily necessary. For example, it would not be prohibited for a soldier to use a non-prohibited laser system to blind an enemy combatant to save the user's life.[390]

6.16 RIOT CONTROL AGENTS

The use of riot control agents is subject to certain prohibitions and restrictions. Riot control agents are widely used by governments for law enforcement purposes (such as crowd control), but are prohibited as a method of warfare.

6.16.1 Definition of Riot Control Agents. *Riot control agents* mean any chemical not listed in a Schedule Annexed to the Chemical Weapons Convention, which can produce rapidly in humans sensory irritation or disabling physical effects that disappear within a short time following termination of exposure.[391] Riot control agents include, for example, tear gas and pepper spray, but generally are understood to exclude the broader class of non-lethal weapons that may sometimes be used for riot control or other similar purposes, such as foams, water cannons, bean bags, or rubber bullets.[392]

The United States does not consider riot control agents to be "chemical weapons,"[393] or otherwise to fall under the prohibition against asphyxiating, poisonous, or other gases, and all analogous liquids, materials, or devices.[394]

6.16.2 Prohibition on Use of Riot Control Agents as a Method of Warfare. It is prohibited to use riot control agents as a method of warfare.[395] The United States has understood this prohibition not to prohibit the use of riot control agents in war in defensive military modes to save lives, such as use of riot control agents:

• in riot control situations in areas under direct and distinct U.S. military control, including controlling rioting POWs;[396]

• in situations in which civilians are used to mask or screen attacks and civilian casualties can be reduced or avoided;[397]

• in rescue missions in remotely isolated areas, of downed aircrews and passengers, and escaping prisoners; and

• in rear echelon areas outside the zone of immediate combat to protect convoys from civil disturbances, terrorists, and paramilitary organizations.

These uses are as articulated in Executive Order 11850.[398] Even though Executive Order 11850 predated the Chemical Weapons Convention (which could have created legal obligations that were inconsistent with Executive Order 11850), interpreting the Chemical Weapons Convention consistent with Executive Order 11850 was a condition of the Senate giving its advice and consent to ratification of the Chemical Weapons Convention.[399] Thus, Executive Order 11850 has remained an important part of U.S. policy on the use of riot control agents.[400]

In addition to being permitted in war in defensive military modes to save lives, it is not prohibited to use riot control agents in military operations outside of war or armed conflict. Specifically, the United States has taken the position that riot control agents may be used in the conduct of:

• peacetime military operations within an area of ongoing armed conflict when the United States is not a party to the conflict;

• consensual peacekeeping operations when the use of force is authorized by the receiving state, including operations pursuant to Chapter VI of the United Nations Charter; and

• peacekeeping operations when force is authorized by the Security Council under Chapter VII of the United Nations Charter.[401]

6.17 HERBICIDES

The United States has renounced, as a matter of national policy, first use of herbicides in war except use, under regulations applicable to their domestic use, for control of vegetation within U.S. bases and installations or around their immediate defensive perimeters.[402]

6.17.1 Definition of Herbicide. An herbicide is a chemical compound that will kill or damage plants. Herbicides that are harmless to human beings are not prohibited under the rule against the use of poison or poisoned weapons.[403]

6.17.2 Chemical Weapons Convention and Herbicides. The Chemical Weapons Convention does not add any new constraints on the use of herbicides.[404] The Chemical Weapons Convention addresses toxic chemicals that cause death, temporary incapacitation, or permanent harm to humans or animals, rather than plants.[405] Moreover, even if an herbicide were such a toxic chemical, its use would likely be for a purpose not prohibited by the Chemical Weapons Convention.[406]

6.17.3 ENMOD Convention and Herbicides. Under certain circumstances, the use of herbicides could be prohibited by the ENMOD Convention.[407] However, the use of herbicides to control vegetation within U.S. bases and installations or around their immediate defensive perimeters has been understood by the United States to be permitted under international law.[408]

6.17.4 Authority Under Domestic Law to Employ Herbicides in War. Use of herbicides in war by the U.S. armed forces requires advance Presidential approval.[409] Additional regulations govern the use of herbicides.

6.18 NUCLEARWEAPONS

There is no general prohibition in treaty or customary international law on the use of nuclear weapons.[410] The United States has not accepted a treaty rule that prohibits the use of nuclear weapons *per se*, and thus nuclear weapons are lawful weapons for the United States. The law of war governs the use of nuclear weapons, just as it governs the use of conventional weapons.[411] For example, nuclear weapons must be directed against military objectives.[412] In addition, attacks using nuclear weapons must not be conducted when the expected incidental harm to civilians is excessive compared to the military advantage expected to be gained.[413]

6.18.1 U.S. Policy on the Use of Nuclear Weapons. The United States has developed national policy on the use of nuclear weapons. For example, the United States has stated that it would only consider the use of nuclear weapons in extreme circumstances to defend the vital interests of the United States or its allies and partners.[414] In addition, the United States has stated that it will not use or threaten to use nuclear weapons against non-nuclear weapons States that are party to the Nuclear Non-Proliferation Treaty and in compliance with their nuclear nonproliferation obligations.[415]

6.18.2 Nuclear Weapons and Arms Control Obligations. Nuclear weapons are regulated by a number of arms control agreements restricting their development, testing, production, proliferation, deployment, use, and, with respect to specific types, possession. Some of these agreements may not apply in times of war. Guidance on nuclear arms control agreements is beyond the scope of this manual.

6.18.3 AP I Provisions and Nuclear Weapons. Parties to AP I have expressed the understanding that the rules relating the use of weapons introduced by AP I were intended to apply exclusively to conventional weapons.[416] Thus, Parties to AP I have understood AP I provisions not to regulate or prohibit the use of nuclear weapons.[417] Although the United States is not a Party to AP I, the United States participated in the diplomatic conference that negotiated AP I based upon this understanding.[418]

6.18.4 Authority to Launch Nuclear Weapons. The authority to launch nuclear weapons generally is restricted to the highest levels of government. The domestic law and procedures concerning nuclear weapons employment are beyond the scope of this manual.

6.19 EXPLOSIVE ORDNANCE

Information should be kept on the use of explosive ordnance in order to reduce the risk posed by explosive remnants of war once the conflict has ended.[419] If possible, explosive ordnance that must be abandoned should be left in a safe and secure manner, and information should be recorded and retained on its abandonment.[420] In addition, generic preventive measures should be taken during the life-cycle of explosive ordnance in order to minimize the occurrence of explosive remnants of war.[421]

6.19.1 Definitions of Explosive Ordnance, Unexploded Ordnance, and Abandoned Ordnance.

6.19.1.1 *Definition of Explosive Ordnance*. "Explosive ordnance" means conventional munitions containing explosives, with the exception of mines, booby-traps, and other devices as defined in the CCW Amended Mines Protocol.[422]

As a factual matter, mines, booby-traps, and other devices may constitute explosive ordnance, but the legal definition of explosive ordnance in the CCW Protocol V on Explosive Remnants of War excludes these items because they are addressed by the CCW Amended Mines Protocol.[423]

The reference to conventional munitions in the definition of "explosive ordnance" means that chemical, biological, and nuclear weapons are not covered by the CCW Protocol V on Explosive Remnants of War.[424]

6.19.1.2 *Definition of Unexploded Ordnance (UXO)*. "Unexploded ordnance" means explosive ordnance that has been primed, fused, armed, or otherwise prepared for use and used in an armed conflict.[425] "Explosive ordnance" is defined in § 6.19.1.1 (Definition of Explosive Ordnance). UXO may have been fired, dropped, launched, or projected, and should have exploded, but failed to do so.[426] For example, UXO does not include weapons that have remained in the armory or arsenal and that have not been prepared for firing.[427] UXO are sometimes called "duds."

6.19.1.3 *Definition of Abandoned Explosive Ordnance (AXO)*. "Abandoned explosive ordnance" means explosive ordnance that has not been used during an armed conflict, that has been left behind or dumped by a party to an armed conflict, and that is no longer under control of the party that left it behind or dumped it.[428] AXO may or may not have been primed, fused, armed, or otherwise prepared for use.[429] For example, mortar rounds in a supply point that is abandoned as a result of a quick departure would constitute AXO.

6.19.2 Using Explosive Ordnance. To the maximum extent possible and as far as practicable, information should be recorded and retained on the use of explosive ordnance during armed conflict in order to reduce the risks posed by explosive remnants of war.[430] The best practice is to record and retain, as accurately as possible, the following information regarding explosive ordnance that may have become unexploded ordnance:

- the location of areas targeted using explosive ordnance;
- the approximate number, type, and nature of explosive ordnance used in these areas; and
- the general location of known and probable unexploded ordnance.[431]

6.19.3 Abandoning Explosive Ordnance. Where it has been necessary to abandon explosive ordnance in the course of operations, the best practice, if possible, is to leave it in a safe and secure manner.[432] In addition, to the maximum extent possible and as far as practicable, information should be recorded and retained on the abandonment of explosive ordnance in order to reduce the risks posed by explosive remnants of war.[433] The best practice is to record as accurately as possible and retain the following information regarding abandoned explosive ordnance:

• the location of AXO;

• the approximate amount of AXO at each specific site; and

• the types of AXO at each specific site.[434]

6.19.4 Storing Records on the Use or Abandonment of Explosive Ordnance. Information regarding the use or abandonment of explosive ordnance should be stored in such a manner as to allow for its subsequent retrieval in accordance with § 6.20.6 (Making Available Information on Used or Abandoned Explosive Ordnance That May Have Become Explosive Remnants of War).

6.19.5 Generic Preventive Measures During the Life-Cycle of Explosive Ordnance Aimed at Minimizing the Occurrence of Explosive Remnants of War. Generic preventive measures aimed at minimizing the occurrence of explosive remnants of war are described in part 3 of the Technical Annex to the CCW Protocol V on Explosive Remnants of War.[435] Such measures include munitions manufacturing management, munitions management, training, transfer, and future production. When producing or procuring explosive ordnance, to the extent possible, and as appropriate, these best practice measures should be implemented and respected during the life cycle of explosive ordnance.[436]

6.19.5.1 *Munitions Manufacturing Management*. The following best practices should be applied during munitions manufacturing management:

• Production processes should be designed to achieve the greatest reliability of munitions.

• Production processes should be subject to certified quality control measures.

• During the production of explosive ordnance, certified quality assurance standards that are internationally recognized should be applied.

• Acceptance testing should be conducted through live-fire testing over a range of conditions or through other validated procedures.

• High reliability standards should be required in the course of explosive ordnance transactions and transfers.[437]

6.19.5.2 *Munitions Management*. In order to ensure the best possible long-term reliability of explosive ordnance, best practice norms and operating procedures with respect to its storage, transport, field storage, and handling should be applied in accordance with the following guidance.[438]

• Explosive ordnance, where necessary, should be stored in secure facilities or appropriate containers that protect the explosive ordnance and its components in a controlled atmosphere, if necessary.[439]

• Explosive ordnance should be transported to and from production facilities, storage facilities, and the field in a manner that minimizes the risk of damage to the explosive ordnance.[440]

• Appropriate containers and controlled environments, where necessary, should be used when stockpiling and transporting explosive ordnance.[441]

• The risk of explosions in stockpiles should be minimized by the use of appropriate stockpile arrangements.[442]

• Appropriate explosive ordnance logging, tracking, and testing procedures should be applied. These procedures should include information on: (1) the date of manufacture of each number, lot, or batch of explosive ordnance; (2) where the explosive ordnance has been; (3) under what conditions the explosive ordnance has been stored; and (4) to what environmental factors the explosive ordnance has been exposed.[443]

• Periodically, stockpiled explosive ordnance should undergo, where appropriate, live-fire testing to ensure that munitions function as desired.[444]

• Sub-assemblies of stockpiled explosive ordnance should, where appropriate, undergo laboratory testing to ensure that munitions function as desired.[445]

• Where necessary, appropriate action (including adjustment to the expected shelf-life of ordnance) should be taken as a result of information acquired by logging, tracking, and testing procedures, in order to maintain the reliability of stockpiled explosive ordnance.[446]

6.19.5.3 *Training*. All personnel who are involved in handling, transporting, and using explosive ordnance should be properly trained through suitable training programs on the munitions that they will be required to handle.[447]

6.19.5.4 *Transfer*. When contemplating transfers of explosive ordnance to another State that did not previously possess that type of explosive ordnance, reasonable efforts should be made to ensure that the receiving State has the capability to store, maintain, and use that explosive ordnance correctly.[448]

6.19.5.5 *Future Production*. When producing, procuring, or developing explosive ordnance, ways and means of improving the reliability of explosive ordnance should be considered with a view to achieving the highest possible reliability.[449]

6.20 EXPLOSIVE REMNANTS OF WAR

There are certain obligations with respect to explosive remnants of war on territory under U.S. control. Most of these obligations are trigged upon the cessation of active hostilities.

6.20.1 Definition of Explosive Remnants of War. Under the CCW Protocol V on Explosive Remnants of War, "explosive remnants of war" means: (1) unexploded ordnance (UXO); and (2) abandoned explosive ordnance (AXO).[450]

6.19.1 "Unexploded ordnance" is defined in §.2 (Definition of Unexploded Ordnance (UXO)). "Abandoned explosive ordnance" is defined in § 6.19.1.3 (Definition of Abandoned Explosive Ordnance (AXO)).

6.20.2 Scope of the Obligations Created by the CCW Protocol V on Explosive Remnants of War. The CCW Protocol V on Explosive Remnants of War applies to explosive remnants of war on the land territory, including internal waters, of States that are Parties to the Protocol.[451] The CCW Protocol V on Explosive Remnants of War applies to situations resulting from armed conflict and occupation.[452]

Certain obligations in the CCW Protocol V on Explosive Remnants of War only apply to explosive remnants of war that were created after the entry into force of the Protocol for the High Contracting Party on whose territory the explosive remnants of war exist.[453] These obligations include obligations discussed in:

- § 6.20.3 (Taking Feasible Precautions to Protect Civilians From Explosive Remnants of War);
- § 6.20.6 (Making Available Information on Used or Abandoned Explosive Ordnance That May Have Become Explosive Remnants of War);
- § 6.20.7 (Clearance, Removal, or Destruction of Explosive Remnants of War From Territory Under a Party's Control); and
- § 6.20.8 (Providing Assistance to Facilitate the Removal of Explosive Remnants of War From a Party's Military Operations in Areas Outside Its Control).

6.20.3 Taking Feasible Precautions to Protect Civilians From Explosive Remnants of War. Parties to the CCW Protocol V on Explosive Remnants of War and parties to an armed conflict shall take all feasible precautions in the territory under their control affected by explosive remnants of war to protect the civilian population, individual civilians, and civilian objects from the risks and effects of explosive remnants of war.[454]

Feasible precautions are those precautions that are practicable or practically possible, taking into account all circumstances ruling at the time, including humanitarian and military considerations.[455]

These precautions may include warnings, risk education to the civilian population, and marking, fencing, and monitoring of territory affected by explosive remnants of war, as set out in Part 2 of the Technical Annex to CCW Protocol V on Explosive Remnants of War.[456]

6.20.3.1 *Best Practice Elements on Warnings and Risk Education*. Warnings are the punctual provision of cautionary information to the civilian population, intended to minimize risks caused by explosive remnants of war in affected territories.[457]

Risk education to the civilian population should consist of risk education programs to facilitate information exchange between affected communities, government authorities, and humanitarian organizations so that affected communities are informed about the threat from explosive remnants of war. Risk education programs are usually long-term activities.[458]

All programs of warnings and risk education should, where possible, take into account prevailing national and international standards, including the International Mine Action Standards.[459]

Warnings and risk education should be provided to the affected civilian population, which comprises civilians living in or around areas containing explosive remnants of war and civilians who transit such areas.[460]

Warnings should be given as soon as possible, depending on the context and the information available. A risk education program should replace a warnings program as soon as possible. Warnings and risk education always should be provided to the affected communities at the earliest possible time.[461]

Third parties, such as international organizations and non-governmental organizations, should be used when there are insufficient resources and skills to deliver efficient risk education.[462]

If possible, additional resources for warnings and risk education should be provided. Such items might include: provision of logistical support, production of risk education materials, financial support, and general cartographic information.[463]

6.20.3.2 *Best Practices on Marking, Fencing, and Monitoring Areas Containing Explosive Remnants of War.* When possible, at any time during the course of a conflict and at the earliest possible time and to the maximum extent possible, the best practice is for areas containing explosive remnants of war to be marked, fenced, and monitored so as to ensure the effective exclusion of civilians, in accordance with the following provisions.[464]

Warning signs based on methods of marking recognized by the affected community should be used in the marking of suspected hazardous areas. Signs and other hazardous area boundary markers should as far as possible be visible, legible, durable, and resistant to environmental effects and should clearly identify which side of the marked boundary is considered safe and which side is considered to be within the area affected by explosive remnants of war.[465]

An appropriate structure should be put in place with responsibility for the monitoring and maintenance of permanent and temporary marking systems, integrated with national and local risk education programs.[466]

6.20.4 Protecting, as Far as Feasible, Humanitarian Missions and Organizations From the Effects of Explosive Remnants of War. Each Party to the CCW Protocol V on Explosive Remnants of War and party to an armed conflict shall protect, as far as feasible, from the effects of explosive remnants of war, humanitarian missions and organizations that are or will be operating in the area under the control of the Party to the CCW Protocol V on Explosive Remnants of War or party to an armed conflict and with that party's consent.[467]

In addition, each Party to the CCW Protocol V on Explosive Remnants of War and party to an armed conflict shall, upon request by such a humanitarian mission or organization, provide, as far as feasible, information on the location of all explosive remnants of war that it is aware of in territory where the requesting humanitarian mission or organization will operate or is operating.[468] Other treaties or certain U.N. Security Council resolutions may provide a higher level of additional protection.[469]

6.20.5 Obligations Under the CCW Protocol V on Explosive Remnants of War That Are Triggered by the Cessation of Active Hostilities. Certain obligations under the CCW Protocol V on Explosive Remnants of War are triggered by the cessation of active hostilities. These obligations include those obligations addressed in:

- § 6.20.6 (Making Available Information on Used or Abandoned Explosive Ordnance That May Have Become Explosive Remnants of War);
- § 6.20.7 (Clearance, Removal, or Destruction of Explosive Remnants of War From Territory Under a Party's Control); and
- § 6.20.8 (Providing Assistance to Facilitate the Removal of Explosive Remnants of War From a Party's Military Operations in Areas Outside Its Control).

The determination that active hostilities have ceased for the purposes of triggering these obligations would be made at the national level.

In addition, peace treaties or similar agreements negotiated in connection with the settlement of armed conflicts may also allocate responsibilities with respect to explosive remnants of war.[470] Nothing in the CCW Protocol V on Explosive Remnants of War precludes future arrangements in connection with the settlement of armed conflicts, or assistance connected thereto, to allocate responsibilities with respect to explosive remnants of war in a manner that respects the essential spirit and purpose of the CCW Protocol V on Explosive Remnants of War.[471]

6.20.6 Making Available Information on Used or Abandoned Explosive Ordnance That May Have Become Explosive Remnants of War. Without delay after the cessation of active hostilities and as far as practicable, Parties to the CCW Protocol V on Explosive Remnants of War and parties to an armed conflict have an obligation, subject to their legitimate security interests, to make available certain information on used or abandoned explosive ordnance, which may have become explosive remnants of war, to certain recipients.[472]

6.20.6.1 *Timing of the Release of Information*. The best practice is for the information to be released as soon as possible, taking into account such matters as any ongoing military and humanitarian operations in the affected areas, the availability and reliability of information, and relevant security issues.[473]

6.20.6.2 *Recipients of the Released Information*. The information shall be made available to the party or parties in control of the affected territory and, upon request, to other relevant organizations that the party providing the information believes are, or will be, undertaking risk education and the marking and clearance, removal, or destruction of explosive remnants of war in the affected area.[474]

The best practice is to release this information to the party or parties in control of the affected territory and to those persons or institutions that the releasing State is satisfied are, or will be, involved in UXO or AXO clearance in the affected area, or in the education of the civilian population on the risks of UXO or AXO.[475]

6.20.6.3 *Mechanism for the Release of Information*. This information may be made available bilaterally (*e.g.*, directly from the United States to the recipient) or through a mutually agreed third party (*e.g.*, the United Nations).[476] In addition, a best practice would be to make use of those mechanisms established internationally or locally for the release of such information, such as through the United Nations Mine Action Service (UNMAS), the Information Management System for Mine Action (IMSMA), and other expert agencies, as considered appropriate by the releasing State.[477]

6.20.6.4 *Best Practices for the Content of Released Information on UXO*. The best practice is for the released information on UXO to contain details on:

• the general location of known and probable UXO;

• the types and approximate number of explosive ordnance used in the targeted areas;

• the method of identifying the explosive ordnance including color, size, and shape, and

other relevant markings; and

• the method for safe disposal of the explosive ordnance.[478]

This information may be drawn from the information recorded and retained when using explosive ordnance.[479]

6.20.6.5 *Best Practices for the Content of Released Information on AXO*. The best practice is for the released information on AXO to contain details on:

• the location of the AXO;

• the approximate number of AXO at each specific site;

• the types of AXO at each specific site;

• the method of identifying the AXO, including color, size, and shape;

• information on type and methods of packing for AXO;

• the state of readiness; and

• the location and nature of any booby traps known to be present in the area of AXO.[480]

This information may be drawn from the information recorded and retained when abandoning explosive ordnance.[481]

6.20.7 Clearance, Removal, or Destruction of Explosive Remnants of War From Territory Under a Party's Control. After the cessation of active hostilities and as soon as feasible, each Party to the CCW Protocol V on Explosive Remnants of War and party to an armed conflict shall mark and clear, remove, or destroy explosive remnants of war in affected territories under its control.[482] This obligation is consistent with long-established customary international law principles of sovereignty and responsibility.[483] In particular, the following measures are to be taken in such territories:

• surveying and assessing the threat posed by explosive remnants of war;

• assessing and prioritizing the needs and practicability in terms of marking and clearance, removal, or destruction;

• marking and clearing, removing, or destroying explosive remnants of war; and

• taking steps to mobilize resources to carry out these activities.[484]

In conducting these activities, international standards, including the International Mine Action Standards, should be taken into account.[485] Areas affected by explosive remnants of war that are assessed to pose a serious humanitarian risk are to be accorded priority status for clearance, removal, or destruction.[486]

Territories under its control include territory over which the State is an Occupying Power. An Occupying Power assumes responsibility for the safety of the civilian population, including responsibilities associated with explosive remnants of war.[487]

6.20.8 Providing Assistance to Facilitate the Removal of Explosive Remnants of War From a Party's Military Operations in Areas Outside Its Control. In cases where a user of explosive ordnance that has become explosive remnants of war does not exercise control of the territory, the user shall, after the cessation of active hostilities, provide where feasible, *inter alia*, technical, financial, material or human resources assistance, bilaterally or through a mutually agreed third party, including, *inter alia*, through the United Nations system or other relevant organizations, to facilitate the marking and clearance, removal, or destruction of such explosive remnants of war.[488]

VII –Wounded, Sick, Shipwrecked, Dead, and the Medical Services

Chapter Contents

7.1 INTRODUCTION

This Chapter addresses the protection of the wounded, sick, shipwrecked, and dead, both on the battlefield and after having fallen into the hands of the enemy during international armed conflict. It also addresses the rights, duties, and liabilities of military medical and religious personnel, and the protection of military medical units, facilities, and transports. Lastly, it addresses the protection of civilians who are wounded, sick, infirm, or maternity cases, and civilian hospitals.

Hospital and safety zones are addressed in Chapter V. [1] Rights and obligations of neutral States with respect to the wounded, sick, and shipwrecked are addressed in Chapter XV. [2] Rules for the protection of medical personel and the wounded, sick, and shipwrecked during non- international armed conflict are addressed in Chapter XVII.[3]

7.1.1 Interpretation and Application of Provisions Relating to Medical Issues in the GWS, GWS-Sea, GPW, and GC. The GWS and GWS-Sea underlie most of the international law rules applicable to the United States for: (1) the protection of the wounded, sick, shipwrecked, and dead; (2) the rights, duties, and liabilities of military medical and religious personnel; and (3) the protection of military medical units, facilities, and transports.

In addition to the rules of the GWS and GWS-Sea, the GPW also has rules on the medical treatment of POWs that may apply to persons who are wounded and sick.[4] The GWS and GWSSea, however, have rules for the protection of the wounded, sick, shipwrecked, and dead who are not POWs (*e.g.*, they are on the battlefield and not in the power of the adverse party, they died without being captured).

The GC has provisions relating to the protection of civilians who are wounded, sick, infirm, or maternity cases, and civilian hospitals. [5] The GC also has rules addressing medical care for civilians in occupied territory.[6]

7.1.1.1 *Interpretation of the GWS and GWS-Sea.* As with the GPW and GC, the provisions of the GWS and GWS-Sea should be interpreted in light of the principles that underlie the treatment of the wounded, sick, shipwrecked, dead, and the medical services, and, in particular, in light of the goal of promoting the care of the wounded and sick.[7] The subsequent practice of States in applying the GWS and GWS-Sea could also assist in interpreting their provisions because States' decades of experience in applying the GWS and GWS-Sea may be very helpful in understanding their requirements.[8]

7.1.1.2 *Special Agreements Under the GWS and GWS-Sea.* Under the GWS and GWS-Sea, States may conclude a variety of special agreements during international armed conflict to facilitate the protection of the wounded, sick, shipwrecked, and others protected by the GWS and GWS-Sea. The GWS and GWS-Sea specifically provide for agreements:

• to entrust to an effective and impartial organization the duties of the Protecting Powers;[9]

• to establish an armistice or suspension of fire, or local arrangement to permit the removal, exchange, and transport of the wounded, sick, and shipwrecked;[10]

• to establish and recognize hospital zones and localities;[11]

• to establish the corresponding ranks for the medical personnel, including the staff of any authorized voluntary aid societies, in order to help determine the senior retained medical officer in POW camps;[12]

• to put neutral observers on board ships to verify the strict observation of provisions of the GWS-Sea or verify the equipment carried on board is exclusively for the treatment of the wounded and sick, or for the prevention of disease;[13]

• to determine the percentage of personnel to be retained, in proportion to the number of POWs, and the distribution of the retained personnel in POW camps;[14]

• to establish flights or additional markings for medical aircraft;[15]

• to approve chartered medical transport ships;[16]

• between the belligerent parties and a neutral Power on the wounded, sick, and shipwrecked who are disembarked onto the neutral Power's territory;[17]

• to use the most modern methods available to facilitate the identification of hospital ships;[18]and

• on the procedure, or to select an umpire who will decide the procedure, to be followed for an inquiry concerning any alleged violation of the GWS or GWS-Sea.[19]

In addition to these agreements expressly provided for in the GWS and GWS-Sea, Parties to the GWS or GWS-Sea may conclude other special agreements for all matters concerning which they may deem it suitable to make separate provision.[20]

No special agreement shall adversely affect the situation of wounded, sick, and shipwrecked persons, of members of the medical personnel, or of chaplains, as defined by the GWS or the GWS-Sea, nor restrict the rights that the GWS or GWS-Sea confer upon them.[21]

Wounded, sick, and shipwrecked persons, as well as medical personnel and chaplains, shall continue to have the benefit of such agreements as long as the GWS or GWS-Sea is applicable to them, except where express provisions to the contrary are contained in these or in subsequent agreements, or where more favorable measures have been taken with regard to them by one or other of the parties to the conflict.[22]

7.2 APPLICATION OF THE PROTECTIONS OF THE GWS AND GWS-SEA

7.2.1 Circumstances and Duration of Application of the GWS and GWS-Sea. In general, the GWS-Sea addresses the protection of persons at sea, while the GWS applies to persons on land. In case of hostilities between land and naval forces of parties to the conflict, the provisions of the GWS-Sea shall apply only to forces on board ship.[23]

For the wounded and sick, the GWS and GWS-Sea provide corresponding protections until these persons no longer suffer from that condition. For example, the GWS would no longer cover a detained member of the armed forces who recovers from his or her wounds, although the GPW would continue to apply to that person because he or she is a POW.[24]

For protected persons who have fallen into the hands of the enemy, the GWS shall apply until their final repatriation.[25]The GWS-Sea does not specify when it ceases to apply, but once persons who are covered by its provisions reach land, the GWS applies. Forces put ashore shall immediately become subject to the provisions of the GWS.[26]If dead persons are landed, the provisions of the GWS shall be applicable.[27]Retained personnel at sea shall be subject, on landing, to the provisions of the GWS.[28]

7.2.2 Non-Renunciation of Rights Secured by the GWS or GWS-Sea. Wounded, sick, and shipwrecked persons, as well as military medical and religious personnel, may in no circumstances renounce in part or in entirety the rights secured to them by the GWS or the GWSSea, or by special agreements, if any, referred to in Article 6 of the GWS or the GWS-Sea.[29]A similar provision of the GPW applies to POWs, and a similar provision of the GC applies to protected persons.[30]

This rule is intended to prevent States from evading their obligations by coercing enemy nationals in their power to waive their rights.[31]The prohibition on the non-renunciation of rights is based on a recognition that: (1) persons in the hands of the enemy are in a vulnerable position; (2) it would be difficult to establish whether they had voluntarily renounced their rights; and (3) an absolute prohibition would best serve the interests of the majority.[32]

This rule does not prohibit States from reassigning their own military medical and religious personnel to non-medical duties or from transferring their own personnel outside of the medical service. Similarly, this rule constitutes an obligation on States and does not directly bind individuals.[33]

7.2.3 Reprisals Prohibited. Reprisals against the wounded, sick, or shipwrecked persons, personnel, buildings, vessels, or equipment protected by the GWS or GWS-Sea are prohibited.[34]

7.3 RESPECT AND PROTECTION OF THEWOUNDED, SICK, AND SHIPWRECKED

Members of the armed forces and other persons mentioned in Article 13 of the GWS and the GWS-Sea, who are wounded, sick, or shipwrecked, shall be respected and protected in all circumstances.[35]Such persons are among the categories of persons placed *hors de combat*; making them the object of attack is strictly prohibited.[36]

7.3.1 Definitions of Wounded, Sick, and Shipwrecked.

7.3.1.1 *Wounded or Sick*. The meanings of "wounded" and "sick" are a matter of common sense and good faith.[37]They include:

- combatants who have been rendered unconscious or otherwise have been incapacitated because of their wounds or sickness;[38]and
- combatants who have surrendered as a consequence of their health.[39]

To merit respect and protection as "wounded" or "sick" under the law of war, combatants must abstain from hostile acts or attempts to evade capture.[40]

7.3.1.2 *Shipwrecked*. For the purpose of applying the protections afforded by the GWS-Sea, the term "shipwreck" means shipwreck from any cause and includes forced landings at sea by or from aircraft.[41]

The shipwrecked may be understood to include those in distress at sea or stranded on the coast who are also helpless.[42]To be considered "shipwrecked," persons must be in need of assistance and care, and they must refrain from any hostile act.[43]For example, shipwrecked persons do not include combatant personnel engaged in amphibious, underwater, or airborne attacks who are proceeding ashore.[44]

7.3.2 Persons Entitled to Protection as Wounded, Sick, or Shipwrecked Under the GWS and GWS-Sea. The GWS and GWS-Sea apply to the wounded, sick, and shipwrecked belonging to the following categories:[45]

- members of the armed forces of a party to the conflict, as well as members of militias or volunteer corps forming part of such armed forces;[46]
- members of other militias and members of other volunteer corps meeting certain conditions;[47]
- members of regular armed forces who profess allegiance to a Government or an authority not recognized by the Detaining Power;[48]
- persons who are authorized to accompany the armed forces;[49]
- members of the crews of the merchant marine and the crews of civil aircraft of the parties to the conflict;[50]and
- participants in a *levée en masse*.[51]

These categories are the same categories of persons to whom the GPW affords POW status if they fall into the hands of the enemy during international armed conflict.[52]Thus, the GWS and GWS-Sea recognize that such persons who fall into enemy hands during international armed conflict shall be POWs, and the provisions of international law concerning POWs (such as the GPW) shall also apply to them.[53]

The categories of persons entitled to POW status under the GPW and protection as wounded, sick, and shipwrecked under the GWS and GWS-Sea were drafted to be the same, but the GWS or GWS-Sea do not refer specifically to Article 4 of the GPW so that the GWS and GWS-Sea could operate as stand-alone legal instruments if a State ratified either the GWS or the GWS-Sea, but not the GPW.[54]

Civilians who are wounded and sick are protected under the GC.[55]

7.3.3 Meaning of "Respect and Protection" of the Wounded, Sick, and Shipwrecked. The wounded, sick, and shipwrecked must be respected and protected at all times. This means that they should not be knowingly attacked, fired upon, or unnecessarily interfered with.[56]

7.3.3.1 *Incidental Harm Not Prohibited.* The respect and protection due to the wounded, sick, and shipwrecked do not prohibit incidental damage or casualties due to their proximity to military objectives or to a justifiable mistake. Combatants who are wounded, sick, or shipwrecked on the battlefield are deemed to have accepted the risk of death or further injury due to their proximity to military operations.[57]Although the presence of the wounded, sick, or shipwrecked on the battlefield does not serve to exempt military objectives from attack due to the risk that such personnel would be incidentally harmed, feasible precautions must be taken to reduce the risk of harm to the wounded, sick, or shipwrecked.[58]

7.3.3.2 *Search and Other Security Measures Not Prohibited.* The wounded, sick, and shipwrecked are not immune from search or other necessary security measures by the enemy.[59]Searches of the wounded, sick, and shipwrecked, and other security measures applied to them, should be consistent with those applicable to other POWs.[60]

7.3.3.3 *Capture of Wounded, Sick, and Shipwrecked Not Prohibited.* The respect and protection afforded the wounded, sick, and shipwrecked do not immunize them from detention.[61]When they fall into enemy hands, they are POWs.[62]

It is not prohibited to capture the wounded and sick, even if they are in the care of military medical units or facilities, or civilian hospitals.[63]However, upon taking custody, the Detaining Power is required to ensure that they continue to receive the necessary medical care.[64]All warships of a belligerent party shall have the right to demand that the wounded, sick, or shipwrecked on board military hospital ships, and hospital ships belonging to relief societies or to private individuals, as well as merchant vessels, yachts, and other craft shall be surrendered, whatever their nationality, provided that the wounded and sick are in a fit state to be moved and that the warship can provide adequate facilities for necessary medical treatment.[65]

In cases of capture at sea, the captor may decide, according to circumstances, whether it is expedient to hold the wounded, sick, or shipwrecked, or to convey them to a port in the captor's own country, to a neutral port, or even to a port in enemy territory. In the last case, POWs thus returned to their home country may not serve for the duration of the war.[66]

7.3.3.4 *Prohibition Against Use in Booby-Traps.* Booby trapping the wounded or sick is expressly prohibited.[67]

7.4 SEARCH, COLLECTION, AND AFFIRMATIVE PROTECTION OF THE WOUNDED, SICK, SHIPWRECKED, AND DEAD

At all times, and particularly after an engagement, parties to the conflict shall, without delay, take all possible measures to search for and collect the wounded and sick on land, to protect them against pillage and ill-treatment, to ensure their adequate care, and to search for the dead and prevent their being despoiled.[68]

The GC provides for the obligation to search for, collect, protect, and care for, civilians who are wounded, sick, shipwrecked, and dead.[69]

7.4.1 GWS-Sea Obligation Regarding the Search, Collection, and Affirmative Protection of the Wounded, Sick, Shipwrecked, and Dead. After each engagement, parties to the conflict shall, without delay, take all possible measures to search for and collect the wounded, sick, and shipwrecked at sea, to protect them against pillage and ill-treatment, to ensure their adequate care, and to search for the dead and prevent their being despoiled.[70]

The obligation in the GWS-Sea to search for, and collect, certain persons is written differently than the comparable obligation in the GWS. Instead of a general obligation to take measures "at all times," in Article 15 of the GWS, in Article 18 of the GWS-Sea, the obligation to search for and collect the wounded, sick, and shipwrecked applies only "after each engagement."

7.4.2 Affirmative Measures to Protect Against Pillage and Ill-Treatment. If practicable, affirmative measures (including, in some cases, the use of force) must be taken to protect the wounded, sick, and shipwrecked, from pillage or ill-treatment by any person, whether military or civilian, seeking to harm them.[71]

7.4.3 Range of Measures to Search for, Collect, and Protect the Wounded, Sick, and Shipwrecked. A range of measures may be taken to fulfill the obligation to search for, collect, and protect the wounded, sick, and shipwrecked. Military forces may directly engage in these activities.[72]

In addition to searching for, collecting, and protecting the wounded, sick, and shipwrecked directly, commanders may take other measures to fulfill this obligation. For example, commanders may request the help of civilian volunteers.[73] As another example, if a warship is unable to collect the shipwrecked after an engagement, it might be able to alert a hospital ship in the vicinity or provide the shipwrecked with a lifeboat.[74]

7.4.3.1 *Armistices and Local Arrangements to Permit the Removal, Exchange, or Transport of the Wounded.* Whenever circumstances permit, an armistice or a suspension of fire shall be arranged, or local arrangements made, to permit the removal, exchange, and transport of the wounded left on the battlefield.[75] Such arrangements may take the form of or include a protected or neutral zone.[76]

Likewise, local arrangements may be concluded between parties to the conflict for the removal or exchange of wounded and sick by land or sea from a besieged or encircled area, or for the passage of medical and religious personnel and equipment on their way to that area.[77]

For example, parties to a conflict may agree to a temporary cease-fire to permit evacuation of the wounded from the fighting area.[78]

7.4.4 Practical Limitations on the Obligation to Search for, Collect, and Take Measures to Protect the Wounded, Sick, and Shipwrecked. The obligations to search for, collect, and take affirmative steps to protect the wounded, sick, and shipwrecked are subject to practical limitations.

Military commanders are to judge what is possible, and to what extent they can commit their personnel to these duties.[79] In some cases, commanders might designate specific units or personnel to engage in such missions.

7.4.4.1 *Risk of Harm.* For example, personnel performing rescue and recovery missions need not place their lives at undue risk in an effort to search for, collect, or protect the wounded, sick, shipwrecked, or dead (*e.g.*, recovery of a body from a minefield, or entry into a disabled enemy armored vehicle that might contain unexploded ordnance or other hazards).

Similarly, a commander of a naval ship need not increase the risk to his or her vessel from threats (*e.g.*, by slowing his or her transit, or by placing his or her ship dead in the water) in order to recover shipwrecked enemy military personnel from a sunken vessel or crashed aircraft.[80]

7.4.4.2 *Military Necessity.* Similarly, the requirements of ongoing military operations may render rescue efforts impractical. For example, during a fast-tempo operation (offensive or defensive), it might not be possible to devote resources to the search and collection of the wounded, sick, and shipwrecked. In other cases, the rescue of enemy personnel may exceed the abilities of the force and its medical personnel. For example, a small patrol operating behind enemy lines or a submarine may not have the capability to receive and care for large numbers of injured personnel.[81]

7.4.5 Collection and Care of the Wounded, Sick, and Shipwrecked by Civilians.

7.4.5.1 *Requests for Civilian Volunteers to Collect and Care for the Wounded, Sick, and Shipwrecked.* The military authorities may appeal to the charity of the inhabitants voluntarily to collect and care for, under their direction, the wounded, sick, and shipwrecked, granting persons who have responded to this appeal the necessary protection and facilities.[82]

Should the adverse party take or retake control of the area, that party shall likewise grant these persons the same protection and the same facilities.[83]

7.4.5.2 *Spontaneous Collection and Care for the Wounded, Sick, and Shipwrecked by Inhabitants and Relief Societies.* The military authorities shall permit the inhabitants and relief societies, even in invaded or occupied areas, spontaneously to collect and care for wounded or sick of whatever nationality.[84] The civilian population shall respect these wounded and sick, and in particular abstain from offering them violence.[85]

7.4.5.3 *Prohibition on Molestation or Conviction for Nursing the Wounded or Sick.* No one shall ever be molested or convicted for having nursed the wounded or sick.[86] For example, persons should not be prosecuted for offenses of aiding the enemy because they tended to wounded members of enemy military forces.

7.4.6 Collection and Care of the Wounded, Sick, and Shipwrecked by Neutral Vessels. The parties to the conflict may appeal to the charity of commanders of neutral merchant vessels, yachts, or other craft, to take on board and care for wounded, sick, or shipwrecked persons, and to collect the dead.[87]

Vessels of any kind responding to this appeal, and those having of their own accord collected wounded, sick, or shipwrecked persons, shall enjoy special protection and facilities to carry out such assistance.[88]

They may, in no case, be captured on account of any such transport; but, in the absence of any promise to the contrary, they shall remain liable to capture for any violations of neutrality they may have committed.[89]

7.5 HUMANE TREATMENT AND CARE OF ENEMY MILITARY WOUNDED, SICK, AND SHIPWRECKED IN THE POWER OF A PARTY TO THE CONFLICT

The wounded, sick, and shipwrecked who are protected by the GWS and GWS-Sea shall be treated humanely and cared for by the party to the conflict in whose power they may be, without any adverse distinction founded on sex, race, nationality, religion, political opinions, or any other similar criteria.[90]Any attempts upon their lives, or violence to their persons, shall be strictly prohibited; in particular, they shall not be murdered or exterminated or subjected to torture or to biological experiments; they shall not willfully be left without medical assistance and care, nor shall conditions exposing them to contagion or infection be created.[91]

7.5.1 Humane Treatment of the Wounded, Sick, and Shipwrecked Without Adverse Distinction. The wounded, sick, and shipwrecked who are protected by the GWS or GWS-Sea who are in the power of a party to the conflict are also POWs. Thus, their humane treatment is required by the GPW as well.[92]

For example, the prohibition on subjecting the wounded, sick, or shipwrecked to biological experiments is also reflected in the GPW.[93]Similarly, the wounded, sick, and shipwrecked who are women shall be treated with all consideration due to their sex.[94]

7.5.2 Medical Care of the Wounded and Sick. Below are general rules for the medical care of the wounded and sick, including both those on the battlefield and those who are POWs. POWs shall be provided necessary medical attention during evacuation, although they may be temporarily kept back in a danger zone if they would face greater risks by being evacuated than by remaining where they are.[95]Additional rules provide for the medical attention and care for POWs who have been interned in POW camps.[96]

7.5.2.1 *Prohibitions Against Willful Medical Neglect or Deliberate Endangerment*. The wounded, sick, and shipwrecked shall not willfully be left without medical assistance and care. In addition, conditions exposing them to contagion or infection shall not be willfully created.

7.5.2.2 *Affirmative Obligation to Provide Adequate Care*. In addition to the prohibitions against willful neglect or deliberate endangerment, the parties to a conflict shall take all possible measures to ensure the adequate care of the wounded, sick, and shipwrecked.[97]The obligation to care for enemy combatants who are wounded and sick is a longstanding law of war obligation.[98]

The obligation to provide medical care incorporates practical considerations; whether resources may be committed to medical care may depend on military necessity, such as the requirements of the mission or the immediate tactical situation.[99]

7.5.2.3 *Priority of Treatment Based on Medical Urgency*. Only urgent medical reasons will authorize priority in the order of treatment to be administered.[100]

For example, in addressing an influx of wounded that include both friends and enemies, doctors should attend to those patients for whom delay might be fatal or at any rate prejudicial, proceeding afterwards to those whose condition is not such as to necessitate immediate attention.[101]

7.5.2.4 *Compulsory Medical Treatment*. The wounded, sick, and shipwrecked, and other POWs, may be ordered to receive medical treatment or care that is warranted by their medical condition. Because POWs are subject to the laws, regulations, and orders in force in the armed forces of the Detaining Power, POWs may be ordered to receive medical treatment just as Detaining Power military personnel may be ordered to do so.[102]However, the wounded, sick, and shipwrecked, and other POWs, may not be subjected to medical or biological experiments, even if Detaining Power military personnel could be ordered to be subjected to such procedures.[103]

The Detaining Power's authority to order POWs to receive medical treatment also derives from the Detaining Power's duty to ensure the well-being of POWs.[104]For example, the Detaining Power's duty to take all sanitary measures necessary to prevent epidemics in the POW camp may require the Detaining Power to order POWs to receive certain measures, such as vaccinations.[105]

AP I, Article 11, paragraph 5 provides for a "right to refuse any surgical operation."[106]

However, this rule could operate in an inhumane manner in some circumstances, and the United States has not accepted it.[107]

7.5.2.5 *Medical Attention Free of Charge*. The Power detaining POWs shall be bound to provide free of charge for the medical attention required by POWs' state of health.[108]

7.5.3 Abandonment of the Wounded or Sick on the Battlefield. In an extreme case, military forces in an international armed conflict may be compelled to abandon wounded or sick to the enemy.[109]The party to the conflict that is compelled to abandon wounded or sick to the enemy shall, as far as military considerations permit, leave with them a part of its medical personnel and material to assist in their care.[110]

Unlike the obligations to provide humane treatment for the wounded and sick who are in custody, the obligation to provide medical personnel and material for the care of abandoned wounded or sick is a strong recommendation, rather than an absolute requirement.[111]

The party that encounters and captures the wounded and sick who have been abandoned by another party to the conflict is not relieved of its duties because the party that has abandoned wounded and sick personnel has not left medical personnel and material to assist in the care of those wounded and sick.[112]On the other hand, the party that encounters and captures the wounded and sick also is not relieved of its duties when the party that has abandoned wounded and sick personnel has left medical personnel and material to assist in the care of those wounded and sick.[113]

In addition to providing for the medical care of wounded and sick abandoned to the enemy, other feasible precautions should be taken to ensure their safety.[114]

7.6 ACCOUNTABILITY INFORMATION FROM THE ENEMY MILITARY WOUNDED, SICK, SHIPWRECKED, AND DEAD

The GWS and GWS-Sea impose certain obligations to gather and report information in order to gain as full an accounting as possible for captured enemy personnel and enemy dead recovered from the battlefield. These obligations complement similar obligations in the GPW to maintain accountability for POWs.[115]

7.6.1 Recording Identifying Information. Parties to the conflict shall record as soon as possible, in respect of each wounded, sick, shipwrecked, or dead person of the adverse party falling into their hands, any particulars that may assist in his or her identification.[116]These records should, if possible, include:[117]

- designation of the Power on which the person depends;
- surname (*i.e.*, last name) and first name or names (*i.e.*, first and middle names);
- army, regimental, personal, or serial number;
- date of birth;
- any other particulars shown on the person's identity card or disc;[118]
- date and place of capture or death; and
- particulars concerning wounds or illness, or cause of death.

The GPW requires the collection of a similar set of information for all POWs.[119]

In order to implement these obligations, as part of their initial processing, the recovering or capturing unit should endeavor to obtain this information from each living POW.[120]POWs who, owing to their physical or mental condition, are unable to state their identity, shall be handed over to the medical service.[121]

Other sources of information should be used to establish the identity of the wounded, sick, shipwrecked, and dead. For example, items associated with the person (*e.g.*, personal effects, a life jacket bearing the name of the ship on which that person served) should be examined, and other members of the person's unit should be questioned. An examination of a deceased person's body may also facilitate identification by gathering information that can be used to identify the person later (*e.g.*, measurements and description of the body and its physical features, examination of the teeth, fingerprints, photograph, DNA samples).[122]This examination should be conducted before burial or cremation.[123]

7.6.2 Forwarding Identifying Information. As soon as possible, the above mentioned identifying information shall be forwarded to the National POW Information Bureau described in Article 122 of the GPW, which shall transmit this information to the Power on which these persons depend through the intermediary of the Protecting Power and of the Central POW Agency.[124]

7.6.2.1 *Military Necessity to Delay Reporting*. As the phrase "as soon as possible" reflects, reporting may be delayed due to practical considerations, including security risks.[125]

7.6.3 U.S. Practice in Reporting Through the ICRC Central Tracing Agency. In U.S. practice, information has been provided to the Theater Detainee Reporting Center, which in turn processes it and reports it to the National Detainee Reporting Center, which has then transmitted it to the ICRC Central Tracing Agency, acting as the Central POW Information Agency.[126]

7.7 TREATMENT AND HANDLING OF ENEMY MILITARY DEAD The GWS and GWS-Sea address the treatment and handling of enemy dead who have been collected from the battlefield.[127]However, the provisions of the GWS and GWS-Sea relating to the treatment and handling of enemy dead may also be interpreted to include wounded and sick combatants who have died shortly after having been collected.[128]The GPW provides other rules relating to POWs who have died while in detention.[129]Thus, certain enemy dead may be addressed by both the GPW and GWS.

Although the provisions of the GWS and GWS-Sea are not applicable to civilian dead, in many cases it may be appropriate and practical to apply those rules by analogy. The GC

provides rules relating to protected persons who have died while interned.[130]

7.7.1 Treatment of Enemy Military Dead. Subject to the tactical situation and sanitation requirements, the handling and burial of enemy military dead must be discharged with the same respect as would be afforded to, or expected for, friendly military dead. The respectful treatment of the dead is one of the oldest rules in the law of war.[131]

7.7.1.1 *No Disrespectful or Degrading Treatment of the Dead.* Enemy military dead must be protected from disrespectful or degrading acts.[132]For example, mutilation or cannibalism of dead bodies is prohibited.[133] In addition, posing with bodies for photographs or leaving a “calling card” on a body are also inconsistent with the respectful treatment of the dead.

7.7.1.2 *Prohibition Against Booby-Trapping Dead Bodies.* The dead may not be booby-trapped.[134]

7.7.2 Handling of Personal Effects Belonging to Enemy Dead. During international armed conflict, personal property recovered from enemy dead becomes the property of the recovering State for the purpose of returning it to the next-of-kin of the deceased.

7.7.2.1 *Measures to Prevent the Dead From Being Despoiled.* At all times, and particularly after an engagement, parties to the conflict shall, without delay, take all possible measures to prevent the dead from being despoiled.[135]This means that the personal effects of deceased persons may not be pillaged or otherwise taken without proper authorization.[136]

Individuals are not entitled to retain the personal effects of a deceased person as “booty”; the individual military member or person accompanying the armed forces who captures or finds enemy property acquires no title or claim.[137]

7.7.2.2 *Forwarding Valuable Articles Found on the Dead.* Parties to the conflict shall collect and forward, through the National POW Information Bureau, one-half of the double identity disc, last wills or other documents of importance to the next of kin, money, and in general all articles of an intrinsic or sentimental value, which are found on the dead.[138]

If there is only a single identity disc, it should remain with the body if the body is to be buried on land, so that the body may subsequently be identified as needed.[139]If the body is to be buried at sea, then the single identity disc itself should be collected and forwarded through the National POW Information Bureau.[140]

These articles, together with unidentified articles, shall be sent in sealed packets, accompanied by statements giving all particulars necessary for the identification of the deceased owners, as well as by a complete list of the contents of the parcel.[141]

The GPW also requires the forwarding of wills and all personal valuables left by POWs, including those who have died.[142]

7.7.3 Preparation and Forwarding of Death Certificates and Lists of the Dead. Parties to the conflict shall prepare and forward to each other, through the National POW Information Bureau, certificates of death or duly authenticated lists of the dead.[143]Where applicable, these certificates or lists shall state the circumstances and reasons for cremation.[144]

The GPW provides further requirements for death certificates and lists of persons who died as POWs.[145]

7.7.4 Handling of Enemy Battlefield Dead – Burial and Cremation.

7.7.4.1 *Examination of Bodies Before Burial or Cremation.* Parties to the conflict shall ensure that burial (whether on land or at sea) or cremation of the dead, carried out individually as far as circumstances permit, is preceded by a careful examination, if possible by a medical examination, of the bodies, with a view to confirming death, establishing identity, and enabling a report to be made.[146]

An examination helps ensure that persons who are still living are not mistakenly buried or cremated. An examination also helps ensure that accountability information can be gathered and forwarded in accordance with other requirements.[147]

7.7.4.2 *Burial on Land.* Parties to the conflict shall further ensure that the dead are honorably interred, if possible according to the rites of the religion to which they belonged, and that their graves are respected, grouped if possible according to the nationality of the deceased, properly maintained, and marked so that they may always be found.[148]The use of collective graves should be avoided.[149]

One-half of the double identity disc, or the identity disc itself if it is a single disc, should remain on the body.[150]

7.7.4.3 *Burial at Sea.* Burial at sea is contemplated by the GWS-Sea. Burial at sea shall be conducted honorably and, if possible, according to the rites of the religion to which the dead belonged.

Where a double identity disc is used, one-half of the disc should remain on the body.[151]

On the other hand, if a single identity disc is used, then it should be collected and forwarded through the National POW Information Bureau along with other valuable articles.[152]

An examination and report are especially important in such cases in order to ensure that accountability information is collected and forwarded.[153]The position (latitude and longitude) and the date should be recorded in the death certificate.[154]

7.7.4.4 *Cremation.* Bodies shall not be cremated except for imperative reasons of hygiene or for motives based on the religion of the deceased.[155]For example, hygiene reasons may include the risk of contamination to recovery personnel or other personnel from disease or exposure to chemical or biological warfare agents. In case of cremation, the circumstances and reasons for cremation shall be stated in detail in the death certificate or on the authenticated list of the dead.[156]

7.7.5 Graves Registration Service. In order to ensure that graves may always be found, parties to the conflict shall organize at the commencement of hostilities an Official Graves Registration Service to allow subsequent exhumations and to ensure the identification of bodies, whatever the site of the graves, and the possible transportation to the home country.[157]The Graves Registration Service shall also identify, record, and respectfully keep the ashes until they can be disposed of in accordance with the wishes of the home country.[158]

The GPW provides additional requirements for the Graves Registration Service with respect to POWs who have died while POWs.[159]

7.7.5.1 *Exchange of Information Regarding Battlefield Graves.* As soon as circumstances permit, and at the latest at the end of hostilities, the Graves Registration Services of the belligerents shall exchange, through the National POW Information Bureaus, lists showing the exact location and markings of the graves, together with particulars of the dead interred therein.[160]

7.8 RESPECT AND PROTECTION OF CATEGORIES OF MEDICAL AND RELIGIOUS PERSONNEL

Certain categories of medical and religious personnel shall be respected and protected during military operations.

7.8.1 Categories of Persons Who Are Entitled to Respect and Protection as Medical and Religious Personnel on the Battlefield. The categories of persons who are entitled to respect and protection as medical and religious personnel during military operations include:

• military medical and religious personnel, including[161]

o medical personnel exclusively engaged in medical duties;[162]

o administrative staff exclusively engaged in support to medical units;[163]and

o chaplains attached to the armed forces;[164]

• auxiliary medical personnel, if they are carrying out their medical duties at the time when they come into contact with the enemy or fall into the enemy's hands;[165]

• authorized staff of voluntary aid societies;[166]

• staff of a recognized aid society of a neutral country;[167]and

• the religious, medical, and hospital personnel of hospital ships and their crews.[168]

7.8.2 Meaning of "Respect and Protection" of Medical and Religious Personnel. The respect and protection accorded medical and religious personnel by the GWS and GWS-Sea mean that they must not knowingly be attacked, fired upon, or unnecessarily prevented from discharging their proper functions.[169]

7.8.2.1 *Incidental Harm Not Prohibited.* The incidental killing or wounding of such personnel, due to their presence among or in proximity to combatant elements actually engaged by fire directed at the latter, gives no just cause for complaint.[170]Medical and religious personnel are deemed to have accepted the risk of death or injury due to their proximity to military operations.[171]Although the presence of medical and religious personnel does not serve to exempt nearby military objectives from attack due to the risk that military medical and religious personnel would be incidentally harmed, feasible precautions must be taken to reduce the risk of harm to military medical and religious personnel.[172]

7.8.2.2 *Search and Other Security Measures Not Prohibited.* Medical and religious personnel are not immune from search or other necessary security measures by the enemy.[173]Searches of medical and religious personnel, and other security measures applied to them, should be consistent with the rules applicable when such actions are taken with respect to POWs.[174]

7.8.2.3 *Capture Not Prohibited.* The respect and protection afforded medical and religious personnel do not immunize them from detention.[175]However, certain classes of medical and religious personnel are, under other rules, exempt from detention or capture.[176]

7.8.3 Loss of Protection for Medical and Religious Personnel From Being Made the Object of Attack. Medical and religious personnel forfeit their protected status and therefore may be made the object of attack if they, outside of their humanitarian duties, participate in hostilities or otherwise commit acts harmful to the enemy.[177]

For example, weapons may not be used by medical and religious personnel against enemy military forces except in self-defense or defense of their patients.[178]Similarly, medical or religious personnel who, during combat, drive vehicles carrying able-bodied combatants or ammunition for able-bodied combatants, forfeit their protection from being made the object of attack by enemy military forces.[179]Hampering or impeding enemy military operations also would be a basis for finding that medical or religious personnel have forfeited protection.[180]Acts that are part of their humanitarian duties, such as caring for the wounded and sick, are not a basis for depriving medical and religious personnel of their protection.[181]

7.8.4 Use of the Distinctive Emblem to Facilitate the Respect and Protection of Military Medical and Religious Personnel. The GWS contemplates the use of the distinctive emblem, including through wearing an armlet bearing the emblem, to facilitate the identification of protected medical and religious personnel, as such. Claiming protected medical status (including feigning such status) in order to deceive the adversary into affording protection so as then to kill or wound the enemy is prohibited.[182]As with other instances of the display of the distinctive emblem, the wearing of the armlet with the distinctive emblem is controlled by the competent military authority, and the armlet may be obscured or removed at the competent military authority's direction.[183]The fact that military medical and religious personnel do not wear an armlet does not deprive them of their protection from being made the object of attack if they are recognized as such.[184]

7.8.4.1 *Wearing of Armlet With the Distinctive Emblem.* The GWS contemplates that

• military medical and religious personnel;[185]

• authorized staff of voluntary aid societies;[186]and

• staff of a recognized aid society of a neutral country;[187]

shall wear, affixed to the left arm, a water-resistant armlet bearing the distinctive emblem, issued and stamped by the military authority.188 [188]In no circumstances may these personnel be deprived of the right to wear the armlet.[189]

7.8.4.2 *Wearing of Armlet With Miniature Distinctive Emblem.* Auxiliary medical personnel shall wear, but only while carrying out medical duties, a white armlet bearing in its center the distinctive sign in miniature; the armlet shall be issued and stamped by the military authority.[190]

7.9 CAPTURED MEDICAL AND RELIGIOUS PERSONNEL

The treatment of captured persons who perform medical or religious duties depends on their legal status. Identification cards are used to help determine the correct status of medical and religious personnel. Retained personnel are subject to certain duties, rights, and privileges during detention.

7.9.1 Different Classes of Persons Performing Medical or Religious Duties. Persons who perform medical or religious duties may be grouped into three categories based on their treatment with respect to detention: (1) medical and religious personnel who are exempt from capture; (2) medical and religious personnel who may be retained; and (3) persons who perform medical and religious duties who are held as POWs.

7.9.1.1 *Medical and Religious Personnel Who Are Exempt From Capture and Detention*. Certain classes of medical and religious personnel are exempt from capture and detention. These classes of personnel include:

• staff of a recognized aid society of a neutral country;[191] and

• the religious, medical, and hospital personnel of hospital ships and their crews during the time they are in service of the hospital ship, whether or not there are wounded and sick on board.[192]

Such individuals may be held temporarily pending determination of their status or until conditions permit their safe return or release.[193]

7.9.1.2 *Medical and Religious Personnel Who May Be Retained*. Certain classes of medical and religious personnel who fall into the hands of the adverse party shall be retained only in so far as the state of health, the spiritual needs, and the number of POWs require.[194]

Personnel who are retained in this way are not considered POWs.[195]

These classes of personnel include:

• military medical and religious personnel, including[196]

o medical personnel exclusively engaged in medical duties;[197]

o administrative staff exclusively engaged in support to medical units;[198] and

o chaplains attached to the armed forces;[199] and

• authorized staff of voluntary aid societies.[200]

7.9.1.3 *Persons Performing Medical or Religious Duties Who Are Held as POWs*. Certain classes of persons who perform medical or religious duties are held as POWs if they fall into the hands of the enemy during international armed conflict. These classes of personnel include:

• auxiliary medical personnel, if they are carrying out their medical duties at the time when they come into contact with the enemy or fall into the enemy's hands;[201]

• members of the armed forces who are trained as medical personnel, but not attached to the medical service;[202]

• members of the armed forces who are ministers of religion without having officiated as chaplains to their own forces;[203] and

• military medical and religious personnel who forfeit their status as retained personnel because they are not exclusively engaged in medical and religious duties.[204]

7.9.2 Use of Identification Card to Help Establish Retained Personnel Status. The 1949 Geneva Conventions contemplate that identification cards will be used to help establish the proper status of persons, including the entitlement to retained personnel status.[205]

Persons entitled to retained personnel status, in addition to wearing the identity disc mentioned in Article 16 of the GWS or Article 19 of the GWS-Sea, shall also carry a special identity card bearing the distinctive emblem.[206]

This card shall be water-resistant and of such size that it can be carried in the pocket. It shall be worded in the national language; shall mention at least the surname and first names, the date of birth, the rank, and the service number of the bearer; and shall state in what capacity the bearer is entitled to the protection of the GWS or GWS-Sea. The card shall bear the photograph of the owner and also either the bearer's signature or fingerprints or both. It shall be embossed with the stamp of the military authority.[207]

The identity card shall be uniform throughout the same armed forces and, as far as possible, of a similar type in the armed forces of the Parties to the GWS and GWS-Sea. The parties to the conflict may be guided by the model, which is annexed, by way of example, to the GWS and GWS-Sea. Parties to the conflict shall inform each other, at the outbreak of hostilities, of the model they are using. Identity cards should be made out, if possible, at least in duplicate, one copy being kept by the home country.[208]

In no circumstances may retained personnel be deprived of their insignia or identity cards nor of the right to wear the armlet.[209] In case of loss, they shall be entitled to receive duplicates of the cards and to have the insignia replaced.[210]

7.9.3 Duties of Retained Personnel. Within the framework of the military laws and regulations of the Detaining Power, and under the authority of its competent service, retained personnel shall continue to carry out, in accordance with their professional ethics, their medical and spiritual duties on behalf of POWs, preferably those of the armed forces to which they themselves belong.[211] This preference for retained personnel to treat the POWs of the armed forces to which the retained personnel belong is also reflected in the provisions of the GPW related to the medical treatment of POWs.[212]

Like POWs, retained personnel may be subject to disciplinary or judicial proceedings for violations of the laws and regulations of the Detaining Power.[213]

If retained personnel refuse to perform their medical duties on behalf of POWs, they are liable to forfeit their status as retained personnel.[214]

7.9.4 Return of Personnel Whose Retention Is Not Indispensable. Personnel whose retention is not indispensable to provide for the health and spiritual needs of POWs shall be returned to the party to the conflict to whom they belong, as soon as a road is open for their return and military requirements permit.[215] As a historical matter, such return has been exceptional while hostilities are ongoing.[216]

7.9.4.1 *Treatment Pending Return.* Pending their return, such personnel whose retention is not indispensable shall not be deemed POWs; nevertheless, they shall at least benefit from all the provisions of the GPW.[217] They shall continue to fulfill their duties under the orders of the adverse party and shall preferably be engaged in the care of the wounded and sick of the party to the conflict to which they themselves belong.[218]

On their departure, they shall take with them the effects, personal belongings, valuables, and instruments belonging to them.[219]

7.9.4.2 *Selection of Personnel for Return.* The selection of personnel for return under Article 30 of the GWS shall be made irrespective of any consideration of race, religion, or political opinion, but preferably according to the chronological order of their capture and their state of health.[220]

7.9.4.3 *Special Agreements on the Percentage of Personnel to Be Retained.* As from the outbreak of hostilities, parties to the conflict may determine by special agreement the percentage of personnel to be retained, in proportion to the number of POWs and the distribution of these medical and religious personnel in the camps.[221]

7.9.5 Rights and Privileges of Retained Personnel.

7.9.5.1 *POW Treatment as a Minimum.* Retained personnel shall receive, as a minimum, the benefits and protection of the GPW.[222]

7.9.5.2 *All Facilities Necessary to Provide for the Medical Care of, and Religious Ministration to, POWs.* Retained personnel shall also be granted all facilities necessary to provide for the medical care of, and religious ministration to, POWs.[223]

7.9.5.3 *Visits of POWs Outside the Camp.* Retained personnel, including chaplains, shall be authorized to visit periodically POWs situated in working detachments or in hospitals outside the camp.[224] For this purpose, the Detaining Power shall place at their disposal the necessary means of transport.[225]

7.9.5.4 *Correspondence by Chaplains.* Chaplains shall be free to correspond, subject to censorship, on matters concerning their religious duties with the ecclesiastical authorities in the country of detention and with international religious organizations.[226] Letters and cards, which they may send for this purpose, shall be in addition to the quota provided for in Article 71 of the GPW.[227]

7.9.5.5 *Proposal of POWs for Examination by Mixed Medical Commissions.* Retained medical personnel may propose wounded and sick POWs for examination by Mixed Medical Commissions with a view towards repatriation or accommodation in a neutral country and the entitlement to attend examinations conducted by Mixed Medical Commissions.[228]

7.9.5.6 *No Other Compulsory Duties.* Although they shall be subject to the internal discipline of the camp in which they are retained, such personnel may not be compelled to carry out any work other than that concerned with their medical or religious duties.[229]

7.9.5.7 *Senior Medical Officer in the Camp.* The senior medical officer in each camp shall be responsible to the camp military authorities for everything connected with the activities of retained medical personnel.[230] For this purpose, parties to the conflict shall agree at the outbreak of hostilities on the subject of the corresponding ranks of the medical personnel, including that of societies mentioned in Article 26 of the GWS.[231]

In all questions arising out of their duties, the senior medical officer, and chaplains, shall have direct access to the military and medical authorities of the camp, who shall grant them the facilities they may require for correspondence relating to these questions.[232]

7.9.5.8 *Agreements on Possible Relief of Retained Personnel.* During hostilities, the parties to the conflict shall agree concerning the possible relief of retained personnel and shall settle the procedure to be followed.[233] This provision arose from limited practice in World War II where retained personnel were repatriated and replaced by medical personnel from their home country.[234]

7.9.6 No Relief of Obligations of the Detaining Power. None of the provisions of Article 28 of the GWS or Article 33 of the GPW shall relieve the Detaining Power of its obligations with regard to POWs from the medical or spiritual point of view.[235] In other words, the fact that the Detaining Power permits and enables retained personnel to care for POWs does not relieve the Detaining Power of its own responsibilities to care for POWs.

7.10 MILITARY MEDICAL UNITS AND FACILITIES

Fixed establishments and mobile medical units of the Medical Service (*i.e.*, military medical units and facilities) may in no circumstances be attacked, but shall at all times be respected and protected by the parties to the conflict.[236] In addition, establishments ashore that are entitled to the protection of the GWS shall be protected from bombardment or attack from the sea.[237]

7.10.1 Meaning of "Respect and Protection" of Military Medical Units and Facilities. The respect and protection accorded by the GWS to military medical units and facilities mean that they must not knowingly be attacked, fired upon, or unnecessarily prevented from discharging their proper functions.[238]

7.10.1.1 *Incidental Harm Not Prohibited.* The incidental harm to military medical units or facilities, due to their presence among or in proximity to combatant elements actually engaged, by fire directed at the latter, gives no just cause for complaint.[239] Commanders are obligated to situate military medical units and facilities such that they are not in danger from attack against military objectives.[240] Military medical units and facilities that are positioned near military objectives are deemed to have accepted the risk of harm due to their proximity to military operations.[241] Although the presence of military medical units and facilities does not serve to exempt nearby military objectives from attack due to the risk that military medical units and facilities would be incidentally harmed, feasible precautions must be taken to reduce the risk of harm to military medical units and facilities.[242]

7.10.1.2 *Search Not Prohibited.* Military medical units and facilities are not immune from search by the enemy.[243] For example, search may be necessary to verify that these units and facilities are not being used, outside their humanitarian duties, to commit acts harmful to the enemy.

7.10.1.3 *Capture Not Prohibited.* The respect and protection afforded military medical units and facilities do not immunize them from capture.[244]

7.10.2 Obligation to Situate Military Medical Units and Facilities Relative to Military Objectives. Responsible authorities shall ensure that military medical establishments and units are, as far as possible, situated in such a manner that attacks against military objectives cannot imperil their safety.[245]

In no case may a military medical unit or facility be used for the purpose of shielding military objectives from attack.[246]

7.10.2.1 *Necessity for the Intermingling of Military Objectives and Military Medical Units and Facilities.* The obligation to situate medical units and facilities so that attacks against military objectives cannot imperil their safety is limited by practical considerations. For example, in order to perform their medical duties effectively, medical units and facilities must, to some degree, be intermingled with military objectives (such as combatants or vehicles that constitute military objectives). In addition, medical units and facilities may be faced with threats of unlawful attacks (*e.g.*, from criminal elements or enemy combatants not abiding by the law of war) that may warrant the placement of medical units and facilities in a more secure location (*e.g.*, within the perimeter of a military base) so that they may be guarded by combatant units. Logistical considerations (*e.g.*, access to roads, water, power, or communications) may also limit the implementation of this obligation.

7.10.2.2 *Implementation of the Obligation to Situate Medical Units and Facilities*. Implementing the obligation to ensure that military medical units and facilities are, as far as possible, situated in such a manner that attacks against military objectives cannot imperil their safety is a responsibility of commanders of military medical units, as well as commanders of combatant units. Commanders of military medical units and facilities should, as far as possible, place their units and facilities so as to protect them from the effects of enemy attacks on military objectives. Commanders of combatant units should also avoid placing their units in proximity to medical units and facilities to the extent feasible. For example, procedures should be established to ensure that, during international armed conflict, non-medical vehicles or personnel do not remain unnecessarily within or near medical facilities.

7.10.3 Loss of Protection of Military Medical Units and Facilities From Being Made the Object of Attack. The protection from being made the object of attack, to which fixed establishments and mobile medical units of the Medical Service are entitled, shall not cease unless they are used to commit, outside their humanitarian duties, acts harmful to the enemy.[247]

7.10.3.1 *Acts Harmful to the Enemy*. Medical establishments and units must refrain from all interference, direct or indirect, in military operations.[248]For example, a hospital may not be used as a shelter for able-bodied combatants or fugitives, as an arms or ammunition depot, or as a military observation post.[249]A medical unit must not be deliberately situated so as to hamper or impede an enemy attack.[250]

Acts that are part of their humanitarian duties, such as caring for the wounded and sick, are not a basis for depriving medical units and facilities of their protection.[251]

7.10.3.2 *Due Warning Before Cessation of Protection*. Protection may, however, cease only after a due warning has been given, naming, in all appropriate cases, a reasonable time limit, and after such warning has remained unheeded.[252]

The obligation to refrain from use of force against a medical unit acting in violation of its mission and protected status without due warning does not prohibit the exercise of the right of self-defense. There may be cases in which, in the exercise of the right of self-defense, a warning is not "due" or a reasonable time limit is not appropriate. For example, forces receiving heavy fire from a hospital may exercise their right of self-defense and return fire.[253]Such use of force in self-defense against medical units or facilities must be proportionate.[254]For example, a single enemy rifleman firing from a hospital window would warrant a response against the rifleman only, rather than the destruction of the hospital.

7.10.3.3 *Conditions Not Depriving Military Medical Units or Facilities of Protection*. The following conditions shall not be considered as depriving a medical unit or establishment of the protection guaranteed by Article 19 of the GWS:

- That the personnel of the unit or establishment are armed, and that they use the arms in their own defense, or in that of the wounded and sick in their charge.
- That in the absence of armed orderlies, the unit or establishment is protected by a picket, by sentries, or by an escort.
- That small arms and ammunition taken from the wounded and sick and not yet handed to the proper service are found in the unit or establishment.
- That personnel and material of the veterinary service are found in the unit or establishment, without forming an integral part thereof; and
- That the humanitarian activities of medical units and establishments or of their personnel extend to the care of civilian wounded or sick.[255]

7.10.3.4 *Arming of Military Medical Units or Facilities*. The GWS contemplates that the personnel of medical units or establishments may be armed, and it does not specifically restrict the weapons that medical units or facilities may have.[256]

Military medical units and facilities may be armed to the extent necessary to enable them to defend themselves or their patients against unlawful attacks.[257]For example, military medical and religious personnel may be equipped with weapons to meet internal security needs, to maintain discipline, to protect staff and patients from criminal threats of violence, and to prevent the theft of medical supplies. On the other hand, medical units or establishments should not be armed such that they would appear to an enemy military force to present an offensive threat.[258] The type of weapon that is necessary for defensive or other legitimate purposes may depend on the anticipated threats against the medical unit or facility.

7.10.3.5 *Use of Weapons in Self-Defense or Defense of the Wounded and Sick*. Although medical personnel may carry arms for self-defense, they may not employ such arms against enemy forces acting in conformity with the law of war.[259] Medical personnel must refrain from all aggressive action and may not use force to resist the advance of enemy forces (*e.g.*, to prevent the search or capture of their unit or the wounded and sick in their care by enemy military forces).[260]

Medical personnel may use force to defend themselves or the wounded and sick in their care from unlawful attacks, such as those from unprivileged belligerents or other persons violating the law of war.[261]

Medical personnel who use their arms in circumstances not justified by the law of war expose themselves to penalties for violation of the law of war and, provided they have been given due warning to cease such acts, may also forfeit the protection of the medical unit or facility of which they form part or that they are protecting.[262]

7.10.3.6 *Temporary Presence of Combatants or Other Military Objectives Within a Medical Unit or Facility*. In some cases, combatants or other military objectives may be temporarily present within a medical unit or facility. For example, a military vehicle that is not protected as medical aircraft or transport may deliver the wounded or sick to a medical facility. Similarly, a commander may visit a hospital to visit wounded or sick members of the commander's unit for the benefit of their morale, or to award medals. The temporary presence of combatants or other military objectives within a medical unit or facility does not automatically constitute an act harmful to the enemy that forfeits its protection from being made the object of attack. More facts would be necessary to establish that the medical unit or facility was somehow being used to commit acts harmful to the enemy, such as if a unit or facility were being used for the purpose of sheltering able-bodied combatants or providing a base from which military operations were being planned or conducted.

7.10.4 Use of the Distinctive Emblem to Facilitate the Identification of Medical Units and Facilities. The GWS contemplates that the distinctive emblem will be used to facilitate the identification of medical units and facilities as such.

7.10.4.1 *Flying of the Distinctive Flag of the GWS Over Medical Units and Facilities*. The distinctive flag of the GWS shall be hoisted only over such medical units and establishments as are entitled to be respected under the GWS, and only with the consent of the military authorities.[263] In mobile units, as in fixed establishments, it may be accompanied by the national flag of the party to the conflict to which the unit or establishment belongs.[264]

Nevertheless, medical units that have fallen into the hands of the enemy shall not fly any flag other than that of the GWS.[265]

7.10.4.2 *Necessary Steps to Make the Distinctive Emblems Clearly Visible*. Parties to the conflict shall take the necessary steps, in so far as military considerations permit, to make the distinctive emblems indicating medical units and establishments clearly visible to the enemy land, air, and naval forces, in order to obviate the possibility of any hostile action.[266]

7.10.5 Capture of Military Medical Units and Facilities. Military medical units and facilities may be captured. Should they fall into the hands of the adverse party, their personnel shall be free to pursue their duties, as long as the capturing Power has not itself ensured the necessary care of the wounded and sick found in such establishments and units.[267]

7.10.5.1 *Disposition of Buildings, Material, and Stores of Captured Medical Units and Facilities*. The material of mobile medical units of the armed forces that fall into the hands of the enemy shall be reserved for the care of wounded and sick.[268]

The buildings, material, and stores of fixed medical establishments of the armed forces shall remain subject to the laws of war (*e.g.*, they may be captured and used by the enemy military force[269]), but may not be diverted from their purpose as long as they are required for the care of wounded and sick.[270] Nevertheless, the commanders of forces in the field may make use of them, in case of urgent military necessity, provided that they make previous arrangements for the welfare of the wounded and sick who are being cared for in them.[271]

The material and stores of mobile medical units and fixed medical establishments of the armed forces that fall into the hands of the enemy shall not be intentionally destroyed.[272]

7.11 GROUND TRANSPORTS OF THE WOUNDED AND SICK, OR OF MEDICAL EQUIPMENT

Ground transports of wounded and sick, or of medical equipment, shall be respected and protected in the same way as mobile medical units.[273]

7.11.1 Protection of Ground Medical Transports on the Same Basis as That of Medical Units. The protection to which these transports are entitled shall not cease unless they are used to commit, outside their humanitarian duties, acts harmful to the enemy.[274]

As with medical units, these ground transports must refrain from all interference, direct or indirect, in military operations in order to retain protection.[275]For example, these ground transports must not be used to transport able-bodied combatants or to carry ammunition to combat forces.[276]

The protection for ground transports of the wounded and sick, or of medical equipment, may cease only after due warning has been given, naming, in all appropriate cases, a reasonable time limit, and after such warning has remained unheeded.[277]

7.11.2 Captured Medical Transport. Should these ground medical transports or vehicles fall into the hands of the adverse party, they shall be subject to the laws of war, on condition that the party to the conflict who captures them shall in all cases ensure the care of the wounded and sick they contain.[278]The reference to the laws of war means that the adverse party may seize and dispose of the property as enemy property.[279]Thus, the adverse party may use or dispose of such transports (including by removing the distinctive emblem and using the vehicle for a hostile purpose), provided that the capturing party ensures the care of the wounded and sick being carried in such transports.[280]

The civilian personnel and all means of transport obtained by requisition shall be subject to the general rules of international law.[281]This rule is stated broadly to account for a variety of situations. For example, if the civilian personnel and vehicles have been requisitioned in occupied territory by the Occupying Power and then captured by a State that is adverse to the Occupying Power, the persons and property would likely belong to the capturing State or a State that is friendly to the capturing State and thus likely would not fall under the GC.[282]On the other hand, the civilian personnel and vehicles could have been requisitioned by a belligerent State within in its own territory and, upon capture by enemy forces, the persons and property likely would fall under the protection of the GC.[283]

7.12 HOSPITAL SHIPS, SICK-BAYS IN WARSHIPS, AND COASTAL RESCUE CRAFT

7.12.1 Types of Hospital Ships and Coastal Rescue Craft.

7.12.1.1 *Military Hospital Ships*. Military hospital ships, that is to say, ships built or equipped by the Powers specially and solely with a view to assisting the wounded, sick, and shipwrecked, to treating them, and to transporting them, may in no circumstances be attacked or captured, but shall at all times be respected and protected on condition that their names and descriptions have been notified to the parties to the conflict ten days before those ships are employed.[284]

The characteristics, which must appear in the notification, shall include registered gross tonnage, the length from stem to stern, and the number of masts and funnels.[285]Many States have employed ships that are equipped or converted into hospital ships, rather than building them specifically as hospital ships.[286]

7.12.1.2 *Commissioned Civilian Hospital Ships*. Hospital ships used by the National Red Cross Societies, by officially recognized relief societies, or by private persons shall have the same protection as military hospital ships and shall be exempt from capture, if the party to the conflict on which they depend has given them an official commission and in so far as the provisions of Article 22 of the GWS-Sea concerning notification have been complied with.[287]These ships must be provided with certificates from the responsible authorities, stating that the vessels have been under their control while fitting out and on departure.[288]

7.12.1.3 *Authorized Neutral Civilian Hospital Ships*. Civilian hospital ships may also be sent from neutral countries.[289]

Hospital ships utilized by National Red Cross Societies, officially recognized relief societies, or private persons of neutral countries shall have the same protection as military hospital ships and shall be exempt from capture, on condition that they have placed themselves under the control of one of the parties to the conflict, with the previous consent of their own governments and with the authorization of the party to the conflict concerned, in so far as the provisions of Article 22 of the GWS-Sea concerning notification have been complied with.[290]

7.12.1.4 *Size of Hospital Ships*. The protection for military hospital ships, commissioned civilian hospital ships, and authorized neutral civilian hospital ships mentioned in Articles 22, 24, and 25 of the GWS-Sea, respectively, shall apply to hospital ships of any tonnage and to their lifeboats, wherever they are operating.[291] Nevertheless, to ensure the maximum comfort and security, the parties to the conflict shall endeavor to use, for the transport of the wounded, sick, and shipwrecked over long distances and on the high seas, only hospital ships of more than 2,000 tons gross.[292] For example, U.S. military hospital ships have adhered to this recommendation by a large margin.[293]

7.12.1.5 *Coastal Rescue Craft and Fixed Coastal Installations Used Exclusively by Them*. Under the same conditions as those provided for in Articles 22 and 24 of the GWSSea, small craft employed by the State, or by the officially recognized lifeboat institutions for coastal rescue operations, shall also be respected and protected, so far as operational requirements permit.[294] The same shall apply so far as possible to fixed coastal installations used exclusively by these craft for their humanitarian missions.[295]

The phrase, "so far as operational requirements permit," has been used to acknowledge the risks incurred by such craft, because of their small size, in a zone of military operations.[296] As provided in Articles 22 and 24 of the GWS-Sea, the names and descriptions of these small craft (including registered gross tonnage, the length from stem to stern, and the number of masts and funnels) must be notified to the parties to the conflict ten days before such craft are employed.[297] In addition, if the craft are not military craft belonging to a party to the conflict, they must be provided an official commission from the party to the conflict on which they depend and with certificates from the responsible authorities.[298]

7.12.2 Duties and Liabilities of Hospital Ships and Coastal Rescue Craft. The following duties and liabilities apply to hospital ships and coastal rescue craft.

7.12.2.1 *Relief and Assistance Without Distinction of Nationality*. The military hospital ships, commissioned civilian hospital ships, authorized neutral civilian hospital ships, and coastal rescue craft described in Articles 22, 24, 25, and 27 of the GWS-Sea, respectively, shall afford relief and assistance to the wounded, sick, and shipwrecked without distinction of nationality.[299] In other words, these vessels should help friendly and enemy personnel without distinction.[300]

7.12.2.2 *No Use for Military Purposes*. Parties to the GWS-Sea undertake not to use these vessels for any military purpose.[301] These vessels must not participate in any way in the armed conflict or the war effort.[302] For example, these vessels may not be used to relay military orders, transport able-bodied combatants or military equipment, or engage in reconnaissance.[303] Such acts constitute acts harmful to the enemy that forfeit a vessel's protection from being made the object of attack and capture.[304]

7.12.2.3 *No Hampering the Movement of Combatants*. Such vessels shall in no way hamper the movement of combatants.[305] Any deliberate attempt by a hospital vessel to hamper the movement of combatants would constitute an act harmful to the enemy, although inadvertent actions might also put those vessels at increased of harm (*e.g.*, if enemy forces mistakenly believe that the vessel is acting deliberately to impede their movements).[306]

For example, hospital ships and coastal rescue craft belonging to one State must not interfere with opposing forces' efforts to capture shipwrecked personnel, nor otherwise interfere with enemy military operations.[307] Therefore, hospital ships and coastal rescue craft should not engage in such activities lest they forfeit the special protection afforded by the GWS-Sea. Special agreements between the opposing forces, however, may provide protection to ships engaged in search, rescue, or other recovery missions.

7.12.2.4 *Merchant Vessels Transformed Into Hospital Ships*. Merchant vessels that have been transformed into hospital ships cannot be put to any other use throughout the duration of hostilities.[308]

7.12.2.5 *Acceptance of the Risk From Proximity to Combat Operations*. During and after an engagement, such vessels will act at their own risk. [309] A hospital ship or coastal rescue craft operating in proximity to combat operations assumes a certain risk of damage as a result of lawful enemy operations, including misidentification by enemy or friendly forces. Although the presence of hospital ships or coastal rescue craft does not serve to exempt nearby military objectives from attack due to the risk that the hospital ships or coastal rescue craft would be incidentally damaged, feasible precautions must be taken to reduce the risk of harm to hospital ships or coastal rescue craft.[310]

7.12.2.6 *Control and Search of Hospital Ships and Coastal Rescue Craft*. The parties to the conflict shall have the right to control and search military hospital ships, commissioned civilian hospital ships, authorized neutral civilian hospital ships, and costal rescue craft mentioned in Articles 22, 24, 25, and 27 of the GWS-Sea, respectively.[311]They may refuse assistance from these vessels, order them off, make them take a certain course, control the use of their wireless and other means of communication, and even detain them for a period not exceeding seven days from the time of interception, if the gravity of the circumstances so requires.[312]For example, a belligerent may delay the passage of a hospital ship belonging to its adversary in order to avoid interference with its military operations.[313]

Parties to a conflict may put a commissioner temporarily on board such vessels whose sole task shall be to see that these orders are carried out.[314]

As far as possible, the parties to the conflict shall enter in the log of the hospital ship, in a language he or she can understand, the orders they have given the captain of the vessel.[315]Parties to the conflict may, either unilaterally or by particular agreements, put on board their ships neutral observers who shall verify the strict observation of the provisions contained in the GWS-Sea.[316]

7.12.2.7 *Use of Secret Codes for Communication.* The GWS-Sea provides that hospital ships may not possess or use a secret code for their wireless or other means of communication.[317]The lack of secret codes for communication was intended to afford belligerents a guarantee that hospital ships were not improperly participating in hostilities.[318]

This rule was formulated before developments in modern communications technology made encryption more common.[319]Substantial practical difficulties have arisen in construing this rule in a way that would prohibit any use of encrypted communications by hospital ships.[320]

In light of the requirement for encryption in modern communications systems (*e.g.*, satellite communications and video teleconference systems) and navigational technology (*e.g.*, global positioning systems) and the goal of ensuring that hospital ships can effectively fulfill their humanitarian mission, the United States has employed hospital ships with the capability to conduct encrypted communications.[321]Other States have not objected to this practice.[322]Some States have taken the same view.[323]Scholarly commentators have also supported this view.[324]Hospital ships may not use encrypted communications for military purposes (*e.g.*, transmitting intelligence data) or in any way that is harmful to an adversary.[325]Commissioners or neutral observers may be placed on board hospital ships to verify compliance with this rule.[326]

7.12.3 Distinctive Marking and Other Identification of Hospital Ships and Coastal Rescue Craft. The ships designated for protection by Articles 22, 24, 25, and 27 of the GWS-Sea (*i.e.*, military hospital ships, commissioned civilian hospital ships, authorized neutral civilian hospital ships, and coastal rescue craft, respectively) shall be distinctively marked as follows: • all exterior surfaces shall be white; and

• one or more dark red crosses, as large as possible, shall be painted and displayed on each side of the hull and on the horizontal surfaces, so placed as to afford the greatest possible visibility from the sea and from the air.[327]

7.12.3.1 *Rules for Flying the National Flag, the Flag of the Party to the Conflict, and the Distinctive Emblem Flag.* All hospital ships must make themselves known by hoisting their national flag and, further, if they belong to a neutral State, the flag of the party to the conflict whose direction they have accepted.[328]A white flag with a red cross shall be flown at the mainmast as high as possible.[329]

Hospital ships that, in accordance with Article 31 of the GWS-Sea, are provisionally detained by the enemy, must haul down the flag of the party to the conflict in whose service they are or whose direction they have accepted.[330]

Coastal lifeboats, if they continue to operate with the consent of the Occupying Power from a base that is occupied, may be allowed, when away from their base, to continue to fly their own national colors along with a flag carrying a red cross on a white ground, subject to prior notification to all the parties to the conflict concerned.[331]

7.12.3.2 *Lifeboats and All Small Craft Used by the Medical Service.* Lifeboats of hospital ships, coastal lifeboats, and all small craft used by the Medical Service shall be painted white with dark red crosses prominently displayed and shall, in general, comply with the identification system prescribed above for hospital ships.[332]

7.12.3.3 *Night-Time Identification Measures.* The above-mentioned ships and craft, which may wish to ensure by night and in times of reduced visibility the protection to which they are entitled, must, subject to the assent of the party to the conflict under whose power they are, take the necessary measures to render their painting and distinctive emblems sufficiently apparent.[333]

7.12.3.4 *Other Emblems Instead of the Red Cross*. All the provisions in Article 43 of the GWS-Sea relating to the red cross apply equally to the other emblems mentioned in Article 41 of the GWS-Sea.[334]

The distinctive emblem of the red crystal also may be used for the same purposes as the red cross.[335]

7.12.3.5 *Agreements to Facilitate the Identification of Hospital Ships*. Parties to the conflict shall at all times endeavor to conclude mutual agreements in order to use the most modern methods available to facilitate the identification of hospital ships.[336]

7.12.4 Rights and Privileges of Hospital Ships and Coastal Rescue Craft. Military hospital ships, commissioned civilian hospital ships, and authorized neutral civilian hospital ships that meet the applicable requirements shall be respected and protected and are exempt from capture. Coastal rescue craft shall also be respected and protected so far as operational requirements permit.

In addition, the following protections apply:

7.12.4.1 *Protection From Attacks From Land*. Hospital ships entitled to the protection of the GWS-Sea shall not be attacked from land.[337]

7.12.4.2 *Authority to Leave Ports Held by the Enemy*. Any hospital ship in a port that falls into the hands of the enemy shall be authorized to leave that port.[338]

7.12.4.3 *Relief From Taxation in Time of War*. In time of war, military hospital ships, commissioned civilian hospital ships, and authorized neutral civilian hospital ships are exempt from all dues and taxes imposed for the benefit of the State in the ports of the Parties to a particular 1904 Hague treaty. Article 1 of this 1904 Hague treaty provides that certain categories of hospital ships listed in an 1899 Hague treaty receive this benefit.[339] These categories of hospital ships listed in the 1899 Hague treaty are the same categories of hospital ships that are used in the GWS-Sea.[340] Some States may be Parties to the GWS-Sea but not to the 1904 Hague treaty that creates this exemption.

7.12.4.4 *Stay in Neutral Ports*. Vessels described in Articles 22, 24, 25, and 27 of the GWS-Sea (*i.e.*, military hospital ships, commissioned civilian hospital ships, authorized neutral civilian hospital ships, and coastal rescue craft, respectively), are not classed as warships as regards their stay in a neutral port.[341] Thus, these vessels are not subject to the restrictions that belligerent warships staying in a neutral port are subject to by neutral Powers.[342]

7.12.5 Protection of Sick-Bays in Warships. Should fighting occur on board a warship, the sick-bays shall be respected and spared as far as possible.[343]

7.12.5.1 *Use of Seized Sick-Bays and Their Equipment*. Sick-bays and their equipment shall remain subject to the laws of warfare, but may not be diverted from their purpose so long as they are required for the wounded and sick.[344] Nevertheless, the commander into whose power they have fallen may, after ensuring the proper care of the wounded and sick who are accommodated therein, apply them to other purposes in case of urgent military necessity.[345]

7.12.6 Forfeiture of Protection of Hospital Ships and Sick Bays. The protection to which hospital ships and sick-bays are entitled shall not cease unless the vessels are used to commit, outside their humanitarian duties, acts harmful to the enemy.[346] In particular, hospital ships and coastal rescue craft may not be used for military purposes nor to hamper the movements of combatants.[347]

7.12.6.1 *Due Warning Before Cessation of Protection*. Protection may, however, cease only after due warning has been given, naming in all appropriate cases a reasonable time limit, and after such warning has remained unheeded.[348]

The obligation to refrain from use of force against a medical vessel or sick-bay acting in violation of its mission and protected status without due warning does not prohibit the exercise of the right of self-defense. There may be cases in which, in the exercise of the right of self-defense, a warning is not "due" or a reasonable time limit is not appropriate. For example, forces receiving heavy fire may exercise their right of self-defense and return fire. Such use of force in self-defense must also be proportionate.[349]

7.12.6.2 *Conditions That Do Not Deprive Hospital Ships and Sick-Bays of Vessels of Their Protection*. The following conditions shall not be considered as depriving hospital ships or sick-bays of vessels of the protection due to them:

• the fact that the crews of ships or sick-bays are armed for the maintenance of order, for their own defense, or for the defense of the sick and wounded;

• the presence on board of apparatus exclusively intended to facilitate navigation or communication;

• the discovery on board hospital ships or in sick-bays of portable arms and ammunition taken from the wounded, sick, and shipwrecked, and not yet handed over to the proper service;

• the fact that the humanitarian activities of hospital ships and sick-bays of vessels or of the crews extend to the care of wounded, sick, or shipwrecked civilians; and • the transport of equipment and of personnel intended exclusively for medical duties, over and above the normal requirements.[350]

7.12.6.3 *Arming of Hospital Ships and Equipping Them With Defensive Devices*. The GWS-Sea contemplates that the crews of hospital ships may be armed and does not specifically restrict the weapons that they may have. The crews of hospital ships may be armed to the extent necessary to enable them to defend themselves or their patients against unlawful attacks. Such arming is viewed as consistent with the ship's humanitarian purpose and the crew's duty to safeguard the wounded and sick. For example, crews of hospital ships may be equipped with weapons to meet internal security needs, to maintain discipline, to protect staff and patients from criminal threats of violence, and to prevent the theft of medical supplies. The type of weapon that is necessary for defensive or other legitimate purposes may depend on the nature of the threats against the hospital ship.[351]For example, hospital ships may be armed with defensive weapon systems, including when necessary certain crew-served weapons, as a prudent anti-terrorism/force protection (AT/FP) measure to defend against small boat attacks.[352]On the other hand, crews of hospital ships should not be armed such that they would appear to an enemy military force to present an offensive threat.

Equipping hospital ships with other defensive devices – such as chaff for protection against over-the-horizon weapons or similar threats – is not prohibited.[353]However, such devices, like weapons, must not be used to commit acts harmful to enemy military forces acting in conformity with the law of war.[354]

7.13 CHARTERED MEDICAL TRANSPORT SHIPS

Ships chartered for medical transport purposes shall be authorized to transport equipment exclusively intended for the treatment of wounded and sick members of armed forces or for the prevention of disease, provided that the particulars regarding their voyage have been notified to the adverse Power and approved by the latter.[355]The adverse Power shall preserve the right to board the carrier ships, but not to capture them or seize the equipment carried.[356]

7.13.1Agreement to Place Neutral Observers on Chartered Medical Transports. By agreement among the parties to the conflict, neutral observers may be placed on board such ships to verify the equipment carried.[357]For this purpose, free access to the equipment shall be given.[358]

7.14 MILITARY MEDICAL AIRCRAFT

Medical aircraft – that is to say, aircraft exclusively employed for the removal of the wounded, sick, and shipwrecked, and for the transport of medical personnel and equipment – shall not be attacked, but shall be respected by the belligerents, while flying at heights and times, and on routes, specifically agreed upon by the belligerents concerned.[359]

7.14.1 Medical Aircraft Not Flying at Heights and Times and on Routes Specifically Agreed Upon by Belligerents Concerned. The use of protected medical aircraft generally depends on an agreement between the belligerents.[360]

However, known medical aircraft, when performing their humanitarian functions, must be respected and protected.[361]Such aircraft does not constitute a military objective that is liable to being made the object of attack.[362]Thus, even if not flying pursuant to an agreement, such aircraft shall not be deliberately attacked or fired upon, if identified as protected medical aircraft. For example, if there is no agreement and a military force happens upon a medical aircraft belonging to an enemy State, the aircraft must not be made the object of attack until all other means of control (such as directing the aircraft to land and submit to search) have been exhausted.[363]

A medical aircraft that is not flying pursuant to a special agreement that seeks to claim protection as medical aircraft shall make every effort to identify itself and to inform the enemy State of its status and its operations, such as its flight times and routes.[364]For example, an unknown aircraft within a theater of military operations would often be reasonably presumed to be a military objective, and the aircraft must take affirmative steps to rebut this presumption.[365]

In order to maintain its entitlement to protection, such aircraft must obey the directions of the enemy State, such as directions to land and to submit to search.

7.14.2 Exclusively Employed for the Removal of the Wounded, Sick, or Shipwrecked, and for the Transport of Medical Personnel and Equipment. In order to receive protection as medical aircraft, the aircraft must be exclusively employed for the removal of wounded, sick, or shipwrecked combatants and for the transport of medical personnel and equipment. For example, such aircraft, while designated or operating as medical aircraft, may not be used also for military purposes, such as to transport able-bodied combatants or to carry ammunition to combat forces.[366]

7.14.2.1 *Conversion of Aircraft To and From Use Under a Special Agreement for Medical Purposes*. If medical aircraft are flying pursuant to an agreement, it is not necessary that the aircraft should have been specially built and equipped for medical purposes.[367]There is no objection to converting ordinary aircraft into medical aircraft for use under a special agreement or to using medical aircraft that have been used under a special agreement for other purposes, provided the distinctive markings are removed.[368]

7.14.2.2 *Search, Rescue, or Other Recovery Excluded*. Medical aircraft may engage in the removal of wounded, sick, or shipwrecked combatants. However, medical aircraft may neither prevent the capture of combatants by enemy military forces nor otherwise interfere with enemy military operations.[369]Therefore, medical aircraft do not include combat search and rescue aircraft, and medical aircraft should not engage in such activities lest they forfeit the special protection afforded by the GWS or GWS-Sea.[370]Special agreements between the opposing forces, however, may provide protection to aircraft engaged in search, rescue, or other recovery missions.

7.14.3 Markings on Medical Aircraft. Medical aircraft shall bear, clearly marked, the distinctive emblem (prescribed in Article 38 of the GWS and Article 41 of the GWS-Sea), together with their national colors, on their lower, upper, and lateral surfaces.[371]They shall be provided with any other markings or means of identification that may be agreed upon between the belligerents upon the outbreak or during the course of hostilities.[372]

7.14.4 Prohibition on Flights Over Enemy or Enemy-Occupied Territory. Unless otherwise agreed, flights over enemy or enemy-occupied territory are prohibited.[373]If flying pursuant to an agreement, however, medical aircraft are to receive protection when flying over enemy or enemy-occupied territory.[374]

7.14.5 Landing and Inspection of Military Medical Aircraft. Medical aircraft shall obey every summons to land (including water landings if the aircraft is equipped for them).[375]In the event of a landing thus imposed, the aircraft with its occupants may continue its flight after examination, if any.[376]

7.14.5.1 *Treatment of Personnel in the Event of Involuntary Landing in Enemy or Enemy-Occupied Territory*. In the event of an involuntary landing in enemy or enemy-occupied territory (including a water landing), the wounded and sick, as well as the crew of the aircraft, shall be POWs.[377]The medical personnel shall be treated according to Article 24 and the Articles following of the GWS, and Articles 36 and 37 of the GWS-Sea.[378]

7.14.6 Belligerent Military Medical Aircraft in Neutral Territory. Subject to conditions set by the neutral Powers, medical aircraft of parties to the conflict may fly over the territory of neutral Powers, land on it in case of necessity, or use it as a port of call.[379]They shall give neutral Powers prior notice of their passage over the said territory, and obey every summons to alight, on land or water.[380]They will be immune from attack only when flying on routes, at heights, and at times specifically agreed upon between the parties to the conflict and the neutral Power concerned.[381]

7.14.6.1 *Conditions Set by Neutral Powers*. The neutral Powers may, however, place conditions or restrictions on the passage or landing of medical aircraft on their territory.[382]

Such possible conditions or restrictions shall be applied equally to all parties to the conflict.[383]

7.14.6.2 *Detention of the Wounded, Sick, or Shipwrecked*. Unless otherwise agreed between the neutral Powers and the parties to the conflict, the wounded, sick, or shipwrecked who are disembarked with the consent of the local authorities on neutral territory by medical aircraft shall be detained by the neutral Power, where so required by international law, in such a manner that they cannot again take part in operations of war.[384]The cost of their accommodation and internment shall be borne by the Power on which they depend.[385]

7.15 DISPLAY OF THE DISTINCTIVE EMBLEM TO FACILITATE IDENTIFICATION

The GWS and GWS-Sea contemplate that the distinctive emblem, usually a red cross on a white background, will be used to facilitate the identification of the persons and objects protected by the GWS and GWS-Sea.[386]It helps identify protected persons and objects (*e.g.*, medical and religious personnel, medical transports, and medical facilities), but does not itself confer on them, or by its absence deprive them of, legal protection. The use of the distinctive emblem to facilitate protection is to take place under the direction of the competent military authority. The misuse of the distinctive emblem is prohibited.

7.15.1 The Distinctive Emblems: Red Cross, Crescent, and Crystal. To facilitate visual identification of medical and religious personnel, and medical units, facilities, transports, and equipment, four distinctive emblems have been recognized in treaties to which the United States is a Party. As displayed below, they are 1) a red cross; 2) a red crescent; 3) a red crystal; and 4) a red lion and sun. Each is displayed on a white background.

The Distinctive Emblems

Red Cross Red Crescent Red Crystal Red Lion and Sun (withdrawn)

7.15.1.1 *Red Cross*. "As a compliment to Switzerland," the heraldic emblem of the red cross on a white ground formed by reversing the Federal colors is retained as the emblem and distinctive sign of the Medical Services of armed forces.[387]The explanation that the red cross is used as a compliment to Switzerland was added to emphasize that the red cross is not intended to have religious significance.[388]The red cross has long been used to identify medical personnel during armed conflict.[389]

7.15.1.2 *Red Crescent, and Red Lion and Sun*. In the case of countries that already use, in place of the red cross, the emblem of the red crescent or the emblem of the red lion and sun on a white ground, those emblems are also recognized by the terms of the GWS and GWS-Sea.[390]

7.15.1.3 *Red Crystal*. AP III recognizes an additional distinctive emblem in addition to, and for the same purposes as, the distinctive emblems of the Geneva Conventions, which, for parties to AP III, shall enjoy equal status to the other emblems.[391]

This additional distinctive emblem, composed of a red frame in the shape of a square on edge on a white ground, shall conform to the illustration in the Annex to AP III.[392]

This distinctive emblem is referred to in AP III as the "third Protocol emblem"[393]but is often called the red crystal. The conditions for use of and respect for the third Protocol emblem are identical to those for the distinctive emblems established by the 1949 Geneva Conventions and, where applicable, AP I and AP II.[394]

AP III further provides for use of the Red Crystal for indicative purposes by national societies of the Parties to AP III. Such uses for indicative purposes by national societies are uses of the red crystal that do not indicate protection under the AP III, the GWS, or the GWS-Sea, but that indicate an association with the Red Cross movement.[395]

7.15.1.4 *State Discretion Among Emblems*. Although the GWS contemplates that the emblem of the red crescent and the emblem of the red sun and lion will be used only in States that already used them before ratifying the 1949 Geneva Conventions, AP III provides that the distinctive emblems of the Geneva Conventions shall enjoy equal status.[396]In addition, AP III provides that the medical services and religious personnel of armed forces of Parties to AP III may, without prejudice to their current emblems, make temporary use of any distinctive emblems where this may enhance protection.[397]

Some States have adopted the red crescent, without objection by other States Parties, even though their use of the red crescent did not predate the adoption of the GWS.[398]Israel ratified the 1949 Geneva Conventions with the reservation that it will use a Red Shield of David as its distinctive sign.[399]

7.15.2 Control of Display by the Competent Military Authority. Under the direction of the competent military authority, the emblem shall be displayed on the flags, armlets, and on all equipment employed in the Medical Service.[400]For example, under military regulations, helmets or vehicles used by the Medical Service might bear the distinctive emblem to facilitate the respect and protection of medical personnel by the enemy.[401]

7.15.2.1 *Removal or Obscuration of the Distinctive Emblem*. The display of the distinctive emblem is under the direction of the competent military authority.[402]Thus, the military command may authorize the removal or obscuring of the distinctive emblem for tactical purposes, such as camouflage.[403]Similarly, it would be appropriate for the distinctive emblem to be removed if it is assessed that enemy forces will fail to respect the emblem and seek to attack medical personnel; display of the emblem in such circumstances would not be considered "feasible" because in that instance it would not result in a humanitarian benefit.[404]

In the practice of the United States, removal or obscuration of the distinctive emblem has generally been controlled by the responsible major tactical commander, such as a brigade commander or higher.[405]

7.15.3 Significance of the Display of the Distinctive Emblem.

7.15.3.1 *Distinctive Emblem Facilitates Protection*. The purpose of the display of the distinctive emblem is to facilitate the identification of protected status.[406]The absence of the distinctive emblem may increase the risk that enemy forces will not recognize the protected status of military medical and religious personnel and other protected persons and objects, and attack them in error.

7.15.3.2 *Distinctive Emblem Does Not in and of Itself Establish the Right to Protection*. The distinctive emblem does not in and of itself establish the right to protection.[407]Rather, the right to protection is established by the fact that the units, facilities, or personnel, have met the applicable requirements for protected status.[408]

Thus, if the distinctive emblem is displayed by forces not entitled to protection, the display of the emblem does not confer protection and those forces may be made the object of attack. On the other hand, if personnel who are entitled to protection are recognized as such, they remain entitled to respect and protection even if the distinctive emblem is not displayed.[409]

7.15.4 Prohibitions on Unauthorized Uses of the Distinctive Emblem. With the exception of certain cases mentioned in Article 44 of the GWS (discussed below), the emblem of the Red Cross on a white ground and the words "Red Cross" or "Geneva Cross" may not be employed, either in time of peace or in time of war, except to indicate or to protect the medical units and establishments, the personnel, and material protected by the GWS and other Conventions dealing with similar matters (*e.g.*, the GWS-Sea).[410]The same shall apply to the emblems mentioned in the second paragraph of Article 38 of the GWS (*i.e.*, the emblem of red crescent and the emblem of the red lion and sun), in respect of the countries that use them.[411]

The distinguishing signs referred to in Article 43 of the GWS-Sea (*i.e.*, certain red crosses on white backgrounds) may only be used, whether in time of peace or war, for indicating or protecting the ships mentioned in Article 43, except as may be provided in any other international Convention (*e.g.*, the GWS) or by agreement between all the parties to the conflict concerned.[412]

In particular, it is prohibited to use the distinctive emblem: (1) while engaging in attacks; (2) in order to shield, favor, or protect one's own military operations; or (3) to impede enemy military operations.[413]For example, using an ambulance marked with the Red Cross to mount a surprise attack against enemy forces would be prohibited.[414]

Certain non-military uses of the distinctive emblem are authorized under the GWS and GWS-Sea: (1) use by National Red Cross Societies and other Authorized Voluntary Aid Societies; (2) use by international Red Cross organizations and their duly authorized personnel; and (3) use by ambulances and free aid stations.

7.15.4.1 *Use by National Red Cross Societies and Other Authorized Voluntary Aid Societies*. The National Red Cross Societies and other Authorized Voluntary Aid Societies designated in Article 26 of the GWS shall have the right to use the distinctive emblem conferring the protection of the GWS only within the framework of paragraph 1 of Article 44 of the GWS.[415]

Furthermore, National Red Cross (Red Crescent, Red Lion and Sun) Societies may, in time of peace, in accordance with their national legislation, make use of the name and emblem of the Red Cross for their other activities that are in conformity with the principles laid down by the International Red Cross Conferences.[416]When those activities are carried out in time of war, the conditions for the use of the emblem shall be such that it cannot be considered as conferring the protection of the GWS; the emblem shall be comparatively small in size and may not be placed on armlets or on the roofs of buildings.[417]This use of the emblem is simply to indicate an association with the Red Cross movement.[418]

7.15.4.2 *Use by International Red Cross Organizations and Their Duly Authorized Personnel*. The international Red Cross organizations and their duly authorized personnel shall be permitted to make use, at all times, of the emblem of the Red Cross on a white ground.[419]

7.15.4.3 *Use by Ambulances and Free Aid Stations*. As an exceptional measure, in conformity with national legislation and with the express permission of one of the National Red Cross (Red Crescent, Red Lion and Sun) Societies, the emblem of the GWS may be employed in time of peace to identify vehicles used as ambulances and to mark the position of aid stations exclusively assigned to the purpose of giving free treatment to the wounded or sick.[420]

7.16 PROTECTION OF CIVILIANS WHO ARE WOUNDED, SICK, INFIRM, OR EXPECTANT MOTHERS

Wounded and sick civilians, as well as the infirm, and expectant mothers, shall be the object of particular protection and respect.[421]The infirm and expectant mothers are given special consideration along with the wounded and sick because they are vulnerable and also do not take part in hostilities.[422]

7.16.1Obligation to Facilitate Efforts to Search for Civilians Who Are Wounded, Killed, Shipwrecked, or Exposed to Grave Danger. As far as military considerations allow, each party to the conflict shall facilitate the steps taken to search for the killed and wounded, to assist the shipwrecked and other persons exposed to grave danger, and to protect them against pillage and ill-treatment.[423]

The obligation with respect to civilians who are killed, wounded, shipwrecked, and exposed to grave danger is framed slightly differently from the corresponding obligation with respect to military personnel who are wounded, sick, and shipwrecked. These differences may be attributed to the fact that civilian authorities would often be responsible for collecting and bringing in civilian casualties.[424]However, as a practical matter, the armed forces may be asked to lead such efforts or to carry out a joint relief operation with civilian authorities.[425]For example, the U.S. armed forces often have undertaken to care for the civilian wounded and sick.[426]

7.17 CIVILIAN HOSPITALS AND THEIR PERSONNEL

During international armed conflict, civilian hospitals organized to give care to the wounded and sick, the infirm, and maternity cases, may in no circumstances be the object of attack, but shall at all times be respected and protected by the parties to the conflict.[427]

7.17.1 Loss of Protection for Civilian Hospitals Used to Commit Acts Harmful to the Enemy. The protection to which civilian hospitals are entitled shall not cease unless they are used to commit, outside their humanitarian duties, acts harmful to the enemy.[428]

7.17.1.1 *Acts Harmful to the Enemy*. Civilian hospitals must avoid any interference, direct or indirect, in military operations, such as the use of a hospital as a shelter for able-bodied combatants or fugitives, as an arms or ammunition store, as a military observation post, or as a center for liaison with combat forces.[429]However, the fact that sick or wounded members of the armed forces are nursed in these hospitals, or the presence of small arms and ammunition taken from such combatants and not yet handed to the proper service, shall not be considered acts harmful to the enemy.[430]

7.17.1.2 *Due Warning Before Cessation of Protection*. In addition, protection for civilian hospitals may cease only after due warning has been given, naming, in all appropriate cases, a reasonable time limit, and after such warning has remained unheeded.[431]

The obligation to refrain from use of force against a civilian medical facility acting in violation of its mission and protected status without due warning does not prohibit the exercise of the right of self-defense. There may be cases in which, in the exercise of the right of self-defense, a warning is not "due" or a reasonable time limit is not appropriate. For example, forces receiving heavy fire from a hospital may exercise their right of self-defense and return fire. Such use of force in self-defense against medical units or facilities must be proportionate. For example, a single enemy rifleman firing from a hospital window would warrant a response against the rifleman only, rather than the destruction of the hospital.[432]

7.17.2 Identification of Protected Civilian Hospitals. The GC provides for civilian hospitals to be identified with certificates and the distinctive emblem.

7.17.2.1 *State-Issued Certificates for Civilian Hospitals*. States that are parties to a conflict shall provide all civilian hospitals with certificates showing that they are civilian hospitals and that the buildings that they occupy are not used for any purpose that would deprive these hospitals of protection in accordance with Article 19 of the GC.[433]

7.17.2.2 *Markings of Civilian Hospitals With the Distinctive Emblem*. Civilian hospitals shall be marked by means of the distinctive emblem provided for in Article 38 of the GWS, but only if authorized by the State.[434]

The parties to the conflict shall, in so far as military considerations permit, take the necessary steps to make the distinctive emblems indicating civilian hospitals clearly visible to the enemy land, air, and naval forces in order to obviate the possibility of any hostile action.[435]

7.17.3 Location of Civilian Hospitals. In view of the dangers to which civilian hospitals may be exposed by being close to military objectives, it is recommended that such hospitals be situated as far as possible from such objectives.[436]

This provision of the GC was drafted because the protection for civilian hospitals in the GC does not confer immunity on military objectives situated close to the hospital or restrict the right to make such military objectives the object of attack.[437]

7.17.4 Protection of Civilian Hospital Personnel.

7.17.4.1 *Civilian Hospital Personnel Who Are Regularly and Solely Engaged.* Persons regularly and solely engaged in the operation and administration of civilian hospitals, including the personnel engaged in the search for, removal and transporting of, and caring for wounded and sick civilians, the infirm, and maternity cases, shall be respected and protected at all times.[438]For example, a surgeon who works regularly in a hospital, but is not exclusively employed there, would not be engaged "solely" in hospital duties and would consequently not be covered by this provision, but could fall under the category of "other civilian hospital personnel."[439]

In occupied territory and in zones of military operations, civilian hospital personnel who are regularly and solely engaged in hospital duties shall be recognizable by means of an identity card certifying their status, bearing the photograph of the holder and embossed with the stamp of the responsible authority, and also by means of a stamped, water-resistant armlet that they shall wear on the left arm while carrying out their duties.[440]This armlet shall be issued by the State and shall bear the emblem provided for in Article 38 of the GWS.[441]

7.17.4.2 *Other Civilian Hospital Personnel.* Other personnel who are engaged in the operation and administration of civilian hospitals shall also be entitled to respect and protection and to wear the armlet while they are employed on such duties.[442]The identity card shall state the duties on which they are employed.[443]

7.17.4.3 *List of Permanent and Temporary Civilian Hospital Personnel.* The management of each hospital shall at all times hold at the disposal of the competent national or occupying authorities an up-to-date list of such personnel.[444]This list should provide the names of all hospital personnel, both permanent and temporary, and specify the duties of each of them.[445]

7.18 LAND AND SEA CIVILIAN HOSPITAL CONVOYS

Convoys of vehicles or hospital trains on land or specially provided vessels on sea, conveying wounded and sick civilians, the infirm, and maternity cases, shall be respected and protected in the same manner as the hospitals provided for in Article 18 of the GC, and shall be marked, with the consent of the State, by the display of the distinctive emblem provided for in Article 38 of the GWS.[446]

The vehicles in a medical convoy need not be used permanently for medical purposes, but while they are in the convoy, they are not to be used for other purposes.[447]

7.18.1 Protection of Civilian Hospital Convoys on the Same Basis as That of Civilian Hospitals. The protection to which these convoys are entitled shall not cease unless they are used to commit, outside their humanitarian duties, acts harmful to the enemy.[448]

As with civilian hospitals, medical convoys must maintain a strict abstention from any direct or indirect participation in a hostile act in order to retain protection.[449]However, the fact that sick or wounded members of the armed forces are transported by these convoys, or the presence of small arms and ammunition taken from such combatants and not yet handed to the proper service, shall not be considered acts harmful to the enemy.[450]

In addition, protection for civilian medical convoys may cease only after due warning has been given, naming, in all appropriate cases, a reasonable time limit, and after such warning has remained unheeded.[451]

7.19 CIVILIAN MEDICAL AIRCRAFT

Aircraft exclusively employed for the removal of wounded and sick civilians, the infirm, and maternity cases, or for the transport of medical personnel and equipment, shall not be attacked, but shall be respected while flying at heights, times, and on routes specifically agreed upon between all the parties to the conflict concerned.[452]

7.19.1 Civilian Medical Aircraft Not Flying at Heights and Times and on Routes Specifically Agreed Upon by Belligerents Concerned. As with medical aircraft for military personnel, the use of protected medical aircraft for civilians generally depends on an agreement between the belligerents.[453]

However, as with medical aircraft for military personnel, known civilian medical aircraft, when performing their humanitarian functions, must be respected and protected, provided they identify themselves as such and comply with the directions of the enemy State.[454]

7.19.2 Marking of Civilian Medical Aircraft With the Distinctive Emblem. The aircraft may be marked with the distinctive emblem provided for in Article 38 of the GWS.[455]

7.19.3 Prohibition of Flight Over Enemy Territory. Unless otherwise agreed, flights over enemy or enemy-occupied territory are prohibited.[456]

7.19.4 Summons to Land. Such aircraft shall obey every summons to land.[457]If the aircraft refuses to obey, it does so at its own risk.[458]

In the event of a landing thus imposed, the aircraft with its occupants may continue its flight after examination, if any.[459]However, if examination reveals that an act "harmful to the enemy" has been committed, *e.g.*, if the aircraft is carrying munitions or has been used for military observation, it loses the benefit of the GC; the aircraft may be seized and the crew and passengers detained.[460]

7.20 AP I PROVISIONS ON THE WOUNDED, SICK, AND SHIPWRECKED Part II of AP I has a number of provisions regarding the wounded, sick, and shipwrecked. These provisions have generally not been discussed in this Chapter because the United States is not a Party to AP I. These provisions, however, are summarized in this section to familiarize service members with them in case U.S. forces are engaged in multinational operations with, or are engaged in armed conflict against, States that are Parties to AP I.

7.20.1 AP I Provisions on General Protection of the Wounded, Sick, and Shipwrecked Persons. Section I of Part II of AP I seeks to elaborate upon and change the protections afforded the wounded, sick, and shipwrecked. For example, AP I provides protection for certain civilian medical units.[461]The obligations in AP I also cover both combatants and civilians. For the purpose of applying AP I, Article 8 of AP I defines "wounded" and "sick" to include persons "whether military or civilian" and "maternity cases, new-born babies and other persons who may be in need of immediate medical assistance or care, such as the infirm or expectant mothers, and who refrain from any act of hostility."[462]On the other hand, in the 1949 Geneva Conventions, the protections for civilians who are wounded and sick are addressed in the GC, and the protections for combatants who are wounded and sick are addressed in the GWS, GWS-Sea, and GPW.[463]

7.20.2 AP I Provisions on Medical Transportation. Section II of AP I has a number of articles that are intended to expand upon the protection afforded to medical transports. For example, under AP I, the respect and protection of medical aircraft of a party to the conflict in and over land areas physically controlled by friendly forces or in and over sea areas not physically controlled by an adverse party are not dependent on any agreement with an adverse party.[464]Similarly, the protection of small craft described in Article 27 of the GWS-Sea is afforded, even if the notification envisaged by that Article has not been made.[465]

7.20.3 AP I Provisions on Missing and Dead Persons. Section III of Part II of AP I has three articles that are intended to help families know the fate of their relatives.[466]For example, as soon as circumstances and the relations between adverse parties permit, the States that are Parties to AP I in whose territories graves may be shall conclude agreements: (1) to facilitate access to gravesites by relatives of the deceased and by representatives of official Graves Registration Services and to regulate the practical arrangements for such access; (2) to protect and maintain such gravesites permanently; and (3) to facilitate the return of the remains of the deceased and of personal effects to the home country upon its request or, unless that country objects, upon the request of the next of kin.[467]

VIII – Detention: Overview and Baseline Rules

Chapter Contents

8.1 INTRODUCTION

8.1.1 Overview of Detention Rules in This Manual and the Scope of Chapter VIII. The internment of POWs is addressed in Chapter IX.

The internment of protected persons in the home territory of a belligerent or in occupied territory is addressed in Chapter X. Detention may occur outside these contexts. For example, the detention of civilians may occur outside a belligerent's home territory but before the establishment of occupation. Similarly, detention may occur in non-international armed conflict or other military operations, such as non-combatant evacuation operations, peace operations, or humanitarian assistance operations. There are no "law-free zones" in which detainees are outside the protection of the law. Where no more specific rules apply, the rules in this Chapter apply as baseline rules addressing the detention of persons during military operations. For example, the following situations are covered by the rules in this Chapter:

- detention of persons who have participated in hostilities or who belong to armed groups, but who are not entitled to POW status or protected person status in international armed conflict;
- detention of persons held for reasons related to a non-international armed conflict; and
- any other detentions.

8.1.2 DoD Policies and Regulations Regarding the Treatment of Detainees. DoD policies and regulations regarding the treatment of detainees provide authoritative guidance for DoD personnel and fill an important role in implementing the requirements of Common Article 3 of the 1949 Geneva Conventions and applicable customary international law.[1]Practitioners are advised to consult all applicable policies and regulations, as these, in many cases, exceed the requirements of international law, U.S. statutes, and Executive Orders. The practice of the U.S. armed forces has been to go beyond a minimalist approach of complying with the baseline rules of Common Article 3 of the 1949 Geneva Conventions and instead to endeavor to enhance conditions so as to ensure that treatment of detainees exceeds the requirements of humane treatment standards under international law, consistent with security concerns.[2]

8.1.3 Ad Hoc Legal Instruments or Frameworks That May Be Applicable to Detention Outside the Context of POW or Civilian Internment. The detention of persons under the law of war, outside the context of the regimes specified in Chapters IX (POWs under the GPW) and X (protected persons under the GC who are interned in a belligerent's home territory or occupied territory) may be addressed in ad hoc legal instruments or frameworks.

Such legal instruments may include, for example, an international agreement between the State conducting detention operations and the State on whose territory the operations are being conducted.[3] As another example, a U.N. Security Council Resolution may provide a basis for detention operations.[4]

8.1.3.1 *Detention Authority*. The legal authority to detain outside the context of the internment of POWs or protected persons may be addressed in ad hoc international instruments. Even if not specifically addressed in an international legal instrument, the legal authority to detain may sometimes be understood as an incident of general authorities conferred by that instrument, such as a U.N. Security Council Resolution authorizing military operations or an agreement authorizing one State to conduct military operations on the territory of another State.[5]

In many cases, the legal authority to detain would be understood as an exercise of the Detaining Power's sovereign rights under international law rather than, or in addition to, authorities arising from an international legal instrument.[6] The authority to detain is often understood as an incident to more general authorities because detention is fundamental to waging war or conducting other military operations (*e.g.*, noncombatant evacuation operations, peacekeeping operations).[7] Detention operations may be militarily necessary to achieve the object of those operations.[8] In addition, the right to use force in self-defense during such operations includes at least a limited right to detain for security reasons.[9] In fact, it may be inhumane to conduct military operations without some provision for those who are detained incident to the operations (*e.g.*, being prepared to conduct detention operations, and provision for transfer of captured persons to coalition partners who are conducting humane detention operations).[10]

8.1.3.2 *Specific Procedures or Restrictions Applicable to Detention Outside the Context of POW or Protected Person Internment*. Ad hoc legal instruments may provide additional procedures or restrictions applicable to detention outside the context of the internment of POWs or protected persons. For example, special agreements between opposing non-State parties to a conflict could be used to bring into force provisions of the GPW or GC for the purposes of that conflict.[11] As another example, agreements among coalition partners regarding the transfer of detainees between one another might prescribe additional requirements.

8.1.4 Sources of Baseline Standards for the Treatment of Detainees. The baseline standards for the treatment of detainees by U.S. armed forces addressed in this Chapter are derived from Common Article 3 of the 1949 Geneva Conventions, applicable customary international law, and applicable U.S. law and national policy. The principle of humanity animates the rules discussed in this Chapter.[12] In addition, analogous provisions of the GPW and the GC may also be helpful in understanding the baseline rules for detention.

8.1.4.1 *Common Article 3 of the 1949 Geneva Conventions*. Although Common Article 3 of the 1949 Geneva Conventions provides that it applies "[i]n the case of armed conflict not of an international character occurring in the territory of one of the High Contracting Parties," Common Article 3 reflects minimum standards for humane treatment that apply to all military operations.[13] Thus, during both international and non-international armed conflict, Common Article 3 reflects a minimum yardstick of humane treatment protections for "all persons taking no active part in hostilities, including members of armed forces who have laid down their arms and those placed *hors de combat* by sickness, wounds, detention, or any other cause."[14]

DoD policy has explicitly incorporated the standards in Common Article 3 as minimum standards.[15]

8.1.4.2 *Article 75 of AP I and Relevant AP II Provisions*. Article 75 of AP I reflects fundamental guarantees for the treatment of persons detained during international armed conflict. Although not a Party to AP I, the United States has stated that the U.S. Government will choose out of a sense of legal obligation to treat the principles set forth in Article 75 as applicable to any individual it detains in an international armed conflict, and expects all other nations to adhere to these principles as well.[16] This statement was intended to contribute to the crystallization of the principles contained in Article 75 as rules of customary international law applicable in international armed conflict.[17]

AP II also contains provisions that expand on Common Article 3's protections for persons held for reasons related to non-international armed conflict. Although the United States has not ratified AP II, President Reagan submitted AP II to the Senate for its advice and consent to ratification, and the Executive Branch has urged the Senate to act as soon as practicable on AP II, noting further that reviews have determined that relevant U.S. military practices are consistent with AP II.[18]

8.1.4.3 *U.S. Domestic Law and National Policy*. In addition to international law, the requirements in this Chapter are also based on the requirements of applicable U.S. statutes, such as the Detainee Treatment Act of 2005, or national policy, such as the requirements stated in applicable Executive Orders.

8.1.4.4 *Analogous GPW and GC Provisions*. In some cases, analogous provisions of the GPW and the GC may also be helpful in understanding the baseline rules for detention. For example, it may be appropriate to apply the principles of the GPW and GC, even when the relevant provisions do not apply as a matter of law.[19]

In addition, as a general matter, analogous provisions of the GPW and GC may be helpful for understanding the requirements in international law for conducting detention operations because the baseline standards for such operations during armed conflict are not more restrictive upon States than the requirements with respect to POWs and civilian internees under the GPW and GC, respectively.[20]

In this Chapter, cross-references to sections that address analogous GPW and GC requirements are used for the purposes described above.

8.1.4.5 *Copenhagen Process on the Handling of Detainees in International Military Operations*. From 2007 to 2012, U.S. representatives participated, along with those of 23 other States and from several international organizations, in a process led by the Government of Denmark to formulate principles and guidelines on the handling of detainees in international military operations.[21]Participants in the Copenhagen Process sought to develop principles to guide the implementation of the existing obligations with respect to detention in international military operations and to promote best practices.[22]The Copenhagen Process Principles and Guidelines is not a text of a legally binding nature.[23]

The Copenhagen Process Principles and Guidelines are intended to apply to international military operations in the context of non-international armed conflicts and peace operations; they are not intended to address international armed conflicts.[24]

8.2 HUMANE TREATMENT OF DETAINEES

Detainees shall in all circumstances be treated humanely and protected against any cruel, inhuman, or degrading treatment. This requirement has been reflected in international law,[25]domestic law,[26]national policy,[27]and DoD policies.[28]Violations of the requirement to treat detainees humanely may be violations of criminal law, although it is not a purpose of this section to address liability under criminal law.

8.2.1 Protection Against Violence, Torture, and Cruel Treatment. Detainees must be protected against violence to life and person, in particular murder of all kinds, mutilation, cruel treatment, torture, and any form of corporal punishment.[29]

8.2.2 Protection Against Humiliating or Degrading Treatment. Detainees must be protected against outrages upon personal dignity, in particular, humiliating and degrading treatment.[30]

8.2.2.1 *Protection Against Rape, Enforced Prostitution, and Other Indecent Assault*. Detainees must be protected against rape, forced prostitution, and other indecent assault.[31]Indecent assault is generally referred to today as sexual assault.[32]

8.2.2.2 *Protection Against Public Curiosity*. Detainees must be protected against insults and public curiosity.[33]For example, displaying detainees publicly with the purpose of exposing them to ridicule and humiliation is prohibited.[34]

8.2.2.3 *DoD Practice of Generally Prohibiting Taking Photographs Without Authorization*. In order to protect detainees against public curiosity, among other reasons, DoD policy has generally prohibited the taking of photographs of detainees except for authorized purposes.[35]

8.2.3 Prohibition Against Biological or Medical Experiments. Medical or biological experiments on detainees are prohibited.[36]

8.2.4 Threats to Commit Inhumane Treatment. Threats to commit the unlawful acts described above (*i.e.*, violence against detainees, or humiliating or degrading treatment, or biological or medical experiments) are also prohibited.[37]This prohibition may be understood to arise separately (*i.e.*, as a distinct prohibition against certain threats), or it may be understood to result when such threats themselves constitute a form of torture or other abuse.[38]

8.2.5 Duty to Protect Detainees. Detainees should be protected not only against unlawful acts by agents of the Detaining Power, but also against unlawful acts from others, including other detainees or the civilian population.[39]

8.2.6 No Adverse Distinction. Detainees shall be treated humanely without any adverse distinction founded on race, color, religion or faith, sex, birth or wealth, national or social origin, political or other opinion, or any other similar criteria.[40]

Special consideration for more vulnerable categories of detainees is consistent with the requirement that detainees must be humanely treated without adverse distinction.[41]

Distinction based on the above criteria may also be made so long as it is not adverse and it is made to advance legitimate interests, such as maintaining order in the camp.[42]

8.3 SECURITY MEASURES

8.3.1 Searches. Although detainees must always be treated humanely, detainees and their property may be searched and secured, when necessary for security reasons and intelligence purposes.[43]

The dignity and honor of the detainee being searched should be protected to the greatest degree possible under the circumstances.[44]The person conducting the search should avoid doing or saying anything likely to be regarded as indecent. In some circumstances, it may be appropriate for a witness to observe the search so as to protect both the person being searched from abuse and the person conducting the search from unfounded accusations of abuse.

8.3.1.1 *Search of Female Detainees*. It is appropriate for female detainees to be searched by female personnel. This practice helps reduce the risk of accusations of indecent behavior.

8.3.2 Search of Detainee Property. Items in the possession of detainees may be removed and searched for security and intelligence purposes.[45]

8.3.3 Securing Detainees With Handcuffs and Other Security Devices. When necessary for security reasons, detainees may be secured temporarily with handcuffs, flex cuffs, blindfolds, or other security devices.[46]

8.3.4 Use of Force to Maintain Order and to Prevent Escape. Physical force is not to be used against a detained person except in circumstances where such force is necessary and proportionate.[47]

As with POWs, the use of weapons against detainees, especially against those who are escaping or attempting to escape, shall constitute an extreme measure, which shall always be preceded by warnings appropriate to the circumstances.[48]For example, detainees should not be fired upon if they are apprehended within the camp limits while making preparations to escape, and there is no risk of escape or harm to anyone.

8.3.4.1 *Use of Non-Lethal Weapons (Including Riot Control Agents)*. Non-lethal weapons may be employed to control rioting detainees or to prevent their escape. In particular, the use of riot control agents to control rioting detainees is not prohibited.[49]If the use of deadly force is warranted and authorized against detainees, there is no legal requirement to employ non-lethal weapons before resort to deadly force.[50]

8.4 INTERROGATION

The law of war does not prohibit interrogation of detainees, but interrogation must be conducted in accordance with the requirements for humane treatment.

8.4.1 Humane Treatment During Interrogation. Interrogation must be carried out in a manner consistent with the requirements for humane treatment, including the prohibitions against torture, cruelty, degrading treatment, and acts or threats of violence.[51]In addition to the legal prohibitions on torture or other illegal methods of interrogation, practical considerations have also strongly counseled against such methods.[52]

8.4.2 Additional U.S. Law and Policy on Interrogation. U.S. law and policy impose additional requirements on the interrogation of detainees.

8.4.2.1 *Only Treatment or Techniques of Interrogation Authorized by and Listed in the U.S. Army Field Manual on Intelligence Interrogation*. No person in the custody or under the effective control of the Department of Defense or under detention in a Department of Defense facility shall be subject to any treatment or technique of interrogation not authorized by and listed in the United States Army Field Manual on Intelligence Interrogation.[53]The Army has published field manuals on interrogation for many years.[54]

8.5 ADEQUATE FOOD, DRINKING WATER, AND CLOTHING

Detainees shall be provided with adequate food, drinking water, and clothing.

8.5.1 Food for Detainees. Daily food rations for detainees shall be sufficient in quantity, quality, and variety to keep detainees in a good state of health or in any event no worse than that afforded the local civilian population.[55]

8.5.1.1 *Accounting for the Customary Diet.* DoD practice has been to take account of the customary diet of the internees.[56]For example, the detainee's cultural and religious requirements have been considered in determining and ensuring the appropriate diet.

8.5.2 Drinking Water. Sufficient drinking water shall be supplied to detainees.[57]

8.5.3 Clothing. As needed, detainees shall be supplied with sufficient clothing, underwear, and footwear that is appropriate for the climate.[58]

8.6 GENERAL CONDITIONS OF DETENTION FACILITIES

Detainees shall be afforded, to the same extent as the local civilian population, safeguards as regards health and hygiene and protection against the rigors of the climate and the dangers of the armed conflict.[59]

8.6.1 Safeguards as Regards Health and Hygiene. The obligation to afford safeguards as regards health and hygiene would include taking sanitary measures necessary to ensure the cleanliness and healthfulness of camps and to prevent epidemics. For example, the DoD practice has been to give detainees medical examinations upon arrival at an internment facility, disinfect them, and provide them with any necessary inoculations. In addition, quarters should be kept free from vermin, and detainees suffering from contagious diseases should be placed in quarantine as needed.[60]

8.6.2 Protection Against the Rigors of Climate. The obligation to afford protection against the rigors of climate would include ensuring that detention facilities are adequately heated, lighted, and protected against dampness.[61]

8.6.3 Protection From the Dangers of Armed Conflict. Detention facilities shall not be located close to the combat zone.[62]

8.6.4 No Prohibition Against Detention Aboard Ships. Provided the above requirements are met, there is no prohibition against the humane detention of persons on ships.[63]The GPW generally requires that POWs be interned on premises located on land.[64]

8.7 SEGREGATION OF DETAINEES

Detainees may be segregated into camps or camp compounds according to their nationality, language, and customs, and the Detaining Power may use other criteria to segregate detainees for administrative, security, intelligence, medical, or law enforcement purposes.[65]

8.7.1 Gender and Family Segregation. Except when men and women of a family are accommodated together, women shall be held in quarters separated from those of men and shall be under the immediate supervision of women.[66]

8.8 MEDICAL ATTENTION

Wounded and sick detainees shall be cared for.[67]They should receive the medical care and attention required by their condition.[68]Medical or biological experiments on detainees are prohibited.[69]

8.8.1 Consent. Medical care should, wherever possible, be undertaken with the consent of the wounded or sick detainee.[70]However, medical actions to preserve the health of the detainee may be justified even where the detainee refuses to provide consent.[71]For example, it is not prohibited to administer vaccinations to detainees in order to preserve their health and to prevent epidemics. Similarly, it is not prohibited to order detainees to be fed, if they undertake a hunger strike.[72]

8.8.2 Blood Donation and Skin Grafting. Detainees may voluntarily consent to give blood for transfusion or skin for grafting for therapeutic purposes; such procedures should take place under conditions consistent with generally accepted medical standards and controls designed for the benefit of both the donor and the recipient.[73]

8.9 ADMINISTRATION AND DISCIPLINE IN DETENTION FACILITIES

Detention facility commanders may establish rules for detainees in the detention facility and conduct disciplinary proceedings, subject to the requirements for humane treatment.[74]

8.9.1 Complaints. Detainees should be permitted to submit, without reprisal, complaints regarding their treatment or conditions of detention to the detaining authorities.[75]Investigatory procedures and a practice of timely and effective investigations of credible complaints of maltreatment also help ensure that allegations regarding treatment or conditions of detention that may arise years after the detention can be addressed.[76]

8.10 CONTACTS WITH THE OUTSIDE WORLD

Subject to security measures, practical considerations, and other military necessities, detainees should be afforded appropriate contact with the outside world, including: (1) receipt of individual or collective relief; (2) correspondence; (3) communication with family; and (4) ICRC access.

8.10.1 Receipt of Individual or Collective Relief. Detainees shall be allowed to receive individual or collective relief.[77]

8.10.2 Correspondence. Detainees shall be allowed to send and receive letters and cards, the number of which may be limited by a competent authority if it deems necessary.[78]

8.10.3 Communication With Family. DoD practice has been, where practicable, to grant detainees the means to communicate with family members (*e.g.*, exchange of letters, phone calls, and video teleconferences with family, family visits).[79]

8.10.4 ICRC Access to Detainees. An impartial humanitarian body, such as the ICRC, may offer its services to the parties to the conflict.[80]

All departments and agencies of the Federal Government shall provide the International Committee of the Red Cross with notification of, and timely access to, any individual detained in any armed conflict in the custody or under the effective control of an officer, employee, or other agent of the U.S. Government or detained within a facility owned, operated, or controlled by a department or agency of the U.S. Government, consistent with Department of Defense regulations and policies.[81]

8.11 RELIGIOUS EXERCISE

Detainees shall be granted free exercise of religion, consistent with the requirements of detention.[82]Detainees' religious practices shall be respected.[83]They shall be allowed to practice their religion and, if requested and appropriate, to receive spiritual assistance from persons, such as chaplains, performing religious functions.[84]

DoD practice has been for detainees to be provided religious materials of their faith (*e.g.*, copies of religious texts) as well as time and other accommodations for religious exercise.[85]

8.12 INTELLECTUAL, PHYSICAL, AND RECREATIONAL ACTIVITIES

Detainees should be provided with reasonable access to intellectual, physical, and recreational activities.[86]Such opportunity is increasingly important as detention occurs for longer periods of time. Access to additional facilities for intellectual, physical, and recreational activities may be made contingent on compliance with camp rules.[87]Education programs may be offered to discourage violent extremism.[88]

8.13 ADEQUATE WORKING CONDITIONS

If made to work, detainees shall have the benefit of working conditions and safeguards similar to those enjoyed by the local civilian population.[89]

8.13.1 Prohibition on Slavery. Slavery and the slave trade in all their forms are prohibited.[90]For example, enforced prostitution is prohibited.[91]

8.14 PROCEDURES FOR DETENTION

8.14.1 Informing of Reasons for Detention. Detainees shall be informed promptly of the reasons for their detention in a language that they understand.[92]

A prompt notification should generally occur within 10 days of detention.[93]The notification should ensure that the detainee adequately understands the general reason for detention (*e.g.*, security reasons, participation in hostilities or association with an armed group that is engaged in hostilities, suspect or witness in law enforcement proceedings).[94]When feasible, more specific information should be provided to the detainee so that he or she may participate in subsequent review procedures from an informed position.[95]

8.14.2 Review of Continued Detention for Security Reasons. DoD practice has been to review periodically the detention of all persons not afforded POW status or treatment.[96]A detainee who has been deprived of liberty for security reasons is to have, in addition to a prompt initial review, the decision to detain reconsidered periodically by an impartial and objective authority that is authorized to determine the lawfulness and appropriateness of continued detention.[97]

The authority conducting the review is not necessarily outside the military, and often would be a military commander.[98]

There is no fixed requirement regarding how often the review must occur, and the period between reviews will depend on a variety of factors, including: (1) operational necessities or resource constraints, such as force protection, the availability of interpreters, or large numbers of detainees; (2) the thoroughness of the review process; and (3) whether there is a true prospect that the legal or factual predicates justifying detention have changed.[99]

8.14.3 Release When the Circumstances Justifying Detention Have Ceased to Exist. Except in cases of arrest or detention for penal offenses, such persons shall be released with the minimum delay possible and in any event as soon as the circumstances justifying the arrest, detention, or internment have ceased to exist.[100]

8.14.3.1 *Participants in Hostilities or Persons Belonging to Armed Groups That Are Engaged in Hostilities*. For persons who have participated in hostilities or belong to armed groups that are engaged in hostilities, the circumstance that justifies their continued detention is the continuation of hostilities.[101]Thus, release of such persons is generally only required after the conflict has ceased.[102]As a matter of policy, release of lawfully detained persons often occurs before the conclusion of hostilities.[103]

However, even after hostilities have ceased, other circumstances may warrant continued detention of such persons. For example, persons who have participated in hostilities on behalf of non-State armed groups might be detained pending law enforcement proceedings and, after a conviction, pursuant to a lawful sentence.[104]

8.14.3.2 *Safe and Orderly Release*. If it is decided to release persons deprived of their liberty, necessary measures to ensure their safety shall be taken by those so deciding.[105]For example, detainees should not be released into a situation in which the detainee would be attacked by hostile elements upon release.[106]

The obligation to release detainees in a safe and orderly manner, as well as operational necessities (*e.g.*, force protection) and logistical constraints (*e.g.*, the limited availability of transport), may make it difficult to release a detainee at the precise moment that the circumstances justifying detention cease.[107]Continued detention in order to facilitate a safe and orderly release may be necessary.[108]

8.14.4 Transfers of Detainees. Because the transfer of a detainee from U.S. custody to another entity or State may implicate certain U.S. legal obligations or policies, detainees should not be transferred to another State or entity without proper authorization.[109]

8.14.4.1 *U.S. Policy Prohibiting Transfers in Cases in Which Detainees Would Likely Be Tortured*. U.S. policy provides that no person shall be transferred to another State if it is more likely than not that the person would be tortured in the receiving country, regardless of whether the person is physically present in the United States or whether such transfer would otherwise be consistent with applicable law.[110]

8.15 NATIONAL ACCOUNTABILITY FOR DETENTION

A proper accounting of detainees is an important part of a State's implementation of the requirements of humane treatment.

8.15.1 Registration of Detainees. The detaining authority should register detainees promptly.[111]Detainees should be registered within a reasonable time, taking into account other essential tasks and resource limitations that may affect the detaining authority's ability to register detainees.[112]

Registration of detainees assists in ensuring that all detainees can be accounted for and that allegations of illegal detention can be addressed.[113]DoD practice has been to register detainees with the National Detainee Reporting Center that is also used to account for the detention of POWs under the GPW and protected persons under the GC.[114]

8.15.2 Records of Detainee Property. DoD practice has been for property in the possession of detainees to be inventoried, and for records of such property to be maintained in order to maintain accountability of it (*e.g.*, to prevent theft) and to ensure its lawful disposition.[115]

8.16 CRIMINAL PROCEDURE AND PUNISHMENT

The passing of sentences and the carrying out of executions without previous judgment pronounced by a regularly constituted court affording all the judicial guarantees that are recognized as indispensable by civilized peoples are prohibited.[116]

The specific rules addressed in this section may be understood as examples of such guarantees.[117]

8.16.1 No Private Rights Created by This Section in Particular. As with all other sections of this manual, Section 8.16 does not create any private rights.[118]The competent prosecuting authorities shall ensure that the requirements reflected in this section are met.

8.16.2 General Limits on Punishments.

8.16.2.1 *Individual Penal Responsibility and No Collective Punishment*. No one shall be convicted of an offense except on the basis of individual penal responsibility.[119]Collective punishments are prohibited. Collective punishments have been specifically prohibited with respect to POWs, protected persons under the GC, and inhabitants of occupied territory, and also in connection with non-international armed conflict.[120]This prohibition includes penalties of any kind inflicted upon persons or groups of persons for acts that these persons have not committed, including administrative penalties.[121]Punishment on the basis of conspiracy, joint criminal enterprise, and other theories of secondary liability are not prohibited by this rule.[122]

8.16.2.2 *No Crime or Punishment Without Prior Law*. No one shall be accused or convicted of a criminal offense on account of any act or omission that did not constitute a criminal offense under the national or international law to which he or she was subject at the time when it was committed; nor shall a heavier penalty be imposed than that which was applicable at the time when the criminal offense was committed; if, after the commission of the offense, provision is made by law for the imposition of a lighter penalty, the offender shall benefit thereby.[123]

8.16.2.3 *No Repetition of Punishment*. No one shall be prosecuted or punished by the same party for an offense in respect of which a final judgment acquitting or convicting that person has been previously pronounced under the same law and judicial procedure.[124]

8.16.2.4 *Limitations on the Death Penalty*. The death penalty shall not be pronounced on persons who were under the age of eighteen years at the time of the offense and shall not be carried out on pregnant women or mothers of young children.[125]

8.16.3 Rights of Defense and Trial Procedure. The procedure shall provide for an accused to be informed without delay of the particulars of the notice alleged against him or her and shall afford the accused before and during his or her trial all necessary rights and means of defense.[126]

8.16.3.1 *Presumption of Innocence*. Anyone charged with an offense is presumed innocent until proved guilty according to law.[127]

8.16.3.2 *Right to Trial in Presence*. Anyone charged with an offense shall have the right to be tried in his or her presence.[128]

8.16.3.3 *No Compulsory Self-Incrimination*. No one shall be compelled to testify against himself or herself or to confess guilt.[129]

8.16.3.4 *Right to Obtain and Examine Witnesses*. Anyone charged with an offense shall have the right to examine, or have examined, the witnesses against him or her and to obtain the attendance and examination of witnesses on his or her behalf under the same conditions as witnesses against him or her.[130]

8.16.3.5 *Right to Have a Public Judgment*. Anyone prosecuted for an offense shall have the right to have the judgment pronounced publicly.[131]

8.16.3.6 *Advice on Appellate Procedure*. A convicted person shall be advised on conviction of his or her judicial and other remedies, and of the time-limits within which they may be exercised.[132]

IX – Prisoners of War (POWs)

Chapter Contents

9.1 INTRODUCTION

This Chapter addresses the treatment of prisoners of war (POWs).

9.1.1 Brief History of POW Law. States have made significant humanitarian advances in the international law relating to the treatment of POWs.[1] Before the modern law of war, POWs were often put to death, held for ransom or as hostages, or sold into slavery.

Gradually, protections for POWs, such as providing for their repatriation at the end of conflict without ransom, were concluded in treaties.[2] The United States concluded bilateral treaties for the humane treatment and protection of POWs, written with a view towards forbidding abuses that occurred during the Revolutionary War.[3] Eventually multilateral treaties for the protection of POWs were concluded, which also drew from the experience of POWs during prior conflicts.[4] War crimes trials after World War II for POW mistreatment recognized that the humane treatment of POWs was required under customary international law.[5] After World War II, the 1949 Geneva Conventions were concluded, and more than 190 States, including the United States, are Parties to the GPW.[6]

9.1.2 Interpretation and Application of the GPW. The GPW underlies most of the international law rules applicable to the United States for the treatment of POWs. The GPW's provisions should be interpreted in light of the principles that underlie POW detention and, in particular, in light of the goal of advancing the humane treatment of POWs.[7]

The subsequent practice of States in applying the GPW could assist in interpreting its provisions because States' decades of experience in applying the GPW may be very helpful in understanding its requirements.[8]

9.1.2.1 *GPW – Notes on Terminology*. "Prisoner of war" (abbreviated POW in this manual) does not refer to any person who is held prisoner in connection with a war; rather, POW is term of art that is defined and used in the GPW.[9]

In the GPW, the *Detaining Power* refers to the State that holds the POW.

In the GPW, the *Power on which the POWs depend* means the State that the POWs served before falling into the power of the enemy. (In some cases, this State might be different from the POW's State of nationality.)

In the GPW, the *Protecting Power* refers to a neutral State that helps implement the GPW.[10]

9.1.2.2 *Special Agreements Under the GPW*. Under the GPW, States may conclude a variety of special agreements during international armed conflict to facilitate the protection of POWs. The GPW specifically provides for agreements:

- to entrust to an effective and impartial organization the duties of the Protecting Powers;[11]
- to establish a system for marking POW camps;[12]
- on how to allocate the profits made by camp canteens in case of a general repatriation;[13]
- to establish the corresponding ranks for their medical personnel and staff of National Red Cross Societies and other Voluntary Aid Services in order to help determine the senior medical officer in POW camps;[14]
- for the relief of retained personnel and the procedure for relief;[15]
- to modify the amount of advances of pay due to POWs based on their military rank during and at the close of hostilities;[16]
- to notify periodically, through the Protecting Power, the account balances of POWs;[17]
- to vary the required information provided for in the list of all appropriate particulars of all POWs whose captivity has been terminated;[18]
- regarding individual parcels and collective relief shipments to POWs;[19]
- for the accommodation in neutral countries of seriously wounded and sick POWs;[20]
- for the direct repatriation or internment in a neutral country of able-bodied POWs who have undergone a long period of captivity;[21]
- to determine the conditions that POWs who are accommodated in a neutral country must fulfill in order to permit their repatriation, and to establish the status of POWs accommodated in a neutral country;[22]
- to determine the equitable apportionment of costs of and a plan for repatriation for when hostilities have ceased;[23]
- to regulate the conditions of the transport of personal effects and the payment of the costs associated with the repatriation of POWs;[24]
- to establish a commission to search for dispersed POWs and to ensure their repatriation with the least possible delay;[25] and
- on the procedure, or to select an umpire who will decide the procedure, to be followed for an inquiry concerning any alleged violation of the GPW.[26]

In addition to the agreements expressly provided for in Articles 10, 23, 28, 33, 60, 65, 66, 67, 72, 73, 75, 109, 110, 118, 119, 122, and 132 of the GPW, Parties to the GPW may conclude other special agreements for all matters concerning which they may deem it suitable to make separate provision.[27]No special agreement shall adversely affect the situation of POWs, as defined by the GPW, nor restrict the rights that it confers upon them.[28]

POWs shall continue to have the benefit of such agreements as long as the GPW is applicable to them, except where express provisions to the contrary are contained in these or subsequent agreements, or where more favorable measures have been taken with regard to them by one or other of the parties to the conflict.[29]

9.1.3 DoD Policies and Regulations for the Treatment of POWs. DoD policies and regulations regarding the treatment of POWs provide authoritative guidance for DoD personnel and fill an important role in implementing the complex requirements of the GPW.[30]Practitioners are advised to consult all applicable policies and regulations, as these, in many cases, exceed the requirements of the GPW, U.S. statutes, and Executive Orders.

9.2 GENERAL PRINCIPLES APPLICABLE TO THE TREATMENT OF POWS Certain principles provide the foundation for the rules governing the treatment of POWs.

9.2.1 Detention Is Non-Punitive in Character. The detention of POWs is not punishment, and is devoid of all penal character.[31]It is intended to disable the POW and to prevent him or her from further participation in hostilities.[32]

9.2.2 Responsibility of the Detaining Power. POWs are in the hands of the enemy Power, and not of the individuals or military units who have captured them.[33]Irrespective of the individual responsibilities that may exist, the Detaining Power is responsible for the treatment given POWs.[34]

9.2.3 Responsibility of POWs. Just as the Detaining Power is responsible for the treatment of POWs, POWs also have certain responsibilities related to their detention. POWs continue to have obligations to the Power on which they depend.[35]POWs are also obliged to obey the laws, regulations, and orders of the Detaining Power.[36]Despite this general obligation, POWs may not be punished in respect of their successful escape, if recaptured by the Detaining Power.[37]

9.2.4 Humane Treatment. Humane treatment is a fundamental principle underlying the GPW; it is always required.[38]Moreover, interpretations of the GPW that provide for humane treatment should be favored over rigid interpretations of the GPW that lead to results that would be detrimental to the welfare of POWs.

9.2.5 Reciprocity in the Treatment of POWs. The rules for the treatment of POWs have long been based on the principle that POWs should be treated as the Detaining Power would want its forces held by the enemy to be treated.[39]Aside from being legally required, the favorable treatment of POWs held by the U.S. armed forces can promote the favorable treatment of members of the U.S. armed forces held by enemy States.[40]

A similar concept may also be found in provisions of the GPW that provide for the treatment of POWs with reference to the treatment of the armed forces of the Detaining Power. For example, POWs shall be quartered under conditions as favorable as those for the forces of the Detaining Power who are billeted in the same area.[41]Similarly, in judicial proceedings, POWs should benefit from the same courts and same procedures as the members of the armed forces of the Detaining Power.[42]

9.3 POW STATUS

9.3.1 POW Status Versus POW Protections. In some cases, the policy of the United States has been to afford detainees certain POW protections even when they may not apply as a matter of law.[43]Certain POW protections may be afforded to an individual without affecting the legal status of that individual or the legal status of a group to which that person belongs.[44]

9.3.2 Persons Entitled to POW Status. During international armed conflict, persons entitled to POW status, if they fall into the power of the enemy, include:[45]

• members of the armed forces of a State that is a party to the conflict, including;[46]

o deserters;[47]and

o military medical and religious personnel not entitled to retained personnel status (*e.g.*, those not exclusively engaged in medical duties at the time of their capture);[48]

• members of certain militia and volunteer corps;[49]

• members of regular armed forces who profess allegiance to a government or authority not recognized by Detaining Power;[50]

• persons authorized to accompany the armed forces;[51]

• members of crews of merchant marine vessels or civil aircraft;[52]and • participants in a *levée en masse*.[53]

9.3.2.1 *Persons Who Are Not Entitled to POW Status*. Certain categories of persons are not entitled to POW status:

• spies, saboteurs, and other persons engaging in similar acts behind enemy lines;54 [54]and

• persons who are nationals of the Detaining Power or its co-belligerents,[55]such as a defector who subsequently is captured by the force from which he or she defected.[56]

9.3.2.2 *Examples of Categories of Persons Who Are Not Necessarily Excluded From POW Status*. The following categories of persons are not necessarily excluded from POW status simply because they belong to one of these categories:

• mercenaries;[57]

• persons who are alleged to have committed war crimes;[58]

• nationals of neutral or non-belligerent States serving in the armed forces of an enemy State;[59]and

• persons whose capture has not been acknowledged by the Power to which they belong.[60]

9.3.3 Persons Entitled to POW Treatment. Certain persons, although not entitled to POW status, shall likewise be treated as POWs under the GPW:

• persons belonging, or having belonged, to the armed forces of an Occupied State if it is deemed necessary to intern them;[61]and

• persons belonging to one of the categories enumerated in Article 4 of the GPW who have been received by neutral or non-belligerent Powers on their territory and whom these Powers are required to intern under international law.[62]

9.3.4 Fallen Into the Power of the Enemy.

9.3.4.1 *Having Fallen*. Persons must have "fallen" into the power of the enemy to receive POW status.[63]

The person need not have been captured to have "fallen" into the power of the enemy; it may be the case that he or she became a prisoner without fighting, *e.g.*, by surrendering.[64]Although defectors serving with their new armed force may be viewed as being in "the power of the enemy," they are not regarded as "hav[ing] fallen into" that condition, since they have voluntarily chosen to switch sides.[65]Thus, defectors are not considered by their new armed forces to be entitled to POW status.[66]

9.3.4.2 *Into the Power*. A person is considered to have fallen into the power of the enemy when that person has been captured by, or surrendered to, members of the military forces, the civilian police, or local civilian defense organizations or other enemy civilians who have taken that person into custody.[67]The person need not be captured by the armed forces of the opposing State.

9.3.4.3 *Of the Enemy*. Persons who have been captured by States that are not adversaries in the armed conflict are not technically in the hands of the enemy.[68]However, persons who are entitled to POW status or treatment and are interned by a neutral State under its duties under international law are generally entitled to POW treatment, as a minimum, under Article 4B(2) of the GPW.[69]Similarly, for example, military forces engaged in peace operations who are detained by the forces of a State that is a party to a conflict should, at a minimum, be afforded POW treatment by analogy.

9.3.5 Treatment and Determination of POW Status in Case of Doubt. Should any doubt arise as to whether persons, having committed a belligerent act and having fallen into the hands of the enemy, belong to any of the categories enumerated in Article 4 of the GPW, such persons shall enjoy the protection of the GPW until such time as their status has been determined by a competent tribunal.[70]

9.3.6 Commencement and Duration of POW Status and Treatment. The GPW applies to persons referred to in Article 4 of the GPW from the time they fall into the power of the enemy until their final release and repatriation.[71]

9.3.7 Non-Renunciation of Rights Secured by the GPW. POWs may in no circumstances renounce, in part or in entirety, the rights secured to them by the GPW, or by special agreements, if any, that are referred to in Article 6 of the GPW.[72]A similar provision of the GC applies to protected persons, and similar provisions of the GWS and GWS-Sea apply to the wounded, sick, shipwrecked, and military medical and religious personnel.[73]

This rule is intended to prevent States from evading their obligations by coercing enemy nationals in their power to waive their rights. The prohibition on the non-renunciation of rights is based on a recognition that: (1) POWs are in a vulnerable position; (2) it would be difficult to establish whether a POW had voluntarily renounced his or her rights; and (3) an absolute prohibition would best serve the interests of the majority of POWs.[74]

The non-renunciation of rights, in particular, prevents a POW from being compelled to serve in the armed forces of the Detaining Power.[75]The non-renunciation of POW rights, however, applies only to those who have fallen into the power of the enemy, and thus would not prevent persons from defecting to the enemy's side before capture.[76]

The non-renunciation of rights by the GPW does not prohibit States from affording POWs the right to refuse repatriation at the conclusion of the conflict.[77]

9.4 NATIONAL-LEVEL GPWIMPLEMENTATION MEASURES

The GPW requires a number of measures at the national level to facilitate protection for POWs.

9.4.1 Appointment or Establishment of Entities. The GPW provides for certain bodies or entities to be appointed or established.

9.4.1.1 *Appointment of a Protecting Power.* The GPW contemplates that a Protecting Power will have been appointed to assist in the implementation of the GPW.[78]A Protecting Power has a variety of roles under the GPW related to the protection of POWs.[79]

9.4.1.2 *Establishment of a National POW Information Bureau.* A National POW Information Bureau shall be established by each of the Powers concerned.[80]

9.4.1.3 *Establishment of a Central POW Information Agency.* A Central POW Information Agency shall be established in a neutral country by the Power concerned jointly.[81]

9.4.1.4 *Appointment of Mixed Medical Commissions.* Mixed Medical Commissions shall be appointed to examine sick and wounded POWs and to make all appropriate decisions regarding them.[82]

9.4.1.5 *Graves Registration Service.* A Graves Registration Service shall be established by each of the Powers concerned.[83]

9.4.2 Notification of Certain Laws or Policies Between Belligerents. Parties to a conflict must notify opposing belligerents or the Protecting Power of certain matters relating to POWs.

9.4.2.1 *Notification of Parole Policy.* Upon the outbreak of hostilities, each party to a conflict shall notify the adverse party or parties of its laws or regulations allowing or forbidding its armed forces to accept parole.[84]U.S. policy has prohibited the acceptance of parole.[85]

9.4.2.2 *Notification of Offenses Punishable by Death.* Information on what offenses are punishable by death should be communicated to the Protecting Power and the POWs.[86]

9.4.2.3 *Location and Marking of POW Camps.* The Detaining Power should provide the Powers concerned useful information regarding the geographic location of POW camps, and agreements on their markings may be reached.[87]

9.4.2.4 *Requirements for the Validity of Wills.* Information on the requirements for the validity of wills should be communicated to the opposing party so that POWs in enemy custody may prepare valid wills.[88]

9.4.2.5 *Communication or Agreement Between the Powers Concerned on Ranks.* Titles and ranks of persons entitled to POW status or treatment shall be communicated to one another by the parties to the conflict.[89]Ranks of persons entitled to retained personnel status or treatment should also be communicated and agreed upon so that the senior medical officer in a POW Camp may be determined.[90]

9.4.2.6 *Arrangements Made for POWs to Write Correspondence and Receive Collective Relief.* Immediately upon POWs falling into its power, the Detaining Power shall inform them and the Powers on which they depend, through the Protecting Power, of the measures taken to carry out the provisions of the Section V of the GPW, which relates to POW correspondence and relief shipments.[91]The Detaining Power shall likewise inform the parties concerned of any subsequent modifications of such measures.[92]

9.4.3 Issue of Identification Cards to Persons Liable to Become POWs. Parties to the GPW must provide persons under their jurisdiction who are liable to become POWs with identity cards.[93]This card must show:

• the owner's surname (*i.e.*, last name), first names (*i.e.*, first and middle names);

• rank, army, regimental, personal or serial number or equivalent information; and • date of birth. In accordance with this requirement, DoD practice has been to issue all members of the U.S. armed forces identification cards that indicate their status under the Geneva Conventions.[94]

An individual's loss of the identification card issued by a State Party is not a basis for denial of POW status.[95]

9.5 HUMANE TREATMENT AND BASIC PROTECTIONS FOR POWS

POWs must at all times be humanely treated.[96]POWs are entitled in all circumstances to respect for their persons and their honor.[97]Likewise, POWs must at all times be protected, particularly against acts of violence or intimidation and against insults and public curiosity.[98]Any unlawful act or omission by the Detaining Power causing death or seriously endangering the health of a POW in its custody is prohibited, and will be regarded as a serious breach of the GPW.[99]

9.5.1 Respect for Their Persons and Honor. POWs are entitled in all circumstances to respect for their persons and their honor.[100]For example, the rape or other indecent assault of POWs is forbidden.[101]

9.5.2 Protection Against Acts of Violence or Intimidation. POWs must at all times be protected, particularly against acts of violence or intimidation.[102]For example, the murder of POWs is forbidden.[103]

9.5.2.1 *Prohibition on Killing of POWs*. A commander of a force may not put enemy prisoners to death because their presence retards the force's movements or diminishes the force's power of resistance by necessitating a large guard, or by reason of the prisoners consuming supplies, or because it appears certain that they will regain their liberty through the impending success of enemy forces. It is likewise unlawful for a commander to kill enemy prisoners in the force's custody on grounds of self-preservation, even in the case of airborne or commando operations, although the circumstances of the operation may make necessary rigorous supervision of and restraint upon the movement of POWs.[104]

Older sources that permitted commanders in dire circumstances to deny quarter do not reflect the current law.[105]

9.5.2.2 *Protection Against Violence by the Civilian Population or Others*. POWs must be protected against violence by the civilian population.1[106]POWs should be protected not only against unlawful acts by the agents of the Detaining Power, but also against violence from other POWs.[107]

9.5.2.3 *No Exposure to a Combat Zone or Use as a Human Shield*. No POW may at any time be sent to or detained in areas where he or she may be exposed to the fire of the combat zone, nor may his or her presence be used to render certain points or areas immune from military operations.[108]

9.5.2.4 *No Physical Mutilation or Medical, Scientific, or Biological Experiments*. In particular, no POW may be subjected to physical mutilation or to medical or scientific experiments of any kind that are not justified by the medical, dental, or hospital treatment of the POW concerned and carried out in his or her interest.[109]The GWS and GWS-Sea also prohibit biological experiments on the wounded, sick, and shipwrecked.[110]

This prohibition was established in the 1949 Geneva Conventions in order to prohibit expressly criminal practices that occurred during World War II and to prevent the wounded, sick, or shipwrecked in captivity from being used as "guinea pigs" for medical experiments.[111]

The prohibition on subjecting the wounded, sick, or shipwrecked to biological experiments does not prevent doctors from trying new treatments that are justified on medical grounds and are employed solely for therapeutic purposes.[112]

POWs may voluntarily consent to give blood for transfusion or skin for grafting for therapeutic purposes; such procedures should take place under conditions consistent with generally accepted medical standards and controls designed for the benefit of both the donor and the recipient.[113]

9.5.3 Protection Against Insults and Public Curiosity. POWs must at all times be protected against insults and public curiosity.[114]For example, organizing a parade of POWs through the civilian population, thereby exposing them to assault, ridicule, and insults, would be prohibited.[115]Displaying POWs in a humiliating fashion on television or on the internet would also be prohibited.[116]For this reason and others, DoD policy has prohibited the taking of photographs of detainees except for authorized purposes.[117]

9.5.4 Reprisals Prohibited. Measures of reprisal against POWs are prohibited.[118]

9.5.5 No Adverse Distinction Based on Race, Nationality, Religion, or Political Belief. Taking into consideration the provisions of the GPW relating to rank and sex, and subject to any privileged treatment that may be accorded to them by reason of their state of health, age, or professional qualifications, all POWs shall be treated alike by the Detaining Power, without any adverse distinction based on race, nationality, religious belief, or political opinions, or any other distinction founded on similar criteria.[119]

Distinction based on nationality, political opinion, or other similar criteria may be made so long as it is not adverse and it is made to advance legitimate interests, such as maintaining order in the camp.[120]

In some cases, the Detaining Power must make distinctions between POWs as to rank or age,[121]state of health,[122]or professional status,[123]as well as to provide specific protection for women.

9.5.6 Due Regard for Women POWs. Women POWs shall be treated with all the regard due to their sex and shall in all cases benefit from treatment as favorable as that granted to men.[124]This principle also applies under the GWS and GWS-Sea to women who are wounded, sick, and shipwrecked.[125]

The GPW provides specifically for separate dormitories for women POWs in camps with both men and women and for separate bathroom facilities for women POWs.[126]The GPW also provides for women POWs undergoing disciplinary punishment or serving a sentence to be confined in separate quarters under the supervision of women.[127]Although the GPW does not specifically provide for this, female POWs should be under the immediate supervision of women.[128]

9.6 SECURITY MEASURES WITH RESPECT TO POWS

Although POWs must always be treated humanely, the confinement and mode of treating a POW may be varied during captivity for security purposes.[129]For example, POWs and their property may be searched and secured, when operationally necessary (*e.g.*, immediately upon capture).[130]

9.6.1 Search of POWs. POWs may be searched for security reasons, intelligence purposes, and other administrative reasons. For example, POWs may be searched for concealed weapons, items of intelligence value, or contraband, such as cameras and cell phones.[131]The dignity and honor of the POW being searched should be protected to the greatest degree possible under the circumstances.[132]The person conducting the search should avoid doing anything unnecessary or saying anything likely to be regarded as indecent or inappropriate. In some circumstances, it may be appropriate for a witness to observe the search so as to protect both the person being searched from abuse and the person conducting the search from unfounded accusations of abuse.

It is appropriate for female POWs to be searched by female personnel of the Detaining Power, if possible.[133]This practice helps reduce the risk of accusations of indecent behavior.

9.6.2 Securing POWs With Handcuffs and Other Security Devices. When necessary for security reasons (*e.g.*, to prevent escape or destruction of documents), POWs may be secured temporarily with handcuffs, flex cuffs, blindfolds, or other security devices.[134]

9.6.3 Search of POW Property. Items in the possession of POWs may be removed and searched for security and intelligence purposes, but certain items of personal or sentimental value must be returned as soon as possible.[135]

9.7 POWEFFECTS AND ARTICLES OF PERSONAL USE

9.7.1 POW Effects and Articles of Personal Use. All effects and articles of personal use, except arms, horses, military equipment, and military documents, shall remain in the possession of POWs, likewise their metal helmets and gas masks and like articles issued for personal protection.[136]Effects and articles used for their clothing or feeding shall likewise remain in their possession, even if such effects and articles are part of their regulation military equipment.[137]

These rules do not prohibit the search of POW effects and articles of personal use.[138]Items that may otherwise be legitimately impounded (*e.g.*, articles of value being withdrawn for reasons of security) are not precluded from being impounded because they may be characterized as an effect or article of personal use or an effect or article used for clothing or feeding.[139]

9.7.2 Identity Documents. At no time should POWs be without identity documents.[140]

The Detaining Power shall supply such documents to POWs who possess none.[141]

Identity documents may be seized when necessary (*e.g.*, intelligence purposes), provided that the Detaining Power supplies a replacement identity document.[142]

9.7.3 Badges, Decorations, and Articles Having Above All a Personal or Sentimental Value. Badges of rank and nationality, decorations, and articles having "above all" a personal or sentimental value may not be taken from POWs.[143]Such items might include a ring, wristwatch, personal correspondence, or family photographs.

This rule does not prohibit the search of items, including a search in order to determine whether the items in question have above all a personal or sentimental value.[144]Items that may otherwise be legitimately impounded (*e.g.*, articles of value being withdrawn for reasons of security) are not precluded from being impounded because of their personal or sentimental value to the POW.[145]

9.7.4 Money and Articles of Value. Sums of money carried by POWs may not be taken away from them except by order of an officer, and after the amount and particulars of the owner have been recorded in a special register and an itemized receipt has been given, legibly inscribed with the name, rank, and unit of the person issuing this receipt.[146]

Sums in the currency of the Detaining Power, or that are changed into such currency at the POW's request, shall be placed to the credit of the POW's account as provided for in Article 64 of the GPW.[147]

The Detaining Power may withdraw articles of value from POWs only for reasons of security; when such articles are withdrawn, the procedure laid down for sums of money impounded shall apply.148 [148]For example, an article of value may be impounded for reasons of security, even if it constitutes an article of personal use or an article having above all a personal or sentimental value.[149]

Such objects, likewise sums taken away in any currency other than that of the Detaining Power and the conversion of which has not been asked for by the owners, shall be kept in the custody of the Detaining Power and shall be returned in their initial shape to POWs at the end of their captivity.[150]

9.7.4.1 *Unexplained Possession of Large Sums of Money by POWs*. The unexplained possession by a POW of a large sum of money justifiably leads to the inference that such funds are not his or her own property and are in fact either property of the enemy government or property that has been looted or otherwise stolen.[151]

9.7.5 No Bartering With POWs for Personal Effects. It is not proper for members of the forces of the Detaining Power to engage in bartering and other transactions with POWs concerning their personal effects.[152]

9.7.6 Accountability for Impounded Personal Effects. In addition to the specific requirements for seized money or articles of value, it is a best practice for other items taken from POWs to be itemized and separated (packaged) with the name, unit, service number of the POW, and other identifying information to permit accountability, safekeeping, and return when appropriate to the POW.

9.8 INTERROGATION OF POWS

The law of war does not prohibit interrogation of POWs, but specifies conditions and limitations for conducting interrogation.

9.8.1 Humane Treatment During Interrogation. Interrogation must be carried out in a manner consistent with the requirements for humane treatment, including the prohibition against acts of violence or intimidation, and insults.[153]

No physical or mental torture, nor any other form of coercion, may be inflicted on POWs to secure from them information of any kind whatever.[154]POWs who refuse to answer may not be threatened, insulted, or exposed to unpleasant or disadvantageous treatment of any kind.[155]Prohibited means include imposing inhumane conditions,[156]denial of medical treatment, or the use of mind-altering chemicals.[157]

9.8.2 Language of Interrogation. The questioning of POWs shall be carried out in a language that they understand.[158]

9.8.3 Additional U.S. Law and Policy on Interrogation. U.S. law and policy impose additional requirements on the interrogation of POWs.[159]

9.8.4 Accountability Information That POWs Are Bound to Provide Upon Questioning. Every POW, when questioned on the subject, is bound to give only his or her surname, first names and rank, date of birth, and army, regimental, personal or serial number, or, failing this, equivalent information.[160]If POWs willfully infringe this rule, they may render themselves liable to a restriction of the privileges accorded to their rank or status.[161]However, POWs who refuse to provide this information may not be coerced or exposed to any other unpleasant or disadvantageous treatment of any kind for failing to respond. The provision of accountability information is important because it allows the Detaining Power to fulfill its obligations under the GPW. For example, the Detaining Power requires this information to establish lists of POWs for evacuation.[162]In addition, the Detaining Power is required to gather further information on POWs to facilitate notification of their families.[163]

9.8.4.1 *POWs Unable to State Their Identity*. POWs who, owing to their physical or mental condition, are unable to state their identity, shall be handed over to the medical service.[164]The identity of such POWs shall be established by all possible means, subject to the prohibition on physical or mental torture, or coercion, as well as the prohibition on threats, insults, or exposure to unpleasant or disadvantageous treatment.[165]

9.9 EVACUATION FROM COMBAT AREAS

POWs shall be evacuated, as soon as possible after their capture, to camps situated in an area far enough from the combat zone for them to be out of danger.[166]POWs shall not be unnecessarily exposed to danger while awaiting evacuation from a fighting zone.[167]

9.9.1 Delay in Evacuation for Medical Reasons. Only those POWs who, owing to wounds or sickness, would run greater risks by being evacuated than by remaining where they are, may be temporarily kept back in a danger zone.[168]

9.9.2 Conditions of Evacuation. The evacuation of POWs shall always be effected humanely and in conditions similar to those for the forces of the Detaining Power in their changes of station.[169]The Detaining Power shall supply POWs who are being evacuated with sufficient food and potable water, and with the necessary clothing and medical attention.[170]The Detaining Power shall take all suitable precautions to ensure their safety during evacuation, and shall establish as soon as possible a list of the POWs who are evacuated.[171]Provided that the requirements of humane treatment are met, the Detaining Power may subject POWs to security measures while evacuating them away from the combat zone. For example, POWs may be handcuffed, flex-cuffed, or blindfolded for security reasons.[172]

9.9.3 Release Under Unusual Conditions Rather Than Evacuation. When persons entitled to protection as POWs have fallen into the power of an adverse party under unusual conditions of combat that prevent their evacuation as provided for in Part III, Section I, of the GPW, they may be released, provided that feasible precautions are taken to ensure their safety.[173]

Release in such circumstances is permissible, but is not required.[174]In some cases, enemy forces may give their parole.[175]

9.9.3.1 *Unusual Conditions of Combat Preventing Evacuation*. Unusual conditions of combat may include airborne operations, commando raids, and long-range reconnaissance patrols.[176]

9.9.3.2 *Feasible Precautions to Ensure Safety*. POWs may be released in these circumstances only if feasible precautions are taken to ensure their safety.[177]For example, a long-range reconnaissance patrol need not make itself militarily ineffective by handing over all its supplies to POWs whom it has captured.[178]However, the patrol should take reasonable steps under the circumstances to help ensure the POWs' safety and survival, such as providing them with a modicum of food and water.[179]

9.9.3.3 *Accountability Measures*. In addition, it is a best practice to take feasible measures to account for the persons captured and released, such as recording: (1) full names, (2) serial numbers, (3) unit affiliation, and (4) the time, date, and location of capture and release.

9.10 TRANSIT OR SCREENING FACILITIES

POWs may be sent to transit or screening facilities before internment in a POW camp. For example, special interrogation centers have been used as screening facilities to interrogate POWs before their internment in the POW camp.[180]

9.10.1 Sending of Capture Cards in Transit or Screening Camps. Even in transit or screening camps, POWs are to be enabled to send capture cards within one week after arrival at a camp.[181]

9.10.2 Transit or Screening Camps Near Military Operations. Facilities within a zone of military operations are often established in order to house POWs captured during military operations in that particular zone that, due to their proximity to the fighting, are not able to meet all the requirements of the GPW.[182]Transit or screening facilities must, however, provide POWs reasonable protection from the effects of combat, basic sanitation, and food and water.[183]

If POWs must, during evacuation, pass through transit camps, their stay in such camps shall be as brief as possible.[184]

9.10.3 Facilities of a Permanent Nature. Transit or screening camps of a permanent kind shall be fitted out under conditions similar to those described in Section II of the GPW for regular internment camps for POWs, and the POWs in those camps shall have the same treatment as in other camps.[185]

9.10.4 Use of Ships for Temporary Detention. POWs may be detained temporarily on board a ship if operational or humanitarian needs dictate, pending a reasonable opportunity to transfer them to a shore facility, or to another vessel for evacuation to a shore facility.[186] For example, they may be temporarily detained on board naval vessels: (a) while being transported between land facilities; or (b) if such action would appreciably improve their safety or health prospects, such as avoidance of exposure to severe environmental or combat conditions, or improved access to medical care for those requiring it. Such limited detention does not violate the requirement for internment of POWs on land.[187]

9.11 GENERAL CONDITIONS IN POWCAMPS: LOCATION, SAFETY, HYGIENE, AND LIVING CONDITIONS

9.11.1 Internment in POW Camps. The Detaining Power may subject POWs to internment.[188] It may impose on them the obligation of not leaving, beyond certain limits, the camp where they are interned, or if that camp is fenced in, of not going outside its perimeter.[189]

9.11.1.1 *Prohibition on Close Confinement*. Subject to the provisions of the GPW relative to penal and disciplinary sanctions, POWs may not be held in close confinement except where necessary to safeguard their health and then only during the continuation of the circumstances that make such confinement necessary.[190]

9.11.2 Parole of POWs. POWs may be partially or wholly released on parole or promise, in so far as is allowed by the laws of the Power on which they depend.[191] Such measures shall be taken particularly in cases where this may contribute to the improvement of their state of health.[192] No POW shall be compelled to accept liberty on parole or promise.[193]

Upon the outbreak of hostilities, each party to the conflict shall notify the adverse party of the laws and regulations allowing or forbidding its own nationals to accept liberty on parole or promise.[194] U.S. policy prohibits U.S. service members from accepting parole or special favors from the enemy.[195]

POWs who are paroled or who have given their promise in conformity with the laws and regulations so notified are bound on their personal honor scrupulously to fulfill, both towards the Power on which they depend and towards the Power that has captured them, the engagements of their paroles or promises.[196] In such cases, the Power on which they depend is bound neither to require nor to accept from them any service incompatible with the parole or promise given.[197]

9.11.3 Location of POW Camps.

9.11.3.1 *Location on Land*. POWs may be interned only in premises located on land.[198] This rule is intended to ensure that POWs are interned in a relatively safe and healthy environment. For example, in prior conflicts, POWs interned on ships were not held in hygienic and humane conditions. Similarly, POWs held on ships faced increased risk from the dangers of war.

Because the purpose of the rule is to provide for the detention of POWs in a relatively safe and healthy environment, detention aboard ship for POWs captured at sea or pending the establishment of suitable facilities on land is nonetheless consistent with the GPW if detention on a ship provides the most appropriate living conditions for POWs.[199] Ships may also be used to transport POWs or for screening.[200]

9.11.3.2 *Use of Penitentiaries as POW Camps*. Except in particular cases that are justified by the interests of the POWs themselves, POWs shall not be interned in penitentiaries.[201] This rule reflects the principle that POW detention is not punitive, but also recognizes that penitentiaries may sometimes provide the most appropriate lodging.[202]

9.11.3.3 *Location Outside the Theater of Military Operations*. The GPW does not prohibit locating POW camps outside the theater of military operations, and there is much State practice in conducting detention operations outside of the theater in which POWs were captured.[203] Although the GC imposes geographic restrictions on the transfer of protected persons held in occupied territory, the GPW does not impose such restrictions on the transfer of POWs from the territory in which they were captured.[204]

Locating POW camps outside the theater of operations may be necessary for security (such as to discourage escape or reduce risk of enemy raids to release POWs) or other military reasons, such as improving the safety of POWs.

9.11.4 Safety of POW Camps.

9.11.4.1 *Avoidance of the Combat Zone*. POW camps should be situated in an area far enough from the combat zone for POWs to be out of danger.[205]

9.11.4.2 *Sharing Information on the Location of POW Camps.* Detaining Powers shall give the Powers concerned, through the intermediary of the Protecting Powers, all useful information regarding the geographic location of POW camps.[206]

9.11.4.3 *Marking of POW Camps.* Whenever military considerations permit, POW camps shall be indicated in the daytime by the letters "PW" or "PG," placed so as to be clearly visible from the air.[207]The Powers concerned may, however, agree upon any other system of marking.[208]No place, other than an internment camp, shall be marked as such.[209]

The caveat that POW camps need to be marked, as such, only when military considerations permit was intended to allow for camps not to be marked, *e.g.*, if marking camps would provide landmarks that would facilitate aerial bombardment by the enemy.[210]

9.11.4.4 *Air-Raid Shelters and Protective Measures for POWs.* POWs shall have shelters against air bombardment and other hazards of war, to the same extent as the local civilian population.[211]With the exception of those engaged in the protection of their quarters against these hazards, they may enter such shelters as soon as possible after the giving of the alarm.[212]Any other protective measure taken in favor of the population shall also apply to POWs.[213]

9.11.4.5 *Fire Precautions.* All precautions must be taken in POW camps against the danger of fire.[214]

9.11.5 Hygiene of POW Camps. POWs may be interned only in premises affording every guarantee of hygiene and healthfulness.[215]POWs interned in unhealthy areas, or where the climate is injurious for them, shall be removed as soon as possible to a more favorable climate.[216]

9.11.5.1 *Necessary Sanitary Measures.* The Detaining Power shall be bound to take all sanitary measures necessary to ensure the cleanliness and healthfulness of camps and to prevent epidemics.[217]For example, the Detaining Power should give POWs medical examinations upon arrival, disinfect them, and provide them with any necessary inoculations.[218]In addition, quarters should be kept free from vermin, and POWs suffering from contagious diseases should be placed in quarantine as needed.[219]

Although the Detaining Power is ultimately responsible for sanitary conditions, the Detaining Power may require POWs to assist in ensuring that POW camps are clean and healthy. For example, POWs may be assigned cleaning duties.[220]Additionally, POWs may be required to take baths or showers if necessary for health reasons.[221]Similarly, POWs may be required to receive inoculations to prevent the spread of contagious diseases.

9.11.5.2 *Sanitary Conveniences (e.g., latrines, bathrooms).* POWs shall have for their use, day and night, conveniences that conform to the rules of hygiene and are maintained in a constant state of cleanliness.[222]The term "sanitary conveniences" should be taken to mean primarily the latrines. Access to the latrines at night is specified in the 1949 Geneva Conventions because lack of access was a frequent problem in POW camps in World War II.[223]In any camps in which women POWs are accommodated, separate conveniences shall be provided for them.[224]

9.11.5.3 *Showers, Baths, Personal Toilet, and Laundry.* Also, apart from the baths and showers with which the camps shall be furnished, POWs shall be provided with sufficient water and soap for their personal toilet and for washing their personal laundry; the necessary installations, facilities, and time shall be granted them for that purpose.[225]For example, the POW camp may have camp laundry facilities that the POWs may use to wash their clothes, or the POWs may have access to a laundry service outside the camp.[226]

Although the frequency with which baths or showers may be taken is not specified, a reasonable opportunity (*e.g.*, considering the available resources, the POWs' cultural practices, the activities in which they are engaged) should be afforded.[227]

9.11.6 Conditions of Living Quarters. POWs shall be quartered under conditions as favorable as those for the forces of the Detaining Power who are billeted in the same area.[228]These conditions shall make allowance for the habits and customs of the POWs and shall in no case be prejudicial to their health.[229]In particular, the dormitories of POWs as regards both total surface and minimum cubic space, and the general installations, bedding, and blankets for POWs, should be as favorable as those of the forces of the Detaining Power in that area and should allow for their habits and customs.[230]

The premises provided for the use of POWs individually or collectively shall be entirely protected from dampness and adequately heated and lighted, in particular between dusk and lights out.[231]

9.11.6.1 *Separate Dormitories for Women*. In any camps in which women POWs, as well as men, are accommodated, separate dormitories shall be provided for women POWs.[232]

Although the GPW refers only to separate dormitories and the quarters as a whole need not necessarily be separated, the Detaining Power may provide separate quarters or even camps for men and women POWs, as appropriate.[233]

9.12 SEGREGATION OF POWS

Although the Detaining Power must treat POWs without any adverse distinction based upon race, nationality, religion, or political belief, or any other distinction based upon similar criteria,[234]the Detaining Power shall assemble POWs in camps or camp compounds according to their nationality, language, and customs, and may use other similar criteria to segregate detainees in order to maintain order in the camps or to impose punishment, or for medical reasons.

9.12.1 Assembling According to Nationality, Language, and Customs. The Detaining Power shall assemble POWs in camps or camp compounds according to their nationality, language, and customs, provided that such POWs shall not be separated from POWs belonging to the armed forces with which they were serving at the time of their capture, except with their consent.[235]For example, unless they otherwise consent, POWs who are serving in the armed forces of a State of which they are not nationals should be quartered with POWs from that State as opposed to with POWs of their nationality. The assembling of POWs on the basis of nationality, language, and customs is intended to ameliorate the possibility of cultural conflicts within POW camps, such as those that occurred during World War I.[236]

9.12.1.1 *Other Permissible Criteria for Segregating POWs*. In addition to nationality, language, and customs, the Detaining Power may segregate POWs along the basis of other criteria to advance legitimate interests, such as maintaining order in camps or imposing punishment, or for medical reasons.[237]For example, it may be appropriate to segregate POWs on the basis of political opinion.[238]POWs who are deserters or who fear attacks by other POWs may also be segregated.[239]In addition, it may be appropriate to segregate POWs based on whether they are connected with potential war crimes proceedings.[240]POWs may be segregated according to their known or suspected security risk level. A combination of classifications may be used to ensure an orderly and secure POW camp.[241]

9.13 FOOD,WATER, TOBACCO, AND CLOTHING FOR POWS

9.13.1 Food for POWs. The basic daily food rations shall be sufficient in quantity, quality, and variety to keep POWs in good health and to prevent loss of weight or the development of nutritional deficiencies.[242]

9.13.1.1 *Accounting for the Habitual Diet*. Account shall also be taken of the habitual diet of the POWs.[243]For example, the POWs' cultural and religious requirements should be considered in determining and ensuring the appropriate diet.[244]The preparation of the food by the POWs themselves also helps conform the food to the POWs' habitual diet.[245]

9.13.1.2 *Additional Food for Certain Groups*. The Detaining Power shall supply POWs who work with such additional rations as are necessary for the labor on which they are employed.[246]

9.13.1.3 *No Collective Disciplinary Measures Affecting Food*. Collective disciplinary measures affecting food are prohibited.[247]The use of food as an inducement to recalcitrant POWs to restore order in a POW camp is permissible.[248]

9.13.1.4 *Messing and Means for Preparing Additional Food*. Adequate premises shall be provided for messing.[249]Supervision of messes by the POWs themselves shall be facilitated in every way.[250]

POWs shall, as far as possible, be associated with the preparation of their meals; they may be employed for that purpose in the kitchens.[251]Furthermore, they shall be given the means of preparing, themselves, the additional food in their possession.[252]For example, the POWs should have access to facilities to prepare food purchased at the canteen or received in relief packages.

9.13.2 Drinking Water. Sufficient drinking water shall be supplied to POWs.[253]

9.13.3 Use of Tobacco. The use of tobacco shall be permitted.[254]The Detaining Power is not required to give tobacco to POWs, but should offer it for sale at the canteen.[255]The Detaining Power may impose reasonable restrictions on smoking to ensure that the camp is a healthful and safe environment.

9.13.4 Clothing for POWs. Clothing, underwear, and footwear shall be supplied to POWs in sufficient quantities by the Detaining Power, which shall make allowance for the climate of the region where the POWs are detained.[256] The regular replacement and repair of these articles shall be assured by the Detaining Power.[257]

9.13.4.1 *Uniforms for POWs*. Uniforms of enemy armed forces captured by the Detaining Power should, if suitable for the climate, be made available to clothe POWs.[258] The Power to which the POWs belong may send uniforms to the POWs through the Protecting Powers or ICRC.[259]

The GPW does not prohibit providing POWs with uniforms of the armed forces of the Detaining Power. If this were necessary, it would be appropriate to remove insignia or alter the uniforms, which would reduce the risk that POWs might be mistaken for members of the armed forces of the Detaining Power.[260]

Similarly, uniforms or other clothing for POWs may contain markings denoting the individuals as POWs because the absence of any such markings may facilitate escape.[261] Any markings on clothing for POWs must be consistent with the principle of respect for the person and honor of the POW and, therefore, may not be humiliating or degrading.[262]

9.13.4.2 *Appropriate Clothing for Work*. POWs who work shall receive appropriate clothing, wherever the nature of the work demands.[263]

9.14 MEDICAL ATTENTION FOR INTERNED POWS

In addition to the rules relating to the general rules applicable to the medical care of the wounded and sick,[264] the following rules also apply to the medical care of POWs who have been interned in POW camps.

The GPW and the GWS contemplate that POWs will benefit from the services of retained medical personnel or POWs who are trained as medical personnel but not attached to the medical service.[265] However, the provision in the GPW and the GWS for such personnel does not relieve the Detaining Power of its obligations to provide for the medical care of POWs.[266]

9.14.1 Access to Medical Attention.

9.14.1.1 *Camp Infirmary*. Every camp shall have an adequate infirmary where POWs may have the attention they require, as well as an appropriate diet.[267] Isolation wards shall, if necessary, be set aside for cases of contagious or mental disease.[268]

9.14.1.2 *Right of POWs to Present Themselves for Examination*. POWs may not be prevented from presenting themselves to the medical authorities for examination.[269] The GPW contemplates that POWs would have daily sick call in which they could present themselves for medical examination, even when undergoing disciplinary punishment.[270]

The right of POWs to present themselves for examination does not mean that POWs must necessarily be examined every day by a doctor, nor does it preclude disciplinary punishment if POWs abusively make groundless requests for medical examination.[271]

9.14.1.3 *Monthly Medical Inspections*. Medical inspections of POWs shall be held at least once a month.[272] They shall include the checking and the recording of the weight of each POW.[273] Their purpose shall be, in particular, to supervise the general state of health, nutrition, and cleanliness of POWs, and to detect contagious diseases, especially tuberculosis, malaria, and venereal disease.[274] For this purpose, the most efficient methods available shall be employed.[275] The GPW gives as an example periodic mass miniature radiography for early detection of tuberculosis, but medical experts should be consulted for the best practices.

9.14.2 Medical Treatment for Interned POWs.

9.14.2.1 *Persons Requiring Special Treatment*. POWs suffering from serious disease, or whose condition necessitates special treatment, a surgical operation, or hospital care, must be admitted to any military or civilian medical unit where such treatment can be given, even if their repatriation is contemplated in the near future.[276] Special facilities shall be afforded for the care to be given to the disabled, in particular to the blind, and for their rehabilitation, pending repatriation.[277]

9.14.2.2 *Treatment by Medical Personnel of the Power on Which They Depend*. POWs shall have the attention, preferably, of medical personnel of the Power on which they depend and, if possible, of their nationality.[278] The preference for POWs to have the attention of the medical personnel of the Power on which they depend is also reflected in the provisions of the GPW and GWS on the duties of retained personnel.[279]

These medical personnel may be retained personnel or persons who are trained as medical personnel but not attached to the medical service in their armed forces.[280]

9.14.2.3 *Certification Recording Treatment.* The detaining authorities shall, upon request, issue to every POW who has undergone treatment an official certificate indicating the nature of his or her illness or injury, and the duration and kind of treatment received.[281] A duplicate of this certificate shall be forwarded to the Central POW Information Agency.[282]

9.14.2.4 *Cost of Treatment.* The costs of treatment, including those of any apparatus necessary for the maintenance of POWs in good health, particularly dentures and other artificial appliances, and eyeglasses, shall be borne by the Detaining Power.[283]

9.14.3 Receipt of Medical Supplies. POWs may receive individual parcels and collective shipments containing medical supplies.[284]However, medical supplies generally are to be sent in collective shipments so that they may be properly administered by the camp doctors.[285]

9.14.4 Inquiry in Certain Cases of Death or Serious Injury. In certain cases of death or serious injury of a POW, an official inquiry shall be conducted by the Detaining Power.[286]

9.15 RELIGIOUS EXERCISE BY POWS

POWs shall enjoy complete latitude in the exercise of their religious duties, including attendance at the service of their faith, on condition that they comply with the disciplinary routine prescribed by the military authorities.[287]

The GPW and the GWS contemplate that POWs will benefit from the services of retained chaplains or POWs who are ministers of religion without having officiated as chaplains to their own forces.[288]However, the provision in the GPW and the GWS for such personnel does not relieve the Detaining Power of its obligations to provide for the exercise of religion by POWs.[289]

9.15.1 Premises for Religious Services for POWs. Adequate premises shall be provided where religious services may be held.[290]The premises where services are held should be sufficiently spacious and clean, and should provide effective shelter to those attending services, but need not be set aside exclusively for religious services.[291]

9.15.2 Chaplains. Chaplains who fall into the hands of the enemy Power and who remain or are retained with a view to assisting POWs shall be allowed to minister to them and to exercise freely their ministry among POWs of the same religion, in accordance with their religious conscience.[292]They shall be allocated among the various camps and labor detachments containing POWs belonging to the same forces, speaking the same language, or practicing the same religion.[293]

9.15.3 Appointment of Other Ordained Ministers or Qualified Laypersons. When POWs do not have the assistance of a retained chaplain or of a POW minister of their faith, a minister belonging to the POWs' or a similar denomination, or in his or her absence a qualified layperson, if such a course is feasible from a confessional point of view, shall be appointed, at the request of the POWs concerned, to fill this office.[294]This appointment, subject to the approval of the Detaining Power, shall take place with the agreement of the community of POWs concerned and, wherever necessary, with the approval of the local religious authorities of the same faith.[295]The person thus appointed shall comply with all regulations established by the Detaining Power in the interests of discipline and military security.[296]

9.16 INTELLECTUAL, PHYSICAL, AND RECREATIONAL ACTIVITIES

While respecting the individual preferences of every POW, the Detaining Power shall encourage the practice of intellectual, educational, and recreational pursuits, and sports and games, among POWs, and shall take the measures necessary to ensure the exercise thereof by providing them with adequate premises and necessary equipment.[297]Adequate premises and necessary equipment may include, for example, musical instruments, theatrical accessories, books, language courses, recreation rooms, or soccer fields.[298]

9.16.1 Voluntariness of the Activities. The Detaining Power should seek to encourage the POWs in engaging in these activities, while respecting their individual preferences. For example, it would be prohibited to compel POWs to listen to propaganda or to punish them if they do not participate.[299]

On the other hand, as reflected in U.S. practice, censoring educational or intellectual materials for security reasons or to refrain from providing POWs with anti-democratic political propaganda would be permissible.[300]Such censorship would be consistent with the general authority of the Detaining Power to censor

POW correspondence and shipments with a view towards deleting or confiscating matter prejudicial to its military security.[301]

9.16.2 Education. The GPW does not provide specifically for the Detaining Power to grant facilities for POWs to continue their studies or to take up new subjects as the GC does for internees.[302]Nonetheless, as the duration of detention increases, the Detaining Power should make an effort to do so as part of its general obligation to encourage intellectual and educational pursuits among POWs.[303]

To advance educational opportunities for POWs, the Detaining Power may seek ways to encourage and assist them in teaching one another.[304]

9.16.3 Opportunities for Physical Exercise and for Being Outdoors. POWs shall have opportunities for taking physical exercise, including sports and games, and for being outdoors.[305]

Sufficient open spaces shall be provided for this purpose in all camps.[306]

9.16.4 Contributions by Sources Apart From the Detaining Power. Apart from the Detaining Power, POWs may receive assistance from a variety of sources that allow them to engage in intellectual, physical, and recreational activities.

For example, relief organizations, including the ICRC, may contribute to ensuring that POWs have opportunities for intellectual, physical, and recreational activities.3[307]In addition, POWs may receive shipments that are intended to allow them to engage in these activities.[308]

Profits from the canteen may also be used for this purpose.[309]

9.17 CANTEENS FOR POWS

Canteens shall be established in all camps, where POWs may procure foodstuffs, soap and tobacco, and ordinary articles for daily use.[310]The purpose is to improve the morale of POWs by offering goods that are beyond the minimum necessities.[311]

9.17.1 Establishment of Canteens. Canteens should be established within a reasonable period of time, such as after more basic camp facilities have been established and similar facilities have been established for U.S. forces in the area. In conflicts of short duration or where POWs are to be transferred to another party to the conflict for longer-term internment, it may be unnecessary to establish a canteen.

9.17.1.1 *"Comfort Pack" Instead of Canteen and Advance of Pay*. Before canteens can be established, the need for a canteen may be mitigated if POWs are provided a free "comfort pack." The "comfort pack" was developed during Operation DESERT STORM for U.S. military personnel during the period before opening of exchanges in which such items were available for purchase. Comfort packs were provided to enemy POWs for similar reasons.[312]

9.17.2 Canteen Stock and Prices. Canteens are like a base or post exchange for POWs. The canteen stock should include foodstuffs, soap, tobacco, and other ordinary articles for daily use, but need not include luxury items. Ordinary articles for daily use may include, but are not limited to, items such as correspondence materials (*e.g.*, stationary, pencils, pens, ink, stamps), supplies for repairing personal effects (*e.g.*, buttons, thread, shoe laces), personal hygiene articles, tobacco, water, soft drinks, other non-alcoholic beverages, and reading materials.[313]

The canteen need not stock alcoholic beverages.[314]

The prices at canteens shall never be in excess of local market prices.[315]

9.17.3 Camp Canteen Management and Profits. Canteen profits shall be used for the benefit of the POWs, and a special fund shall be created for this purpose.[316]The POW Representative shall have the right to collaborate in the management of the canteen and of this fund.[317]The wishes expressed by the POWs should be considered to the extent consistent with camp regulations, and the profits should be used whenever needed to improve the conditions for POWs.[318]

9.17.3.1 *Use of Canteen Profits for Working Pay of the POW Representative*. Canteen profits should also be used for working pay of the POW Representative, of his or her advisers, if any, and of his or her assistants.[319]

9.17.3.2 *Disposition of Canteen Profits Upon Camp Closure*. When a camp is closed, the credit balance of the special fund shall be handed to an international welfare foundation, to be employed for the benefit of POWs of the same nationality as those who have contributed to the fund.[320]In case of a general repatriation, the profits shall be kept by the Detaining Power, subject to any agreement to the contrary between the Powers concerned.[321]

9.18 FINANCIAL RESOURCES OF POWS

9.18.1 Detaining Power Regulations on the Maximum Amount of Money in POW Possession. Upon the outbreak of hostilities, and pending an arrangement on this matter with the Protecting Power, the Detaining Power may determine the maximum amount of money in cash or in any similar form that POWs may have in their possession.[322] Any amount in excess, which was properly in their possession and which has been taken or withheld from them, shall be placed to their account, together with any monies deposited by them, and shall not be converted into any other currency without their consent.[323]

9.18.2 Purchases Outside the POW Camp. If POWs are permitted to purchase services or commodities outside the camp "against payment in cash," such payments shall be made by the POW himself or herself or by the camp administration who will charge them to the accounts of the POWs concerned.[324] The Detaining Power will establish the necessary rules in this respect.[325]

9.18.3 Advance of Pay. The Detaining Power shall grant all POWs a monthly advance of pay, the amount of which shall be fixed by conversion into the currency of the Detaining Power.[326]

This money allows POWs to purchase items at the canteen. This money may be paid in scrip or vouchers that can be used only in the POW camp, to prevent POWs from having or hoarding currency that might facilitate their escape.

Since the promulgation of the GPW, advances of pay to POWs by the Detaining Power have been infrequent.[327]

9.18.3.1 *GPW Specified Amounts for Monthly Advances of Pay*. Article 60 of the GPW specifies the following amounts for monthly advances of pay:[328]

- Category I: POWs ranking below sergeants: eight Swiss francs.[329]
- Category II: Sergeants and other non-commissioned officers, or POWs of equivalent rank: twelve Swiss francs.[330]
- Category III: Warrant officers and commissioned officers below the rank of major or POWs of equivalent rank: fifty Swiss francs.[331]
- Category IV: Majors, lieutenant-colonels, colonels, or POWs of equivalent rank: sixty Swiss francs.[332]
- Category V: General officers or POWs of equivalent rank: seventy-five Swiss francs.[333]

"POWs of equivalent rank" refers to POWs, such as persons authorized to accompany the armed forces, who are not members of the armed forces and thus do not have a military rank.[334]

9.18.3.2 *Special Agreement to Modify the Amount of Advances of Pay*. The parties to the conflict concerned may, by special agreement, modify the amount of advances of pay due to POWs of the categories specified in Article 60 of the GPW.[335]

9.18.3.3 *Procedure for Limiting Amounts Drawn From Accounts Pending Special Agreement*. If the amounts indicated in Article 60 of the GPW would be unduly high compared with the pay of the Detaining Power's armed forces or would, for any reason, seriously embarrass the Detaining Power, then, pending the conclusion of a special agreement with the Power on which the POWs depend to vary the amounts indicated above, the Detaining Power:[336]

- shall continue to credit the accounts of the POWs with the amounts indicated in the first paragraph of Article 60 of the GPW;[337]
- may temporarily limit the amount made available from these advances of pay to POWs for their own use, to sums that are reasonable, but that, for Category I, shall never be inferior to the amount that the Detaining Power gives to the members of its own armed forces.[338]

The reasons for any limitations will be given without delay to the Protecting Power.[339]

9.18.3.4 *Reimbursement by the Power on Which the POWs Depend*. Advances of pay, issued to POWs in conformity with Article 60 of the GPW, shall be considered as made on behalf of the Power on which they depend.[340]

Such advances of pay, as well as all payments made by the Power on which the POWs depend under the third paragraph of Article 63 of the GPW (payments made by the POW in their own country[341]) and Article 68 of the GPW (claims by POWs for compensation for personal loss[342]), shall form the subject of arrangements between the Powers concerned, at the close of hostilities.[343]

9.18.4 Supplementary Pay From the Power on Which the POWs Depend. The Detaining Power shall accept for distribution as supplementary pay to POWs sums that the Power on which the POWs depend may forward to them, on condition that the sums to be paid shall be the same for each POW of the same category, shall be payable to all POWs of that category depending on that Power, and shall be placed in their separate accounts, at the earliest opportunity, in accordance with the provisions of Article 64 of the GPW.[344]Such supplementary pay shall not relieve the Detaining Power of any obligation under the GPW.[345]

9.18.5 Remittances and Other Money Transfers. POWs shall be permitted to receive remittances of money addressed to them individually or collectively.[346]

Every POW shall have at his or her disposal the credit balance of his or her account as provided for in the Article 64 of the GPW, within the limits fixed by the Detaining Power, which shall make such payments as are requested.[347]Subject to financial or monetary restrictions that the Detaining Power regards as essential, POWs may also have payments made abroad.[348]In this case, payments addressed by POWs to dependents shall be given priority.[349]

9.18.5.1 *Payments Made in the POW's Own Country*. In any event, and subject to the consent of the Power on which they depend, POWs may have payments made in their own country, as follows:

• The Detaining Power shall send to the Power on which the POWs depend, through the Protecting Power, a notification giving all the necessary particulars concerning the POWs, the beneficiaries of the payments, and the amount of the sums to be paid, expressed in the Detaining Power's currency.[350]

• This notification shall be signed by the POWs and countersigned by the camp commander.[351]

• The Detaining Power shall debit the POWs' accounts by a corresponding amount; the sums thus debited shall be placed by it to the credit of the Power on which the POWs depend.[352]

To apply the foregoing provisions, the Detaining Power may usefully consult the Model Regulations in Annex V of the GPW.[353]

9.18.6 POW Accounts. The Detaining Power shall hold an account for each POW, showing at least the following:[354]

• the amounts due to the POW or received by him or her as advances of pay, as working pay or derived from any other source;

• the sums in the currency of the Detaining Power that were taken from him or her;

• the sums taken from the POW and converted at his or her request into the currency of that Power;

• the payments made to the POW in cash, or in any other similar form;

• the payments made on his or her behalf and request; and

• the sums transferred for payments made in a POW's own country under the third paragraph of Article 63 of the GPW.[355]

Every item entered in the account of a POW shall be countersigned or initialed by the POW, or by the POW Representative acting on his or her behalf.[356]

POWs shall at all times be afforded reasonable facilities for consulting and obtaining copies of their accounts, which may likewise be inspected by the representatives of the Protecting Powers at the time of visits to the camp.[357]

9.18.6.1 *Cash in POW Hands Taken at the Time of Capture*. Cash that was taken from POWs, in accordance with Article 18 of the GPW, at the time of their capture, and that is in the currency of the Detaining Power, shall be placed to their separate accounts, in accordance with the Article 64 of the GPW.[358]

The amounts, in the currency of the Detaining Power, due to the conversion of sums in other currencies that are taken from the POWs at the same time, shall also be credited to their separate accounts.[359]

9.18.6.2 *Personal Account in the Event of Transfers to Another POW Camp*. When POWs are transferred from one camp to another, their personal accounts will follow them.[360]In case of transfer from one Detaining Power to another, the monies that are their property and are not in the currency of the Detaining Power will follow them.[361]POWs shall be given certificates for any other monies standing to the credit of their accounts.[362]

9.18.6.3 *Periodic Notification of the Amount of the Accounts of POWs*. The parties to the conflict concerned may agree to notify to each other at specific intervals, through the Protecting Power, the amount of the accounts of the POWs.[363]

9.18.6.4 *Statements of Credit Balance on Termination of Captivity*. On the termination of captivity, through the release of a POW or the POW's repatriation, the Detaining Power shall give him or her a statement, signed by an authorized officer of that Power, showing the credit balance then due to the POW.[364]

The Detaining Power shall also send through the Protecting Power to the government upon which the POWs depend lists giving all appropriate particulars of all POWs whose captivity has been terminated by repatriation, release, escape, death, or any other means, and showing the amount of their credit balances.[365]Such lists shall be certified on each sheet by an authorized representative of the Detaining Power.[366]

These rules may be varied by mutual agreement between any two parties to the conflict.[367]

9.18.6.5 *Responsibility for Credit Balance Due From the Detaining Power on the Termination of Captivity*. The Power on which the POWs depend shall be responsible for settling with each POW any credit balance due to him or her from the Detaining Power on the termination of the POW's captivity.[368]

9.18.7 Claims by POWs in Respect of Personal Loss. Any claim by a POW for compensation in respect of personal effects, monies, or valuables impounded by the Detaining Power under Article 18 of the GPW and not forthcoming on his or her repatriation, or in respect of loss alleged to be due to the fault of the Detaining Power or any of its servants, shall likewise be referred to the Power on which the POW depends.[369]Nevertheless, any such personal effects required for use by the POW while in captivity shall be replaced at the expense of the Detaining Power.[370]

The Detaining Power will, in all cases, provide the POW with a statement, signed by a responsible officer, showing all available information regarding the reasons why such effects, monies, or valuables have not been restored to him or her.[371]A copy of this statement will be forwarded to the Power on which he or she depends through the Central POW Information Agency provided for in Article 123 of the GPW.[372]

9.19 POWLABOR

The Detaining Power may use the labor of POWs who are physically fit, taking into account their age, sex, rank, and physical aptitude, and with a view particularly to maintaining them in a good state of physical and mental health.[373]

9.19.1 Determining the Appropriate Labor Assignment. In determining whether labor should be compelled, as well as the appropriate labor assignment for a POW, the POW's age, gender, rank, and physical aptitude should be considered.[374]

9.19.1.1 *Labor Assignment – Non-Commissioned Officers*. Noncommissioned officers shall only be required to do supervisory work.[375]Noncommissioned officers who are not required to do supervisory work may ask for other suitable work, which shall, so far as possible, be found for them.[376]

9.19.1.2 *Labor Assignment – Officers and Persons of Equivalent Status*. If officers or persons of equivalent status ask for suitable work, it shall be found for them, so far as possible, but they may in no circumstances be compelled to work.[377]

9.19.1.3 *Labor Assignment – POWs Who May Not Be Compelled to Do Other Work*. Retained personnel and persons treated like retained personnel (*e.g.*, POWs trained as medical personnel who are directed to provide medical care for fellow POWs), however, may not be compelled to carry out any work other than that concerned with their medical or religious duties.3[378]POW Representatives and their assistants also shall not be required to perform any other work, if the accomplishment of their duties is thereby made more difficult.[379]

9.19.2 Types of Labor Assignments.

9.19.2.1 *Unhealthy or Dangerous Labor*. Unless the POW is a volunteer, he or she may not be employed on labor that is of an unhealthy or dangerous nature.[380]For example, the removal of landmines or similar devices is considered dangerous labor.[381]"Similar devices" may be understood to include booby-traps and other devices that are similar to landmines.[382]

When employing POWs on dangerous or unhealthy activities, the Protecting Power or the ICRC should assist in verifying the voluntariness of the employment, and the suitability of the POW for such hazardous activities (*e.g.*, any special skills or training in relation to the hazards) should be considered.[383]

9.19.2.2 *Prohibited Labor Assignments*. No POW shall be assigned to labor that would be looked upon as humiliating for a member of the Detaining Power's own forces.[384]

9.19.2.3 *Labor Assignments That May Be Compelled.* POWs, like other enemy nationals, may not be compelled to take part in operations of war directed against their own country.[385] POWs may be compelled to do only such work as is included in the following classes:

• POW camp administration, installation, and maintenance;

• agriculture;

• industries connected with the production or the extraction of raw materials, and manufacturing industries, with the exception of metallurgical, machinery, and chemical industries;

• public works and building operations having no military character or purpose;

• transport and handling of stores not of a military character or purpose;

• commercial businesses, including arts and crafts;

• domestic services; and

• public utilities having no military character or purpose.[386]

Should the above provisions be infringed, POWs shall be allowed to exercise their right of complaint, in conformity with Article 78 of the GPW.[387]

9.19.3 Suitable Work Conditions. POWs must be granted suitable working conditions, especially as regards accommodation, food, clothing, and equipment.[388] For example, POWs must be given additional rations or appropriate clothing, as necessary, for their work.[389]

POWs' working conditions shall not be inferior to those enjoyed by nationals of the Detaining Power employed in similar work.[390] Account shall also be taken of climatic conditions.[391]

9.19.3.1 *Application of National Legislation for the Protection of Labor and Regulations for Worker Safety.* The Detaining Power, in utilizing the labor of POWs, shall ensure that in areas in which such POWs are employed, the national legislation concerning the protection of labor, and, more particularly, the regulations for the safety of workers, are duly applied.[392] For example, labor conditions for POWs working in the United States would be determined by U.S. Federal labor and safety laws. Conditions of labor shall in no case be rendered more arduous by disciplinary measures.[393] For example, the Detaining Power may not lower safety standards or disregard the requirements for protective equipment as a punishment for misbehavior.[394]

9.19.3.2 *Training and Means of Protection.* POWs shall receive training and be provided with the means of protection suitable to the work they will have to do and similar to those accorded to the nationals of the Detaining Power.[395] Subject to the provisions of Article 52 of the GPW, POWs may be submitted to the normal risks run by these civilian workers.[396]

9.19.3.3 *Limits on the Duration of Labor.* The duration of the daily labor of POWs, including any travel time, shall not be excessive, and must in no case exceed that permitted for civilian workers in the district, who are nationals of the Detaining Power and employed on the same work.[397]

POWs must be allowed, in the middle of the day's work, a rest of not less than one hour.[398] This rest will be the same as that to which workers of the Detaining Power are entitled, if the latter is of longer duration.[399] POWs shall be allowed, in addition, a rest of 24 consecutive hours every week, preferably on Sunday or the day of rest in their country of origin.[400]

Furthermore, every POW who has worked for one year shall be granted a rest of eight consecutive days, during which his or her working pay shall be paid to him or her.[401] If methods of labor, such as piece work, are employed, the length of the working period shall not be rendered excessive thereby.[402] Piece work refers to the practice of assigning and compensating for tasks as opposed to compensating based on the amount of time worked.[403]

Thus, the limits on the number of hours of daily work cannot be circumvented by assigning work on the basis of tasks completed as opposed to requiring POWs to work a certain number of hours.[404]

9.19.4 Working Pay. The working pay due to POWs shall be fixed in accordance with the provisions of Article 62 of the GPW.[405]

POWs shall be paid a fair working rate of pay by the detaining authorities directly.[406]

The rate shall be fixed by these authorities, but shall at no time be less than one-fourth of one Swiss franc for a full working day.[407] The Detaining Power shall inform POWs, as well as the Power on which they depend, through the intermediary of the Protecting Power, of the rate of daily working pay that it has fixed.[408]

Working pay shall likewise be paid by the detaining authorities to POWs permanently detailed to duties, or to a skilled or semi-skilled occupation, in connection with the administration, installation, or maintenance of camps, and to the POWs who are required to carry out spiritual or medical duties on behalf of their comrades.[409]

Working pay of POW Representatives, their advisers, if any, or assistants, also is provided for in the GPW.[410]

9.19.5 Medical Fitness to Work. The fitness of POWs for work shall be periodically verified by medical examinations, at least once a month.[411]The examinations shall have particular regard to the nature of the work that the POWs are required to do.[412]

If any POW considers himself or herself incapable of working, he or she shall be permitted to appear before the medical authorities of the camp.[413]Physicians or surgeons may recommend that the POWs who are, in their opinion, unfit for work, be exempted therefrom.[414]

9.19.6 Accidents or Disease in Connection With Work. POWs who sustain accidents in connection with work, or who contract a disease in the course or in consequence of their work, shall receive all the care their condition may require.[415]The Detaining Power shall furthermore deliver to such POWs a medical certificate enabling them to submit their claims to the Power on which they depend, and shall send a duplicate to the Central POW Information Agency provided for in Article 123 of the GPW.[416]

In certain cases, POWs who meet with accidents in connection with work would receive the benefit of the provisions of the GPW as regards repatriation or accommodation in a neutral country.[417]

9.19.6.1 *Claims for Compensation From Injury or Other Disability Arising Out of Work.* Any claim by a POW for compensation in respect of any injury or other disability arising out of work shall be referred to the Power on which he or she depends, through the Protecting Power.[418]In accordance with Article 54 of the GPW, the Detaining Power will, in all cases, provide the POW concerned with a statement showing the nature of the injury or disability, the circumstances in which it arose, and the particulars of medical or hospital treatment given for it.[419]This statement will be signed by a responsible officer of the Detaining Power and the medical particulars certified by a medical officer.[420]

9.19.7 Labor Detachments. The organization and administration of labor detachments shall be similar to those of POW camps.[421]

Every labor detachment shall remain under the control of, and administratively part of, a POW camp.[422]The military authorities and the commander of the POW camp shall be responsible, under the direction of their government, for the observance of the provisions of the GPW in labor detachments that are dependent upon the POW camp.[423]

The camp commander shall keep an up-to-date record of the labor detachments dependent on the camp, and shall communicate it to the delegates of the Protecting Power, of the ICRC, or of other agencies giving relief to POWs, who may visit the camp.[424]

9.19.8 POWs Working for Private Persons. The treatment of POWs who work for private persons, even if the latter are responsible for guarding and protecting them, shall not be inferior to that which is provided for by the GPW.[425]The Detaining Power, the military authorities, and the commander of the camp to which such POWs belong shall be entirely responsible for the maintenance, care, treatment, and payment of the working pay of such POWs.[426]

Such POWs shall have the right to remain in communication with the POW Representatives in the camps on which they depend.[427]

9.20 POWCORRESPONDENCE AND RELIEF SHIPMENTS

POWs may send and receive mail and relief shipments, subject to security requirements. In addition, POWs benefit from certain postage exemptions.

9.20.1 Capture, Health, or Transfer Card. Immediately upon capture, or not more than one week after arrival at a camp, even if it is a transit camp, likewise in case of sickness or transfer to a hospital or to another camp, every POW shall be enabled to write directly to his or her family, on the one hand, and to the Central POW Information Agency provided for in Article 123 of the GPW, on the other hand, a card similar, if possible, to the model annexed to the GPW (GPW Annex IV B), informing his or her relatives of his or her capture, address, and state of health.[428]

These cards shall be forwarded as rapidly as possible and may not be delayed in any manner.[429]

9.20.2 POW Correspondence Rights and Quota. POWs shall be allowed to send and receive letters and cards.[430]

9.20.2.1 *POW Correspondence Quota*. If the Detaining Power deems it necessary to limit the number of letters and cards sent by each POW, this number shall not be less than two letters and four cards monthly, exclusive of the capture cards provided for in Article 70 of the GPW (GPW Annex IV B), and conforming as closely as possible to the models annexed to the GPW (GPW Annex IV C).[431]In addition, the following correspondence also does not count toward this quota:

• formal requests and complaints;[432]

• correspondence between a duly appointed POW Representative at a principal camp and a POW Representative at a labor camp;[433]

• authorized correspondence by chaplains;[434]and • correspondence by the senior medical officer in a camp with the competent authorities of the camp regarding the duties of retained medical personnel.[435]

Further limitations on POW correspondence may be imposed only if the Protecting Power is satisfied that it would be in the interests of the POWs concerned to do so owing to difficulties of translation caused by the Detaining Power's inability to find sufficient qualified linguists to carry out the necessary censorship.[436]If limitations must be placed on the correspondence addressed to POWs, they may be ordered only by the Power on which the POWs depend, possibly at the request of the Detaining Power.[437]Such letters and cards must be conveyed by the most rapid method at the disposal of the Detaining Power.[438]

9.20.2.2 *No Delay in Correspondence for Disciplinary Reasons*. Cards and letters may not be delayed or retained for disciplinary reasons.[439]

9.20.2.3 *More Rapid Means of Correspondence in Appropriate Cases*. POWs who have been without news for a long period, or who are unable to receive news from their next of kin or to give them news by the ordinary postal route, as well as those who are at a great distance from their homes, shall be permitted to send telegrams, the fees being charged against the POWs' accounts with the Detaining Power or paid in the currency at their disposal.[440]They shall likewise benefit from this measure in cases of urgency.[441]

More modern means of communication, such as email, should be considered for POW correspondence, as appropriate.[442]

9.20.2.4 *Language of POW Correspondence*. As a general rule, the correspondence of POWs shall be written in their native language.[443]The parties to the conflict may allow correspondence in other languages.[444]

9.20.2.5 *Handling of Sacks of POW Mail*. Sacks containing POW mail must be securely sealed and labeled so as clearly to indicate their contents, and must be addressed to "offices of destination."[445]

9.20.3 Receipt of Individual and Collective Relief Shipments for POWs. POWs shall be allowed to receive by post, or by any other means, individual parcels or collective shipments containing, in particular, foodstuffs, clothing, medical supplies, and articles of a religious, educational, or recreational character, which may meet their needs, including books, devotional articles, scientific equipment, examination papers, musical instruments, sports outfits, and materials allowing POWs to pursue their studies or their cultural activities.[446]Such shipments shall in no way free the Detaining Power from the obligations imposed upon it by virtue of the GPW.[447]

The only limits that may be placed on these shipments shall be those proposed by the Protecting Power in the interest of the POWs themselves, or by the ICRC or any other organization giving assistance to the POWs, in respect of their own shipments only, on account of exceptional strain on transport or communications.[448]

9.20.3.1 *Separation of Books From Other Relief Supplies*. Parcels of clothing and foodstuffs may not include books.[449]Books should be separated because the time required for censorship of books would likely delay the distribution of other relief supplies.[450]

9.20.3.2 *Medical Supplies in Relief Shipments*. Medical supplies shall, as a rule, be sent in collective parcels.[451]This rule is meant to ensure that medical supplies can be administered by the camp doctors and not by the POWs themselves without medical supervision.[452]This rule, however, would not prohibit, in exceptional cases, for example, special medication being sent from family members to a POW.[453]

9.20.3.3 *Special Agreements Concerning Relief Shipments*. The conditions for the sending of individual parcels and collective shipments shall, if necessary, be the subject of special agreements between the Powers concerned, which may in no case delay the receipt by the POWs of relief supplies.[454]

9.20.3.4 *Collective Relief for POWs*. In the absence of special agreements between the Powers concerned on the conditions for the receipt and distribution of collective relief shipments, the rules and regulations concerning collective relief that are annexed to the GPW shall be applied (GPW Annex III).[455]

The special agreements referred to above shall in no case restrict the right of POW Representatives to take possession of collective relief shipments intended for POWs, to proceed to their distribution, or to dispose of them in the interest of the POWs.[456]Nor shall such agreements restrict the right of representatives of the Protecting Power, the ICRC, or any other organization giving assistance to POWs and responsible for the forwarding of collective shipments, to supervise their distribution to the recipients.[457]

9.20.4 Exemptions From Postal and Shipping Charges. Relief shipments for POWs and mail sent by POWs or to them benefit from certain exemptions under international law.

9.20.4.1 *Exemption From Dues for Relief Shipments*. All relief shipments for POWs shall be exempt from import, customs, or other dues.[458]

9.20.4.2 *GPW Postal Dues Exemption*. Under the GPW, correspondence, relief shipments, and authorized remittances of money addressed to POWs or sent by them through the post office, either directly or through the National POW Information Bureau provided for in Article 122 of the GPW and the Central POW Information Agency provided for in Article 123 of the GPW, shall be exempt from any postal dues, both in the countries of origin and destination, and in intermediate countries.[459]

9.20.4.3 *Exemption From Postal Charges Under the Universal Postal Convention*. Under the Universal Postal Convention, with the exception of air surcharges for airmail, letter-post items, postal parcels, and monetary articles addressed to or sent by POWs, either directly or through the National POW Information Bureau and the Central POW Information Agency, shall be exempt from all postal charges.[460]This exemption also applies to belligerents captured and interned in neutral countries.[461]

Parcels shall be admitted free of postage up to a weight of 5 kg.[462]The weight limit shall be increased to 10 kg in the case of parcels the contents of which cannot be split up and of parcels addressed to a camp or the POW Representatives there for distribution to the POWs.[463]

Items exempt from postal charges and POW parcels shall bear "Service des prisonniers de guerre" ("Prisoner-of-war Service) and a translation in another language, if appropriate.[464]Postal administrations shall not be liable for the loss of, theft from, or damage to parcels in the case of POW parcels.[465]

9.20.4.4 *Costs of Transporting Relief Shipments Outside the Post Office*. If relief shipments intended for POWs cannot be sent through the post office by reason of weight or for any other cause, the cost of transportation shall be borne by the Detaining Power in all the territories under its control.4[466]The other Parties to the GPW shall bear the cost of transport in their respective territories.[467]

In the absence of special agreements between the Parties concerned, the costs connected with transport of such shipments, other than costs covered by the above exemption, shall be charged to the senders.[468]

9.20.4.5 *Rates Charged for Telegrams*. The Parties to the GPW shall endeavor to reduce, so far as possible, the rates charged for telegrams sent by POWs, or addressed to them.[469]

9.20.5 Special Means of Transport of Shipments to POWs. Should military operations prevent the Powers concerned from fulfilling their obligation to ensure the transport of the shipments referred to in Articles 70, 71, 72, and 77 of the GPW, the Protecting Powers concerned, the ICRC, or any other organization duly approved by the parties to the conflict may undertake to ensure the conveyance of such shipments by suitable means (railway wagons, motor vehicles, vessels, or aircraft, etc.).[470]For this purpose, the Parties to the GPW shall endeavor to supply them with such transport, and to allow its circulation, especially by granting the necessary safe-conducts.[471]

Such transport may also be used to convey:

- correspondence, lists, and reports exchanged between the Central POW Information Agency referred to in Article 123 of the GPW and the National POW Information Bureau

referred to in Article 122 of the GPW; and

- correspondence and reports relating to POWs that the Protecting Powers, the ICRC, or any other body assisting the POWs exchange either with their own delegates or with the parties to the conflict.[472]

These provisions in no way detract from the right of any party to the conflict to arrange other means of transport, if it should so prefer, nor preclude the granting of safe-conducts, under mutually agreed conditions, to such means of transport.[473]

In the absence of special agreements, the costs occasioned by the use of such means of transport shall be borne proportionally by the parties to the conflict whose nationals are benefited thereby.[474]Expenses in setting up the special transport system are not specifically addressed by the GPW, and presumably would be addressed by an agreement between the body that takes the initiative in establishing the system and the Powers concerned.[475]

9.20.6 Censorship and Security Review of POW Correspondence and Shipments. The Detaining Power may examine and censor all communications sent to or by POWs, including correspondence, telegrams, parcels, newspapers, periodicals, and books, with a view to deleting or confiscating matter prejudicial to its military security.[476]The Detaining Power's general right to conduct censorship is an implicit assumption in the GPW, as it is in the GC.[477]POWs' complaints, petitions, and reports may be subject to security review and censorship to ensure that they are not misused.[478]

The censoring of correspondence addressed to POWs or sent by them shall be done as quickly as possible.[479]Mail shall be censored only by the State from which the mail is sent and the State that receives the mail, and only once by each State.[480]

9.20.6.1 *Examination of Consignments*. The examination of consignments intended for POWs shall not be carried out under conditions that will expose the goods contained in them to deterioration; except in the case of written or printed matter, it shall be done in the presence of the addressee, or of a fellow POW duly delegated by him or her.[481]

The delivery to POWs of individual or collective consignments shall not be delayed under the pretext of difficulties of censorship.[482]

9.20.6.2 *Prohibitions of Correspondence*. Any prohibition of correspondence ordered by parties to the conflict, either for military or political reasons, shall be only temporary and its duration shall be as short as possible.[483]

9.21 PRIVATE LEGAL MATTERS OF POWS

9.21.1 Civil Capacity. POWs shall retain their full civil capacity that they enjoyed at the time of their capture.[484]Their retention of their full civil capacity at the time of capture means that the law that applied to them before becoming a POW would continue to apply to them.[485]

The Detaining Power may not restrict the exercise, either within or without its own territory, of the rights such capacity confers, except in so far as the captivity requires.[486]Thus, POWs may, to the extent consistent with their detention, continue to exercise the civil capacity that they would have in their own country under that country's law.[487]For example, to the degree permitted by captivity, POWs may take legal steps that they were able to take before captivity, such as disposing of property, making wills, giving consent to marriage, voting, or executing a power of attorney.[488]

9.21.2 Legal Documents and Assistance. The Detaining Powers shall provide all facilities for the transmission, through the Protecting Power or the Central POW Information Agency provided for in Article 123 of the GPW, of instruments, papers, or documents intended for POWs or sent by them, especially powers of attorney and wills.[489]In all cases, the Detaining Power shall facilitate the preparation and execution of such documents on behalf of POWs; in particular, they shall allow them to consult a lawyer and shall take what measures are necessary for the authentication of their signatures.[490]

9.21.2.1 *Preparation of Wills*. Wills of POWs shall be drawn up so as to satisfy the conditions of validity required by the legislation of their country of origin, which will take steps to inform the Detaining Power of its requirements in this respect.[491]For States with federal systems like the United States in which the requirements of wills may vary according to jurisdictions within the State, the will should satisfy those provincial or state law requirements. At the request of the POW, the will shall be transmitted without delay to the Protecting Power; a certified copy shall be sent to the Central POW Information Agency.[492]

9.22 INTERNAL DISCIPLINE OF POWCAMPS

9.22.1 POW Camp Commander. Every POW camp shall be put under the immediate authority of a responsible commissioned officer belonging to the regular armed forces of the Detaining Power.[493]For example, a non-commissioned officer, a contractor, a civilian, a member of a paramilitary force, or an enemy POW may not be the camp commander.[494]The POW camp commander has disciplinary powers over POWs.[495]

9.22.2 POWs' Status With Respect to Their Armed Forces. Although POWs are not subject to the judicial or disciplinary procedures of the Power to which they belong while they are POWs, POWs remain subject to the law, disciplinary authority, and regulations of the Power on which they depend.[496]For example, while they are POWs, they may not conduct disciplinary proceedings against one another; however, they are liable to punishment for violations committed during captivity, once they have been released and repatriated.[497]

9.22.3 Saluting Between POWs and Officers of the Detaining Power. POWs, with the exception of officers, must salute and show to all officers of the Detaining Power the external marks of respect provided for by the regulations applying in their own forces.[498]POWs need not salute Detaining Power non-commissioned officers or persons of equivalent status to officers belonging to the Detaining Power.[499]

The form and conditions for saluting and showing external marks of respect are to be determined by the regulations applying in the armed forces to which the POWs belong.[500]Other external marks of respect may include, for example, standing at attention when an officer enters the room.[501]

Officer POWs are bound to salute only Detaining Power officers of higher rank and the camp commander, regardless of his or her rank.[502]

Although not required by the GPW, Detaining Power officers should return salutes as a matter of courtesy.[503]Members of the armed forces of the Detaining Power are not required to salute POWs of superior rank.[504]

9.22.4 Rank and Age of POWs. POWs shall be treated with the regard due to their rank and age.[505]POWs shall be permitted to wear their badges of rank, nationality, or decorations.[506]

For example, rank insignia or similar devices may be worn by POWs. The rank of POWs also determines the amount of the advances of pay that they are entitled to receive.[507]In addition, POWs awarded disciplinary punishment may not be deprived of the prerogatives of rank.[508]

9.22.4.1 *Communication of Titles and Ranks*. Upon the outbreak of hostilities, the parties to the conflict shall communicate to one another the titles and ranks of all the persons mentioned in Article 4 of the GPW, in order to ensure equality of treatment between POWs of equivalent rank.[509]Titles and ranks that are subsequently created shall form the subject of similar communications.[510]

9.22.4.2 *Recognition of Promotions in Rank*. Promotions in rank of POWs that have been accorded to POWs and that have been duly notified by the Power on which the POWs depend shall be recognized.[511]

9.22.5 Posting of Convention and Camp Orders. In every POW camp the text of the GPW and its Annexes, and the contents of any applicable special agreements, shall be posted, in the POWs' own language, in places where all may read them.[512]The POWs' own language means an official language of the State to which they belong.[513]Copies shall be supplied, on request, to the POWs who cannot have access to the posted copy.[514]

Regulations, orders, notices, and publications of every kind relating to the conduct of POWs shall be issued to them in a language that they understand.[515]Such regulations, orders, and publications shall be posted in the same manner as the GPW, and copies shall be handed to the POW Representative.[516]Every order and command addressed to POWs individually must likewise be given in a language that they understand.[517]

9.22.6 Use of Force to Maintain Order and to Prevent Escape. The use of weapons against POWs, especially against those who are escaping or attempting to escape, shall constitute an extreme measure, which shall always be preceded by warnings appropriate to the circumstances.[518]For example, POWs should not be fired upon if they are apprehended within the camp limits while making preparations to escape and there is no risk of escape or harm to anyone.[519]

9.22.6.1 *Use of Non-Lethal Weapons (Including Riot Control Agents)*. Non-lethal weapons may be employed to control rioting POWs or to prevent their escape.[520]In particular, the use of riot control agents to control rioting POWs is not prohibited.[521]

If the use of deadly force is warranted and authorized against POWs, there is no legal requirement to employ non-lethal weapons before resort to deadly force.[522]

9.23 POWREQUESTS, COMPLAINTS, AND REPORTS ABOUT CONDITIONS OF CAPTIVITY

9.23.1 POW Right to Make Requests and Complaints. POWs shall have the right to make known, to the military authorities in whose power they are, their requests regarding the conditions of captivity to which they are subjected.[523]The commander of a POW camp may issue regulations that establish procedures about how POWs are to make these requests.[524]POWs shall also have the unrestricted right to apply to the representatives of the

Protecting Powers, either through their POW Representative or, if they consider it necessary, directly, in order to draw their attention to any points on which they may have complaints to make regarding their conditions of captivity.[525]

9.23.1.1 *Making Requests and Complaints to the ICRC*. Such complaints may also be made to the delegates of the ICRC who enjoy the same prerogatives of access as the representatives of the Protecting Power.[526] In the past, the ICRC has been able to take appropriate measures besides merely forwarding the complaint to the Power on which the POWs depend, including measures on a confidential basis that help improve the situations of POWs.[527]

9.23.2 Communication of Requests and Complaints. These requests and complaints shall neither be limited nor considered to be part of the correspondence quota referred to in Article 71 of the GPW.[528] They must be transmitted immediately.[529]

9.23.2.1 *Review and Censorship of Requests and Complaints by the Detaining Power*. Complaints and requests by the POWs that are to be transmitted to the representatives of the Protecting Power may be reviewed and censored for security reasons by the Detaining Power, although the right of the POWs to make complaints about their conditions of captivity to the representatives of the Protecting Power must not be restricted.[530] In addition, security review and censorship must be conducted in accordance with the general rules on censoring POW correspondence.[531]

9.23.2.2 *No Punishment for Making Complaints*. Even if the complaints are recognized to be unfounded, they may not give rise to any punishment.[532] This rule departs from the usual practice in armed forces, in which excessive use of the right of complaint within a State's armed forces may be punished.[533] The GPW provision against punishment for unfounded complaints arose after POWs were tortured for complaining to the representatives of the Protecting Power.[534] In any event, it could be contrary to POWs' interests to abuse this right by making groundless complaints because complaints that are justified might not, as a result, receive the appropriate attention.[535]

9.23.3 Periodic Reports by the POW Representatives. POW Representatives may send periodic reports on the situation in the camps and the needs of the POWs to the representatives of the Protecting Powers.[536] POW Representatives have discretion about how and when to submit their reports.[537] The Detaining Power may subject these reports to security review and censorship to ensure that these reports are not misused.[538]

9.24 POW REPRESENTATIVES

The GPW contemplates that certain POWs will serve as representatives of the POWs before the Detaining Power, the Protecting Power, and other entities. These persons have certain prerogatives that enable them to carry out their duties to further the well-being of the POWs. In the GPW, these persons are called "prisoners' representatives." This manual refers to them as POW Representatives.

9.24.1 General Qualifications of the POW Representative. In all cases, the POW Representative must have the same nationality, language, and customs as the POWs whom he or she represents.[539] Thus, POWs distributed in different sections of a camp, according to their nationality, language, or customs, shall have for each section their own POW Representative.[540]

In practice, it is advisable for the POW Representative to be a POW as opposed to a retained person. Medical personnel have their own representative for issues related to their activities.[541] In addition, the duties of the POW Representative and the duties of retained personnel could interfere with one another, and special provision is made in the GPW to ensure that other duties not interfere with the duties of retained personnel or the duties of the POW Representative.[542]

9.24.2 Procedure for Selecting POW Representatives, Advisers, and Assistants. The procedure for selecting POW Representatives, and their advisers (POWs who are officers) and assistants (POWs who are not officers), depends on the rank of the POWs in the camp.

9.24.2.1 *Camps With Officer POWs*. In camps for officers and persons of equivalent status or in mixed camps, the senior officer among the POWs shall be recognized as the camp POW Representative.[543] A mixed camp refers to camps composed of both officers and other ranks.[544]

The senior officer is the officer of highest rank, not the oldest officer.[545] The senior officer is also a member of the armed forces, as opposed to a person of equivalent status (*e.g.*, such as a person authorized to accompany the armed forces who is of senior status).[546]

In camps for officers, the POW Representative shall be assisted by one or more advisers chosen by the officers.[547] These advisers may assist the POW Representative if his or her health makes it difficult to carry out the duties of the POW Representative and also may assist the POW Representative in gauging the wishes of other POWs.[548]

In mixed camps, the POW Representative's assistants shall be chosen from among the POWs who are not officers and shall be elected by them.[549]

9.24.2.2 *Labor Camps With Officers Who Carry Out Camp Administration Duties for POWs*. Officer POWs of the same nationality shall be stationed in labor camps for POWs, for the purpose of carrying out the camp administration duties for which the POWs are responsible.[550] These officers may be elected as POW Representatives under the first paragraph of Article 79 of the GPW.[551] In such a case, the assistants to the POW Representatives shall be chosen from among those POWs who are not officers.[552]

9.24.2.3 *Places Without Officer POWs*. In all places where there are POWs, except in those where there are officers, the POWs shall freely elect POW Representatives by secret ballot, every six months, and also in case of vacancies.[553] These POW Representatives shall be eligible for re-election.[554]

9.24.2.4 *Approval of Elected POW Representatives by the Detaining Power*. Every elected POW Representative must be approved by the Detaining Power before he or she has the right to commence his or her duties.[555] Where the Detaining Power refuses to approve a POW elected by fellow POWs, it must inform the Protecting Power of the reason for such refusal.[556]

9.24.3 Duties of POW Representatives. The POW Representatives are entrusted with representing the POWs before the military authorities, the Protecting Powers, the ICRC, and any other organization that may assist them.[557] POW Representatives shall further the physical, spiritual, and intellectual well-being of POWs.[558]

The POW Representative's general duties imply that the POW Representatives will undertake a variety of activities to ensure that POWs receive proper treatment by the Detaining Power and to advance their welfare, even activities that are not specified in the GPW as constituting their duties.

9.24.3.1 *POW Representatives and Mutual Assistance Among POWs*. In particular, where the POWs decide to organize among themselves a system of mutual assistance, this organization will be within the province of the POW Representative, in addition to the special duties entrusted to him or her by other provisions of the GPW.[559]

9.24.3.2 *Other Specific Duties*. In addition to the general duties that the GPW imposes on the POW Representative, the GPW also imposes specific duties on POW Representatives in three areas: (1) relief activities; (2) relations between the POWs and the authorities; and (3) verification that the guarantees provided under GPW are respected.[560] The POW Representative's duties in relation to relief activities include:

• assisting in the transport of the POWs' community property and luggage in cases of transfers of POWs;[561]

• managing the distribution of collective relief;[562]

• receiving parcels and remittances of money for POWs undergoing disciplinary punishment, and handing over to the infirmary any perishable goods;[563] and

• signing receipts for relief supplies or materials to the relief society or organization making the shipment.[564]

The POW Representative's duties in facilitating relations between the POWs and the authorities include:

• remaining in communication with POWs who work for private employers;[565]

• transmitting requests and complaints, and sending periodic reports on the needs of POWs;[566] and

• interviewing with delegates of the Protecting Powers.[567]

The POW Representative's duties in verifying that the guarantees provided under GPW are respected include:

• collaborating in the management of the canteen and its special fund;[568]

• receiving copies of regulations, orders, notices, and publications relating to the conduct of POWs issued by the Detaining Power for communication to POWs;[569]

• countersigning for entries entered in a POW's account on his or her behalf, or for notifications of payments sent by POWs to their own country;[570]

• monitoring decisions announced from any disciplinary proceeding against a POW;[571]

• receiving notifications of judicial proceedings instituted against POWs, and of any judgment or sentences pronounced upon them;[572]and

• proposing repatriation for certain POWs and participating in Mixed Medical Commissions.[573]

9.24.4 Prerogatives of POW Representatives. In order to accomplish their duties, POW Representatives are afforded certain prerogatives.

9.24.4.1 *Limitations on Other Work Assignments*. POW Representatives shall not be required to perform any other work, if the accomplishment of their duties is thereby made more difficult.[574]Officers, in any event, may not be required to work.[575]

9.24.4.2 *Assistants for the POW Representatives*. In addition to the advisers and assistants for the POW Representatives that are provided for in camps with officers, POW Representatives may appoint from among the POWs such assistants as they may require.[576]Assistants, like the POW Representative, should be exempted from any other work to the extent that the accomplishment of their duties would thereby be made more difficult.[577]

9.24.4.3 *Material Facilities, Including Certain Freedom of Movement*. All material facilities shall be granted to POW Representatives, particularly a certain freedom of movement necessary for the accomplishment of their duties (inspection of labor detachments, receipt of supplies, etc.).[578]This freedom of movement does not mean complete freedom.[579]

9.24.4.4 *Visiting Premises Where POWs Are Detained and Consultation With POWs*. POW Representatives shall be permitted to visit premises where POWs are detained, and every POW shall have the right to consult freely with his or her POW Representative.[580]The premises that may be visited will include the kitchen, infirmary, and other annexes to the POW Camp.[581]

9.24.4.5 *Working Pay of POW Representatives, Advisers, and Assistants*. The working pay of POW Representatives, advisers, if any, and assistants, shall be paid out of the fund maintained by canteen profits.[582]The scale of this working pay shall be fixed by the POW Representative and approved by the camp commander.[583]If there is no such fund, the detaining authorities shall pay these POWs a fair working rate of pay.[584]

9.24.4.6 *Facilities for Communication*. All facilities shall likewise be accorded to POW Representatives for communication by post and telegraph with the detaining authorities, the Protecting Powers, the ICRC and their delegates, the Mixed Medical Commissions, and the bodies that give assistance to POWs.[585]

POW Representatives of labor detachments shall enjoy the same facilities for communication with the POW Representatives of the principal camp.[586]Such communications shall not be restricted, nor considered as forming a part of the quota mentioned in Article 71 of the GPW.[587]

The facilities to be granted do not, however, include freedom from censorship, but correspondence may not be withheld.[588]

9.24.4.7 *No Responsibility for Offenses Committed by POWs by Reason of Their Duties*. POW Representatives shall not be held responsible, simply by reason of their duties, for any offenses committed by POWs.[589]For example, POW Representatives may not be held responsible for escape attempts by other POWs, unless they personally participated in such activities.[590]

9.24.4.8 *Turnover for Successor POW Representatives Upon Transfer*. POW Representatives who are transferred shall be allowed a reasonable time to acquaint their successors with current affairs.[591]

9.24.5 Dismissal of POW Representative. In case of dismissal, the reason for the dismissal shall be communicated to the Protecting Power.[592]

The GPW does not provide a procedure for POWs to dismiss POW Representatives aside from not re-electing them.[593]However, the Detaining Power may withdraw its approval of a POW Representative and request that POWs hold new elections.[594]

9.25 POW ESCAPES

POWs who escape successfully are not liable to any punishment in respect of their previous escape. A number of rules limit the punishment of POWs who do not escape successfully. Notifications of POW escape and recapture should also be made. POWs punished as a result of an unsuccessful escape may be subjected to special surveillance.

9.25.1 No Punishment for Successful Escape. POWs who have made good their escape in the sense of Article 91 of the GPW, and who are recaptured, shall not be liable to any punishment in respect of their previous escape.[595]

In this way, POWs who have escaped successfully are treated similarly to persons who have engaged in espionage and returned safely to friendly lines.[596]Escaping POWs must not kill or wound the enemy by resort to perfidy.[597]

9.25.1.1 *Types of Successful Escapes*. Under Article 91 of the GPW, the escape of a POW shall be deemed to have succeeded when the POW has:

• joined the armed forces of the Power on which he or she depends, or those of an allied Power;

• left the territory under the control of the Detaining Power, or of an ally of the Detaining Power; or

• joined a ship flying the flag of the Power on which he or she depends, or of an allied

Power, in the territorial waters of the Detaining Power, this ship not being under the control of the Detaining Power.[598]

The general principle is that the POW must have actually gone beyond the reach of the Detaining Power.[599]Thus, for example, a POW who escapes from the territory of the Detaining Power to the territory of one of the Detaining Power's allies will not be deemed to have escaped successfully. On the other hand, if the POW reaches neutral territory or the high seas, he or she will have escaped successfully.[600]

The situation of POWs who have successfully escaped into neutral territory is addressed under the law of neutrality.[601]

9.25.2 Unsuccessful Escapes. POWs who do not escape successfully retain their entitlement to POW status upon recapture. In particular, the wearing of civilian clothes does not deny escaping POWs their status as POWs.[602]

A number of rules limit the punishment of POWs who do not escape successfully. By limiting the punishment in respect of the act of escape, the GPW recognizes that POWs may legitimately try to escape from their captors.[603]In some cases, POWs may even be under an obligation to escape.[604]For example, U.S. military personnel have a duty to make every effort to escape captivity.[605]

9.25.2.1 *Handover of a Recaptured POW to Competent Military Authority*. A POW who is recaptured shall be handed over without delay to the competent military authority.[606]For example, if recaptured by private persons or by civilian law enforcement authorities, POWs should be handed over to competent military authority without delay. This rule makes certain practices used by Germany during World War II clearly unlawful.[607]

9.25.2.2 *Only Disciplinary Punishments in Respect of an Act of Escape*. A POW who attempts to escape and is recaptured before having made good his or her escape in the sense of Article 91 of the GPW shall be liable only to a disciplinary punishment in respect of this act, even if it is a repeated offense.[608]

9.25.2.3 *Disciplinary Punishment for Certain Offenses Committed in Connection With Escape*. In conformity with the principle stated in Article 83 of the GPW (*i.e.*, leniency in favor of disciplinary rather than judicial proceedings), offenses committed by POWs with the sole intention of facilitating their escape and that do not entail any violence against life or limb, such as offenses against public property, theft without intention of self-enrichment, the drawing up or use of false papers, or the wearing of civilian clothing, shall occasion disciplinary punishment only.[609]

For example, if a POW steals food, money, or means of transport; wears civilian clothing; or fabricates false documents in order to facilitate his or her escape and is caught before escaping successfully, such acts may only incur disciplinary punishment.

9.25.2.4 *Escape Not an Aggravating Circumstance in Respect of a Judicial Offense Committed During Escape*. Escape or attempt to escape, even if it is a repeated offense, shall not be deemed an aggravating circumstance if the POW is subjected to trial by judicial proceedings in respect of an offense committed during his or her escape or attempt to escape.[610]

For example, an escaping POW who kills or injures a Detaining Power guard while escaping could be liable to judicial punishment for that offense. However, the circumstance of escape shall not be deemed to aggravate the sentence of the POW, even if the POW is one who frequently attempts to escape.

9.25.3 Liability of POWs for Other POWs' Escape or Attempts to Escape. POWs who aid or abet an escape or an attempt to escape are liable on this count to disciplinary punishment only.[611]

Collective punishment of POWs for an escape attempt by other POWs is also prohibited.[612]

9.25.4 Special Surveillance of POWs as a Result of Unsuccessful Escape. Despite Article 88 of the GPW, which articulates the principle that POWs who have served disciplinary or judicial sentences may not be treated differently from other POWs,[613]POWs punished as a result of an unsuccessful escape may be subjected to special surveillance.[614]Such surveillance: (1) must not affect the state of their health; (2) must be undergone in a POW camp; and (3) must not entail the suppression of any of the safeguards granted to them by the GPW.[615]

9.25.5 Notification of Escape and Recapture. If a POW escapes, the State on which the POW depends should be notified by the Detaining Power's National POW Information Bureau through the intermediary of the Protecting Powers and Central Tracing Agency.[616]If an escaped POW is recaptured, the Power on which the POW depends shall be notified of the recapture in the manner prescribed in Article 122 of the GPW (*i.e.*, by the Detaining Power's National POW Information Bureau through the intermediary of the Protecting Powers and Central Tracing Agency), provided notification of the POW's escape has been made.[617]

9.26 GENERAL PRINCIPLES APPLICABLE TO POWDISCIPLINE

In the GPW, the rules on POW discipline are provided under three sub-headings:

• General principles applicable at all times.[618]

• "Disciplinary measures," which means punishment by the commander of the POW camp, and is equivalent to summary disposal by a commanding officer.[619]

• "Judicial proceedings," which means trial and punishment by a court having jurisdiction to try POWs, and is equivalent to trial by court-martial.[620]

This section addresses the general principles applicable to both disciplinary and judicial proceedings against POWs.

9.26.1 POWs Subject to the Laws, Regulations, and Orders in Force in the Armed Forces of the Detaining Power. A POW shall be subject to the laws, regulations, and orders in force in the armed forces of the Detaining Power; the Detaining Power shall be justified in taking judicial or disciplinary measures in respect of any offense committed by a POW against such laws, regulations, or orders.[621]For example, POWs in custody of the U.S. armed forces are subject to the Uniform Code of Military Justice.[622]However, no proceedings or punishments contrary to the provisions of Chapter III of the GPW shall be allowed.[623]

POWs also remain subject to the laws of the State to which they claim allegiance, and may be prosecuted by that State following release and repatriation for misconduct committed during their captivity.[624]

9.26.2 Leniency in Favor of Disciplinary Rather Than Judicial Proceedings. In deciding whether proceedings in respect of an offense alleged to have been committed by a POW shall be judicial or disciplinary, the Detaining Power shall ensure that the competent authorities exercise the greatest leniency and adopt, wherever possible, disciplinary rather than judicial measures.[625]

This requirement does not compel the competent authorities to choose disciplinary over judicial proceedings in any particular case.

9.26.3 Trial by Military Courts. A POW shall be tried only by a military court, unless the existing laws of the Detaining Power expressly permit the civil courts to try a member of the armed forces of the Detaining Power in respect of the particular offense alleged to have been committed by the POW.[626]

In no circumstances whatever shall a POW be tried by a court of any kind that does not offer the essential guarantees of independence and impartiality as generally recognized and, in particular, the procedure of which does not afford the accused the rights and means of defense provided for in Article 105 of the GPW.[627]

9.26.4 Retention of Benefits of the GPW Even if Prosecuted for Pre-Capture Acts. POWs prosecuted under the laws of the Detaining Power for acts committed prior to capture shall retain, even if convicted, the benefits of the GPW.[628]For example, POWs prosecuted for pre-capture acts are entitled to the fair trial protections provided by the GPW.[629]POWs' fair trial protections could not be ensured if the mere allegation of a war crime could deny them POW status.[630]

9.26.5 Repetition of Punishment. No POW may be punished more than once for the same act, or on the same charge.[631]

9.26.6 Prohibited Penalties. POWs may not be sentenced by the military authorities and courts of the Detaining Power to any penalties except those provided for in respect of members of the armed forces of the Detaining Power who have committed the same acts.[632] The following punishments are expressly prohibited:

- collective punishments for individual acts;[633]
- corporal punishment;
- imprisonment in premises without daylight;
- any form of torture or cruelty; and
- deprivation of rank or of the right to wear badges.[634]

The courts and authorities of the Detaining Power, in reaching decisions on punishment, must at all times remember that the accused:

- does not owe any allegiance to the Detaining Power and may be, for example, under a duty to escape;[635] and
- is in its power through circumstances beyond his or her control so that, for example, depression brought on by captivity should not be regarded as self-induced. Courts and disciplinary authorities shall have the discretion to reduce the penalty below the minimum prescribed for members of the armed forces of the Detaining Power.[636]

9.26.7 Treatment of POWs Undergoing Punishment. Officers, non-commissioned officers, and men and women who are POWs undergoing a disciplinary or judicial punishment, shall not be subjected to more severe treatment than that applied in respect of the same punishment to members of the armed forces of the Detaining Power of equivalent rank.[637]

A woman POW shall not be awarded or sentenced to a punishment more severe, or treated while undergoing punishment more severely, than either a female or male member of the armed forces of the Detaining Power dealt with for a similar offense.[638]

POWs who have served disciplinary or judicial sentences may not be treated differently from other POWs.[639] However, POWs who have been punished as a result of an attempted escape may nonetheless be subjected to special surveillance.[640]

9.27 DISCIPLINARY PROCEEDINGS AND PUNISHMENT

In addition to the general principles applicable to POW punishment, the following rules address disciplinary proceedings and punishment.

9.27.1 POW Camp Authorities Who May Order Disciplinary Punishment. Apart from judicial authorities or superior military authorities, only the camp commander, the officer acting in the commander's place, or an officer to whom the commander has delegated his or her disciplinary powers, may order disciplinary punishment.[641] A camp commander's power to order disciplinary punishment of POWs may not be delegated to POWs.[642]

9.27.2 Confinement of POWs Pending Disposal of Offense. A POW accused of an offense against discipline shall not be kept in confinement pending his or her hearing (*e.g.*, during the investigation of the accusation), unless (1) a member of the armed forces of the Detaining Power would be so confined if he or she were accused of a similar offense, or (2) if confinement is otherwise essential in the interests of camp order and discipline.[643]

Any period spent in confinement awaiting disposal of an offense against discipline shall be reduced to the absolute minimum and shall not exceed fourteen days.[644]

The provisions of Articles 97 and 98 of the GPW shall apply to a POW's confinement while awaiting the disposal of offenses against discipline.[645] Thus, the conditions of any confinement before the hearing should be at least as good as that afforded POWs who are confined as part of disciplinary punishment.[646]

9.27.3 Rights of POWs in Disciplinary Proceedings. Before any disciplinary award is pronounced, the accused shall be given (1) precise information regarding the offenses of which he or she is accused; (2) an opportunity to explain his or her conduct; and (3) an opportunity to defend himself or herself.[647] He or she shall be permitted, in particular, to call witnesses and to have recourse, if necessary, to the services of a qualified interpreter.[648]

The decision shall be announced to the accused POW and to the POW Representative.[649]

9.27.4 Record of Disciplinary Punishments. A record of disciplinary punishments shall be maintained by the camp commander and shall be open to inspection by representatives of the Protecting Power.[650]

9.27.5 Types of Disciplinary Punishments and Procedures for Dispensing Punishment. In no case shall disciplinary punishments be inhuman, brutal, or dangerous to the health of POWs.[651] The disciplinary punishments applicable to POWs are the following:

• a fine that shall not exceed 50 percent of the advances of pay and working pay that the POW would receive under Articles 60 and 62 of the GPW during a period of not more than thirty days;

• discontinuance of privileges granted over and above the treatment provided for by the GPW;

• fatigue duties, not exceeding two hours daily;

o "Fatigue duties" refers to details of extra-duty chores (*e.g.*, such as policing of the POW camp grounds, kitchen duty).[652]

o The punishment of fatigue duties shall not be applied to officers.[653] In addition, fatigue details should meet normal standards for working conditions and must not be made more arduous as a disciplinary measure.[654]

• confinement.[655]

The duration of any single punishment shall in no case exceed thirty days.[656] The maximum of thirty days may not be exceeded, even if the POW is answerable for several acts when punishment is awarded, regardless of whether such acts are related.[657]

Any period of confinement awaiting the hearing of a disciplinary offense or the award of disciplinary punishment shall be deducted from an award pronounced against a POW.[658] The period between the pronouncing of an award of disciplinary punishment and its execution shall not exceed one month.[659] When a POW is awarded a further disciplinary punishment, a period of at least three days shall elapse between the execution of any two of the punishments, if the duration of one of these punishments is ten days or more.[660]

9.27.6 Conditions for POWs Serving Disciplinary Punishments. Conditions for disciplinary punishments must comply with the requirements for humane treatment.[661] In addition, the following rules apply:

9.27.6.1 *Premises Where Disciplinary Punishments Are to Be Served.* POWs shall not in any case be transferred to penitentiary establishments (prisons, penitentiaries, convict prisons, etc.) to undergo disciplinary punishment.[662]

Officers and persons of equivalent status shall not be lodged in the same quarters as noncommissioned officers or enlisted personnel.[663]

Women POWs undergoing disciplinary punishment shall be confined in separate quarters from male POWs and shall be under the immediate supervision of women.[664]

9.27.6.2 *Retention of the Benefits of the GPW While Undergoing Disciplinary Punishment.* A POW undergoing confinement as a disciplinary punishment shall continue to receive the benefits of the GPW, except insofar as these benefits are necessarily rendered inapplicable by the mere fact that the POW is confined.[665]

In no case, however, may POWs undergoing disciplinary punishments be deprived of the benefits of the provisions of Articles 78 and 126 of the GPW.[666] Thus, POWs undergoing disciplinary punishments may not be deprived of the right to make requests and complaints, or to deal with representatives of the Protecting Power (including representatives of an impartial humanitarian organization performing the functions of the Protecting Power).[667]

9.27.6.3 *Disciplinary Punishment - Retention of Prerogatives of Rank.* A POW awarded disciplinary punishment may not be deprived of prerogatives attached to his or her rank.[668] The prerogatives referred to are those provided by the GPW, such as the right not to be required to work or the right to wear the insignia of their rank.[669]

9.27.6.4 *Attendance at Daily Medical Inspection and Medical Attention.* POWs awarded disciplinary punishment shall be allowed, on their request, to be present at the daily medical inspection.[670] They shall receive the attention that their state of health requires and, if necessary, shall be removed to the camp infirmary or to a hospital.[671]

9.27.6.5 *Exercise and Access to the Open Air.* POWs awarded disciplinary punishment shall be allowed to exercise and to stay in the open air at least two hours daily.[672]

9.27.6.6 *Reading, Writing, Correspondence, and Packages.* POWs awarded disciplinary punishment shall have permission to read and write, and to send and receive letters.[673] Parcels and remittances of money, however, may be withheld from them until the completion of the punishment; such parcels and remittances of money shall meanwhile be entrusted to the POW Representative, who will hand over to the infirmary the perishable goods contained in such parcels.[674]

9.27.7 Disciplinary Punishment and Repatriation or Accommodation in a Neutral Country. No POW on whom a disciplinary punishment has been imposed and who is eligible for repatriation or for accommodation in a neutral country may be kept back because he or she has not undergone his or her punishment.[675]

9.28 JUDICIAL PROCEEDINGS AND PUNISHMENT

In addition to the general principles applicable to POW punishment, the following rules address judicial proceedings and punishment.

9.28.1 No Trial or Sentence Without Pre-Existing Law. No POW may be tried or sentenced for an act that is not forbidden by the law of the Detaining Power or by international law in force at the time that act was committed.[676]

9.28.2 Same Courts and Same Procedures. A POW can be validly sentenced only if the sentence has been pronounced by the same courts according to the same procedure as in the case of members of the armed forces of the Detaining Power, and if, furthermore, the provisions of Chapter III of the GPW have been observed.[677]For example, evidence laws used in the trial of a POW will be the same as those applicable in the trial of a member of the Detaining Power's military forces.

9.28.3 Notification of Proceedings, Investigation of Offenses, and Confinement Before Trial.

9.28.3.1 *Notification of Proceedings.* In any case in which the Detaining Power decides to institute judicial proceedings against a POW, the Detaining Power shall notify the Protecting Power and the POW Representative as soon as possible and at least three weeks before the opening of the trial.[678]This period of three weeks shall run from the day on which such notification reaches the Protecting Power at the address previously indicated by the latter to the Detaining Power.[679]This notification shall contain the following information: • surname and first names of the POW, his or her rank, his or her army, regimental, personal or serial number, his or her date of birth, and his or her profession or trade, if any;

• place of internment or confinement;

• specification of the charge or charges on which the POW is to be arraigned, giving the legal provisions applicable; and

• designation of the court that will try the case, and the date and place fixed for the opening of the trial.[680]

If no evidence is submitted at the opening of a trial that the notification referred to above was received by the Protecting Power, by the POW, and by the POW Representative concerned at least three weeks before the opening of the trial, then the trial may not take place and must be adjourned.[681]

9.28.3.2 *Rapid Investigations.* Judicial investigations relating to a POW shall be conducted as rapidly as circumstances permit so that his or her trial may take place as soon as possible.[682]

9.28.3.3 *Pre-Trial Confinement.* A POW shall not be confined while awaiting trial unless a member of the armed forces of the Detaining Power would be so confined if he or she were accused of a similar offense, or if it is essential to do so in the interests of national security.[683]

In no circumstances, however, may a POW be confined for more than three months.[684]

Any period spent by a POW in confinement awaiting trial shall be deducted from any sentence of imprisonment passed upon him or her and taken into account in fixing any penalty.[685]

Articles 97 and 98 of the GPW apply to a POW's confinement while awaiting trial.[686]

Thus, the conditions of any confinement before trial should be at least as good as that afforded POWs who are confined as part of disciplinary punishment.[687]

9.28.4 Rights of Defense and Trial Procedure.

9.28.4.1 *No Moral or Physical Coercion to Induce Admissions of Guilt.* No moral or physical coercion may be exerted on a POW in order to induce him or her to admit himself or herself guilty of the act of which he or she is accused.[688]

9.28.4.2 *Opportunity to Present Defense and Assistance of Counsel.* No POW may be convicted without having had an opportunity to present his or her defense and to have the assistance of a qualified advocate or counsel.[689]

In judicial proceedings against a POW, the POW shall be entitled to: • assistance by one of his or her POW comrades;

• defense by a qualified advocate or counsel of his or her own choice;

• the calling of witnesses; and

• if he or she deems necessary, the services of a competent interpreter.[690]

The POW shall be advised of these rights by the Detaining Power in due time before the trial to enable him or her to exercise them.[691]

Failing a choice by the POW, the Protecting Power shall find him or her an advocate or counsel, and shall have at least one week at its disposal for this purpose.[692]The Detaining Power shall deliver to the Protecting Power, on request, a list of persons qualified to present the defense.[693]Failing a choice of an advocate or counsel by the POW or the Protecting Power, the Detaining Power shall appoint a competent advocate or counsel to conduct the defense.[694]

9.28.4.3 *Facilities for Defense Counsel*. The advocate or counsel conducting the defense on behalf of the POW shall have at his or her disposal a period of two weeks at least before the opening of the trial, as well as the necessary facilities to prepare the defense of the accused.[695]He or she may, in particular, freely visit the accused and interview him or her in private.[696]He or she may also confer with any witnesses for the defense, including POWs.[697]

He or she shall have the benefit of these facilities until the term of appeal or petition has expired.[698]

9.28.4.4 *Rights of the Accused to Particulars of Charge*. Particulars of the charge or charges on which the POW is to be arraigned, as well as the documents that are generally communicated to the accused by virtue of the laws in force in the armed forces of the Detaining Power, shall be communicated to the accused POW in a language that he or she understands, and in good time before the opening of the trial.[699]The same communication in the same circumstances shall be made to the advocate or counsel conducting the defense on behalf of the POW.[700]

9.28.4.5 *Right of the Protecting Power to Attend Trial*. Representatives of the Protecting Power shall be entitled to attend the trial of the case unless, exceptionally, the trial is held *in camera* in the interest of State security.[701]In such a case, the Detaining Power shall advise the Protecting Power accordingly.[702]

9.28.4.6 *Notification of Trial Outcome*. Any judgment and sentence pronounced upon a POW shall be immediately reported to the Protecting Power and the POW Representative concerned in the form of a summary communication, which shall also indicate whether the POW has the right of appeal with a view to the quashing of the sentence or the reopening of the trial.[703]This communication shall also be sent to the accused POW in a language he or she understands, if the sentence was not pronounced in his or her presence.[704]

9.28.5 Appeals and Notice of Final Conviction or Death Sentence.

9.28.5.1 *Appeals*. Every POW shall have, in the same manner as the members of the armed forces of the Detaining Power, the right of appeal or petition from any sentence pronounced upon him or her, with a view to the quashing or revising of the sentence or the reopening of the trial.[705]

He or she shall be fully informed of his or her right to appeal or petition and of the time limit within which he or she may do so.[706]

The Detaining Power shall immediately communicate to the Protecting Power the decision of the POW to use or to waive his or her right of appeal.[707]

9.28.5.2 *Notification of Final Conviction or Death Sentence*. If a POW is finally convicted or if a sentence pronounced on a POW in the first instance is a death sentence, the Detaining Power shall as soon as possible address to the Protecting Power a detailed communication containing:

• the precise wording of the finding and sentence;

• a summarized report of any preliminary investigation and of the trial, emphasizing in particular the elements of the prosecution and the defense; and

• notification, where applicable, of the establishment where the sentence will be served.[708]

These communications shall be sent to the Protecting Power at the address previously made known to the Detaining Power.[709]

9.28.6 Death Sentences. POWs and the Protecting Powers shall be informed, as soon as possible, of the offenses that are punishable by death sentence under the laws of the Detaining Power.[710]Other offenses shall not thereafter be made punishable by the death penalty without the concurrence of the Power on which the POWs depend.[711]

The death sentence may not be pronounced on a POW unless the attention of the court has, in accordance with the second paragraph of Article 87 of the GPW, been particularly called to the fact:

- that since the accused is not a national of the Detaining Power, he or she is not bound to it by any duty of allegiance; and
- that he or she is in the power of the Detaining Power as the result of circumstances independent of his or her own will.[712]

If the death penalty is pronounced on a POW, the sentence shall not be executed before the expiration of a period of at least six months from the date when the Protecting Power receives, at an indicated address, the detailed communication provided for in Article 107 of the GPW.[713]

9.28.7 Conditions for POWs Serving Judicial Punishments.

9.28.7.1 *Establishments and Conditions Where Sentences Are Served.* A sentence pronounced on a POW after a conviction has become duly enforceable shall be served in the same establishments and under the same conditions as in the case of members of the armed forces of the Detaining Power.[714]These conditions shall in all cases conform to the requirements of health and humanity.[715]

9.28.7.2 *Confinement of Women POWs Who Have Been Sentenced.* A woman POW on whom such a sentence has been pronounced shall be confined in separate quarters and shall be under the supervision of women.[716]

9.28.7.3 *Complaints and Access to ICRC and the Protecting Power During Confinement.* In any case, POWs sentenced to a penalty depriving them of their liberty shall retain the benefit of the provisions of Articles 78 and 126 of the GPW.[717]Thus, they may make requests and complaints and deal with representatives of the Protecting Power or the ICRC.[718]

9.28.7.4 *Correspondence, Relief Parcels, Exercise, Medical Care, and Spiritual Assistance.* Furthermore, POWs sentenced to a penalty depriving them of their liberty shall be entitled to receive and send correspondence, to receive at least one relief parcel monthly, to take regular exercise in the open air, to have the medical care required by their state of health, and to have the spiritual assistance they may desire.[719]

9.28.7.5 *Prohibited Penalties.* Penalties to which POWs may be subjected shall be in accordance with the provisions of the third paragraph of Article 87 of the GPW.[720]Thus, collective punishment for individual acts, corporal punishment, imprisonment in premises without daylight, and, in general, any form of torture or cruelty, are forbidden for POWs serving judicial penalties.[721]

9.29 TRANSFER OF POWS FROM THE POWCAMP

9.29.1Determining Whether to Transfer POWs. The Detaining Power, when deciding upon the transfer of POWs, shall take into account the interests of the POWs themselves, more especially so as not to increase the difficulty of their repatriation.[722]

Sick or wounded POWs shall not be transferred as long as their recovery may be endangered by the journey, unless their safety imperatively demands it.[723]

If the combat zone draws closer to a camp, the POWs in that camp shall not be transferred unless their transfer can be carried out in adequate conditions of safety, or unless they are exposed to greater risks by remaining on the spot than by being transferred.[724]

9.29.2 Conditions for the Transfer of POWs. The transfer of POWs shall always be effected humanely and in conditions not less favorable than those under which the forces of the Detaining Power are transferred.[725]Account shall always be taken of the climatic conditions to which the POWs are accustomed, and the conditions of transfer shall in no case be prejudicial to their health.[726]

The Detaining Power shall supply POWs during transfer with sufficient food and drinking water to keep them in good health, and shall supply them with the necessary clothing, shelter, and medical attention.[727]The Detaining Power shall take adequate precautions, especially in case of transport by sea or by air, to ensure their safety during transfer, and shall draw up a complete list of all transferred POWs before their departure.[728]

9.29.3 Procedures for Transfer of POWs From the Camp.

9.29.3.1 *Notification of Transfer*. In the event of transfer, POWs shall be officially advised of their departure and of their new postal address.[729] Such notifications shall be given in time for them to pack their luggage and inform their next of kin.[730]

9.29.3.2 *Baggage*. POWs shall be allowed to take with them their personal effects and the correspondence and parcels that have arrived for them.[731] The weight of such baggage may be limited, if the conditions of transfer so require, to what each POW can reasonably carry, which shall in no case be more than 25 kilograms (approximately 55 pounds) per person.[732]

9.29.3.3 *Forwarding of Mail*. Mail and parcels addressed to their former camp shall be forwarded to POWs without delay.[733]

9.29.3.4 *Disposition of Community Property and Remaining Property*. The camp commander shall take, in agreement with the POW Representative, any measures needed to ensure the transport of the POWs' community property and of the luggage they are unable to take with them in consequence of restrictions imposed by virtue of the second paragraph of Article 48 of the GPW (*i.e.*, restrictions on the weight of baggage).[734]

9.29.3.5 *Costs of Transfers*. The costs of transfers shall be borne by the Detaining Power.[735]

9.30 TRANSFER OF POWS TO THE CUSTODY OF ANOTHER DETAINING POWER

The GPW specifies a number of rules that apply to the transfer of POWs by the Detaining Power to the custody of another Detaining Power.

9.30.1 Requirements for Transfer to the Custody of Another Detaining Power. POWs may only be transferred by the Detaining Power to a Power that is a Party to the GPW and after the Detaining Power has satisfied itself of the willingness and ability of such receiving Power to apply the GPW.[736] U.S. policy may prescribe additional requirements.[737] To ensure accountability, a POW should not be transferred before his or her formal processing and submission of all required information to the National POW Information Bureau.[738]

9.30.2 Responsibility When Transferred to the Custody of Another Detaining Power. When POWs are transferred under such circumstances, responsibility for the application of the GPW rests on the Power accepting them while they are in its custody.[739]

Nevertheless, if that Power fails to carry out the provisions of the GPW in any important respect, the Power by whom the POWs were transferred shall, upon being notified by the Protecting Power, take effective measures to correct the situation or shall request the return of the POWs.[740] Such requests must be complied with.[741]

9.30.3 Special Agreements on Transfers to the Another Detaining Power. Special agreements may be concluded with respect to POW transfers to the custody of another Detaining Power. U.S. policy has been to require a written agreement or arrangement before transferring POWs to a coalition partner.[742] For example, the receiving State may agree to inspections by the transferring State to verify compliance with the GPW.[743]

9.31 NATIONAL ACCOUNTING OF THE DETENTION OF POWS

A proper accounting of the detention of POWs is an important responsibility of the Detaining Power.[744] In order to provide as full an accounting as possible, and in order to minimize misidentification of POWs in its custody, the Detaining Power shall obtain detailed information about each POW for forwarding through its National POW Information Bureau to the Protecting Powers and the Central POW Information Agency.

9.31.1 Accountability Information That the Detaining Power Should Collect. The following information regarding POWs should be collected by the Detaining Power and given to its National POW Information Bureau for forwarding to the Powers concerned through the intermediary of the Protecting Powers and the Central POW Information Agency. This information shall make it possible quickly to advise the next of kin concerned.[745]

This information includes, in respect of each POW:

- indication of the Power on which he or she depends;
- surname (*i.e.*, last name) and first names (*i.e.*, first and middle names);
- army, regimental, personal, or serial number;
- rank;
- place and full date of birth;
- first name of the father and maiden name of the mother;

- name and address of the person to be informed; and
- the address to which correspondence for the POW may be sent.[746]

In addition to this information, information regarding transfers, releases, repatriations, escapes, admissions to a hospital, and deaths shall also be collected and transmitted.[747] Likewise, information regarding the state of health of POWs who are seriously ill or seriously wounded shall be supplied regularly, every week if possible.[748]

In addition to notices of death, wills and records of death of POWs may be forwarded in accordance with Article 120 of the GPW.[749] Lists of graves and the particulars of POWs interred in cemeteries and elsewhere may also be forwarded in accordance with Article 120 of the GPW.[750]

9.31.1.1 *Difference Between the Accountability Information the Detaining Power Must Seek and the Information the POW Must Provide.* Although POWs are required to provide full name, rank, date of birth, and service (or equivalent) number, and POWs are required to show their identity card upon demand, POWs are not required to inform the Detaining Power of their places of birth, the first name of fathers, the maiden name of mothers, or the names and addresses of persons to be informed of capture.[751]

9.31.2 National POW Information Bureau. Upon the outbreak of a conflict and in all cases of occupation, each of the parties to the conflict shall institute an official National POW Information Bureau for POWs who are in its power.[752] Neutral or non-belligerent Powers, who may have received within their territory persons belonging to one of the categories referred to in Article 4 of the GPW, shall take the same action with respect to such persons.[753]

The Power concerned shall ensure that the National POW Information Bureau is provided with the necessary accommodation, equipment, and staff to ensure its efficient working.[754] It shall be at liberty to employ POWs in the National POW Information Bureau under the conditions laid down in Section III of the GPW dealing with work by POWs.[755]

Within the shortest possible period, each of the parties to the conflict shall give its National POW Information Bureau the information referred to in the fourth, fifth, and sixth paragraphs of Article 122 of the GPW regarding any enemy person belonging to one of the categories referred to in Article 4 who has fallen into its power.[756] Neutral or non-belligerent Powers shall take the same action with regard to persons belonging to such categories whom they have received within their territory.[757]

All written communications made by the National POW Information Bureau shall be authenticated by a signature or a seal.[758]

The National POW Information Bureau: (1) receives and forwards certain information to the Powers concerned through the intermediary of the Protecting Powers and the Central POW Information Agency; (2) replies to inquiries; and (3) collects personal valuables left by POWs.

9.31.2.1 *Receiving and Forwarding Certain Information.* The National POW Information Bureau receives certain information regarding POWs from other departments of the State to which it belongs.[759]

The National POW Information Bureau shall immediately forward such information by the most rapid means to the Powers concerned through the intermediary of the Protecting Powers and likewise of the Central POW Information Agency provided for in Article 123 of the GPW.[760]

9.31.2.2 *Replying to Inquiries Regarding POWs.* The National POW Information Bureau shall also be responsible for replying to all inquiries sent to it concerning POWs, including those who have died in captivity; it will make any inquiries necessary to obtain the information that is asked for if the information is not in its possession.[761] Responses to inquiries must be made consistent with the protection of POWs against insults and public curiosity.[762]

9.31.2.3 *Collection of Personal Valuables Left by POWs.* The National POW Information Bureau shall furthermore be charged with collecting all personal valuables including sums in currencies other than that of the Detaining Power and documents of importance to the next of kin that have been left by POWs who have been repatriated or released, or who have escaped or died, and shall forward these valuables to the Powers concerned.[763] Such articles shall be sent by the National POW Information Bureau in sealed packets, which shall be accompanied by statements giving clear and full particulars of the identity of the person to whom the articles belonged, and by a complete list of the contents of the parcel.[764] Other personal effects of such POWs shall be transmitted under arrangements agreed upon between the parties to the conflict concerned.[765]

9.31.3 Central POW Information Agency. The GPW contemplates that a Central POW Information Agency shall be created in a neutral country.[766] This organization may be the same as that provided for in the GC.[767]

The function of the Central POW Information Agency shall be to collect all the information it may obtain through official or private channels respecting POWs, and to transmit it as rapidly as possible to the country of origin of the POWs or to the Power on which they depend.[768] The Central POW Information Agency shall receive from the parties to the conflict all facilities for effecting such transmissions.[769] Parties to the GPW, and in particular those States whose nationals benefit from the services of the Central POW Information Agency, are requested to give the Central POW Information Agency the financial aid it may require.[770]

The provisions of the GPW establishing the Central POW Information Agency do not restrict the humanitarian activities of the ICRC and of the relief societies described in Article 125 of the GPW.[771]

The GPW contemplates that the ICRC shall, if it deems necessary, propose to the Powers concerned the organization of a Central POW Information Agency.[772] The role of the Central POW Information Agency has been performed by the ICRC Central Tracing Agency, which has also performed this role for protected persons under the GC.

9.31.4 U.S. Practice in Reporting to the ICRC Central Tracing Agency. For the United States, the National POW Information Bureau, which has been established and managed by the Secretary of the Army, has been referred to at different times as the National Prisoner of War Information Center (NPWIC) in Operation DESERT STORM in 1991,[773] and as the National Detainee Reporting Center (NDRC) in Operation IRAQI FREEDOM in 2003. The NPWIC/NDRC has performed the same function for certain protected persons under the GC.

In general, the Theater Prisoner of War Information Center (TPWIC) has received reports from all subordinate POW facilities, regardless of Military Service, consolidated them and reconciled any discrepancies, and then forwarded its report to the NPWIC/NDRC. The NPWIC/NDRC, on behalf of the United States, has further reviewed and reconciled reports as needed, and forwarded its report to the ICRC Central Tracing Agency, which has acted as the Central POW Information Agency. The release of POW accountability information outside this process (such as directly from field components to ICRC field representatives) has been avoided so as to reduce the likelihood of erroneous, conflicting, or duplicate reporting.

9.31.5 Exemption of National POW Information Bureau and Central POW Information Agency From Certain Charges. The National POW Information Bureau and the Central POW Information Agency shall enjoy free postage for mail, all the exemptions provided for in Article 74 of the GPW,[774] and, further, so far as possible, exemption from telegraphic charges or, at least, greatly reduced rates.[775]

In addition, the National POW Information Bureau and the Central POW Information Agency shall also enjoy exemption from postal charges in respect of letter-post items, postal parcels, and monetary articles that concern POWs, which they send or receive, either directly or as intermediaries under the conditions laid down in the Universal Postal Convention.[776] Such items exempt from postal charges should be marked appropriately.[777]

9.32 ROLE OF THE PROTECTING POWER IN THE GPW

The Protecting Power is an organ for ensuring implementation of the GPW. The Protecting Power has extensive duties under the GPW. It transmits information between belligerents, monitors Detaining Power compliance with the GPW, and takes an active role in promoting the welfare of POWs.

9.32.1 Transmission of Information Between Belligerents. The Protecting Power helps transmit information between belligerents, including:

- information concerning the geographic location of camps;[778]
- financial information, including working rates of pay, notifications of payment, lists of credit balances, and claims for compensation;[779]
- notification of arrangements made for POW correspondence and relief shipments;[780]
- legal documents of POWs;[781]
- death certificates of POWs;[782]
- accountability information for the detention of POWs;[783] and
- official translations of the GPW, and laws and regulations that parties to the conflict have adopted to implement it.[784]

9.32.2 Monitoring or Inspecting Compliance With the GPW. The Protecting Power monitors or inspects the Detaining Power's compliance with the GPW, including by:

- visiting all places where POWs may be;[785]
- receiving complaints and requests by POWs, and periodic reports from the POW Representatives;[786]
- communicating with POW Representatives, and receiving the rationale in cases in which the Detaining Power refuses to approve an elected POW Representative or chooses to dismiss a POW Representative;[787]
- inspecting records of disciplinary punishments;[788]
- receiving communications and subsequent reports of death or serious injury of a POW in certain cases;[789]
- receiving the reasons for the limitations imposed by the Detaining Power on advances of pay;[790]
- inspecting POW accounts;[791]
- receiving records of labor detachments;[792] and
- observing trials of POWs if security permits and receiving notification of: (1) judicial proceedings, (2) trial outcome, (3) whether the POW has decided to appeal, and (4) final conviction or death sentence.[793]

9.32.3 Actively Working to Improve the Welfare of POWs. In many other instances, the GPW contemplates that the Protecting Power will actively work to improve the welfare of POWs, such as by:

- helping the Detaining Power determine the maximum amount of money that POWs have in their possession;[794]
- approving any limitation on correspondence beyond two letters and four cards per month;[795]
- proposing any necessary limitations on relief shipments;[796]
- supervising the distribution of collective relief;[797]
- setting up special means of transport to ensure the conveyance of relief shipments to POWs;[798]
- finding POWs defense counsel when necessary;[799]
- notifying the transferring State in the event that a receiving State fails to carry out its obligations;[800]
- participating in the appointment of members of a Mixed Medical Commission;[801] and
- lending its good offices to assist in dispute resolution.[802]

9.33 ACCESS TO POWS BY THE PROTECTING POWERS, ICRC, AND RELIEF ORGANIZATIONS

9.33.1 Access by Protecting Powers. Representatives or delegates of the Protecting Powers shall have permission to go to all places where POWs may be, particularly to places of internment, imprisonment, and labor, and shall have access to all premises occupied by POWs; they also shall be allowed to go to the places of departure, passage, and arrival of POWs who are being transferred.[803] They shall be able to interview POWs, and in particular POW Representatives, without witnesses, either personally or through an interpreter.[804]

Representatives and delegates of the Protecting Powers shall have full liberty to select the places they wish to visit.[805] The duration and frequency of these visits shall not be restricted.[806] Visits may not be prohibited except for reasons of imperative military necessity, and then only as an exceptional and temporary measure.[807]

9.33.1.1 *Participation of Compatriots of POWs in Visits*. The Detaining Power and the Power on which these POWs depend may agree, if necessary, that compatriots of these POWs be permitted to participate in the visits.[808]

9.33.1.2 *ICRC Delegates Enjoying the Same Prerogatives of Access*. The delegates of the ICRC shall enjoy the same prerogatives as those of the representatives and delegates of the Protecting Powers described in Article 126 of the GPW.[809]

The appointment of such delegates shall be submitted to the approval of the Power detaining the POWs to be visited.[810]

9.33.2 Access by Relief Societies and Other Organizations. Subject to the measures that the Detaining Powers may consider essential to ensure their security or to meet any other reasonable need, the representatives of religious organizations, relief societies, or any other organization assisting POWs shall receive from these Powers, for themselves and their duly accredited agents, all necessary facilities for visiting the POWs; for distributing relief supplies and material, from any source, intended for religious, educational, or recreative purposes; and for assisting them in organizing their leisure time within the POW camps.[811] Such societies or organizations may be constituted in the territory of the Detaining Power or in any other country, or they may have an international character.[812]

9.33.2.1 *Limitations on the Number of Relief Organizations*. The Detaining Power may limit the number of societies and organizations whose delegates are allowed to carry out their activities in its territory and under its supervision, on condition, however, that such limitation shall not hinder the effective operation of adequate relief to all POWs.[813]The special position of the ICRC in this field shall be recognized and respected at all times.[814]

9.33.2.2 *Receipts for Relief Consignments*. As soon as relief supplies or material intended for religious, educational, or recreative purposes are handed over to POWs, or very shortly afterwards, receipts for each consignment, signed by the POW Representative, shall be forwarded to the relief society or organization making the shipment.[815]At the same time, receipts for these consignments shall be supplied by the administrative authorities responsible for guarding the POWs.[816]

9.34 DEATH OF POWS

This section addresses rules relating to the death of POWs. The GWS and GWS-Sea also have rules relating to the treatment of the dead; however, the GWS and GWS-Sea provisions are broader in application, *e.g.*, they address members of the armed forces who have died on the battlefield and thus were not held as POWs.[817]

9.34.1 Transmittal of Wills. The Detaining Power has an obligation to help POWs with their wills.[818]In all cases, after death, POWs' wills shall be transmitted without delay to the Protecting Power, and a certified copy shall be sent to the Central POW Information Agency.[819]

9.34.2 Death Certificates. Death certificates in the form given in Annex IV D to the GPW, or lists certified by a responsible officer, of all persons who die as POWs shall be forwarded as rapidly as possible to the National POW Information Bureau established in accordance with Article 122 of the GPW.[820]

The death certificates or certified lists shall show:

• particulars of identity, as set out in the third paragraph of Article 17 of the GPW, that are listed on the POW's identity card;

o such as, surname; first names; rank; date of birth; and army, regimental, personal, or serial number or equivalent information;[821]

• the date and place of death;

• the cause of death;

• the date and place of burial;

• where applicable, the fact of, and reasons for, cremation;[822]and

• all particulars necessary to identify the graves or inurnment/columbarium location.[823]

9.34.3 Burial or Cremation and Inurnment.

9.34.3.1 *Medical Examination Before Burial or Cremation*. The burial or cremation of a POW shall be preceded by a medical examination of the body with a view to confirming death and enabling a report to be made and, where necessary, establishing identity.[824]For example, if a POW's identity was not established before death, and the identity cannot be established, the report should include information that will enable the Power on which the POW depended to establish his or her identity (*e.g.*, examination of teeth, fingerprints, photograph, DNA samples).[825]

9.34.3.2 *Burial*. The detaining authorities shall ensure that POWs who have died in captivity are honorably buried, if possible according to the rites of the religion to which they belonged.[826]

Wherever possible, deceased POWs who depended on the same Power shall be interred in the same place.[827]Deceased POWs shall be buried in individual graves unless unavoidable circumstances require the use of collective graves.[828]For example, an epidemic or military operations may require the Detaining Power to undertake collective burials in the interest of public health because individual graves are not possible due to lack of time and resources.[829]

9.34.3.3 *Cremation*. Bodies of POWs may be cremated only for imperative reasons of hygiene, on account of the religion of the deceased, or in accordance with his or her express wish to this effect.[830]In case of cremation, the fact shall be stated and the reasons given in the death certificate of the deceased.[831]

9.34.4 Maintenance and Records of Graves and Ashes. The detaining authorities shall ensure that the graves of POWs who have died in captivity are respected, suitably maintained, and marked so as to be found at any time.[832]

9.34.4.1 *Records Held by the Graves Registration Service.* In order that graves may always be found, all particulars of burials and graves shall be recorded with a Graves Registration Service established by the Detaining Power.[833] Lists of graves and particulars of the POWs interred in cemeteries and elsewhere shall be transmitted to the Power on which such POWs depended.[834] Responsibility for the care of these graves and for records of any subsequent moves of the bodies shall rest on the Power controlling the territory, if that Power is a Party to the GPW.[835]

The Graves Registration Service shall also identify, record, and respectfully keep the ashes until they can be disposed of in accordance with the wishes of the home country.[836]

The GWS also provides for the Graves Registration Service to record graves of persons belonging to enemy military forces who died but were never held as POWs.[837]

9.34.5 Inquiries Into Death or Serious Injury of POWs in Certain Cases. Every death or serious injury of a POW caused or suspected to have been caused by a sentry, another POW, or any other person, as well as any death the cause of which is unknown, shall be immediately followed by an official inquiry by the Detaining Power.[838] A communication on this subject shall be sent immediately to the Protecting Power.[839]

Serious injury, in many cases, may be understood to mean an injury that requires inpatient treatment in a hospital or infirmary.[840]

Statements shall be taken from witnesses, especially from those who are POWs, and a report including such statements shall be forwarded to the Protecting Power.[841]

If the inquiry indicates the guilt of one or more persons, the Detaining Power shall take all measures for the prosecution of the person or persons responsible.[842]

9.35 EXCHANGES AND OTHER RELEASE AND REPATRIATION OF POWS DURING HOSTILITIES

9.35.1 Exchange of POWs During Hostilities Through Cartel Agreements. The exchange of POWs, other than those whose repatriation is required by the GPW, may be effected by agreement between the belligerents.[843] Such agreements have been called cartels.[844] The modern practice in this area has been infrequent.[845]

As a general principle, the exchange of POWs, other than those whose repatriation is required by the GPW, is an act of convenience to both belligerents.[846]

The conditions for exchange are generally prescribed in the cartel, and exchanges need not be on the basis of number-for-number, or rank-for-rank.[847]

9.35.2 Release and Repatriation of POWs Not on the Basis of Exchange. Even when not required by the GPW or done on the basis of an exchange, a Detaining Power may release and repatriate POWs during hostilities to the Power to which they belong.[848]

Such release and repatriation of POWs during hostilities may be accomplished between governments through agreements. In any case, POWs subject to release and repatriation should be interviewed by representatives of the Protecting Power or the ICRC to ensure that their return is voluntary.[849]

9.36 DIRECT REPATRIATION AND ACCOMMODATION IN NEUTRAL COUNTRIES DURING HOSTILITIES The GPW provides for the direct repatriation or accommodation in neutral countries of certain wounded and sick POWs during hostilities. POWs who meet with accidents shall, unless the injury is self-inflicted, have the benefit of the provisions of the GPW as regards repatriation or accommodation in a neutral country.[850]

States may also conclude agreements with a view to the direct repatriation or internment in a neutral country of able-bodied POWs who have undergone a long period of captivity.[851]

9.36.1 Direct Repatriation of Seriously Wounded, Injured, or Sick POWs. Parties to a conflict have an obligation to send back to their own country, regardless of number or rank, certain seriously wounded and seriously sick POWs, after having cared for them until they are fit enough to travel.[852] Before the 1929 GPW, such repatriations occurred in accordance with specific agreements.[853] Under the GPW, such repatriations have been conducted as part of exchanges or on a unilateral basis.[854]

No sick or injured POW who is eligible for repatriation under the first paragraph of Article 109 of the GPW may be repatriated against his or her will during hostilities.[855]

9.36.1.1 *Categories of POWs Eligible for Direct Repatriation*. The following categories of POWs shall be repatriated directly:

• incurably wounded and sick whose mental or physical fitness seems to have been gravely diminished;

• wounded and sick who, according to medical opinion, are not likely to recover within one year, whose condition requires treatment, and whose mental or physical fitness seems to have been gravely diminished; and

• wounded and sick who have recovered, but whose mental or physical fitness seems to have been gravely and permanently diminished.[856]

9.36.2 Accommodation in Neutral Countries. Throughout the duration of hostilities, parties to the conflict shall endeavor, with the cooperation of the neutral Powers concerned, to make arrangements for the accommodation in neutral countries of certain sick and wounded POWs as discussed below.[857]This provision of the GPW reflected prior State practice.[858]

9.36.2.1 *Categories of Wounded, Injured, or Sick to Be Accommodated in a Neutral State*. The following categories of POWs may be accommodated in a neutral country: • wounded, injured, or sick whose recovery may be expected within one year of the date of the wound or the beginning of the illness, if treatment in a neutral country might increase the prospects of a more certain and speedy recovery; and

• POWs whose mental or physical health, according to medical opinion, is seriously threatened by continued captivity, but whose accommodation in a neutral country might remove such a threat.[859]

9.36.3 Agreements to Intern POWs in Neutral Territory. The Detaining Power, the Power on which the POWs depend, and a neutral Power agreed upon by these two Powers, shall endeavor to conclude agreements that will enable POWs to be interned in the territory of that neutral Power until the close of hostilities.[860]For example, if the Detaining Power were unable to comply with the minimum standards in the GPW for the treatment of POWs, then it would seek to negotiate such agreements with the Power on which the POWs depend and a neutral Power.[861]

9.36.3.1 *Repatriation From a Neutral State*. The conditions that POWs accommodated in a neutral country must fulfill in order to permit their repatriation shall be fixed, as shall likewise their status, by agreement between the Powers concerned.[862]In general, POWs who have been accommodated in a neutral country, and who belong to the following categories, should be repatriated:

• those whose health has deteriorated so far as to fulfill the conditions for direct repatriation;[863]and

• those whose mental or physical powers remain, even after treatment, considerably impaired.[864]

9.36.4 Lack of Special Agreements to Determine Cases of Disablement or Sickness Entailing Direct Repatriation or Accommodation in a Neutral Country. If no special agreements are concluded between the parties to the conflict concerned to determine the cases of disablement or sickness entailing direct repatriation or accommodation in a neutral country, such cases shall be settled in accordance with the principles laid down in the Model Agreement concerning direct repatriation and accommodation in neutral countries of wounded and sick POWs and in the Regulations concerning Mixed Medical Commissions annexed to the GPW (GPW Annex II).[865]

9.36.5 Mixed Medical Commissions. Upon the outbreak of hostilities, Mixed Medical Commissions shall be appointed to examine sick and wounded POWs, and to make all appropriate decisions regarding them.[866]The appointment, duties, and functioning of these Commissions shall be in conformity with the provisions of the Regulations annexed to the GPW (GPW Annex II).[867]

9.36.5.1 *Examination Not Necessary for Manifestly Seriously Injured or Sick POWs*. POWs who, in the opinion of the medical authorities of the Detaining Power, are manifestly seriously injured or seriously sick may be repatriated without having to be examined by a Mixed Medical Commission.[868]

9.36.5.2 *POWs Entitled to Present Themselves for Examination Before the Commission*. Besides those who are designated by the medical authorities of the Detaining Power, wounded or sick POWs belonging to the categories listed below shall be entitled to present themselves for examination by the Mixed Medical Commissions provided for in Article 112 of the GPW:

• Wounded and sick proposed by a physician or surgeon who is of the same nationality, or a national of a party to the conflict allied with the Power on which these POWs depend, and who exercises his or her functions in the camp;

• Wounded and sick proposed by their POW Representative; and

• Wounded and sick proposed by the Power on which they depend, or by an organization duly recognized by this Power and giving assistance to the POWs.[869]

POWs who do not belong to one of the three foregoing categories may nevertheless present themselves for examination by Mixed Medical Commissions, but shall be examined only after those belonging to the three foregoing categories.[870]

9.36.5.3 *Observers at the Examination.* The physician or surgeon of the same nationality as the POWs who present themselves for examination by the Mixed Medical Commission, and the POW Representative of these POWs, shall have permission to be present at the examination.[871]

9.36.6 Costs of Repatriation or Transportation to a Neutral Country During Hostilities. The costs of repatriating POWs or of transporting them to a neutral country are borne, from the frontiers of the Detaining Power, by the Power on which the POWs depend.[872]

9.36.7 Employment of Persons Who Have Been Repatriated. No repatriated person may be employed on active military service.[873]

9.36.7.1 *No Repatriated Person.* The rule against employment on active military service only applies to POWs repatriated by the Detaining Power pursuant to Articles 109 and 110 of the GPW.[874]For example, able-bodied POWs who are voluntarily repatriated by the unilateral decision of the Detaining Power during hostilities may be employed on active military service.[875]

9.36.7.2 *Active Military Service.* Not all military service is "active military service." Repatriated persons may remain employed on military service.[876]What "active military service" they may engage in after repatriation may be the subject of agreements pursuant to Articles 109 and 110 of the GPW. Absent such agreements, active military service may be understood as broadly covering any participation, whether direct or indirect, in armed operations against the former Detaining Power or its allies, but to exclude medical work or strictly administrative duties.[877]

9.37 RELEASE AND REPATRIATION AFTER HOSTILITIES

POWs shall be released and repatriated without delay after the cessation of active hostilities.[878]

9.37.1 Agreements on POW Release and Repatriation. The release and repatriation of POWs generally have been addressed by peace treaties or other agreements among belligerents at the end of the war.[879]For example, agreements may address the apportionment of costs or commissions to search for dispersed POWs.[880]

In the absence of stipulations to the above effect in any agreement concluded between the parties to the conflict with a view to the cessation of hostilities, or failing any such agreement, each of the Detaining Powers shall itself establish and execute without delay a plan of repatriation in conformity with the principle laid down in paragraph 1 of the Article 118 of the GPW.[881]Thus, once active hostilities have ceased, the Detaining Powers must release and repatriate POWs, even if there is no specific agreement regarding the release and repatriation of POWs, or even if a formal peace agreement ending the war has not yet been concluded.[882]During the process of releasing and repatriating POWs, it is proper to expect that each party's conduct with respect to the repatriation of POWs will be reasonable and broadly commensurate with the conduct of the other.[883]For example, it would not be reasonable to expect that a State would release all of the POWs it holds without assurance that its own personnel held by the enemy will also be released.[884]

9.37.2 Cessation of Active Hostilities. According to Lauterpacht, the phrase "cessation of active hostilities" probably does not refer to a situation that leaves open the possibility of a resumption of struggle, but to a situation in which it is out of the question for hostilities to resume.[885]It is the complete end of the fighting with clearly no probability of resumption of hostilities in the near future.[886]The cessation of active hostilities may also be understood as describing the point in time when belligerents feel sufficiently at ease about the future that they are willing to release and repatriate all POWs.[887]The cessation of active hostilities may result from a capitulation or agreement, but such an agreement is not required if there is no prospect that hostilities will resume.

9.37.3 Without Delay. The GPW provides that release and repatriation take place without delay. This requirement, however, does not affect the practical arrangements that must be made to ensure that repatriation takes place in a safe and orderly manner in accordance with the requirements of the GPW.[888]

For example, the availability of manpower and transportation to conduct the repatriation, continuing military operations in other theaters, or the ability of countries to receive POWs in large numbers, may all increase the time required to conduct the release and repatriation.[889]The security situation in the country in which POWs are to be released may also be an important factor.[890]In addition, time may be needed based on the medical condition of the POWs or in order to conduct interviews to ensure that repatriation is voluntary.

9.37.4 POWs Whom the Detaining Power Is Not Required to Repatriate.

9.37.4.1 *Persons Illegally Enrolled in Enemy Forces and Deserters*. The obligation to repatriate does not apply to persons who have been illegally enrolled in the armed forces of the enemy (for example, the inhabitants of occupied territories who have been forced to enlist in the military of the Occupying Power), or to persons who have deserted to the other side.[891]

9.37.4.2 *POWs Who Resist Repatriation*. The Detaining Power is not required to repatriate forcibly POWs who do not wish to be repatriated.

Although the GPW provides that POWs may not renounce the rights secured to them by the GPW,[892]this principle is not violated by the POW rejecting repatriation and requesting asylum, if it is established in a satisfactory manner that the POW is making an informed, voluntary, and personal choice.[893]Thus, the GPW does not itself change accepted principles of international law under which asylum is applicable to POWs, and the Detaining Power may, but is not required to, grant asylum.[894]The policy of the United States has been not to conduct forcible repatriation of POWs and, in particular, not to transfer any person when torture is more likely than not to result.[895]

The voluntariness of POWs' decisions may be established through interviews by an impartial intermediary, such as a neutral State or the ICRC.[896]

9.37.4.3 *POWs Undergoing Criminal Proceedings for an Indictable Offense*. The repatriation of a POW may not be delayed on the basis that the POW has not undergone disciplinary proceedings or punishment.[897]

POWs against whom criminal proceedings for an indictable offense are pending may be detained until the end of such proceedings, and, if necessary, until the completion of the punishment.[898]The same shall apply to POWs already convicted for an indictable offense.[899]In fact, in most instances, as the Supreme Court has observed, the practical administration of the system of military justice under the law of war would fail if such authority were thought to end with the cessation of hostilities because only after their cessation could the greater number of offenders and the principal ones be apprehended and tried.[900]

Parties to the conflict shall communicate to each other the names of any POWs who are detained until the end of the proceedings or until punishment has been completed.[901]

The GPW does not prohibit the extradition of POWs to other Parties to the GPW to face criminal charges.[902]

POWs held after the cessation of active hostilities on this basis, even if they have been convicted, remain entitled to the benefits of the GPW until they are released and repatriated.[903]

9.37.5 Commissions to Search for Dispersed POWs. By agreement between the parties to the conflict, commissions shall be established for the purpose of searching for dispersed POWs and of ensuring their repatriation with the least possible delay.[904]

9.37.6 Costs of Repatriation at the Close of Hostilities. The costs of repatriation of POWs shall in all cases be equitably apportioned between the Detaining Power and the Power on which the POWs depend.[905]

If the two Powers are contiguous, then the Power on which the POWs depend shall bear the costs of repatriation from the frontiers of the Detaining Power.[906]

If the two Powers are not contiguous, the Detaining Power shall bear the costs of transport of POWs over its own territory as far as its frontier or its port of embarkation nearest to the territory of the Power on which the POWs depend.[907]The Parties concerned shall agree between themselves on the equitable apportionment of the remaining costs of the repatriation, but the conclusion of such agreement shall in no circumstances justify any delay in the repatriation of the POWs.[908]

9.38 PROCEDURE ON RELEASE AND REPATRIATION AFTER HOSTILITIES

9.38.1 Conditions During Release and Repatriation. Repatriation shall be effected in conditions similar to those laid down in Articles 46 through 48 of the GPW for the transfer of POWs, having regard to the provisions of Article 118 of the GPW and to those of Article 119 of the GPW.[909]

9.38.1.1 *Conditions of Movement.* The conditions of repatriation must be humane and not less favorable than those under which the forces of the Detaining Power are transferred.[910] Adequate precautions must be taken for the health and safety of POWs, including provision of sufficient food, drinking water, clothing, shelter, and medical attention.[911]

9.38.2 Notification of Repatriation Plan and Departure. POWs should be informed of the plan for their repatriation.[912] For example, this may take place through notices posted in the camps and provided to the POW Representatives, or a public announcement.[913]

POWs who are to be moved must be identified and listed before departure, and both the National POW Information Bureau and the POWs themselves officially notified in advance, so that they can pack their baggage and inform their next of kin.[914]

9.38.3 Personal Property. On repatriation, any articles of value impounded from POWs under Article 18 of the GPW,[915] and any foreign currency that has not been converted into the currency of the Detaining Power, shall be restored to them.[916] Articles of value and foreign currency that, for any reason whatever, are not restored to POWs on repatriation, shall be sent to the National POW Information Bureau set up under Article 122 of the GPW.[917]

POWs shall be allowed to take with them their personal effects and any correspondence and parcels that have arrived for them.[918] The weight of such baggage may be limited, if the conditions of repatriation so require, to what each POW can reasonably carry.[919] Each POW shall in all cases be authorized to carry at least 25 kilograms (approximately 55 pounds).[920] The other personal effects of the repatriated POW shall be left in the charge of the Detaining Power, which shall have them forwarded to him or her as soon as it has concluded an agreement to this effect, regulating the conditions of transport and the payment of the costs involved, with the Power on which the POW depends.[921]

9.38.4 POW Accounts and Financial Matters. POW accounts should be addressed in accordance with Article 66 of the GPW.[922]

In addition, the Powers concerned should make arrangements concerning advances of pay, compensation payments, and remittances.[923]

9.38.5 Canteen Profits. The profits of the canteen should be disposed of in accordance with Article 28 of the GPW.[924]

9.39 CODE OF CONDUCT FOR U.S. ARMED FORCES

The Code of Conduct is a moral code designed to provide U.S. military personnel with a standard of conduct.[925] Although designed for situations in which members of the U.S. armed forces may be held as POWs, the spirit and intent of the Code apply to service members subjected to other types of hostile detention, such as hostage scenarios.[926]

9.39.1 Text of the Code of Conduct and Discussion. The Code of Conduct was written to be consistent with the requirements imposed by the GPW. Violations of the Code of Conduct are not criminally punishable as such, but conduct that violates the Code of Conduct may also violate the Uniform Code of Military Justice.

9.39.1.1 *Code of Conduct – Article I.*

I am an American, fighting in the forces which guard my country and our way of life. I am prepared to give my life in their defense.

9.39.1.2 *Code of Conduct – Article II.*

I will never surrender of my own free will. If in command, I will never surrender the members of my command while they still have the means to resist. Under Article II of the Code of Conduct, the responsibility and authority of a commander never extends to the surrender of command, even if isolated, cut off, or surrounded, while the unit has a reasonable power to resist, break out, or evade to rejoin friendly forces.[927]

Under the Uniform Code of Military Justice, a U.S. commander who shamefully surrenders to the enemy any command or place that is his or her duty to defend is subject to punishment.[928] In addition, any person subject to the Uniform Code of Military Justice who compels or attempts to compel a commander of any place, vessel, aircraft, or other military property, or of any body of members of the armed forces, to give it up to an enemy or to abandon it, or who strikes the colors or flag to an enemy without proper authority, is subject to punishment.[929]

Article II of the Code of Conduct is consistent with the authority of commanders under the law of war to conclude capitulation agreements for the surrender of forces under their command.[930]

9.39.1.3 *Code of Conduct – Article III.*

If I am captured I will continue to resist by all means available. I will make every effort to escape and to aid others to escape. I will accept neither parole nor special favors from the enemy.

By limiting the punishment of POWs with respect to an unsuccessful escape, the GPW acknowledges that POWs may have a duty to seek to escape, as reflected in Article III of the Code of Conduct.[931]

Although the GPW contemplates that POWs may accept parole,[932]U.S. policy, as reflected in Article III of the Code of Conduct, prohibits accepting parole.

9.39.1.4 *Code of Conduct – Article IV.*

If I become a prisoner of war, I will keep faith with my fellow prisoners. I will give no information [n]or take part in any action which might be harmful to my comrades. If I am senior, I will take command. If not, I will obey the lawful orders of those appointed over me and will back them up in every way.

Article IV of the Code of Conduct exemplifies the spirit of comradeship that many POWs have exhibited in past conflicts.[933]It reflects a higher standard of conduct than that punishable under Article 105 of the Uniform Code of Military Justice regarding misconduct towards fellow POWs.[934]The senior POW in command is not necessarily the POW Representative.[935]In POW camps containing only enlisted personnel, the POW Representative is elected.[936]

9.39.1.5 *Code of Conduct – Article V.*

When questioned, should I become a prisoner of war, I am required to give name, rank, service number and date of birth. I will evade answering further questions to the utmost of my ability. I will make no oral or written statements disloyal to my country and its allies or harmful to their cause.

Article V of the Code of Conduct reflects the obligations of the POW to provide accountability information.[937]The Detaining Power may have further obligations to collect additional accountability information relating to POWs.[938]

Article V of the Code of Conduct also underscores one aspect of Article 104 of the Uniform Code of Military Justice regarding aiding the enemy by specifically forbidding oral or written statements disloyal to the United States or harmful to its cause.[939]

9.39.1.6 *Code of Conduct – Article VI.* I will never forget that I am an American, fighting for freedom, responsible for my actions, and dedicated to the principles which made my country free. I will trust in my God and in the United States of America.

9.39.2 Background on the U.S. Code of Conduct. Although misconduct by POWs was not unique to the Korean War, the Code of Conduct was introduced after public awareness of instances of misconduct by U.S. service members during their captivity as POWs during the Korean War. The Secretary of Defense established a committee that recommended that a unified and purposeful standard of conduct for Americans who become POWs be promulgated.[940]

President Dwight D. Eisenhower promulgated the Code of Conduct on August 17, 1955.[941]A review of the Code of Conduct after the return of U.S. POWs from the Vietnam War confirmed the value of the Code of Conduct.[942]

X – Civilians in the Hands of a Party to the Conflict

Chapter Contents

10.1 INTRODUCTION

This Chapter addresses protection of civilians in the hands of a party to an international armed conflict. In particular, the Chapter addresses civilians who are interned in the home territory of a belligerent or occupied territory.

Rules for the protection of civilians in the conduct of hostilities are addressed in Chapter V. Rules for the protection of civilians that are specific to occupied territory are addressed in Chapter XI, while rules that apply to both civilians in occupied territory and civilians in a belligerent's home territory are addressed in this Chapter.

10.1.1 Interpretation and Application of the GC. The GC underlies most of the international law rules applicable to the United States for the treatment of civilians in the hands of a party to the conflict.

As with the GPW, the GC's provisions should be interpreted in light of the principles that underlie the treatment of civilians and, in particular, in light of the goal of advancing the humane treatment of protected persons.[1] The subsequent practice of States in applying the GC could also assist in interpreting its provisions because States' decades of experience in applying the GC may be very helpful in understanding its requirements.[2]

10.1.1.1 *GC – Notes on Terminology*. In the GC, *protected persons* does not simply refer to persons who are "civilians" in the sense of not being members of the military; rather, *protected persons* refers to persons who are protected by the GC in connection with international armed conflict or occupation.[3]

In the GC, *internees* refers to protected persons who are interned.

In the GC, the *Detaining Power* refers to the State that holds the internee.

In the GC, the *Protecting Power* refers to a neutral State that helps implement the GC.[4]

10.1.1.2 *Special Agreements Under the GC*. Under the GC, States may conclude a variety of special agreements during international armed conflict to facilitate the protection of civilians. The GC specifically provides for agreements:

- to entrust to an effective and impartial organization the duties of the Protecting Powers;[5]
- to establish and recognize civilian hospital zones and localities;[6]
- to establish and determine the conditions of neutralized zones;[7]
- for the removal of wounded, sick, infirm, and aged persons, children, and maternity cases, from besieged or encircled areas or for the passage of medical and religious personnel on their way to such areas;[8]
- to settle practical details regarding the departure of protected persons from a belligerent's home territory;[9]
- to establish the conditions for the sending of individual parcels and collective relief shipments;[10]
- for the release, repatriation, the return to places of residence or the accommodation in a neutral country of certain classes of internees, in particular children, pregnant women and mothers with infants and young children, wounded and sick, and internees who have been detained for a long time;[11] and
- on the procedure, or to select an umpire who will decide the procedure, to be followed for an inquiry concerning any alleged violation of the GC.[12]

In addition to the agreements expressly provided for in Articles 11, 14, 15, 17, 36, 108, 109, 132, 133, and 149 of the GC, Parties to the GC may conclude other special agreements for all matters concerning which they may deem it suitable to make separate provision.[13] No special agreement shall adversely affect the situation of protected persons, as defined by the GC, nor restrict the rights that it confers upon them.[14]

Protected persons shall continue to have the benefit of such agreements as long as the GC is applicable to them, except where express provisions to the contrary are contained in these agreements or in subsequent agreements, or where more favorable measures have been taken with regard to them by one or other of the parties to the conflict.[15]

10.1.2 DoD Policies and Regulations for the Treatment of Internees. DoD policies and regulations regarding the treatment of internees provide authoritative guidance for DoD personnel and fill an important role in implementing the complex requirements of the GC.[16]

Practitioners are advised to consult all applicable policies and regulations, as these, in many cases, exceed the requirements of the GC, U.S. statutes, and Executive Orders.

10.2 NATIONAL-LEVEL GC IMPLEMENTATION MEASURES

The GC requires a number of measures at the national level to facilitate protection for protected persons.

10.2.1 Appointment or Establishment of Entities. The GC provides for certain bodies or entities to be appointed or established.

10.2.1.1 *Appointment of a Protecting Power*. The GC contemplates that a Protecting Power has been appointed to assist in the implementation of the GC.[17] A Protecting Power has a variety of roles under the GC related to the protection of protected persons.[18]

10.2.1.2 *Establishment of a National Protected Person Information Bureau*. A National Protected Person Information Bureau shall be established by each of the parties to the conflict.[19]

10.2.1.3 *Establishment of a Central Information Agency for Protected Persons*. A Central Information Agency for protected persons shall be established in a neutral country.[20]

10.2.2 Notification of Certain Laws or Policies Between Belligerents. Parties to a conflict must notify opposing belligerents or the Protecting Power of certain matters relating to protected persons.

10.2.2.1 *Location and Marking of Places of Internment Camps*. The Detaining Power shall provide the Powers concerned useful information regarding the geographic location of places of internment, and agreements on their markings may be reached.[21]

10.2.2.2 *Notification of Measures Taken for Implementing GC Provisions That Address Relations Between Internees and the Exterior*. Immediately upon interning protected persons, the Detaining Power shall inform them, the Power to which they owe allegiance, and their Protecting Power of the measures taken for executing the provisions of Chapter VIII of Section IV of the GC, which address the relations between internees and the exterior (*e.g.*, correspondence, relief shipments, family visits).[22]The Detaining Power shall likewise inform the Parties concerned of any subsequent modifications of such measures.[23]

10.3 PROTECTED PERSON STATUS

In general, the GC uses the concept of *protected person* to define the individuals who are entitled to receive its protections. Principally, *protected persons* include persons of enemy nationality living in the territory of a belligerent State and the inhabitants of occupied territories.[24]Even if persons are not protected persons under the GC, other rules may be applicable to them. For example, certain baseline rules apply to the treatment of all detainees, including those who are not protected persons or POWs.[25]

10.3.1 Protected Person – Notes on Terminology. *Protected persons* is used in the GC as a term to refer to persons who are protected by the GC in connection with international armed conflict or occupation.

Protected persons is often understood as a term that is specific to the GC. In a few places, however, the GPW, the GWS, and the GWS-Sea also refer to "protected persons" or "persons protected."[26]In such cases, "protected persons" and "persons protected" refer to persons protected by that respective convention, as opposed to *protected persons* for purposes of the GC.

Although the GC addresses the protection of civilians, *protected persons* may include certain unprivileged belligerents. Certain rights and privileges that protected persons who are unprivileged belligerents receive, however, are subject to derogation for security reasons.[27]

10.3.2 Persons Who Receive Protected Person Status Under the GC. Subject to certain exceptions addressed below, persons protected by the GC are those who, at a given moment and in any manner whatsoever, find themselves, in case of a conflict or occupation, in the hands of a party to the conflict or Occupying Power of which they are not nationals.[28]

Because the definition of protected person is framed negatively in terms of persons who are in the hands of a State of which they are not nationals, the definition of protected person allows for the possibility that persons without any nationality may be protected persons.[29]

10.3.2.1 *"Find Themselves"*. The persons must "find themselves" in the hands of a party to the conflict or Occupying Power, which suggests an element of happenstance or coincidence.[30]For example, nationals of a neutral or non-belligerent State who travel to an occupied State to fight the Occupying Power cannot be said to have "found themselves" within that occupied territory within the meaning of Article 4 of the GC.

10.3.2.2 *In Occupied Territory or the Home Territory of a Party to the Conflict*. To be entitled to the protections provided under the GC for "protected persons," a person must be located in either (1) occupied territory or (2) the home territory of a party to the conflict.[31]

The references in the GC that limit the scope of protections for protected persons to those "in the territory of a Party to the conflict" refer to the *home territory* of the party to the conflict in whose hands the protected person finds himself or herself.[32]For example, Section II of Part III of the GC, which applies to "Aliens in the Territory of a Party to the Conflict," addresses the protection of aliens present on the home territory of a party to the conflict in relation to that State.

10.3.2.3 *Not Protected by the GWS, GWS-Sea, or the GPW*. Persons protected by the GWS, the GWS-Sea, or the GPW shall not be considered as *protected persons* within the meaning of the GC.[33]For example, persons who receive POW status under the GPW are not considered as *protected persons* within the meaning of the GC.[34]

10.3.2.4 *Unprivileged Belligerents Not Per Se Excluded From Protected Person Status*. The fact that a person has engaged in hostile or belligerent conduct does not *per se* exclude that person from protected person status under the GC.[35]

Certain rights and privileges of the GC that a protected person who has engaged in such conduct would otherwise receive are, however, subject to derogation for security reasons.[36]

10.3.3 Categories of Nationals Specifically Excluded From the Definition of Protected Person Under the GC. As discussed in this subsection, the GC specifically excludes certain classes of persons from the definition of protected person.[37]

The provisions of Part II of the GC, which address the general protection of populations against certain consequences of war, are, however, wider in application, as defined in Article 13 of the GC.[38]These provisions cover the whole of the populations in conflict without any adverse distinction based, in particular, on race, nationality, religion, or political opinion, and are intended to alleviate the sufferings caused by war.[39]Thus, persons excluded by reason of their nationality from the definition of protected person may nonetheless receive the protections of Part II of the GC.

10.3.3.1 *A State's Own Nationals*. In general, a State's own nationals who are in its hands are not *protected persons* under the GC.[40]This exclusion from the definition of "protected person" in the GC is consistent with a traditional principle of international law—the GC "does not interfere in a State's relations with its own nationals."[41]However, the GC does limit the arrest, prosecution, conviction, or deportation of nationals of any Occupying Power who, before the outbreak of hostilities, have sought refuge in the territory of the occupied State.[42]

10.3.3.2 *Nationals of a State That Is Not Bound by the GC*. Nationals of a State that is not bound by the GC are not protected by it.[43]This provision of the GC reflects a legal principle underlying the GC: a State must accept the burdens of the GC in order to receive its benefits.[44]

10.3.3.3 *Nationals of a Neutral State or Co-Belligerent State While Normal Diplomatic Representation Exists*. Nationals of a neutral State who find themselves in the territory of a belligerent State, and nationals of a co-belligerent State, shall not be regarded as protected persons while the State of which they are nationals has normal diplomatic representation in the State in whose hands they are.[45]

These persons were omitted from the GC's definition of "protected person" in order to avoid creating complications or inconsistencies in procedures should both the GC and the law applicable to normal diplomatic representation apply.[46]

Nationals of a neutral State in occupied territory, however, are regarded as protected persons under the GC.[47]

10.3.4 Commencement and Duration of Protected Person Status. As with the general application of the GC, protected persons shall receive their protections from the outset of any conflict or occupation mentioned in Article 2 of the GC.[48]In general, the application of protected person status also ceases when the GC ceases to apply.

In the territory of parties to the conflict, the application of the GC shall cease on the general close of military operations.[49]In most cases, the general close of military operations will be the final end of all fighting between all those concerned.[50]

In the case of occupied territory, the application of the GC shall cease one year after the general close of military operations; however, the Occupying Power shall be bound, for the duration of the occupation, to the extent that such Power exercises the functions of government in such territory, by the provisions of the following Articles of the GC: 1 through 12, 27, 29 through 34, 47, 49, 51, 52, 53, 59, 61 through 77, and 143.[51]

Protected persons whose release, repatriation, or re-establishment may take place after such dates shall meanwhile continue to benefit from the protection of the GC.[52]"Reestablishment" refers to protected persons who cannot be repatriated because, for example, they would be liable to persecution in their own country or because their homes have been destroyed.[53]

10.3.5 State Responsibility for Its Agents' Treatment of Protected Persons. A party to the conflict is responsible for the treatment accorded to protected persons by its agents, irrespective of any individual responsibility that may be incurred.[54]

10.3.6 Non-Renunciation of Rights Secured by the GC. Protected persons may in no circumstances renounce in part or in entirety the rights secured to them by the GC or by special agreements, if any, referred to in Article 7 of the GC. [55]A similar provision of the GPW applies to POWs, and similar provisions of the GWS and GWS-Sea apply to the wounded, sick, shipwrecked, and military medical and religious personnel.[56]

10.4 DEROGATION FOR SECURITY REASONS

Certain rights and privileges of the GC that protected persons receive are subject to derogation for security reasons.

10.4.1 Derogation in Home Territory. Where, in the territory of a party to the conflict, the latter is satisfied that an individual protected person is definitely suspected of or engaged in activities hostile to the security of the State, such individual person shall not be entitled to claim such rights and privileges under the GC as would, if exercised in the favor of such individual person, be prejudicial to the security of such State.[57]

10.4.2 Derogation in Occupied Territory. Where in occupied territory an individual protected person is detained as a spy or saboteur, or as a person under definite suspicion of activity hostile to the security of the Occupying Power, such person shall, in those cases where absolute military security so requires, be regarded as having forfeited rights of communication under the GC.[58]For example, it may be necessary to keep the fact of detention secret temporarily so as not to compromise an ongoing operation against a conspiracy or network of spies.[59]

The derogation provisions relating to the home territory of a belligerent are not applicable in occupied territory, even though the occupied territory may arguably be characterized as the home territory of the opposing belligerent (*i.e.*, the country being occupied).

10.4.3 Derogation in Other Areas. To the extent that the rights and privileges of protected persons afforded by the GC are applied outside the home territory of a party to the conflict or outside occupied territory, it would be reasonable for such rights and privileges similarly to be subject to derogation. Thus, if a party to the conflict is satisfied that an individual protected person is definitely suspected of or engaged in activities hostile to the security of the State in other contexts, such individual person similarly could be deemed not entitled to claim such rights and privileges under the GC as would, if exercised in favor of such individual person, be prejudicial to the security of such State.[60]

10.4.4 Limits on Derogation. In each case, such persons shall nevertheless be treated with humanity, and in case of trial, shall not be deprived of the rights of fair and regular trial prescribed by the GC.[61]No derogation from the provision of humane treatment and the rights of fair and regular trial is permitted.[62]

Such persons shall also be granted the full rights and privileges of a protected person under the GC at the earliest date consistent with the security of the State or Occupying Power, as the case may be.[63]

10.5 HUMANE TREATMENT AND OTHER BASIC PROTECTIONS FOR PROTECTED PERSONS

Protected persons are entitled, in all circumstances, to respect for their persons, their honor, their family rights, their religious convictions and practices, and their manners and customs. They shall at all times be humanely treated, and shall be protected especially against all acts of violence or threats of violence, and against insults and public curiosity.[64]

10.5.1 Protection Against Violence or Threats. Protected persons must at all times be protected, particularly against acts or threats of violence. For example, the murder of protected persons is forbidden.

Protected persons should be protected not only against unlawful acts by the agents of the Detaining Power, but also against violence from other protected persons, such as during internment or, in the case of protected persons in a belligerent's home territory, violence from members of the civilian population.

10.5.1.1 *Measures of Physical Suffering, Extermination, or Other Brutality*. The Parties to the GC specifically agree that each of them is prohibited from taking any measure of such a character as to cause the physical suffering or extermination of protected persons in their hands.[65]This prohibition applies not only to murder, torture, corporal punishment, mutilation, and medical or scientific experiments not necessitated by the medical treatment of a protected person, but also to any other measures of brutality whether applied by civilian or military agents.[66]

10.5.1.2 *Protection for Women Against Rape or Other Indecent Assault*. Women shall be especially protected against any attack on their honor, in particular against rape, enforced prostitution, or any form of indecent assault.[67]

Although the GC provides special protection for women against these offenses, all individuals, including children and men, should also be protected against these offenses. Indecent assault is generally referred to today as sexual assault.

10.5.1.3 *Presence May Not Be Used to Render Places Immune From Military Operations*. The presence of a protected person may not be used to render certain points or areas immune from military operations.[68]

10.5.1.4 *Taking of Hostages*. The taking of hostages is prohibited.[69]

10.5.2 Protection Against Insults and Public Curiosity. Internees must at all times be protected against insults and public curiosity. For example, displaying internees in a humiliating fashion on television or on the internet would be prohibited. [70]For this reason and others, DoD policy has prohibited the taking of photographs of detainees except for authorized purposes.[71]

10.5.3 Other Prohibited Measures.

10.5.3.1 *No Physical or Moral Coercion*. No physical or moral coercion shall be exercised against protected persons, in particular to obtain information from them or from third parties.[72]For example, protected persons in occupied territory may not be compelled to provide information about their State's military defenses.[73]Other requirements, including the requirements of U.S. law and policy, would apply to the interrogation of protected persons.[74]The GC's prohibition against the use of coercion does not apply to those measures implicitly or explicitly authorized by the GC (such as the use of force to prevent one internee from harming another) or other lawful measures that are otherwise consistent with the GC (such as the use of force to prevent internees from escaping internment).[75]

10.5.3.2 *Collective Penalties and Measures of Intimidation or Terrorism*. No protected person may be punished for an offense he or she has not personally committed.[76]

Collective penalties and likewise all measures of intimidation or of terrorism are prohibited.[77]

Collective penalties are prohibited as a general matter.[78]

10.5.3.3 *Pillage Against Protected Persons*. Pillage is prohibited.[79]In addition to this specific prohibition in the GC, pillage is prohibited as a general matter.[80]

10.5.4 Reprisals Against Protected Persons and Their Property. Reprisals against protected persons and their property are prohibited.[81]

10.5.5 No Adverse Distinction Based on Race, Religion, or Political Opinion. Without prejudice to the provisions of the GC relating to their state of health, age, and sex, all protected persons shall be treated with the same consideration by the party to the conflict in whose power they are, without any adverse distinction based, in particular, on race, religion, or political opinion.[82]

Distinctions are permitted, and in some cases required, for humanitarian reasons.[83]For example, the GC provides for special treatment for children during internment.[84]

Distinction based on religion, political opinion, or other similar criteria may also be made so long as it is not adverse and it is made to advance legitimate interests, such as maintaining order in a place of internment.[85]

10.5.6 Facility for Applying to the Protecting Powers and Assistance Organizations Such as the ICRC. Protected persons shall have every facility for making application to the Protecting Powers, the ICRC, the National Red Cross (Red Crescent, Red Lion and Sun) Society of the country where they may be, as well as to any organization that might assist them.[86]

These several organizations shall be granted all facilities for that purpose by the authorities, within the bounds set by military or security considerations.[87]

Apart from the visits of the delegates of the Protecting Powers and of the ICRC, provided for by Article 143 of the GC, the Detaining or Occupying Powers shall facilitate as much as possible visits to protected persons by the representatives of other organizations whose object is to give spiritual aid or material relief to such persons.[88]

10.6 MEASURES OF CONTROL AND SECURITY

The parties to the conflict may take such measures of control and security in regard to protected persons as may be necessary as a result of the war.[89]For example, in a belligerent's home territory, measures of control are normally taken with respect to, at the very least, persons known to be active or reserve members of a hostile army, persons who would be liable to service in the enemy forces, and persons who it is expected would furnish information or other aid to a hostile State.[90]

These measures may include, for example, requiring protected persons: (1) to register with and report periodically to the police authorities; (2) to carry identity cards or special papers; (3) to refrain from carrying weapons; (4) to refrain from changing their place of residence without permission; (5) to refrain from accessing certain areas; (6) to have an assigned residence; and (7) to be interned.[91]

This section addresses searches, interrogation, and general rules for measures of control and security applied to protected persons, which must be applied consistent with humane treatment requirements.[92]

10.6.1 General Authority of a Belligerent to Search and Secure Protected Persons and Their Property. The measures of control and security that a belligerent may take in regard to protected persons include searching and securing protected persons and their property when necessary as a result of the war (*e.g.*, for security reasons and intelligence purposes).[93]

10.6.1.1 *Searches of Protected Persons*. The dignity and honor of the protected person being searched should be protected to the greatest degree possible under the circumstances.[94]The person conducting the search should avoid doing or saying anything likely to be regarded as indecent. In some circumstances, it may be appropriate for a witness to observe the search so as to protect both the person being searched from abuse and the person conducting the search from unfounded accusations of abuse.

10.6.1.2 *Search of Female Protected Persons*. It is appropriate for female protected persons to be searched by female personnel, if possible. This practice helps reduce the risk of accusations of indecent behavior.

A woman internee shall not be searched except by a woman.[95]

10.6.1.3 *Securing Protected Persons With Handcuffs and Other Security Devices*. When necessary for security reasons, protected persons may be secured temporarily with handcuffs, flex cuffs, blindfolds, or other security devices.[96]

10.6.1.4 *Search of the Property of Protected Persons*. When necessary as a result of the war, items in the possession of protected persons may be removed and searched, but property should be returned as soon as possible, absent a legal basis for the seizure of the property.[97]

10.6.2 Interrogation of Protected Persons. The law of war does not prohibit interrogation of protected persons, but specifies conditions and limitations for conducting interrogation.

10.6.2.1 *Humane Treatment During Interrogation*. Interrogation must be carried out in a manner consistent with the requirements for humane treatment, including the prohibition against acts of violence or intimidation, and insults.[98]

No physical or moral coercion shall be exercised against protected persons to obtain information from them or from third parties.[99]

10.6.2.2 *Additional U.S. Law and Policy on Interrogation*. U.S. law and policy impose additional requirements on the interrogation of protected persons.[100]

10.6.3 No Measures of Control More Severe Than Assigned Residence or Internment. If the measures of control mentioned in the GC for protected persons are considered to be inadequate by the Power imposing them, that Power may not have recourse to any other measure of control more severe than that of assigned residence or internment, in accordance with the provisions of Articles 42 and 43 of the GC.[101]

Similarly, if the Occupying Power considers it necessary, for imperative reasons of security, to take safety measures concerning protected persons, it may, at the most, subject them to assigned residence or to internment.[102]

10.6.4 Support to Persons Who Are Assigned Residence Should Be Guided by Internment Standards. In applying the provisions of the second paragraph of Article 39 of the GC to the cases of persons required to leave their usual places of residence by virtue of a decision placing them in assigned residence elsewhere, the Detaining Power shall be guided as closely as possible by the standards of welfare set forth in Part III, Section IV of the GC.[103]In other words, if the State assigns residence to a protected person in its home territory that results in the protected person being unable to support himself or herself and his or her dependents, the State shall provide support to that protected person and his or her dependents, being guided as closely as possible by the standards of welfare set forth in Part III, Section IV of the GC, which provides regulations for the treatment of internees.

Similarly, in occupied territory, protected persons made subject to assigned residence and thus required to leave their homes shall enjoy the full benefit of Article 39 of the GC.1[104]For example, the internment standards in the GC should also be a guide for support to protected persons and their dependents who are subject to assigned residence in occupied territory.

10.6.5 Refugees Who Do Not Enjoy the Protection of Any Government. In applying the measures of control mentioned in the GC to protected persons in its home territory, the Detaining Power shall not treat as enemy aliens exclusively on the basis of their nationality *de jure* of an enemy State, refugees who do not, in fact, enjoy the protection of any government.1[105]For example, a refugee who is not actually protected by his or her government should not be automatically subject to control measures simply on the basis of enemy nationality, but may be subject to any other recognized control measure if there are additional reasons for such action.[106]

10.7 GENERAL TREATMENT OF PROTECTED PERSONS IN A BELLIGERENT'S HOME TERRITORY

10.7.1 Regulation of the Situation of Protected Persons, in Principle, by Provisions Applicable to Aliens in Time of Peace. With the exception of special measures authorized by the GC, in particular by Articles 27 and 41 of the GC (*i.e.*, measures of control and security, such as internment), the situation of protected persons shall continue to be regulated, in principle, by the provisions concerning aliens in time of peace.[107]In any case, the following rights shall be granted to them:

- They shall be enabled to receive the individual or collective relief that may be sent to them (*e.g.*, by their home country or national or international relief organizations, such as the ICRC).
- They shall, if their state of health so requires, receive medical attention and hospital treatment to the same extent as the nationals of the State concerned.
- They shall be allowed to practice their religion and to receive spiritual assistance from ministers of their faith.
- If they reside in an area particularly exposed to the dangers of war, they shall be authorized to move from that area to the same extent as the nationals of the State concerned.
- Children under fifteen years, pregnant women, and mothers of children under seven years shall benefit from any preferential treatment to the same extent as the nationals of the State concerned.[108]

10.7.2 Opportunity for Employment and Other Support. Protected persons who, as a result of the war, have lost their gainful employment, shall be granted the opportunity to find paid employment.[109]That opportunity shall, subject to security considerations and to the provisions of Article 40, be equal to that enjoyed by the nationals of the Power in whose territory they are.[110]

10.7.2.1 *Support When Measures of Control Are Applied.* Where a party to the conflict applies to a protected person methods of control that result in that person being unable to support himself or herself, and especially if such a person is prevented for reasons of security from finding paid employment on reasonable conditions, that party shall ensure his or her support and that of his or her dependents.[111]

10.7.2.2 *Receipt of Allowances.* Protected persons may in any case receive allowances from their home country, the Protecting Power, or the relief societies referred to in Article 30 of the GC (*e.g.*, the ICRC).[112]

10.7.3 Compulsory Work for Protected Persons in a Belligerent's Home Territory. Protected persons may be compelled to work only to the same extent as nationals of the party to the conflict in whose territory they are.[113]

If protected persons are of enemy nationality, they may only be compelled to do work that is normally necessary to ensure the feeding, sheltering, clothing, transport, and health of human beings and that is not directly related to the conduct of military operations.[114]

In the cases mentioned in the two preceding paragraphs, protected persons compelled to work shall have the benefit of the same working conditions and of the same safeguards as national workers, in particular as regards wages, hours of labor, clothing and equipment, previous training, and compensation for occupational accidents and diseases.[115]

If the above provisions are infringed, protected persons shall be allowed to exercise their right of complaint in accordance with Article 30 of the GC.[116]

10.7.4 Cancellation of Restrictive Measures After Hostilities. In so far as they have not been previously withdrawn, restrictive measures taken regarding protected persons shall be cancelled as soon as possible after the close of hostilities.[117]Restrictive measures affecting their property shall be cancelled, in accordance with the law of the Detaining Power, as soon as possible after the close of hostilities.[118]

10.8 EXPULSION FROM AREAS WITHIN A BELLIGERENT'S HOME TERRITORY AND DEPARTURE AND TRANSFERS OF PROTECTED PERSONS FROM A BELLIGERENT'S HOME TERRITORY

Expulsion From Areas Within a Belligerent's Home Territory.

10.8.1 At the outbreak of hostilities, a State may expel or bar the citizens or subjects of the enemy State from its seaports, the area surrounding airbases, airports, and fortified places, areas of possible attack, and the actual or contemplated theaters of operation. When such expulsion is decreed, the persons expelled should be given such reasonable notice, consistent with public safety, as will enable them to arrange for the collection, disposal, and removal of their goods and property, and for the settlement of their personal affairs.[119]

Although such expulsion must be implemented in a humane manner, the provisions of the GC addressing a State's internment or assigned residence of civilians of an enemy State in its home territory (Articles 41-45), including the transfer of internees to other countries, do not apply to such expulsion.[120]

10.8.2 Departures of Protected Persons From a Belligerent's Home Territory. All protected persons who may desire to leave the territory at the outset of or during a conflict shall be entitled to do so, unless their departure is contrary to the national interests of the State.[121]The applications of such persons to leave shall be decided in accordance with regularly established procedures, and the decision shall be taken as rapidly as possible.[122]Those persons permitted to leave may provide themselves with the necessary funds for their journey and take with them a reasonable amount of their effects and articles of personal use.[123]

If any such person is refused permission to leave the territory, he or she shall be entitled to have such refusal reconsidered as soon as possible by an appropriate court or administrative board designated by the Detaining Power for that purpose.[124]

Upon request, representatives of the Protecting Power shall, unless reasons of security prevent it, or the persons concerned object, be furnished with the reasons for refusal of any request for permission to leave the territory and be given, as expeditiously as possible, the names of all persons who have been denied permission to leave.[125]

Departures permitted under Article 35 of the GC shall be carried out in satisfactory conditions as regards safety, hygiene, sanitation, and food.[126]All costs in connection therewith, from the point of exit in the territory of the Detaining Power, shall be borne by the country of destination or, in the case of accommodation in a neutral country, by the Power whose nationals are benefited.[127]The practical details of such movements may, if necessary, be settled by special agreements between the Powers concerned.[128]

The foregoing shall not prejudice such special agreements as may be concluded between parties to the conflict concerning the exchange and repatriation of their nationals in enemy hands.[129]

Protected persons who are confined pending proceedings or serving a sentence involving loss of liberty shall be treated humanely during their confinement.[130]As soon as they are released, they may ask to leave the territory in conformity with Articles 35 and 36 of the GC.[131]

10.8.3 Transfers of Protected Persons From a Belligerent's Home Territory.

10.8.3.1 *Requirements for Transfer*. Protected persons shall not be transferred to a Power that is not a Party to the GC.[132]This provision shall in no way constitute an obstacle to the repatriation of protected persons, or to their return to their country of residence after the cessation of hostilities.[133]

Protected persons may be transferred by the Detaining Power only to a Power that is a Party to the GC and after the Detaining Power has satisfied itself of the willingness and ability of such receiving Power to apply the GC.[134]

In no circumstances shall a protected person be transferred to a country where he or she may have reason to fear persecution for his or her political opinions or religious beliefs.[135]In addition, U.S. policy provides that no detainee shall be transferred to another State if it is more likely than not that the detainee would be tortured in the receiving country.[136]

The provisions of Article 45 of the GC do not constitute an obstacle to the extradition, under extradition treaties concluded before the outbreak of hostilities, of protected persons accused of offenses against ordinary criminal law.[137]

10.8.3.2 *Responsibility Upon Transfer*. If protected persons are transferred under such circumstances, responsibility for the application of the GC rests on the Power accepting them, while they are in its custody.[138]Nevertheless, if that Power fails to carry out the provisions of the GC in any important respect, the Power by which the protected persons were transferred shall, upon being so notified by the Protecting Power, take effective measures to correct the situation or shall request the return of the protected persons.[139]Such request must be complied with.[140]

10.9 INTERNMENT

The parties to the conflict shall not intern protected persons, except in accordance with the provisions of Articles 41, 42, 43, 68, and 78 of the GC.[141]

10.9.1 Principles Underlying the Internment of Protected Persons Compared to Those Underlying POW Internment. In some respects, the principles underlying the internment of protected persons are similar to the principles underlying the internment of POWs.[142]For example, the internment of protected persons is non-punitive, the Detaining Power is responsible for the treatment of internees in its custody, and humane treatment is required.

However, the GC recognizes that the internment of protected persons is quite different in character from that of POWs by requiring the separation of internees from POWs.[143]Protected persons interned for security reasons, in theory, have not participated in hostilities.[144]Thus, their internment shall cease when the reasons that have necessitated it have ceased, which may occur before the end of the conflict.[145]In practice, however, internment for security reasons may involve persons who have participated in hostilities, and such persons' continued detention for the duration of the conflict may be justified in order to prevent their further participation in the conflict.

On the other hand, internees are not members of the armed forces and, thus, in certain respects, have not earned the special privileges that POWs have earned.[146]For example, although internees receive allowances, internees do not receive specified advances of pay like POWs.[147]Similarly, internees who have successfully escaped do not benefit from the immunity from punishment applicable to POWs who have successfully escaped.[148]

10.9.2 Procedure for Internment or Assigned Residence on Home Territory.

10.9.2.1 *Internment or Assigned Residence Only if Absolutely Necessary*. The internment or placing in assigned residence of protected persons may be ordered only if the security of the Detaining Power makes it absolutely necessary.[149]

10.9.2.2 *Voluntary Internment*. If any person, acting through the representatives of the Protecting Power, voluntarily demands internment, and if his or her situation renders this step necessary, he or she shall be interned by the Power in whose hands he or she may be.[150]

10.9.2.3 *Reconsideration and Periodic Review*. Any protected person who has been interned or placed in assigned residence shall be entitled to have such action reconsidered as soon as possible by an appropriate court or administrative board designated by the Detaining Power for that purpose.[151]If the internment or placing in assigned residence is maintained, the court or administrative board shall periodically, and at least twice yearly, give consideration to his or her case, with a view to the favorable amendment of the initial decision, if circumstances permit.[152]

10.9.2.4 *Notification to the Protecting Power*. Unless the protected persons concerned object, the Detaining Power shall, as rapidly as possible, give the Protecting Power the names of any protected persons who have been interned or subjected to assigned residence, or who have been released from internment or assigned residence.[153]The decisions of the courts or boards mentioned in the first paragraph of Article 43 of the GC shall also, subject to the same conditions, be notified as rapidly as possible to the Protecting Power.[154]

10.9.3 Procedure for Internment or Assigned Residence in Occupied Territory. The requirements in the GC for the procedure for assigned residence or internment of protected persons in occupied territory are not as elaborate as those applicable in a belligerent's home territory because of concerns that it would not be practicable to implement the requirements applicable in a belligerent's home territory in view of the disorder that might prevail in occupied territory.[155]

10.9.3.1 *Internment or Assigned Residence for Imperative Reasons of Security*. If the Occupying Power considers it necessary, for imperative reasons of security, to take safety measures concerning protected persons, it may, at the most, subject them to assigned residence or to internment.[156]

Decisions regarding such assigned residence or internment shall be made according to a regular procedure to be prescribed by the Occupying Power in accordance with the provisions of the GC.[157]This procedure shall include the right of appeal for the parties concerned.[158]Appeals shall be decided with the least possible delay.[159]In the event of the decision being upheld, it shall be subject to periodic review, if possible every six months, by a competent body set up by the Occupying Power.[160]"Competent bodies" to review the internment or assigned residence of protected persons may be created with advisory functions only, leaving the final decision to a high official of the Government.[161]

10.9.3.2 *Internment for Minor Offenses Directed Against the Occupying Power*. Protected persons who commit an offense that is solely intended to harm the Occupying Power, but that does not constitute an attempt on the life or limb of members of the occupying forces or administration, nor a grave collective danger, nor seriously damage the property of the occupying forces or administration or the installations used by them, shall be liable to internment or simple imprisonment, provided the duration of such internment or imprisonment is proportionate to the offense committed.[162]Furthermore, internment or imprisonment shall, for such offenses, be the only measure adopted for depriving protected persons of liberty.[163]The courts provided for under Article 66 of the GC may at their discretion convert a sentence of imprisonment to one of internment for the same period.[164]

10.9.4 Maintenance of Internees. Parties to the conflict who intern protected persons shall be bound to provide free of charge for their maintenance, and to grant them also the medical attention required by their state of health.[165]No deduction from the allowances, salaries, or credits due to the internees shall be made for the repayment of these costs.[166]The Detaining Power shall provide for the support of those dependent on the internees, if such dependents are without adequate means of support or are unable to earn a living.[167]

10.9.5 Release As Soon As Reasons for Internment No Longer Exist. Each interned person shall be released by the Detaining Power as soon as the reasons that necessitated his or her internment no longer exist.[168]

10.9.6 Agreements for the Release, Return, or Accommodation in a Neutral Country of Certain Classes of Internees. The parties to the conflict shall endeavor during the course of hostilities to conclude agreements for the release, the repatriation, the return to places of residence, or the accommodation in a neutral country of certain classes of internees, in particular children, pregnant women, and mothers with infants and young children, wounded and sick, and internees who have been detained for a long time.[169]

10.10 SEGREGATION OF INTERNEES

10.10.1 Segregation From Other Types of Detainees. Internees shall be accommodated and administered separately from POWs and from persons deprived of liberty for any other reason.[170]This rule reflects the difference in character between the internment of internees from other types of detention or confinement.[171]

10.10.2 Accommodation According to Nationality, Language, and Customs. The Detaining Power shall, as far as possible, accommodate the internees according to their nationality, language, and customs.[172]Unlike the comparable rule in the GPW, this provision has the qualification "as far as possible" because of the competing interest in leaving internees near their families and because POWs are generally automatically grouped by nationality as a consequence of the circumstances of their capture.[173]Internees who are nationals of the same country shall not be separated merely because they speak different languages.[174]

10.10.3 Families Kept Together. Throughout the duration of their internment, members of the same family, and in particular parents and children, shall be lodged together in the same place of internment, except when separation of a temporary nature is necessitated for reasons of employment or health or for the purposes of enforcement of the provisions of Chapter IX (Penal and Disciplinary Sanctions) of Section IV of the GC.[175]Keeping families together during internment is intended to improve the morale and welfare of internees.[176]

Internees may request that their children who are left at liberty without parental care shall be interned with them.[177]For example, two parents who are interned together may request that their children be interned with them, but one of the parents would not be able to request the internment of a child being cared for by a parent who was not being interned.[178]

Wherever possible, interned members of the same family shall be housed in the same premises and given separate accommodation from other internees, together with facilities for leading a proper family life.[179]

10.11 SAFETY, HYGIENE, AND LIVING CONDITIONS IN PLACES OF INTERNMENT

10.11.1 Safety of Places of Internment.

10.11.1.1 *Avoidance of Particularly Dangerous Areas*. The Detaining Power shall not set up places of internment in areas particularly exposed to the dangers of war.[180]

10.11.1.2 *Sharing Information on the Location of Places of Internment*. Detaining Powers shall give enemy Powers, through the intermediary of the Protecting Powers, all useful information regarding the geographic location of places of internment.[181]

10.11.1.3 *Marking of Internment Camps*. Whenever military considerations permit, internment camps shall be indicated by the letters "IC," placed so as to be clearly visible in the daytime from the air.[182]The Powers concerned may, however, agree upon any other system of marking.[183]No place, other than an internment camp, shall be marked as such.[184]The caveat that internment camps need to be marked, as such, only when military considerations permit was intended to allow for camps not to be marked, *e.g.*, if such identification would enable an Enemy Power to instigate a revolt in the camp, provide weapons to the internees, enable escape attempts, etc.[185]

10.11.1.4 *Air-Raid Shelters and Protective Measures for Internees*. In all places of internment exposed to air raids and other hazards of war, shelters adequate in number and structure to ensure the necessary protection shall be installed.[186]In case of alarms, the internees shall be free to enter such shelters as quickly as possible, excepting those who remain for the protection of their quarters against these hazards.[187]Any protective measures taken in favor of the population shall also apply to the internees.[188]

10.11.1.5 *Fire Precautions*. All due precautions must be taken in places of internment against the danger of fire.[189]

10.11.2 Hygiene of Places of Internment. In no case shall permanent places of internment be situated in unhealthy areas, or in districts the climate of which is injurious to the internees.[190]In all cases where the district in which a protected person is temporarily interned is in an unhealthy area, or has a climate that is harmful to his or her health, he or she shall be removed to a more suitable place of internment as rapidly as circumstances permit.[191]

10.11.2.1 *Sanitary Conveniences (e.g., latrines, bathrooms)*. Internees shall have for their use, day and night, sanitary conveniences that conform to the rules of hygiene and are constantly maintained in a state of cleanliness.[192]The term "sanitary conveniences" should be taken to mean primarily the latrines.[193]Access to the latrines at night is specified in the 1949 Geneva Conventions because this was a frequent problem in POW camps in World War II.[194]

In certain circumstances, women internees shall be provided separate sanitary conveniences for their use.[195]

10.11.2.2 *Showers, Baths, Personal Toilet, and Laundry*. Internees shall be provided with sufficient water and soap for their daily personal toilet and for washing their personal laundry; installations and facilities necessary for this purpose shall be granted to them.[196]

Showers or baths shall also be available.[197]The necessary time shall be set aside for washing and for cleaning.[198]Although the frequency of baths or showers is not specified in the GC, a reasonable opportunity (*e.g.*, considering the available resources, the internees' cultural practices, the activities in which they are engaged) should be afforded.[199]

10.11.3 Conditions of Buildings and Quarters. The Detaining Power is bound to take all necessary and possible measures to ensure that protected persons shall, from the outset of their internment, be accommodated in buildings or quarters that afford every possible safeguard in regard to hygiene and health, and provide efficient protection against the rigors of the climate and the effects of the war.[200]

The premises shall be fully protected from dampness and adequately heated and lighted, in particular between dusk and lights out.[201]The sleeping quarters shall be sufficiently spacious and well ventilated, and the internees shall have suitable bedding and sufficient blankets, with account being taken of the climate and of the age, sex, and state of health of the internees.[202]

10.11.3.1 *Separate Sleeping Quarters and Latrines for Women*. Whenever it is necessary, as an exceptional and temporary measure, to accommodate women internees who are not members of a family unit in the same place of internment as men, women internees shall be provided separate sleeping quarters and sanitary conveniences for their use.[203]

10.12 CANTEENS FOR INTERNEES

Canteens shall be installed in every place of internment, except where other suitable facilities are available.[204]As with canteens for POWs, the purpose of the canteen is to increase the morale and comfort of the internees.[205]

10.12.1 Establishment of the Canteen. Although canteens for POWs are required, canteens for civilian internees are not required if other suitable facilities, such as local shops, are available.[206]

If other suitable facilities are not available, and before canteens can be established, the need for a canteen may be mitigated if internees are provided a free "comfort pack."[207]

10.12.2 Canteen Stock and Prices. The purpose of the canteens shall be to enable internees to make purchases, at prices not higher than local market prices, of foodstuffs and articles of everyday use, including soap and tobacco, such as would increase their personal wellbeing and comfort.[208]

Internees are not entitled to more favorable treatment than the population at large with respect to canteen facilities, and are equally subject to regulations, such as those pertaining to rationing, which are applied to the population generally.[209]

10.12.3 Canteen Management and Profits. Canteen profits shall be credited to a welfare fund to be set up for each place of internment, and administered for the benefit of the internees attached to such place of internment.[210]The Internee Committee shall have the right to check the management of the canteen and of the fund.[211]

When a place of internment is closed down, the balance of the welfare fund shall be transferred to the welfare fund of a place of internment for internees of the same nationality, or, if such a place does not exist, to a central welfare fund that shall be administered for the benefit of all internees remaining in the custody of the Detaining Power.[212]In case of a general release, the profits shall be kept by the Detaining Power, subject to any agreement to the contrary between the Powers concerned.[213]

10.13 FOOD,WATER, TOBACCO, AND CLOTHING FOR INTERNEES

10.13.1 Food for Internees. Daily food rations for internees shall be sufficient in quantity, quality, and variety to keep internees in a good state of health and prevent the development of nutritional deficiencies.[214]

10.13.1.1 *Accounting for the Customary Diet*. Account shall also be taken of the customary diet of the internees.[215]For example, the internee's cultural and religious requirements should be considered in determining and ensuring the appropriate diet.

10.13.1.2 *Additional Food for Certain Groups*. Internees who work shall receive additional rations in proportion to the kind of labor that they perform.[216]Expectant and nursing mothers, and children under fifteen years of age, shall be given additional food, in proportion to their physiological needs.[217]

10.13.1.3 *No Disciplinary Reductions in Food Rations*. The reduction in food rations as a disciplinary punishment is specifically prohibited.[218]

10.13.1.4 *Means for Preparing Additional Food*. Internees shall be given the means by which they can prepare for themselves any additional food in their possession.[219]

10.13.2 Drinking Water. Sufficient drinking water shall be supplied to internees.[220]

10.13.3 Use of Tobacco. The use of tobacco shall be permitted.[221]The Detaining Power is not required to give tobacco to internees, but should offer it for sale at the canteen, if one is established.[222]The Detaining Power may impose reasonable restrictions on smoking to ensure that the camp is a healthful and safe environment.

10.13.4 Clothing of Internees. When taken into custody, internees shall be given all facilities to provide themselves with the necessary clothing, footwear, and change of underwear, and, later on, to procure further supplies, if required.[223]Should any internees not have sufficient clothing, account being taken of the climate, and be unable to procure any, it shall be provided free of charge to them by the Detaining Power.[224]

10.13.4.1 *No Humiliating Clothing*. The clothing supplied by the Detaining Power to internees and the outward markings placed on their own clothes shall not be ignominious nor expose them to ridicule.[225]For example, clothing provided to them should not resemble convicts' uniforms or be intended to humiliate them.[226]

10.13.4.2 *Clothing for Work*. Workers shall receive suitable working outfits, including protective clothing, whenever the nature of their work so requires.[227]

10.14 MEDICAL ATTENTION FOR INTERNEES

Parties to the conflict who intern protected persons shall be bound to grant them the medical attention required by their state of health.[228]

10.14.1 Access to Medical Attention.

10.14.1.1 *Camp Infirmary*. Every place of internment shall have an adequate infirmary, under the direction of a qualified doctor, where internees may have the attention they require, as well as an appropriate diet.[229]Isolation wards shall be set aside for cases of contagious or mental diseases.[230]

10.14.1.2 *Right of Internees to Present Themselves for Examination*. Internees may not be prevented from presenting themselves to the medical authorities for examination.[231]The GC appears to contemplate that the internees would have daily sick call in which they could present themselves for medical examination, because it provides for internees who are undergoing disciplinary punishment to request to attend this inspection.[232]The right of internees to present themselves for examination does not mean that the internees must necessarily be examined every day by a doctor, nor does it preclude disciplinary punishment if internees abusively (on the basis of the doctor's report) make groundless requests for medical examination.[233]

10.14.1.3 *Monthly Medical Inspections*. Medical inspections of internees shall be made at least once a month.[234]Their purpose shall be, in particular, to supervise the general state of health, nutrition, and cleanliness of internees, and to detect contagious diseases, especially tuberculosis, malaria, and venereal diseases.[235]Such inspections shall include, in particular, the checking of weight of each internee and, at least once a year, radioscopic examination.[236]

10.14.2 Compulsory Medical Treatment. Because internees are subject to the laws in force in the territory in which they are detained,[237]internees may be ordered to undergo medical treatment to the extent that other persons in that territory may be so ordered.

10.14.3 Medical Treatment for Internees.

10.14.3.1 *Persons Requiring Special Treatment*. Maternity cases and internees suffering from serious diseases, or whose condition requires special treatment, a surgical operation, or hospital care, must be admitted to any institution where adequate treatment can be given and shall receive care not inferior to that provided for the general population.[238]

10.14.3.2 *Treatment by Medical Personnel of Their Own Nationality*. Internees shall, for preference, have the attention of medical personnel of their own nationality.[239]

10.14.3.3 *Certificate Recording Treatment*. The medical authorities of the Detaining Power shall, upon request, issue to every internee who has undergone treatment an official certificate showing the nature of his or her illness or injury, and the duration and nature of the treatment given.[240]A duplicate of this certificate shall be forwarded to the Central Information Agency for protected persons.[241]

10.14.3.4 *Cost of Medical Treatment*. Treatment, including the provision of any apparatus necessary for the maintenance of internees in good health, particularly dentures and other artificial appliances and spectacles, shall be free of charge to the internee.[242]

10.14.4 Receipt of Medical Shipments. Internees may receive individual parcels and collective shipments containing medical supplies.[243]However, medical supplies are generally to be sent in collective shipments so that they may be properly administered by the camp doctors.[244]

10.14.5 Inquiry in Certain Cases of Death or Serious Injury. In certain cases of death or serious injury of an internee, an official inquiry shall be conducted by the Detaining Power.[245]

10.15 RELIGIOUS EXERCISE BY INTERNEES

Internees shall enjoy complete latitude in the exercise of their religious duties, including attendance at the services of their faith, on condition that they comply with the disciplinary routine prescribed by the detaining authorities.[246]

10.15.1 Premises for Religious Services for Internees. The Detaining Power shall place at the disposal of interned persons, of whatever denomination, premises suitable for the holding of their religious services.[247]The premises where services are held should be sufficiently spacious and clean, and should provide effective shelter to those attending services, but need not be set aside exclusively for religious services.[248]

10.15.2 Interned Ministers of Religion. Ministers of religion who are interned shall be allowed to minister freely to the members of their community.[249]For this purpose, the Detaining Power shall ensure their equitable allocation among the various places of internment in which there are internees speaking the same language and belonging to the same religion.[250]Should such ministers be too few in number, the Detaining Power shall provide them with the necessary facilities, including means of transport, for moving from one place to another, and they shall be authorized to visit any internees who are hospitalized.[251]

10.15.2.1 *Correspondence by Ministers of Religion.* Ministers of religion shall be at liberty to correspond on matters concerning their ministry with the religious authorities in the country of detention and, as far as possible, with the international religious organizations of their faith.[252]Such correspondence shall not be considered as forming a part of the quota mentioned in Article 107 of the GC.[253]It shall, however, be subject to the provisions of Article 112 of the GC.[254]In other words, such correspondence may be censored, but such censorship shall be done as quickly as possible.[255]

10.15.3 Appointment of Other Ministers or Qualified Laypersons. When internees do not have at their disposal the assistance of ministers of their faith, or should these latter be too few in number, the local religious authorities of the same faith may appoint, in agreement with the Detaining Power, a minister of the internees' faith or, if such a course is feasible from a denominational point of view, a minister of a similar religion or a qualified layperson.[256]The latter shall enjoy the facilities granted to the ministry he or she has assumed.[257]Persons so appointed shall comply with all regulations laid down by the Detaining Power in the interests of discipline and security.[258]

10.16 INTELLECTUAL, PHYSICAL, AND RECREATIONAL ACTIVITIES

The Detaining Power shall encourage intellectual, educational, and recreational pursuits, and sports and games among internees, while leaving them free to take part in them or not.[259]It shall take all practicable measures to ensure the exercise thereof, in particular by providing suitable premises for such purposes.[260]

10.16.1 Voluntariness of the Activities. The Detaining Power should seek to encourage the internees in engaging in these activities, while leaving them free to take part or not. For example, it would be prohibited to compel internees to listen to propaganda, such as by punishing them or giving them harsher conditions of confinement if they do not participate. On the other hand, censoring educational or intellectual materials for security reasons or to refrain from providing internees with anti-democratic political propaganda would be permissible.[261]

10.16.2 Education. All possible facilities shall be granted to internees to continue their studies or to take up new subjects.[262]The education of children and young people shall be ensured; they shall be allowed to attend schools either within the place of internment or outside.[263]

To advance educational opportunities for internees, the Detaining Power may seek ways to encourage and assist them in teaching one another.[264]

10.16.3 Opportunities for Physical Exercise, Sports, and Outdoor Games. Internees shall be given opportunities for physical exercise, sports, and outdoor games.[265]For this purpose, sufficient open spaces shall be set aside in all places of internment.[266]Special playgrounds shall be reserved for children and young people.[267]

10.16.4 Contributions by Sources Apart From the Detaining Power. Apart from the Detaining Power, internees may receive assistance from a variety of sources that allow them to engage in intellectual, physical, and recreational activities.

For example, relief organizations, including the ICRC, may contribute to the ensuring that internees have opportunities for intellectual, physical, and recreational activities.[268]In addition, internees may receive shipments that are intended to allow them to engage in these activities.[269]Lastly, the profits from the canteen may be used in this area.[270]

10.17 INTERNEE LABOR

The rules described in this section govern the use of internee labor. Additional rules apply to the use of the labor of protected persons depending on whether they are present in a belligerent's home territory or in occupied territory.[271]

10.17.1 General Rules on Internee Labor. The Detaining Power shall not employ internees as workers, unless the internees so desire.[272]

Employment that, if undertaken under compulsion by a protected person not in internment, would involve a breach of Articles 40 or 51 of the GC, and employment on work that is of a degrading or humiliating character, are in any case prohibited.[273]

After a working period of six weeks, internees shall be free to give up work at any moment, subject to eight days' notice.[274]

10.17.2 Compulsory Work for the Benefit of the Internee Community. The first two paragraphs of Article 95 of the GC constitute no obstacle to the right of the Detaining Power to employ interned doctors, dentists, and other medical personnel in their professional capacity on behalf of their fellow internees, or to employ internees for administrative and maintenance work in places of internment and to detail such persons for work in the kitchens or for other domestic tasks, or to require such persons to undertake duties connected with the protection of internees against aerial bombardment or other war risks.[275]Thus, although internees generally may not be compelled to work, there is an exception for tasks that benefit the internee community as a whole.[276]

No internee may, however, be required to perform tasks for which he or she is, in the opinion of a medical officer, physically unsuited.[277]

In addition, members of the Internee Committee may not be compelled to do other work, if the accomplishment of their duties is thereby rendered more difficult.[278]

10.17.2.1 *Wages and Working Conditions for Internees Permanently Detailed for Work on Behalf of the Internee Community*. Internees permanently detailed for categories of work mentioned in the third paragraph of Article 95 of the GC shall be paid fair wages by the Detaining Power.[279]The working conditions and the scale of compensation for occupational accidents and diseases to internees thus detailed shall not be inferior to those applicable to work of the same nature in the same district.[280]

10.17.3 Working Conditions. The Detaining Power shall take entire responsibility for all working conditions, for medical attention, for the payment of wages, and for ensuring that all employed internees receive compensation for occupational accidents and diseases.[281]The standards prescribed for said working conditions and for compensation shall be in accordance with national laws and regulations, and with existing practice; they shall in no case be inferior to those obtaining for work of the same nature in the same district.[282]

In addition, internees who work shall receive additional rations in proportion to the kind of labor that they perform.[283]

10.17.4 Wage Agreement With Internees. Wages for work done shall be determined on an equitable basis by special agreements between the internees, the Detaining Power, and, if the case arises, employers other than the Detaining Power, due regard being paid to the obligation of the Detaining Power to provide for free maintenance of internees, and for the medical attention that their state of health may require.[284]

10.17.5 Labor Detachments of Internees. All labor detachments shall remain part of, and dependent upon, a place of internment.[285]The competent authorities of the Detaining Power and the commandant of a place of internment shall be responsible for the observance of the provisions of the GC in a labor detachment.[286]

The commandant shall keep an up-to-date list of the labor detachments subordinate to him or her, and shall communicate it to the delegates of the Protecting Power, of the ICRC, and of other humanitarian organizations who may visit the places of internment.[287]

10.18 INTERNEES' ARTICLES OF PERSONAL USE

10.18.1 Internees' Articles of Personal Use. Internees shall be permitted to retain articles of personal use.[288]

Internees, however, are not entitled to retain items that could be put to a use prejudicial to the legitimate interests of the Detaining Power, such as items that may pose a risk to security (*e.g.*, knives or devices that may be used as weapons) or items that may facilitate escape (*e.g.*, flares, flare pistols, compasses, maps, radios, or cell phones).[289]

10.18.2 Articles of Personal or Sentimental Value. Articles that have "above all" a personal or sentimental value may not be taken away.[290]Items that may otherwise be legitimately impounded (*e.g.*, for security reasons) are not precluded from being impounded because of their personal or sentimental value to the internee.[291]

10.18.3 Family or Identity Documents for Internees. Family or identity documents in the possession of internees may not be taken away without a receipt being given.[292]At no time shall internees be left without identity documents.[293]If they have none, they shall be issued special documents drawn up by the detaining authorities, which will serve as their identity papers until the end of their internment.[294]

10.19 FINANCIAL RESOURCES OF INTERNEES

10.19.1 Money and Valuables in the Internee's Possession. Monies, checks, bonds, etc., and valuables in the possession of internees may not be taken from them, except in accordance with established procedure.[295]Detailed receipts shall be given for such valuables.[296]The amounts shall be paid into the account of every internee, as provided for in Article 98 of the GC. [297]Such amounts may not be converted into any other currency unless legislation in force in the territory in which the owner is interned so requires or the internee gives his or her consent.[298]

10.19.1.1 *Pocket Money for Purchases*. Internees may keep on their persons a certain amount of money, in cash or in the form of purchase coupons, to enable them to make purchases.[299]For example, internees could use this money or scrip to purchase items at the canteen.[300]

For security reasons, the amount of money retained by internees generally will be small.[301]

10.19.2 Internee Allowances. All internees shall receive regular allowances, sufficient to enable them to purchase goods and articles, such as tobacco, toilet requisites, etc.[302]Such allowances may take the form of credits or purchase coupons.[303]

Furthermore, internees may receive allowances from the Power to which they owe allegiance, the Protecting Powers, the organizations that may assist them, or their families, as well as the income on their property in accordance with the law of the Detaining Power.[304]

The amount of allowances granted by the Power to which an internee owes allegiance shall be the same for each category of internees (*e.g.*, infirm, sick, pregnant women), but may not be allocated by that Power or distributed by the Detaining Power on the basis of discrimination between internees that is prohibited by Article 27 of the GC.[305]

10.19.3Internee Accounts. The Detaining Power shall open a regular account for every internee, to which shall be credited:

- allowances named in Article 98 of the GC;
- wages earned;
- remittances received; and
- such sums taken from the internee as may be available under the legislation in force in the territory in which he or she is interned.[306]

10.19.3.1 *Drawing Upon Accounts*. Internees may draw from their accounts the amounts necessary for their personal expenses, within the limits fixed by the Detaining Power.[307]

Internees shall be granted all facilities consistent with the legislation in force in such territory to make remittances to their families and to other dependents.[308]

10.19.3.2 *Account Statement*. Internees shall at all times be afforded reasonable facilities for consulting and obtaining copies of their accounts.[309]A statement of accounts shall

be furnished to the Protecting Power on request, and shall accompany the internee in case of transfer.[310]

10.19.3.3 *Settlement of Account on Release or Repatriation*. On release or repatriation, internees shall be given all articles, monies, or other valuables taken from them during internment and shall receive in currency the balance of any credit to their accounts kept in accordance with Article 98 of the GC, with the exception of any articles or amounts withheld by the Detaining Power by virtue of its legislation in force.[311]If the property of an internee is so withheld, the owner shall receive a detailed receipt.[312]

10.20 ADMINISTRATION OF PLACES OF INTERNMENT AND DISCIPLINE

10.20.1 Responsible Officer and Staff. Every place of internment shall be put under the authority of a responsible officer, chosen from the regular military forces or the regular civil administration of the Detaining Power.[313]

The officer in charge of the place of internment must have in his or her possession a copy of the GC in the official language, or one of the official languages, of his or her country and shall be responsible for its application.[314]The staff in control of internees shall be instructed in the provisions of the GC and of the administrative measures adopted to ensure its application.[315]

10.20.2 Posting of Convention and Camp Orders. The text of the GC and the texts of special agreements concluded under the GC shall be posted inside the place of internment, in a language that the internees understand, or shall be in the possession of the Internee Committee.[316]

Regulations, orders, notices, and publications of every kind shall be communicated to the internees and posted inside the places of internment, in a language that the internees understand.[317]

Every order and command addressed to internees individually must likewise be given in a language that they understand.[318]

10.20.3 Humane Disciplinary Regime. The disciplinary regime in places of internment shall be consistent with humanitarian principles, and shall in no circumstances include regulations imposing on internees any physical exertion dangerous to their health or involving physical or moral victimization.[319]

10.20.3.1 *Prohibition Against Tattooing or Imprinting Signs or Markings on the Body*. Identification by tattooing or imprinting signs or markings on the body is prohibited.[320]

10.20.3.2 *Prohibition Against Prolonged Standing, Roll-Calls, Punishment Drill, Military Drill and Manuevers, or the Reduction of Food Rations*. In particular, prolonged standing and roll-calls, punishment drill, military drill and maneuvers, or the reduction of food rations are prohibited.[321]

10.20.4 Use of Force to Maintain Order and to Prevent Escape. As with POWs, the use of weapons against protected persons, especially against those who are escaping or attempting to escape, shall constitute an extreme measure, which shall always be preceded by warnings appropriate to the circumstances.[322]For example, internees should not be fired upon if they are apprehended within the camp limits while making preparations to escape, and there is no risk of escape or harm to anyone. It should be noted that, although the GC has no provision corresponding to Article 42 of the GPW, the above principle is equally valid in the internee context.

10.20.4.1 *Use of Non-Lethal Weapons (Including Riot Control Agents)*. Nonlethal weapons may be employed to control rioting protected persons or to prevent their escape. In particular, the use of riot control agents to control rioting protected persons is not prohibited.[323]

If the use of deadly force is warranted and authorized against protected persons, there is no legal requirement to employ non-lethal weapons before resort to deadly force.[324]

10.21 INTERNEE PETITIONS, COMPLAINTS, AND REPORTS ABOUT CONDITIONS OF INTERNMENT

10.21.1 Internee Right to Present Petitions and Complaints. Internees shall have the right to present to the authorities in whose power they are, any petition with regard to the conditions of internment to which they are subjected.[325]The commandant of an internment facility may issue regulations that establish procedures about how internees are to exercise this right to petition.[326]Internees shall also have the right to apply without restriction through the Internee Committee or, if they consider it necessary, directly to the representatives of the Protecting Power, in order to indicate to them any points on which they may have complaints to make with regard to the conditions of internment.[327]

10.21.1.1 *Making Requests and Complaints to the ICRC*. Such complaints may also be made to the delegates of the ICRC, who enjoy the same prerogatives of access to internees as the representatives of the Protecting Power.[328]In the past, the ICRC has been able to take appropriate measures besides merely forwarding the complaint to the Power on which the POWs depend, including measures on a confidential basis that help to improve the situation of internees.[329]

10.21.2 Communication of Petitions and Complaints. Such petitions and complaints shall be transmitted forthwith and without alteration, and even if the latter are recognized to be unfounded, they may not occasion any punishment.[330]

10.21.2.1 *Review and Censorship of Requests and Complaints by the Detaining Power*. Although petitions and complaints shall be transmitted forthwith and without alteration, the Detaining Power has the right to examine and censor petitions and complaints for security reasons.[331]Communications for transmittal may be examined to determine whether they legitimately contain petitions or complaints.[332]Security review and censorship must be conducted in accordance with the general rules on censoring internee correspondence.[333]

10.21.2.2 *No Punishment for Making Complaints*. Even if the petitions and complaints are recognized to be unfounded, they may not occasion any punishment.[334]In any event, as in the case of POWs, it could be contrary to internees' interests to abuse this right by making groundless complaints because complaints that are justified might not, as a result, receive the appropriate attention.[335]

10.21.3 Periodic Reports by the Internee Committees. Periodic reports on the situation in places of internment and as to the needs of the internees may be sent by the Internee Committees to the representatives of the Protecting Powers.[336]Internee Committees have discretion about how and when to submit their reports.[337]The Detaining Power may subject these reports to security review and censorship to ensure that these reports are not misused.[338]

10.22 INTERNEE COMMITTEES

10.22.1 Election of Internee Committees. In every place of internment, the internees shall freely elect, by secret ballot every six months, the members of a Committee empowered to represent them before the Detaining and the Protecting Powers, the ICRC, and any other organization that may assist them.[339]The members of the Committee shall be eligible for reelection. [340]

10.22.1.1 *Approval of Internee Committee Members by the Detaining Authorities*.

Internees so elected shall enter upon their duties after their election has been approved by the detaining authorities.[341]The reasons for any refusals or dismissals shall be communicated to the Protecting Powers concerned.[342]

10.22.2 Duties of Internee Committees. Internee Committees shall further the physical, spiritual, and intellectual well-being of the internees.[343]Internee Committees have duties similar to the duties of POW Representatives.[344]

The Internee Committee's general duties imply that the Internee Committee will undertake a variety of activities to ensure that internees receive proper treatment by the Detaining Power and to advance their welfare, even activities that are not specified in the GC as constituting their duties.

10.22.2.1 *Internee Committees and Mutual Assistance Among the Internees*. In case the internees decide, in particular, to organize a system of mutual assistance among themselves, this organization would be within the competence of the Internee Committees, in addition to the special duties entrusted to them under other provisions of the GC.[345]This provision of the GC is based on experience during World War II, when camp leaders allowed POWs to assist one another.[346]

10.22.2.2 *Other Specific Duties*. In addition to the general duties that the GC imposes on the Internee Committees, the GC also imposes specific duties on the Internee Committees in the following matters:

• managing the distribution of collective relief;[347]

• assisting in the transport of the internees' community property and luggage in cases of transfers of internees;[348]

• receiving parcels and remittances of money for internees undergoing disciplinary punishment, and handing over to the infirmary any perishable goods;[349]

• checking on the management of the canteen and its special fund;[350]

• transmitting complaints and petitions, and making periodic reports on the needs of internees;[351]

• retaining a copy of the GC and any applicable special agreements under the GC, if a copy is not publicly posted;[352]

• monitoring decisions announced from any disciplinary proceeding against an internee;[353]

and

• receiving information about all judicial proceedings against internees whom they represent, and of their result.[354]

10.22.3 Prerogatives of Internee Committees. In order to accomplish their duties, members of Internee Committtees are afforded certain prerogatives.

10.22.3.1 *Limitations on Other Work Assignments*. Members of Internee Committees shall not be required to perform any other work, if the accomplishment of their duties is thereby rendered more difficult.[355]Because internees are generally exempted from compulsory work, this means that members of Internee Committees are generally exempted from maintenance work.[356]

10.22.3.2 *Assistants for the Members of Internee Committtees*. Members of Internee Committees may appoint from among the internees such assistants as they may require.[357]

10.22.3.3 *Material Facilities, Including Certain Freedom of Movement*. All material facilities shall be granted to members of Internee Committees, particularly a certain freedom of movement necessary for the accomplishment of their duties (*e.g.*, visits to labor detachments, receipt of supplies).[358]For example, members of Internee Committees may need to go to stations or other arrival points where relief supplies are sent.[359]

10.22.3.4 *Facilities for Communication*. All facilities shall likewise be accorded to members of Internee Committees for communication by post and telegraph with the detaining authorities, the Protecting Powers, the ICRC and their delegates, and the organizations that give assistance to internees.[360]Internee Committee members in labor detachments shall enjoy similar facilities for communication with their Internee Committee in the principal place of internment.[361] Such communications shall not be limited nor considered as forming a part of the quota mentioned in Article 107 of the GC.[362]

Censorship of communications with the Protecting Powers, ICRC, etc., is not prohibited, but the Detaining Power should ensure that delays do not occur to the disadvantage of the internees.[363]

10.22.3.5 *Turnover for Successor Members of the Internee Committee Upon Transfer*. Members of Internee Committees who are transferred shall be allowed a reasonable time to acquaint their successors with current affairs.[364]

10.23 INTERNEE CORRESPONDENCE AND RELIEF SHIPMENTS

Internees may send and receive mail and relief shipments, subject to security requirements. In addition, internees benefit from certain postage exemptions.

10.23.1 Internment, Health, or Transfer Card. As soon as he or she is interned, or at the latest not more than one week after his or her arrival in a place of internment, and likewise in cases of sickness or transfer to another place of internment or to a hospital, every internee shall be enabled to send direct to his or her family, on the one hand, and to the Central Information Agency for protected persons, on the other, an internment card that is similar, if possible, to the model annexed to the GC (GC Annex III.I), informing his or her relatives of his or her detention, address, and state of health.[365]

These cards shall be forwarded as rapidly as possible and may not be delayed in any way.[366]

10.23.2 Internees' Correspondence Rights and Quota. Internees shall be allowed to send and receive letters and cards.[367]

10.23.2.1 *Internee Correspondence Quota*. If the Detaining Power deems it necessary to limit the number of letters and cards sent by each internee, this number shall not be fewer than two letters and four cards monthly; these letters and cards shall be drawn up so as to conform as closely as possible to the models annexed to the GC (GC Annexes III.II & III.III).[368]

If limitations must be placed on the correspondence addressed to internees, they may be ordered only by the Power to which such internees owe allegiance, possibly at the request of the Detaining Power.[369] If the Detaining Power can censor and transport more than the minimum stated, in a reasonable time-frame, it should permit more correspondence.[370]

10.23.2.2 *No Delay in Correspondence for Disciplinary Reasons*. Such letters and cards must be conveyed with reasonable dispatch; they may not be delayed or retained for disciplinary reasons.[371]

10.23.2.3 *More Rapid Means of Correspondence in Appropriate Cases*. Internees who have been a long time without news, or who find it impossible to receive news from their relatives, or to give them news by the ordinary postal route, as well as those who are at a considerable distance from their homes, shall be allowed to send telegrams, the charges being paid by them in the currency at their disposal.[372] Internees shall likewise benefit from this provision in cases that are recognized to be urgent.[373]

More modern means of communication, such as email, may be considered for internee correspondence, as appropriate.[374]

10.23.2.4 *Language of Internee Correspondence*. As a rule, internees' mail shall be written in their own language.[375] The parties to the conflict may authorize correspondence in other languages.[376]

10.23.3 Receipt of Individual and Collective Relief Shipments for Internees. Internees shall be allowed to receive, by post or by any other means, individual parcels or collective shipments containing in particular foodstuffs, clothing, medical supplies, as well as books and objects of a devotional, educational, or recreational character that may meet their needs.[377] Such shipments shall in no way free the Detaining Power from the obligations imposed upon it by virtue of the GC.[378]

Should military necessity require the quantity of such shipments to be limited, due notice thereof shall be given to the Protecting Power and to the ICRC, or to any other organization giving assistance to the internees and responsible for the forwarding of such shipments.[379] Such a case of military necessity may be, for example, when military operations could be hindered through the blocking of means of communication by large consignments of relief supplies.[380]

10.23.3.1 *Separation of Books From Other Relief Supplies*. Parcels of clothing and foodstuffs may not include books.[381] Books should be separated because the time required for censorship of books would likely delay the distribution of other relief supplies.[382]

10.23.3.2 *Medical Supplies in Relief Shipments*. Medical relief supplies shall, as a rule, be sent in collective parcels.[383] This rule is meant to ensure that medical supplies can be administered by the camp doctors and not by the internees themselves without medical advice.[384]

However, this rule would not prohibit, in exceptional cases, for example, special medication being sent from family members to an internee.[385]

10.23.3.3 *Special Agreements Concerning Relief Shipments*. The conditions for the sending of individual parcels and collective shipments shall, if necessary, be the subject of special agreements between the Powers concerned, which may in no case delay the receipt by the internees of relief supplies.[386]

10.23.3.4 *Collective Relief for Internees*. In the absence of special agreements between parties to the conflict regarding the conditions for the receipt and distribution of collective relief shipments, the regulations concerning collective relief that are annexed to the GC (GC Annex II) shall be applied.[387]

The special agreements provided for above shall in no case restrict the right of Internee Committees to take possession of collective relief shipments intended for internees, to undertake their distribution, and to dispose of them in the interests of the recipients.[388]

Nor shall such agreements restrict the right of representatives of the Protecting Powers, the ICRC, or any other organization giving assistance to internees and responsible for the forwarding of collective shipments, to supervise their distribution to the recipients.[389]

10.23.4 Exemption From Postal and Shipping Charges for Shipments To and From Internees. Relief shipments for internees and mail sent by internees or to them benefit from certain exemptions in international law.

10.23.4.1 *Exemption From Dues for Relief Shipments*. All relief shipments for internees shall be exempt from import, customs, and other dues.[390]

10.23.4.2 *GC Postal Dues Exemption*. Under the GC, all matter sent by mail, including relief parcels sent by parcel post and remittances of money, addressed from other countries to internees or sent by them through the post office, either directly or through the National Protected Person Information Bureau and the Central Information Agency for protected persons, shall be exempt from all postal dues both in the countries of origin and destination, and in intermediate countries.[391]

To this end, in particular, the exemption provided by the Universal Postal Convention of 1947 and by the agreements of the Universal Postal Union in favor of civilians of enemy nationality detained in camps or civilian prisons, shall be extended to the other interned persons protected by the GC.[392] The countries not signatory to the above-mentioned agreements shall be bound to grant freedom from charges in the same circumstances.[393]

10.23.4.3 *Exemption From Postal Charges Under the Universal Postal Convention*. Under the Universal Postal Convention, with the exception of air surcharges for airmail, letter-post items, postal parcels, and monetary articles originating in other countries and addressed to, or sent by internees, either directly or through the National Protected Person Information Bureau and the Central Information Agency for protected persons, shall be exempt from all postal charges.[394]

Parcels shall be admitted free of postage up to a weight of 5 kg.[395] The weight limit shall be increased to 10 kg in the case of parcels the contents of which cannot be split up and of parcels addressed to a camp or the Internee Committee there for distribution to the internees.[396]

Items exempt from postal charges and internee parcels shall bear "*Service des internés civils*" (Civilian internees service) and a translation in another language if appropriate.[397]

Postal administrations shall not be liable for the loss of, theft from, or damage to parcels in the case of internee parcels.[398]

10.23.4.4 *Costs of Transporting Relief Shipments Outside the Post Office*. The cost of transporting relief shipments that are intended for internees and that, by reason of their weight or any other cause, cannot be sent through the post office, shall be borne by the Detaining Power in all the territories under its control.[399] Other Powers that are Parties to the GC shall bear the cost of transport in their respective territories.[400]

Costs connected with the transport of such shipments that are not covered by the above paragraphs shall be charged to the senders.[401]

10.23.4.5 *Charges for Telegrams*. The Parties to the GC shall endeavor to reduce, so far as possible, the charges for telegrams sent by internees, or addressed to them.[402]

10.23.5 Special Means of Transport of Shipments to Internees. Should military operations prevent the Powers concerned from fulfilling their obligation to ensure the conveyance of the mail and relief shipments provided for in Articles 106, 107, 108, and 113 of the GC, the Protecting Powers concerned, the ICRC, or any other organization duly approved by the parties to the conflict, may undertake the conveyance of such shipments by suitable means (*e.g.*, rail, motor vehicles, vessels, or aircraft).[403]For this purpose, the Parties to the GC shall endeavor to supply them with such transport, and to allow its circulation, especially by granting the necessary safe-conducts.[404]

Such transport may also be used to convey:

• correspondence, lists, and reports exchanged between the Central Information Agency for protected persons and the National Protected Person Information Bureau; and

• correspondence and reports relating to internees that the Protecting Powers, the ICRC, or any other organization assisting the internees exchange either with their own delegates or with the parties to the conflict.[405]

These provisions in no way detract from the right of any party to the conflict to arrange other means of transport if it should so prefer, nor preclude the granting of safe-conducts, under mutually agreed conditions, to such means of transport.[406]

The costs occasioned by the use of such means of transport shall be borne, in proportion to the importance of the shipments, by the parties to the conflict whose nationals are benefited thereby.[407]Expenses in setting up the special transport system are not specifically addressed by the GC, and presumably would be addressed by an agreement between the body that takes the initiative in establishing the system and the Powers concerned.[408]

10.23.6 Censorship and Security Review of Internee Correspondence and Shipments. The Detaining Power may examine and censor all communications sent to or by internees, including correspondence, telegrams, parcels, newspapers, periodicals, and books, with a view to deleting or confiscating matter prejudicial to its military security.[409]The Detaining Power's general right to conduct censorship is an implicit assumption in the GC.[410]Internees' complaints, petitions, and reports may be subject to security review and censorship to ensure that they are not misused.[411]

The censoring of correspondence addressed to internees or sent by them shall be done as quickly as possible.[412]

10.23.6.1 *Examination of Consignments*. The examination of consignments intended for internees shall not be carried out under conditions that will expose the goods in them to deterioration.[413]It shall be done in the presence of the addressee, or of a fellow-internee duly delegated by him or her.[414]This measure is intended to prevent any misappropriation; the fellow-internee would generally be a member of the Internee Committee.[415]

The delivery to internees of individual or collective consignments shall not be delayed under the pretext of difficulties of censorship.[416]

10.23.6.2 *Limitation on Prohibition of Correspondence*. Any prohibition of correspondence ordered by the parties to the conflict for either military or political reasons shall be only temporary, and its duration shall be as short as possible.[417]

10.24 PRIVATE LEGAL MATTERS OF INTERNEES

10.24.1 Civil Capacity. Internees shall retain their full civil capacity and shall exercise such attendant rights as may be compatible with their status.[418]Their civil capacity continues to be governed by the laws that applied to them before they were interned (*e.g.*, the domestic laws of their home country or the laws governing the place where they were detained).[419]

10.24.2 Legal Documents and Assistance. The Detaining Power shall provide all reasonable facilities for the transmission, through the Protecting Power or the Central Information Agency for protected persons, or as otherwise required, of wills, powers of attorney, letters of authority, or any other documents intended for internees or sent by them.[420]In all cases, the Detaining Power shall facilitate the execution and authentication in due legal form of such documents on behalf of internees, in particular by allowing them to consult a lawyer.[421]

10.24.2.1 *Safekeeping of Wills*. In particular, the wills of internees shall be received for safe-keeping by the responsible authorities, *e.g.*, the commandant of the place of internment or, on the responsibility of the commandant, a public notary.[422] Upon the internee's death, the will shall be sent to a person whom the internee had designated.[423]

10.24.3 Management of Internee Property. Because internment is not punishment, enabling internees to manage their property is of major importance.[424]

The Detaining Power shall afford internees all facilities to enable internees to manage their property, provided such facilities are not incompatible with the conditions of internment and the law that is applicable.[425] The law that is applicable may include the Detaining Power's war regulations relating to enemy property.[426]

For the purpose of enabling internees to manage their property, the Detaining Power may give internees permission to leave the place of internment in urgent cases and if circumstances allow.[427] In general, however, the internee would act only by delegating his or her powers.[428]

10.24.4 Court Cases in Which Internees Are Parties. In all cases in which an internee is a party to proceedings in any court, the Detaining Power shall, if the internee so requests, cause the court to be informed of the internee's detention and shall, within legal limits, ensure that all necessary steps are taken to prevent the internee from being in any way prejudiced, by reason of his or her internment, as regards the preparation and conduct of his or her case, or as regards the execution of any judgment of the court.[429]

10.25 INTERNEES AND VISITS OF FAMILY AND FRIENDS

Every internee shall be allowed to receive visitors, especially near relatives, at regular intervals and as frequently as possible.[430] The frequency of the visits will depend on practical considerations, including security needs.[431] The visitors need not be family members, but may be, for example, close friends.[432]

As far as is possible, internees shall be permitted to visit their homes in urgent cases, particularly in cases of death or serious illness of relatives.[433]

10.26 INTERNEE ESCAPES

Similar to the rules on POW escape,[434] a number of provisions of the GC limit punishment of internees with respect to escape. However, unlike POWs,[435] internees are not afforded any protection from punishment for successful escapes.

10.26.1 Punishment for the Escape Itself. Internees who are recaptured after having escaped, or when attempting to escape, shall be liable only to disciplinary punishment in respect of this act, even if it is a repeated offense.[436]

10.26.1.1 *Punishment for Aiding and Abetting an Escape*. Internees who aid and abet an escape, or an attempt to escape, shall be liable on this count to disciplinary punishment only.[437]

10.26.2 Offenses Committed in Connection With, or During, Escape.

10.26.2.1 *Escape Not an Aggravating Circumstance for Offenses Committed During Escape*. Escape, or attempt to escape, even if it is a repeated offense, shall not be deemed an aggravating circumstance in cases where an internee is prosecuted for offenses committed during his or her escape.[438]

10.26.2.2 *Leniency in Favor of Disciplinary Punishments for Offenses Committed in Connection With Escape*. The parties to the conflict shall ensure that the competent authorities exercise leniency in deciding whether punishment inflicted for an offense shall be of a disciplinary or judicial nature, especially in respect of acts committed in connection with an escape, whether successful or not.[439] Thus, offenses that are committed in connection with an escape that are not serious should be punished in disciplinary as opposed to judicial proceedings.[440]

This requirement does not compel the competent authorities to choose disciplinary over judicial proceedings in any particular case.[441]

10.26.3 Special Surveillance of Internees Who Have Previously Escaped or Attempted to Escape. Although paragraph 3 of Article 118 of the GC provides that internees who have served disciplinary or judicial sentences shall not be treated differently from other internees,[442] internees punished as a result of escape or attempt to escape may be subjected to special surveillance, on condition that such surveillance: (1) does not affect the state of their health; (2) is exercised in a place of internment; and (3) does not entail the abolition of any of the safeguards granted by the GC.[443]

10.26.4 Notification of Escape and Recapture. If an internee escapes, the State of which the internee was a national or in which the national resided should be notified by the Detaining Power's National Protected Person Information Bureau through the intermediary of the Protecting Powers and Central Information Agency for protected persons.[444]

10.27 GENERAL PROVISIONS APPLICABLE TO BOTH JUDICIAL AND DISCIPLINARY SANCTIONS REGARDING INTERNEES

This section addresses the rules that apply to both judicial and disciplinary sanctions regarding internees.

10.27.1 Applicability of the Laws in Force to Internees. Subject to the provisions of Chapter IX of the GC, the laws in force in the territory in which they are detained will continue to apply to internees who commit offenses during internment.[445]

10.27.2 Only Disciplinary Punishments for Internee-Specific Offenses. If general laws, regulations, or orders declare acts committed by internees to be punishable, whereas the same acts are not punishable when committed by persons who are not internees, such acts shall entail disciplinary punishments only.[446]

10.27.3 No Repetition of Punishment. No internee may be punished more than once for the same act, or on the same count.[447]

10.27.4 Penalties Imposed on Internees. The courts or authorities shall, in passing sentence, take as far as possible into account the fact that the defendant is not a national of the Detaining Power.[448] They shall be free to reduce the penalty prescribed for the offense with which the internee is charged and shall not be obliged, to this end, to apply the minimum sentence prescribed.[449]

10.27.4.1 *Prohibited Penalties*. Imprisonment in premises without daylight and, in general, all forms of cruelty without exception are forbidden.[450]

Collective punishments of internees are also prohibited.[451]

10.27.4.2 *Deduction From Penalty for Duration of Preventive Detention*. The duration of preventive detention undergone by an internee shall be deducted from any disciplinary or judicial penalty involving confinement to which he or she may be sentenced.[452]

10.27.5 Treatment of Internees Who Have Served Sentences. Internees who have served disciplinary or judicial sentences shall not be treated differently from other internees.[453]

10.28 DISCIPLINARY PROCEEDINGS AND PUNISHMENT

In addition to the general principles applicable to the punishment of internees, the following rules address disciplinary proceedings and punishment.

10.28.1 Internment Camp Authorities Who May Order Disciplinary Punishment. Without prejudice to the competence of courts and higher authorities, disciplinary punishment may be ordered only by the commandant of the place of internment, or by a responsible officer or official who replaces him or her, or to whom the commandant has delegated his or her disciplinary powers.[454] The delegation of disciplinary authorities by the commandant, however, does not absolve the commandant of his or her own responsibility and of his or her duty of supervision.[455]

10.28.2 Confinement of Internees Awaiting Trial. Acts that constitute offenses against discipline shall be investigated immediately.[456] This rule shall be applied, in particular, in cases of escape or attempt to escape.[457] Recaptured internees shall be handed over to the competent authorities as soon as possible.[458] Competent authorities are those directly responsible for the person concerned before the escape and who are entitled to direct that the internee receive a disciplinary punishment.[459]

In cases of offenses against discipline, confinement awaiting trial shall be reduced to an absolute minimum for all internees, and shall not exceed fourteen days.[460] Although the words "awaiting trial" are used in the GC, the GC does not require that this period of investigation before the disciplinary proceedings have the complicated character of pre-trial discovery and involve motions, for example, as in the judicial context.[461]

The provisions of Articles 124 and 125 of the GC shall apply to internees who are in confinement awaiting trial for offenses against discipline.[462] Thus, the conditions of any confinement before the trial should be at least as good as those afforded internees who are confined as part of disciplinary punishment.[463]

10.28.3 Rights of Internees in Disciplinary Proceedings. Before any disciplinary punishment is awarded, the accused internee shall be given: (1) precise information regarding the offenses of which he or she is accused; (2)

an opportunity to explain his or her conduct; and (3) an opportunity to defend himself or herself.[464] The internee shall be permitted, in particular, to call witnesses and to have recourse, if necessary, to the services of a qualified interpreter.[465] The decision shall be announced in the presence of the accused and of a member of the Internee Committee.[466]

10.28.4 Record of Disciplinary Punishments. A record of disciplinary punishments shall be maintained by the commandant of the place of internment and shall be open to inspection by representatives of the Protecting Power.[467]

10.28.5 Types of Disciplinary Punishments and Procedures for Dispensing Punishment. In no case shall disciplinary penalties be inhuman, brutal, or dangerous for the health of internees.[468]

In addition, prolonged standing and roll-calls, punishment drill, military drill and maneuvers, and the reduction of food rations are prohibited.[469]

Account shall be taken of the internee's age, sex, and state of health.[470] The disciplinary punishments applicable to internees shall be the following:

- a fine that shall not exceed 50 percent of the wages that the internee would otherwise receive under Article 95 of the GC during a period of not more than thirty days;
- discontinuance of privileges granted over and above the treatment provided for by the GC;
- fatigue duties, not exceeding two hours daily, in connection with the maintenance of the place of internment; and
- confinement.[471]

The duration of any single punishment shall in no case exceed a maximum of thirty consecutive days, even if the internee is answerable for several breaches of discipline when his or her case is dealt with, regardless of whether such breaches are connected.[472]

The duration of confinement awaiting trial shall in any case be deducted from any sentence of confinement.[473]

The period elapsing between the time of award of a disciplinary punishment and its execution shall not exceed one month.[474] When an internee is awarded a further disciplinary punishment, a period of at least three days shall elapse between the execution of any two of the punishments, if the duration of one of these punishments is ten days or more.[475]

10.28.6 Conditions for Internees Serving Disciplinary Punishments. Conditions for disciplinary punishments must comport with the requirements for humane treatment. In addition, the following rules apply:

10.28.6.1 *Premises Where Disciplinary Punishments Are to Be Served.* Internees shall not, in any case, be transferred to penitentiary establishments (*e.g.*, prisons, penitentiaries, convict prisons) to undergo disciplinary punishments.[476]

The premises in which disciplinary punishments occur shall conform to sanitary requirements; they shall, in particular, be provided with adequate bedding.[477] Internees undergoing punishment shall be enabled to keep themselves in a state of cleanliness.[478]

Women internees undergoing disciplinary punishment shall be confined in separate quarters from male internees and shall be under the immediate supervision of women.[479]

10.28.6.2 *Exercise and Access to the Open Air.* Internees awarded disciplinary punishment shall be allowed to exercise and to stay in the open air at least two hours daily.[480]

Exercising and staying in the open air should be given as an opportunity and should not be compelled.[481]

10.28.6.3 *Attendance at Daily Medical Inspection and Medical Attention.* Internees awarded disciplinary punishment shall be allowed, if they so request, to be present at daily medical inspections.[482] They shall receive the attention that their state of health requires and, if necessary, shall be removed to the infirmary of the place of internment or to a hospital.[483]

10.28.6.4 *Reading, Writing, Correspondence, and Packages.* Internees awarded disciplinary punishment shall have permission to read and write, and to send and receive letters.[484] Parcels and remittances of money, however, may be withheld from them until the completion of their punishment; such consignments shall meanwhile be entrusted to the Internee Committee, who will hand over to the infirmary the perishable goods contained in the parcels.[485]

No internee given a disciplinary punishment may be deprived of the benefit of the provisions of Articles 101 and 143 of the GC.[486] Thus, internees undergoing disciplinary punishments may not be deprived of the right to make requests and complaints, or to deal with representatives of the Protecting Power (including representatives of an impartial humanitarian organization performing the functions of the Protecting Power).[487]

10.29 JUDICIAL PROCEEDINGS REGARDING PROTECTED PERSONS IN OCCUPIED TERRITORY OR INTERNEES IN A BELLIGERENT'S HOME TERRITORY

In addition to the general principles applicable to the punishment of internees, the following rules address judicial proceedings against internees in a belligerent's home territory and protected persons in occupied territory.

The provisions of Articles 71 through 76 of the GC shall apply, by analogy, to proceedings against internees who are in the national territory of the Detaining Power.[488] Articles 71-76 of the GC address the penal procedures for protected persons in occupied territory. Thus, this section has been written with a view towards describing both situations.

10.29.1 Information to Internee Committees. Internee Committees shall be informed of all judicial proceedings instituted against internees whom they represent, and of their result.[489]

10.29.2 Pre-Trial Matters.

10.29.2.1 *Prompt Information of the Charges and Brought to Trial as Rapidly as Possible*. Accused persons who are prosecuted by the Occupying or Detaining Power shall be promptly informed, in writing, in a language that they understand, of the particulars of the charges preferred against them, and shall be brought to trial as rapidly as possible.[490]

10.29.2.2 *Notification of Proceeding in Respect of Charges Involving the Death Penalty or Imprisonment for Two or More Years*. The Protecting Power shall be informed of all proceedings instituted by the Occupying or Detaining Power against protected persons in respect of charges involving the death penalty or imprisonment for two years or more, and shall be enabled, at any time, to obtain information regarding the state of such proceedings.[491]

Furthermore, the Protecting Power shall be entitled, on request, to be furnished with all particulars of these and of any other proceedings instituted by the Occupying or Detaining Power against protected persons.[492]

The notification to the Protecting Power, as provided for in the second paragraph of Article 71 of the GC, shall be sent immediately, and shall in any case reach the Protecting Power three weeks before the date of the first hearing.[493] Unless, at the opening of the trial, evidence is submitted that the provisions of Article 71 of the GC are fully complied with, the trial shall not proceed.[494] The notification shall include the following particulars:

- description of the accused;
- place of residence or detention;
- specification of the charge or charges (with mention of the penal provisions under which it is brought);
- designation of the court that will hear the case; and
- place and date of the first hearing.[495]

10.29.3 Right of Defense and Trial Procedure. Accused persons shall have the right to present evidence necessary to their defense and may, in particular, call witnesses.[496] They shall have the right to be assisted by a qualified advocate or counsel of their own choice, who shall be able to visit them freely and shall enjoy the necessary facilities for preparing the defense.[497]

10.29.3.1 *Additional Provision for Defense Counsel and Interpreter*. Failing a choice by the accused, the Protecting Power may provide him or her with an advocate or counsel.[498] Where the accused person is facing a serious charge and the Protecting Power is not functioning, the Occupying or Detaining Power, subject to the consent of the accused, shall provide an advocate or counsel.[499] Accused persons shall, unless they freely waive such assistance, be aided by an interpreter, both during preliminary investigation and during the hearing in court.[500] They shall have the right at any time to object to the interpreter and to ask for the interpreter to be replaced.[501]

10.29.3.2 *Attendance by the Protecting Power in the Proceedings*. Representatives of the Protecting Power shall have the right to attend the trial of any protected person, unless the hearing, as an exceptional measure, has to be held *in camera* in the interests of the security of the Occupying or Detaining Power, which shall then notify the

Protecting Power.[502] A notification in respect of the date and place of trial shall be sent to the Protecting Power.[503]

10.29.3.3 *Regular Trial.* No sentence shall be pronounced by the competent courts of the Occupying or Detaining Power except after a regular trial.[504]

10.29.3.4 *Notification of Judgment Involving Sentence of Death or Imprisonment.* Any judgment involving a sentence of death, or imprisonment for two years or more, shall be communicated, with the relevant grounds, as rapidly as possible to the Protecting Power.[505] The notification shall contain a reference to the notification made under Article 71 of the GC and, in the case of sentences of imprisonment, the name of the place where the sentence is to be served.[506] A record of judgments other than those referred to above shall be kept by the court and shall be open to inspection by representatives of the Protecting Power.[507] Any period allowed for appeal in the case of sentences involving the death penalty, or imprisonment of two years or more, shall not run until notification of judgment has been received by the Protecting Power.[508]

10.29.4 Right of Appeal. A convicted person shall have the right of appeal provided for by the laws applied by the court.[509] The convicted person shall be fully informed of his or her right to appeal or petition, and of the time limit within which he or she may do so.[510]

The penal procedure in Section III of the GC shall apply, as far as it is applicable, to appeals.[511] Where the laws applied by the court make no provision for appeals, the convicted person shall have the right to petition against the finding and sentence to the competent authority of the Occupying or Detaining Power.[512]

10.29.5 Death Sentences. In no case shall persons condemned to death be deprived of the right of petition for pardon or reprieve.[513]

No death sentence shall be carried out before the expiration of a period of at least six months from the date of receipt by the Protecting Power of the notification of the final judgment confirming such death sentence, or of an order denying pardon or reprieve.[514]

The six-month period of suspension of the death sentence prescribed in Article 75 of the GC may be reduced in individual cases in circumstances of grave emergency involving an organized threat to the security of the Occupying or Detaining Power or its forces, provided always that the Protecting Power is notified of such reduction and is given reasonable time and opportunity to make representations to the competent occupying or domestic authorities in respect of such death sentences.[515]

10.29.6 Conditions for Accused or Convicted Protected Persons Who Are Detained. If found in occupied territory, protected persons accused of offenses shall be detained in the occupied country, and, if convicted, they shall serve their sentences therein.[516] This provision does not preclude the appearance of convicted persons before appeal courts outside the occupied territory, or the possibility of protected persons receiving medical care outside the occupied territory.[517]

Accused or convicted protected persons who are detained shall, if possible, be separated from other detainees and shall enjoy conditions of food and hygiene that will be sufficient to keep them in good health, and that will be at least equal to those obtaining in prisons in that country.[518]

10.29.6.1 *Medical Attention and Spiritual Assistance.* Accused or convicted protected persons who are detained shall receive the medical attention required by their state of health.[519] They shall also have the right to receive any spiritual assistance that they may require.[520]

10.29.6.2 *Separate Confinement for Women and Special Treatment for Minors.* Women shall be confined in separate quarters and shall be under the direct supervision of women.[521] Proper regard shall be paid to the special treatment due to minors.[522]

10.29.6.3 *Visits by the Protecting Power and ICRC and Right to Receive Relief Parcel.* Accused or convicted protected persons who are detained shall have the right to be visited by delegates of the Protecting Power and the ICRC, in accordance with the provisions of Article 143 of the GC.[523] Accused or convicted protected persons who are detained also shall have the right to receive at least one relief parcel monthly.[524]

10.30 TRANSFERS OF INTERNEES FROM THE PLACE OF INTERNMENT

The rules in this section apply to transfers of internees from the place of internment. Additional rules apply to the transfer of protected persons from the belligerent's home territory.[525] Individual or mass forcible transfers of protected persons from occupied territory are prohibited.[526]

10.30.1 Determining Whether to Transfer Internees. When making decisions regarding the transfer of internees, the Detaining Power shall take their interests into account and, in particular, shall not do anything to increase the difficulties of repatriating them or returning them to their own homes.[527]

Sick, wounded, or infirm internees, and maternity cases, shall not be transferred if the journey would be seriously detrimental to them, unless their safety imperatively so demands.[528] If the combat zone draws close to a place of internment, the internees in that place shall not be transferred unless their removal can be carried out in adequate conditions of safety, or unless they are exposed to greater risks by remaining on the spot than by being transferred.[529]

10.30.2 Conditions for the Transfer of Internees. The transfer of internees shall always be effected humanely.[530] As a general rule, it shall be carried out by rail or other means of transport, and under conditions at least equal to those obtaining for the forces of the Detaining Power in their changes of station.[531] If, as an exceptional measure, such removals have to be effected on foot, they may not take place unless the internees are in a fit state of health, and may not in any case expose them to excessive fatigue.[532] This provision seeks to prevent abuses such as the forced "death marches" that occurred during World War II.[533] When considering movement by foot, the Detaining Power must assess the fitness of each individual, not the condition of the majority.[534]

The Detaining Power shall supply internees during transfer with drinking water and food sufficient in quantity, quality, and variety to maintain them in good health, and also with the necessary clothing, adequate shelter, and the necessary medical attention.[535] The Detaining Power shall take all suitable precautions to ensure their safety during transfer, and shall establish before their departure a complete list of all internees transferred.[536]

10.30.3 Procedures for the Transfer of Internees. In the event of transfer, internees shall be officially advised of their departure and of their new postal address.[537] Such notification shall be given in time for them to pack their luggage and inform their next of kin.[538]

10.30.3.1 *Baggage*. Internees shall be allowed to take with them their personal effects, and the correspondence and parcels that have arrived for them.[539] The weight of such baggage may be limited if the conditions of transfer so require, but in no case may be limited to less than 25 kilograms (approximately 55 pounds) per internee.[540]

10.30.3.2 *Forwarding of Mail*. Mail and parcels addressed to their former place of internment shall be forwarded to internees without delay.[541]

10.30.3.3 *Disposition of Community Property and Remaining Property*. The commandant of the place of internment shall take, in agreement with the Internee Committee, any measures needed to ensure the transport of the internees' community property and of the luggage the internees are unable to take with them in consequence of restrictions that may be imposed on baggage that internees may take with them.[542]

10.31 NATIONAL ACCOUNTING FOR PROTECTED PERSONS IN ITS POWER

A proper accounting of the detention or assigned residence of protected persons is an important responsibility of States in relation to protected persons in their power. In order to provide as full an accounting as possible, and in order to minimize misidentification, a State must obtain detailed information about certain protected persons in its power for forwarding through its National Protected Person Information Bureau to the Central Information Agency for protected persons.

10.31.1 Accountability Information That Parties to a Conflict Should Collect. Each of the parties to the conflict shall, within the shortest possible period, give its National Protected Person Information Bureau information of any measure taken by it concerning any protected persons who are: (1) kept in custody for more than two weeks; (2) subjected to assigned residence; or (3) interned.[543]

The information received by the National Protected Person Information Bureau and transmitted by it shall be of such a character as to make it possible to identify the protected person exactly and to advise his or her next of kin quickly.[544] The information in respect of each person shall include at least:

- surname (*i.e.*, last name) and first names (*i.e.*, first and middle names);
- place and date of birth;
- nationality;
- last residence;
- any distinguishing characteristics;

• first name of the father and the maiden name of the mother;

• the date, place, and nature of the action taken with regard to the individual;

• the address at which correspondence may be sent to him or her; and

• the name and address of the person designated by the protected person to be informed of information concerning him or her.[545]

In addition to this information, information regarding all changes pertaining to these protected persons, such as transfers, releases, repatriations, escapes, admittances to hospitals, births, and deaths, shall also be collected and transmitted.[546]

In addition to admittances to hospitals, information regarding the state of health of internees who are seriously ill or seriously wounded must be supplied to the National Protected Person Information Bureau regularly and, if possible, every week.[547]

In addition to notices of death, copies of records of death of internees shall be forwarded in accordance with Article 129 of the GC.[548]Lists of graves of deceased internees shall be forwarded, as soon as circumstances permit, in accordance with Article 130 of the GC.[549]

10.31.2 National Protected Person Information Bureau. Upon the outbreak of an international armed conflict and in all cases of occupation of the territory of one State by another State, each of the parties to the conflict shall establish an official National Protected Person Information Bureau responsible for receiving and transmitting information in respect of the protected persons who are in its power.[550]The State has discretion about the nature, composition, and working methods of the National Protected Person Information Bureau.[551]

All communications in writing made by any National Protected Person Information Bureau shall be authenticated by a signature or a seal.[552]

The National Protected Person Information Bureau: (1) receives and forwards certain information to the Powers concerned, through the intermediary of the Protecting Powers and the Central Information Agency for protected persons; (2) replies to inquiries; (3) withholds certain information from the Powers concerned, but not the Central Information Agency for protected persons, in the interest of protected persons or their relatives; (4) collects personal valuables left by certain protected persons; and (5) in occupied territory, has a special section that is responsible for identifying children whose identity is in doubt.

10.31.2.1 *Receiving and Forwarding Certain Information.* The National Protected Person Information Bureau receives certain information regarding protected persons from other departments of the State to which it belongs.[553]

Each National Protected Person Information Bureau shall immediately forward information concerning protected persons by the most rapid means to the Powers of whom these persons are nationals, or to Powers in whose territory they resided, through the intermediary of the Protecting Powers and likewise through the Central Information Agency for protected persons.[554]

10.31.2.2 *Replying to Inquiries Regarding Protected Persons.* The National Protected Person Information Bureau shall also reply to all inquiries that may be received regarding protected persons.[555]Responses to inquiries must be made consistent with the protection of protected persons against insults and public curiosity.[556]

10.31.2.3 *Withholding of Information When Detrimental to the Person or His or Her Relatives.* The National Protected Person Information Bureau shall transmit information concerning a protected person unless its transmission might be detrimental to the person concerned or to his or her relatives.[557]Even in such a case, the information may not be withheld from the Central Information Agency for protected persons, which, upon being notified of the circumstances, will take the necessary precautions indicated in Article 140 of the GC.[558]This rule seeks to protect persons who have fled persecution from their home governments, as well as the family members of such persons.[559]

10.31.2.4 *Collection of Personal Valuables Left by Protected Persons.* Each National Protected Person Information Bureau shall be responsible for collecting all personal valuables left by protected persons mentioned in Article 136 of the GC, in particular those who have been repatriated or released, or who have escaped or died.[560]The National Protected Person Information Bureau shall forward these valuables to those concerned, either directly or, if necessary, through the Central Information Agency for protected persons.[561]Personal valuables may include wills, articles of sentimental value, or sums of money.[562]Such articles shall be sent by the National Protected Person Information Bureau in sealed packets, which shall be accompanied by statements giving clear and full identity particulars of the person to whom the articles belonged, and by a complete list of the contents of the parcel.[563] Detailed records shall be maintained of the receipt and sending of all such valuables.[564]

10.31.2.5 *Special Section for the Identification of Children in Occupied Territory*. In occupation, a special section of the National Protected Person Information Bureau shall be responsible for taking all necessary steps to identify children whose identity is in doubt.[565]

Particulars of their parents or other near relatives should always be recorded if available.[566]

10.31.3 Central Information Agency for Protected Persons. A Central Information Agency for protected persons, in particular for internees, shall be created in a neutral country.[567]

This agency may be the same as that provided for in Article 123 of the GPW.[568]

The function of the Central Information Agency for protected persons shall be to collect all information of the type set forth in Article 136 of the GC, which it may obtain through official or private channels, and to transmit it as rapidly as possible to the countries of origin or of residence of the persons concerned, except in cases where such transmissions might be detrimental to the persons whom this information concerns, or to their relatives.[569] The Central Information Agency for protected persons shall receive from the parties to the conflict all reasonable facilities for effecting such transmissions.[570] Parties to the GC, and in particular those States whose nationals benefit from the services of the Central Information Agency for protected persons, are requested to give the Central Information Agency for protected persons the financial aid it may require.[571]

The provisions of the GC establishing the Central Information Agency for protected persons do not restrict the humanitarian activities of the ICRC and of the relief societies described in Article 142 of the GC.[572]

The GC contemplates that the ICRC shall, if it deems necessary, propose to the Powers concerned the organization of a Central Information Agency for protected persons.[573] In fact, the role of the Central Information Agency for protected persons has been performed in many conflicts by the ICRC Central Tracing Agency, which has also performed this role for POWs.[574]

10.31.4 U.S. Practice in Reporting to the ICRC Central Tracing Agency. For the United States, the functions of the National Protected Person Information Bureau have been performed by the same entity that performs the functions of the National POW Information Bureau: the National Prisoner of War Information Center or the National Detainee Reporting Center.[575] This organization has provided information to the ICRC Central Tracing Agency, acting as the Central Information Agency for protected persons under the GC.

10.31.5 Exemption of National Protected Person Information Bureau and Central Information Agency for Protected Persons From Certain Charges. The National Protected Person Information Bureau and the Central Information Agency for protected persons shall enjoy free postage for mail, all the exemptions provided for in Article 110 of the GC,[576] and, further, so far as possible, exemption from telegraphic charges or, at least, greatly reduced rates.[577]

The National Protected Person Information Bureau and the Central Information Agency for protected persons shall also enjoy exemption from postal charges in respect of letter-post items, postal parcels, and monetary articles that concern internees, which they send or receive, either directly or as intermediaries, under the conditions laid down in the Universal Postal Convention.[578] Such items exempt from postal charges should be marked appropriately.[579]

10.32 ROLE OF THE PROTECTING POWER IN THE GC

The Protecting Power is an organ for ensuring implementation of the GC. The Protecting Power has extensive duties under the GC. It transmits information between belligerents, monitors compliance with the GC, and takes an active role in promoting the welfare of protected persons.

10.32.1 Transmission of Information Between Belligerents. The Protecting Power helps transmit information between belligerents, including:

- useful information regarding the geographic location of places of internment;[580]
- accountability information concerning measures that States have taken with respect to protected persons in their power;[581]
- legal documents of internees;[582]
- copies of the records of the death of internees;[583]

• notification of measures taken by the Detaining Powers, immediately upon interning protected persons, for executing the provisions of the GC that address relations between internees and the exterior;[584] and

• official translations of the GC and of laws and regulations that parties to the conflict have adopted to implement it.[585]

10.32.2 Monitoring or Inspecting Compliance With the GC. The Protecting Power monitors or inspects the Detaining Power's compliance with the GC, including by:

• visiting places where protected persons are and interviewing them;[586]

• receiving petitions and complaints from internees and periodic reports from the Internee Committees;[587]

• communicating with members of the Internee Committees and receiving the reasons for any refusal or dismissal of a member of an Internee Committee;[588]

• receiving up-to-date lists of labor detachments so its representatives may visit them;[589]

• inspecting records of disciplinary punishments maintained by the commandant of the place of internment;[590]

• receiving notice if military necessity should require that the quantity of individual parcels or collective shipments be limited;[591]

• receiving statements of internee accounts;[592] and

• receiving communications and subsequent reports of death or serious injury of an internee in certain cases.[593]

10.32.3 Actively Working to Improve the Welfare of Protected Persons. In other instances, the GC contemplates that the Protecting Power will actively work to improve the welfare of protected persons, such as by:

• providing internees with allowances;[594]

• setting up special means of transport to ensure the conveyance of relief shipments to internees;[595] and

• lending its good offices with a view towards settling disagreements as to application or interpretation of the provisions of the GC between parties to the conflict.[596]

10.33 ACCESS TO INTERNEES BY PROTECTING POWERS, ICRC, AND OTHER RELIEF ORGANIZATIONS

10.33.1 Access by Protecting Powers. Representatives or delegates of the Protecting Powers shall have permission to go to all places where protected persons are, particularly to places of internment, detention, and work.[597] They shall have access to all premises occupied by protected persons and shall be able to interview the latter without witnesses, personally or through an interpreter.[598] Such visits may not be prohibited except for reasons of imperative military necessity, and then only as an exceptional and temporary measure.[599] Their duration and frequency must not be restricted.[600]

Such representatives and delegates shall have full liberty to select the places they wish to visit.[601]

10.33.1.1 *Participation of Compatriots of Internees in Visits*. The Detaining or Occupying Power, the Protecting Power, and, when the occasion arises, the Power of origin of the persons to be visited, may agree that compatriots of the internees shall be permitted to participate in the visits.[602]

10.33.1.2 *ICRC Delegates Enjoying the Same Prerogatives of Access*. The delegates of the ICRC shall also enjoy the prerogatives as those of the representatives and delegates of the Protecting Powers described in Article 143 of the GC.[603]

The appointment of such delegates shall be submitted to the approval of the Power governing the territories where they will carry out their duties.[604]

10.33.2 Access by Relief Societies and Other Organizations. Subject to the measures that the Detaining Powers may consider essential to ensure their security, or to meet any other reasonable need, the representatives of religious organizations, relief societies, or any other organizations assisting the protected persons shall receive from these Powers all facilities for visiting the protected persons; for distributing relief supplies and material from any source intended for educational, recreational, or religious purposes; or for assisting them in organizing their leisure time within the places of internment.[605] Such societies or organizations may be constituted in occupied territory, in the territory of the Detaining Power, or in any other country, or they may have an international character.[606]

10.33.2.1 *Limitations on the Number of Relief Organizations*. The Detaining Power may limit the number of societies and organizations whose delegates are allowed to carry out their activities in its territory and under its supervision, on condition, however, that such limitation shall not hinder the supply of effective and adequate relief to all protected persons.[607]The special position of the ICRC in this field shall be recognized and respected at all times.[608]

10.34 DEATH OF INTERNEES

10.34.1 Transmittal of Wills. The Detaining Power has an obligation to help internees with their wills, such as by receiving them for safe-keeping.[609]In the event of the death of an internee, his or her will shall be transmitted without delay to a person whom he or she has previously designated.[610]

10.34.2 Death Certificates. Deaths of internees shall be certified in every case by a doctor, and a death certificate shall be made out, showing the causes of death and the conditions under which it occurred.[611]

An official record of the death, duly registered, shall be drawn up in accordance with the procedure relating thereto in force in the territory where the place of internment is situated, and a duly certified copy of such record shall be transmitted without delay to the Protecting Power and to the Central Information Agency for protected persons.[612]

10.34.3 Burial or Cremation and Inurnment. The detaining authorities shall ensure that internees who die while interned are honorably buried, if possible according to the rites of the religion to which they belonged.[613]

Deceased internees shall be buried in individual graves unless unavoidable circumstances require the use of collective graves.[614]For example, an epidemic or military operations may require the Detaining Power to undertake collective burials in the interest of public health because individual graves are not possible due to lack of time and resources.[615]

10.34.3.1 *Cremation*. Bodies of internees may be cremated only for imperative reasons of hygiene, on account of the religion of the deceased, or in accordance with his or her expressed wish to this effect.[616]In case of cremation, the fact shall be stated and the reasons given in the death certificate of the deceased.[617]The ashes shall be retained for safe-keeping by the detaining authorities and shall be transferred as soon as possible to the next of kin on their request.[618]

10.34.4 Maintenance and Records of Graves and Ashes. The detaining authorities shall ensure that the graves of internees who die while interned are respected, properly maintained, and marked in such a way that they can always be recognized.[619]Ideally, the markings on the graves should contain at least the surname, first name, and date of birth of the dead person in detail and in a durable fashion.[620]

As soon as circumstances permit, and not later than the close of hostilities, the Detaining Power shall forward lists of graves of deceased internees to the Powers on whom the deceased internees depended, through the National Protected Person Information Bureau.[621]Such lists shall include all particulars necessary for the identification of the deceased internees, as well as the exact location of their graves.[622]

10.34.5 Inquiries Into Death or Serious Injury of Internees in Certain Cases. Every death or serious injury of an internee, caused or suspected to have been caused by a sentry, another internee, or any other person, as well as any death the cause of which is unknown, shall be immediately followed by an official inquiry by the Detaining Power.[623]A communication on this subject shall be sent immediately to the Protecting Power.[624]

Serious injury, in many cases, may be understood to mean a wound that requires inpatient treatment at a hospital or infirmary.[625]

The evidence of any witnesses shall be taken, and a report including such evidence shall be prepared and forwarded to the Protecting Power.[626]

If the inquiry indicates the guilt of one or more persons, the Detaining Power shall take all necessary steps to ensure the prosecution of the person or persons responsible.[627]For example, depending on the circumstances, prosecutions could take place in the civil or military courts of the Detaining Power or in accordance with the laws in force in the territory where they happen to be interned.[628]

10.35 RELEASE, RETURN, REPATRIATION OF INTERNEES AFTER THE CLOSE OF HOSTILITIES

10.35.1 Cessation of Internment After the Close of Hostilities. Internment shall cease as soon as possible after the close of hostilities.[629]The GC, however, does not preclude internment after the close of hostilities.[630]For example, internment may be necessary during occupation after hostilities have ended.

10.35.1.1 *Close of Hostilities*. The phrase "close of hostilities" should be understood in the same sense as the phrase "cessation of active hostilities" in the GPW.[631]

10.35.1.2 *Internees Subject to Penal Proceedings or Punishment.* Internees in the territory of a party to the conflict, against whom penal proceedings are pending for offenses not exclusively subject to disciplinary penalties, may be detained until the close of such proceedings and, if circumstances require, until the completion of the penalty.[632]The same shall apply to internees who have been previously sentenced to a punishment depriving them of liberty.[633]

10.35.2 Return to the Last Place of Residence or Repatriation. The Parties to the GC shall endeavor, upon the close of hostilities or occupation, to ensure the return of all internees to their last place of residence, or to facilitate their repatriation.[634]

10.35.2.1 *Internees Who Resist Repatriation.* Internees who resist repatriation need not be forcibly repatriated.[635]A similar rule applies to POWs.[636]U.S. policy provides that no person shall be transferred to another State if it is more likely than not that the person would be tortured in the receiving country.[637]

10.35.3 Committees to Search for Dispersed Internees. By agreement between the Detaining Power and the Powers concerned, committees may be set up after the close of hostilities, or after the close of the occupation of territories, to search for dispersed internees.[638]

10.35.4 Costs of Returning Internees. The Detaining Power shall bear the expense of returning released internees to the places where they were residing when interned, or, if it took them into custody while they were in transit or on the high seas, the cost of completing their journey or of their return to their point of departure.[639]

Where a Detaining Power refuses permission to reside in its territory to a released internee who previously had his or her permanent domicile there, such Detaining Power shall pay the cost of this internee's repatriation.[640]If, however, the internee elects to return to his or her country on his or her own responsibility or in obedience to the Government of the Power to which he or she owes allegiance, the Detaining Power need not pay the expenses of his or her journey beyond the point of his or her departure from its territory.[641]

The Detaining Power need not pay the costs of repatriation of an internee who was interned at his or her own request.[642]

If internees are transferred in accordance with Article 45 of the GC, the transferring and receiving Powers shall agree on the portion of the above costs to be borne by each.[643]

The foregoing discussion shall not prejudice such special agreements as may be concluded between parties to the conflict concerning the exchange and repatriation of their nationals in enemy hands.[644]

XI – Military Occupation

Chapter Contents

11.1 INTRODUCTION

This Chapter addresses military occupation. The GC provides specific rules for the internment of protected persons in occupation, which are addressed in Chapter X. Military occupation is a temporary measure for administering territory under the control of invading forces, and involves a complicated, trilateral set of legal relations between the Occupying Power, the temporarily ousted sovereign authority, and the inhabitants of occupied territory.[1]

The law of belligerent occupation seeks to account for both military and humanitarian imperatives. The Occupying Power's primary objective in conducting military occupation is to further the purpose of the war in which the occupying forces are engaged and to ensure the maintenance and security of those forces, but the Occupying Power is also bound to provide for the interests and welfare of the civilian population of the occupied territory.[2]The Occupying Power has obligations related to the maintenance of public order and safety, and the protection of civilians and property in occupied territory.[3]

11.1.1 Military Occupation – Notes on Terminology.

11.1.1.1 *Military Occupation, Military Government, Belligerent Occupation, and Martial Law*. The practice of conducting military occupation is very old, and the law of military occupation has long been part of the law of war.[4]Military occupation is also called belligerent occupation. The conduct of military occupation has also been characterized as an exercise of "military government"[5]or "martial law."[6]

This manual uses the terms "military occupation," "belligerent occupation," and "occupation" to refer to situations governed by the law of belligerent occupation.

11.1.1.2 *Occupying Power, Occupant, and Occupying State*. Treaties or other legal documents may refer to the State conducting the occupation as the "Occupying Power,"

"Occupying State," or "occupant." This manual uses the term "Occupying Power" to refer to a State conducting an occupation.

In some cases, multiple States may conduct an occupation together, and each State may be considered an Occupying Power.[7]

11.1.2 Military Occupation and Bodies of International Law.

11.1.2.1 *Military Occupation and Other Law of War Rules*. In general, the law of war rules that are not specific to the occupation of enemy territory continue to apply to situations that are addressed by those rules that may arise during occupation. For example, the rules for the conduct of hostilities and the humane treatment of detainees apply also to combat operations and detention operations, respectively, during occupation. However, the fact of occupation triggers the application of additional rules specific to occupation.[8]

11.1.2.2 *Occupation and the Hague IV Regulations*. Articles 42 through 56 of the Hague IV Regulations address military occupation. When the GC and the Hague IV Regulations are both applicable, the provisions of the GC regarding occupation supplement the provisions of the Hague IV Regulations regarding occupation.[9]

When the 1954 Hague Cultural Property Convention and the Hague IV Regulations are both applicable, the provisions of the 1954 Hague Cultural Property Convention regarding occupation supplement the provisions of the Hague IV Regulations regarding occupation.[10]

Although in some cases the Hague IV Regulations would not be applicable as a matter of treaty law because belligerent States might not be Parties to Hague IV, many of the provisions in Articles 42 through 56 of the Hague IV Regulations reflect customary international law.[11]For example, Article 42 of the Hague IV Regulations, which provides a standard for when the law of belligerent occupation applies, is regarded as customary international law.[12]

11.1.2.3 *Occupation and the 1949 Geneva Conventions*. Common Article 2 of the 1949 Geneva Conventions provides that "[t]he Convention shall also apply to all cases of partial or total occupation of the territory of a High Contracting Party, even if the said occupation meets with no armed resistance."[13]

Articles 47 through 78 of the GC specifically address occupied territories.[14]Other provisions of the GC also apply to occupied territory. For example, Articles 27 through 34 of the GC contain provisions common to the territories of parties to the conflict and to occupied territories.[15]

11.1.2.4 *Occupation and the 1954 Hague Cultural Property Convention*. The 1954 Hague Cultural Property Convention applies to all cases of partial or total occupation of the territory of a High Contracting Party, even if this occupation meets with no armed resistance.[16]The 1954 Hague Cultural Property Convention and the Regulations for the Execution of the 1954 Hague Cultural Property Convention also have provisions that explicitly address occupation.[17]

11.1.2.5 *Occupation and U.N. Security Council Resolutions*. The U.N. Security Council may call upon Occupying Powers to comply with existing international law.[18]

Acting under the Charter of the United Nations, the U.N. Security Council may also establish authorities or limitations that might interact with those otherwise applicable under occupation law.[19]For example, a U.N. Security Council authorization may provide additional authority for an Occupying Power to govern occupied territory.[20]

In cases of conflict with obligations in other treaties or under customary international law, such obligations provided by the U.N. Security Council may prevail.[21]Thus, such authority may be used to take action that would not otherwise be permissible under the law of belligerent occupation.

11.1.2.6 *Occupation and the ICCPR and Other Human Rights Treaties*. It has been the U.S. view that the International Covenant on Civil and Political Rights (ICCPR) does not create obligations for an Occupying Power with respect to occupied territory because a contracting State's obligations under the ICCPR only extend to persons within its territory and subject to its jurisdiction.[22]Although persons within occupied territory are subject to the jurisdiction of the Occupying Power for certain purposes, they are not within the Occupying Power's national territory. This limit to the scope of the ICCPR's obligations was proposed by the United States to preclude the creation of obligations for States with respect to territories that they occupied, such as post-World War II Germany and Japan.[23]

In addition, the law of belligerent occupation is specially crafted to address the situation of belligerent occupation. Thus, in cases of apparent conflict with other provisions of law that are not intended to address the situation of belligerent occupation, there might be a presumption that such provisions would not conflict with occupation law, or that occupation law likely would prevail when addressing the situation of belligerent occupation.[24]

Other States, such as some coalition partners, may interpret their human rights treaty obligations to create obligations for their military operations outside their home territory in the context of belligerent occupation.[25]

Subject to the Occupying Power's authority to change local law,[26]an occupied State's domestic law that has been enacted pursuant to its human rights treaty obligations or that meets the requirements of the occupied State's human rights treaty obligations may continue to apply during an occupation.

11.1.3 Application of Occupation Law to Situations Not Constituting Belligerent Occupation. The law of belligerent occupation is only applicable as a matter of law when certain conditions have been met.[27]The law of belligerent occupation generally does not apply to (1) mere invasion; (2) liberation of friendly territory; (3) non-international armed conflict; or (4) post-war situations (except for certain provisions of the GC). Nonetheless, it may be appropriate to apply rules from the law of belligerent occupation in such situations.

11.1.3.1 *Mere Invasion*. Mere physical presence of a belligerent's military forces in the territory of its enemy alone does not constitute military occupation. Such presence might not constitute an effective and firm possession of enemy territory, or the belligerent State might lack the intention to displace and substitute the enemy State's authority with its own.[28]

Guidance for the U.S. armed forces has sought, to the extent possible, to apply the rules of the law of occupation as a matter of policy to areas through which U.S. forces are passing, and even on the battlefield.[29]

11.1.3.2 *Liberation of Friendly Territory*. The law of belligerent occupation does not apply to the liberation of friendly territory.[30]Indeed, a belligerent occupation presupposes that the Occupying Power is hostile in relation to the State whose territory is being occupied.[31]

The administration of liberated territory may be conducted in accordance with a civil affairs agreement between the co-belligerent States.[32]In the absence of such an agreement, a military government may be established in the area as a provisional and interim measure.[33]The law of belligerent occupation may provide appropriate rules to apply by analogy, pending an agreement with the lawful government.[34]

11.1.3.3 *Occupation and Non-International Armed Conflict*. The law of belligerent occupation does not address non-international armed conflict as such because a belligerent occupation presupposes that the Occupying Power is hostile in relation to the State whose territory is being occupied.[35]A State's military forces controlling its own territory would not be regarded as conducting an occupation; similarly, foreign forces conducting operations with the consent of the territorial State would also not be regarded as conducting an occupation.[36]

However, the law of belligerent occupation may be applicable to a non-international armed conflict when a non-State party to the conflict has been recognized as a belligerent.[37]In addition, a non-international armed conflict could be regarded as taking place in the context of, or alongside, an occupation.[38]

11.1.3.4 *Occupation and Post-War Situations*. Before the GC was adopted and entered into force, the law of belligerent occupation was not applicable when a state of hostilities had completely ceased.[39]The general inapplicability of the law of belligerent occupation to post-war situations may be viewed as resulting from the law of belligerent occupation presupposing that a hostile relationship exists between the invading force's State and the State of the occupied territory.[40]For example, if the State of the occupied territory unconditionally surrendered and any international armed conflict completely ended, the law of belligerent occupation was not applicable.[41]

The GC, however, continues to apply in occupied territory until one year after the general close of military operations, and the Occupying Power is bound, for the duration of the occupation, to the extent that such State exercises the functions of government in such territory, by the provisions of certain articles of the GC.[42]Additionally, protected persons under the GC who remain in the custody of the Occupying Power following the end of occupation retain that protection until their release, repatriation, or re-establishment.[43]

In addition, it may be appropriate to adhere to the rules of belligerent occupation or to apply those rules by analogy even if not applicable as a matter of law in post-conflict situations.[44]The rules of belligerent occupation may reflect fundamental safeguards applicable to a wider range of situations.[45]

11.2 WHEN MILITARY OCCUPATION LAW APPLIES

The law of military occupation applies when a military occupation exists in fact. Even if the requirements of the law of belligerent occupation do not apply as a matter of law, general law of war principles and rules, such as those for the conduct of hostilities, continue to apply.[46]

11.2.1 Military Occupation as a Fact. Military occupation is a question of fact.[47]The legal consequences arising from the fact of occupation (*i.e.*, that this fact is the basis for both rights and duties) illustrates how the law of war may be viewed as both permissive and restrictive in nature.[48]

The fact of occupation is the basis for the Occupying Power to exercise authority over the occupied territory.[49]The fact of occupation, as a requirement for the exercise of authority over the occupied territory, prevents a State from simply claiming the authorities of military government over an enemy territory without actually controlling such territory.[50]

The fact of occupation also imposes certain duties on the Occupying Power with respect to occupied territory.[51]The fact of occupation, as a requirement for triggering the duties of an Occupying Power, means that a State that does not, in fact, occupy an area, does not incur the obligations of an Occupying Power. Once an occupation exists in fact, regardless of whether the invasion was lawful or unlawful under *jus ad bellum*, the rights and duties of the Occupying Power and the population in relation to each other apply.[52]This application of the law of belligerent occupation is an example of how *jus in bello* rules and *jus ad bellum* rules generally operate independently of one another.[53]

11.2.2 Standard for Determining When Territory Is Considered Occupied. Territory is considered occupied when it is actually placed under the authority of the hostile forces.[54]

This standard for when the law of belligerent occupation applies is reflected in Article 42 of the Hague IV Regulations and is regarded as customary international law.[55]

11.2.2.1 *"Actually Placed" – Effectiveness of Occupation.* Military occupation must be actual and effective; that is, the organized resistance must have been overcome, and the Occupying Power must have taken measures to establish its authority.[56]

It is sufficient that the occupying force can, within a reasonable time, send detachments of forces to enforce its authority within the occupied district. Military occupation does not require the presence of military forces in every populated area, although the occupying force must, *inter alia*, control the most important places.[57]

The type of forces used to maintain the authority of the Occupying Power is not material. For example, the occupation might be maintained by permanently based units or mobile forces, either of which would be able to send detachments of forces to enforce the authority of the Occupying Power within the occupied district.[58]However, air superiority alone would not constitute an effective occupation.

Similarly, as long as the occupation is effective, there is no precise number of forces that are considered necessary to constitute an effective occupation. The number of forces necessary to maintain effective occupation will depend on various considerations, such as the disposition of the inhabitants, the number and density of the population, the nature of the terrain, and similar factors.[59]

An occupation may be effective despite the existence of areas in the enemy State that are temporarily controlled by enemy forces or pockets of resistance. For example, the fact that a defended location (such as a city or town) still controlled by enemy forces exists within an area declared occupied by the Occupying Power does not render the occupation of the remainder of the area invalid, provided that continued resistance in such place or defended zone does not render the occupier unable to exercise control over the remainder of the occupied territory.[60]

Similarly, an occupation may continue to be effective despite intermittent insurgent attacks or temporary seizures of territory by resistance forces.[61]

11.2.2.2 *"Under the Authority" – Suspension and Substitution of Governmental Authority.* Occupation also requires the suspension of the territorial State's authority and the substitution of the Occupying Power's authority for the territorial State's authority.[62]

The territorial State must be rendered incapable of publicly exercising its authority in the territory, and the Occupying Power must substitute its authority for that of the territorial State.[63]

Invading forces in possession of the territory must have taken measures to establish their authority.[64]For example, such measures may include establishing its own governmental authority for that area and making regulations for the conduct of temporary government.[65] The suspension and substitution of authority may take place with local authorities continuing to administer territory subject to the paramount authority of the Occupying Power.[66] On the other hand, routine measures necessary to provide for unit security (*e.g.*, warning private persons not to threaten or interfere with military operations) would not necessarily constitute measures to establish authority over enemy territory. The substitution of authority by the Occupying Power may be shown by a proclamation of occupation, although such a proclamation is not required.[67]

11.2.2.3 *"Of the Hostile Army" – Belligerent Occupation Applies to Enemy Territory*. Occupation occurs when territory is actually placed under the authority of the hostile army. Thus, the existence of an occupation presupposes a hostile relationship between the invading force's State and the State of the occupied territory, although the occupation need not be met with armed resistance.[68]

For example, the law of belligerent occupation would not apply to the use of military forces to control a State's own territory, such as in cases of domestic emergency, insurrection, or non-international armed conflict.[69]

Similarly, the law of belligerent occupation would not apply to the liberation of friendly territory that was previously occupied by the enemy.[70]

The requirement of a hostile relationship between the invading force's State and the State of the occupied territory also prevents the law of belligerent occupation from applying to post conflict situations (except for certain provisions of the GC).[71]

11.2.3 Scope of Occupied Territory. The occupation extends only to the territory where such authority has been established and can be exercised.[72]

For example, a State may remain in control of part of its territory while the remainder of its territory is under occupation. The 1949 Geneva Conventions apply in cases of partial or total occupation of a State.[73]

The end of occupation in one part of occupied territory does not end the occupation in other parts of the occupied territory where the Occupying Power maintains its authority.

11.2.4 Proclamation of Occupation. Due to the special relations established between the civilian population of the occupied territory and the Occupying Power, the fact of military occupation and the territory over which it extends should be made known to the citizens of the occupied territory and to other States.

However, there is no specific legal requirement that the Occupying Power issue a proclamation of military occupation.[74]

The question of whether territory is occupied does not depend upon the issuance of any particular proclamation or other instrument. Territory may be occupied even though no proclamation of occupation has been issued. Conversely, the issuance of proclamation of occupation does not empower a State with the rights of an Occupying Power if it does not control such territory in fact. A proclamation may help fix the date of the beginning of the occupation.[75] The proclamation may also advise inhabitants of occupied territory of the rules with which they must comply.[76] In particular, the proclamation may be used to advise inhabitants of changes to law, including penal law.[77]

The general practice of the United States has been to make the fact of occupation known by proclamation or similar notice.[78] In addition to giving notice to the inhabitants of occupied territory, notice may be given to other governments.[79] In some cases, the U.N. Security Council has recognized the fact of occupation.[80]

11.3 END OF OCCUPATION AND DURATION OF GC OBLIGATIONS

The status of belligerent occupation ends when the conditions for its application are no longer met. Certain GC obligations with respect to occupied territory continue for the duration of the occupation after the general close of military operations.

11.3.1 End of Occupation. Belligerent occupation ceases when the conditions for its application are no longer met.[81] In particular, as discussed below, the status of belligerent occupation ceases when the invader no longer factually governs the occupied territory or when a hostile relationship no longer exists between the State of the occupied territory and the Occupying Power.[82]

Belligerent occupation ends when the Occupying Power no longer has effectively placed the occupied territory under its control.[83] For example, an uprising by the local population may prevent the Occupying Power from actually enforcing its authority over occupied territory. Similarly, the Occupying Power's expulsion or complete withdrawal from the territory would also suffice because the former Occupying Power generally would not be able to control sufficiently the occupied territory.

Belligerent occupation also may end when a hostile relationship no longer exists between the Occupying Power and the State of the occupied territory (although, as discussed in the following subsection, certain GC obligations may continue to apply).[84] For example, if a new, independent government of the previously occupied territory assumes control of the territory and consents to the presence of the previously occupying forces, then such a situation would no longer be considered a belligerent occupation. Similarly, if a peace treaty legitimately transfers the territory to the sovereignty of the Occupying Power, then the Occupying Power would no longer be characterized as such. However, an Occupying Power is not permitted, under the law of belligerent occupation, to annex occupied territory.[85]

11.3.2 Duration of GC Obligations in the Case of Occupied Territory. In the home territory of parties to the conflict, the application of the GC shall cease on the general close of military operations.[86]

In the case of occupied territory, the application of the GC shall cease one year after the general close of military operations; however, the Occupying Power shall be bound, for the duration of the occupation, to the extent that such State exercises the functions of government in such territory, by the provisions of the following Articles of the GC:

- 1 through 12 (general provisions and common articles, *e.g.*, the Protecting Power continues to function, and the derogation for security reasons continues to apply);
- 27, 29 through 34 (humane treatment);
- 47 (preserves rights as against change by annexation or arrangement with the local authorities so long as occupation lasts);
- 49 (transfers, evacuation, and deportation);
- 51, 52 (prohibitions against certain compulsory service and protection of workers);
- 53 (respect for property);
- 59, 61 through 63 (facilitating relief programs);
- 64 through 77 (criminal proceedings); and
- 143 (access by Protecting Powers and the ICRC).[87]

The one-year time limit for the cessation of the application of the GC (apart from the provisions that continue to apply to the extent that the Occupying Power exercises the functions of government in occupied territory) was proposed to account for situations like those of Germany and Japan after World War II.[88] AP I provides that the 1949 Geneva Conventions and AP I shall cease to apply, in the case of occupied territories, on the termination of the occupation;[89] coalition partners that are Occupying Powers and Parties to AP I would be bound by this rule.

In any case, individuals entitled to GC protection who remain in the custody of the Occupying Power following the end of occupation retain that protection until their release, repatriation, or re-establishment.[90] In addition, it may be appropriate following the end of occupation to continue to apply by analogy certain rules from the law of belligerent occupation, even if such rules do not apply as a matter of law.[91]

11.4 LEGAL POSITION OF THE OCCUPYING POWER

Military occupation of enemy territory involves a complicated, trilateral set of legal relations between the Occupying Power, the temporarily ousted sovereign authority, and the inhabitants of occupied territory.[92] The fact of occupation gives the Occupying Power the right to govern enemy territory temporarily, but does not transfer sovereignty over occupied territory to the Occupying Power.

11.4.1 Right of the Occupying Power to Govern the Enemy Territory Temporarily. The right to govern the territory of the enemy during its military occupation is one of the incidents of war.[93] By the fact of occupation (*i.e.*, the Occupying Power's established power over occupied territory), the Occupying Power is conferred the authority to exercise some of the rights of sovereignty.[94] The exercise of these sovereign rights also results from the necessity of maintaining law and order, indispensable both to the inhabitants and to the occupying force, and the failure or inability of the legitimate government to exercise its functions, or the undesirability of allowing it to do so.[95]

11.4.2 Limitations on the Power of the Occupying Power Stemming From Its Lack of Sovereignty Over Occupied Territory. Belligerent occupation in a foreign war, being based upon the possession of enemy territory, necessarily implies that the sovereignty of the occupied territory is not vested in the Occupying Power.[96]Occupation is essentially provisional.[97]

Because sovereignty is not vested in the Occupying Power, the fact of military occupation does not authorize the Occupying Power to take certain actions. For example, the Occupying Power is not authorized by the fact of belligerent occupation to annex occupied territory or to create a new State.[98]In addition, the Occupying Power may not compel the inhabitants of occupied territory to become its nationals or otherwise to swear allegiance to it.[99]

Similarly, in view of the provisional nature of belligerent occupation, the authority of the Occupying Power under occupation law has been interpreted as being subject to limitations on the ability of the Occupying Power to alter institutions of government permanently or change the constitution of a country.[100]

11.5 DUTY OF THE OCCUPYING POWER TO ENSURE PUBLIC ORDER AND SAFETY

The authority of the legitimate power having in fact passed into the hands of the Occupying Power, the latter shall take all the measures in its power to restore, and ensure, as far as possible, public order and safety, while respecting, unless absolutely prevented, the laws in force in the country.[101]This principle has been recognized as customary international law.[102]

The Occupying Power has a general duty to maintain public order and to provide for the preservation of rights of the inhabitants, including rights to their private property.[103]

11.5.1 Authority Conferred by the Occupying Power's Duty to Ensure Public Order and Safety. The Occupying Power's duty to take all the measures in its power to restore and ensure, as far as possible, public order and safety also provides it authority take such actions. For example, the Occupying Power may enact provisions to maintain the orderly government of the territory.[104]

11.5.2 Duty to Respect, Unless Absolutely Prevented, the Laws in Force in the Country. The duty to respect, unless absolutely prevented, the laws in force in the country prohibits the Occupying Power from arbitrarily exercising its authority to suspend, repeal, or change the municipal law applicable to occupied territory.[105]

The duty to respect, unless absolutely prevented, the laws in force in the country has been interpreted not to apply to local administrative laws, such as regulations, executive orders, ordinances, and decrees.[106]

11.6 PROTECTION OF THE POPULATION OF AN OCCUPIED TERRITORY

Under the law of belligerent occupation, the Occupying Power has certain duties with respect to the population of an occupied territory, including protected persons under the GC.

11.6.1 General Protections, Including Humane Treatment, of the Population of an Occupied Territory. The population of an occupied territory, like other protected persons under the GC, are entitled, in all circumstances, to respect for their persons, their honor, their family rights, their religious convictions and practices, and their manners and customs.[107]They shall at all times be humanely treated, and shall be protected especially against all acts of violence or threats of violence, and against insults and public curiosity.

Other provisions for the humane treatment of protected persons set forth in Articles 27 through 34 of the GC apply to the population of an occupied territory. For example, women must be especially protected against any attack on their honor, in particular against rape, enforced prostitution, or any form of indecent assault.[108]Reprisals against protected persons and their property are prohibited.[109]The taking of hostages is prohibited.[110]

In addition, protected persons in occupied territory shall have every facility for making application to the Protecting Powers, the ICRC, the National Red Cross (Red Crescent, Red Lion and Sun) Society of the country where they may be, as well as to any organization that might assist them.[111]

11.6.2 Overview of Additional Protections for the Population That Are Specific to Occupation. There are a number of protections for the population of occupied territory that are specific to occupation. For example, specific provision exists for the protection of children in occupied territory.[112]Specific constraints exist on the authority of the Occupying Power to punish protected persons, direct their movement, or compel them to perform labor.[113]Provision also is made with respect to: (1) food and medical supplies of the population; (2) public health and hygiene; (3) spiritual assistance; and (4) relief efforts and consignments.[114]

The following prohibitions also apply with respect to the inhabitants of occupied territory:

11.6.2.1 *Prohibition on Compelling Inhabitants of Occupied Territory to Swear Allegiance to the Hostile State*. It is forbidden to compel the inhabitants of occupied territory to swear allegiance to the hostile State.[115]

11.6.2.2 *Prohibition Against General Penalties in Occupied Territory*. No general penalty, pecuniary or otherwise, shall be inflicted upon the population on account of the acts of individuals for which they cannot be regarded as jointly and severally responsible.[116]Such penalties are prohibited, even if authorized under the law of the occupied territory. Collective penalties are prohibited as a general matter.[117]

11.6.3 Occupying Power's Duty to Respect the Rights of Protected Persons Secured by the GC. The Occupying Power has certain obligations to respect the rights of protected persons secured by the GC. Protected persons who are in occupied territory shall not be deprived, in any case or in any manner whatsoever, of the benefits of the GC by any change introduced, as the result of the occupation of a territory, into the institutions or government of the occupied territory, nor by any agreement concluded between the authorities of the occupied territories and the Occupying Power, nor by any annexation by the latter of the whole or part of the occupied territory.[118]For example, an Occupying Power may not purport to annex occupied territory in order to avoid its responsibilities as an Occupying Power.[119]

The Occupying Power may not enter into any special agreements that would adversely affect the situation of protected persons, as defined by the GC, nor restrict the rights that the GC confers upon them.[120]

The Occupying Power may also not evade its responsibilities through the purported renunciation by protected persons of the rights secured to them by the GC and by any special agreements referred to in Article 7 of the GC.[121]

In certain cases, a protected person's rights of communication under the GC may be forfeited for security reasons.[122]

11.6.4 Citizens of Neutral States in Occupied Territory. Citizens of neutral States residing within an occupied territory are generally treated the same as other residents of occupied territory.[123]

11.7 AUTHORITY OF THE OCCUPYING POWER OVER INHABITANTS

The Occupying Power's authority over inhabitants of occupied territory derives from its war powers and from its duty to ensure public order and safety in occupied territory. The Occupying Power, as a belligerent State, may take such measures of control and security in regard to protected persons as may be necessary as a result of the war.[124]In addition, the Occupying Power may take measures necessary to fulfill its duty to ensure public order and safety.[125]

11.7.1 Inhabitants' Obedience to the Occupying Power. It is the duty of the inhabitants to carry on their ordinary peaceful pursuits, to behave in an absolutely peaceful manner, to take no part whatever in the hostilities carried on, to refrain from all injurious acts toward the forces or in respect to their operations, and to render strict obedience to the orders of the occupant.[126]Subject to the restrictions imposed by international law, the Occupying Power may demand and enforce from the inhabitants of occupied territory such obedience as may be necessary for the security of its forces, for the maintenance of law and order, and for the proper administration of the country.[127]

The inhabitant's obedience to the Occupying Power is generally distinguished from a duty of allegiance.[128]The inhabitant's duty of allegiance to his or her State of nationality is not severed.[129]The inhabitants, however, are not bound to obey their State of nationality.[130]

11.7.1.1 *Enforcement of Obedience*. The Occupying Power's enforcement of obedience must comply with the law of war. For example, measures prohibited by the law of war, such as collective punishments, acts of torture, or reprisals, may not be used.

11.7.2 Censorship and Other Regulation of the Media. Under the law of belligerent occupation,[131]for the purposes of security, an Occupying Power may establish censorship or regulation of any or all forms of media (*e.g.*, press, radio, television) and entertainment (*e.g.*, theater, movies), of correspondence, and of other means of communication.[132]For example, an Occupying Power may prohibit entirely the publication of newspapers that pose a threat to security, or it may prescribe regulations for the publication or circulation of newspapers of other media for the purpose of fulfilling its obligations to restore public order.[133]

The Occupying Power is not required to furnish facilities for postal service, but may take charge of them itself, especially if the officials of the occupied district fail to act or to obey its orders.[134]

11.7.3 Control of Means of Transportation. An Occupying Power is entitled to exercise authority over all means of public and private transportation, whether land, waterborne, or air, within the occupied territory, and may seize them and regulate their operation.[135]

11.7.4 Limits on Safety Measures. If the Occupying Power considers it necessary, for imperative reasons of security, to take safety measures concerning protected persons, it may, at the most, subject them to assigned residence or to internment.[136]The Occupying Power must comply with certain procedural requirements (*e.g.*, appeal, periodic review) when conducting assigned residence or internment of protected persons in occupied territory based on imperative reasons of security.[137]The Occupying Power also has an obligation to support persons who are assigned residence and thus are required to leave their homes, consistent with the standards for the treatment of internees.[138]

11.8 ADMINISTRATION OF OCCUPIED TERRITORY

11.8.1 Paramount Authority of the Occupying Power Over Government Functions in Occupied Territory. The functions of the hostile government—whether of a general, provincial, or local character—continue only to the extent they are sanctioned by the Occupying Power.[139]

11.8.2 Continued Performance of Duties by Civil Servants and Other Officials of Local Governments. The Occupying Power may, while retaining its paramount authority, permit the government of the country to perform some or all of its normal functions.[140]It may, for example, call upon the local authorities to administer designated rear areas, subject to the guidance and direction of the Occupying Power.[141]Such action is consistent with the status of occupation, so long as there exists the firm possession of territory and the purpose to maintain paramount authority.[142]Similarly, for example, courts are generally to continue the ordinary administration of justice during occupation.[143]

The compulsion of civil servants and other officials of local governments to continue to perform their duties must be justified by military necessity and consistent with applicable provisions of the GC.[144]

11.8.3 Local Governments Under Duress or Surrogate Governments. The restrictions placed upon the authority of a belligerent State cannot be avoided by a system of using a puppet government, central or local, to carry out acts that would be unlawful if performed directly by the Occupying Power. Acts induced or compelled by the Occupying Power are nonetheless its acts.[145]

11.8.4 Responsibility of the Occupying Power for Treatment Accorded to Protected Persons by Its Agents. An Occupying Power, like belligerent States, is responsible for the treatment accorded to protected persons by its agents, irrespective of any individual responsibility that may be incurred.[146]The responsibility of the Occupying Power for its agents includes those agents that are not its nationals, such as locally recruited agents of the nationality of the State whose territory is occupied.[147]

11.8.5 Immunity of Occupation Personnel From Local Law. Military and civilian personnel of the occupying forces and occupation administration and persons accompanying them are not subject to the local law or to the jurisdiction of the local civil or criminal courts of the occupied territory, unless expressly made subject thereto by a competent officer of the Occupying Power.[148]

It is important for the Occupying Power to ensure that an appropriate system of substantive law applies to such persons and that tribunals are in existence to deal with civil litigation to which they are parties and with offenses committed by them.[149]In the past, provost courts have been used for these purposes.[150]In recent practice, the Uniform Code of Military Justice and the Military Extraterritorial Jurisdiction Act, among others, have been used to address offenses committed by military and civilian personnel in the context of occupation.[151]

11.8.6 Civilian or Military Nature of the Occupation Government. It is immaterial whether the government over an enemy's territory consists in a military or civil or mixed administration. Its character is the same and the source of its authority the same. It is a government imposed by force, and the legality of its acts is determined by the law of war.[152]

For example, the governing authority established by the Occupying Power may be composed of civilian personnel.[153]

11.9 LOCAL LAW AND LEGISLATION

11.9.1General Continuation of Municipal Law of the Occupied Territory as Between Inhabitants. In general, the municipal law of the occupied territory (*i.e.*, the ordinary domestic civil and criminal law) and their administration remain in full force so far as the inhabitants of occupied territory are concerned, unless changed by the Occupying Power.[154] For example, the penal laws of the occupied territory generally continue in force.[155]

As a foreign State and as the paramount authority in the occupied territory, the Occupying Power is not bound by the municipal law of the occupied territory.[156] For example, the Occupying Power would not be bound by a municipal law that had been enacted by the State of the occupied territory to prevent trading with the enemy.[157] Similarly, the Occupying Power's personnel are immune from the application of local law.[158]

11.9.2 Authority for the Occupying Power to Suspend, Repeal, or Change the Municipal Law Applicable to Occupied Territory. The duty of the Occupying Power to respect, unless absolutely prevented, the laws in force in the country prohibits it from arbitrarily exercising its authority to suspend, repeal, or change the municipal law applicable to occupied territory.[159] As with other authorities under the law of war, the Occupying Power must use its power with respect to the municipal law of occupied territory in good faith and not for the purpose of oppressing the population.[160]

The Occupying Power may subject the population of the occupied territory to provisions: (1) that are essential to enable the Occupying Power to fulfill its obligations under the GC; (2) to maintain the orderly government of the territory; and (3) to ensure the security of the Occupying Power, of the members and property of the occupying forces or administration, and likewise of the establishments and lines of communication used by them.[161]

Broader authority for the Occupying Power to change the laws of the occupied territory may be provided by the U.N. Security Council.[162]

After the Occupying Power's occupation has ended, its changes to the municipal law of the occupied territory may be changed, subject to the requirements of applicable international law.

11.9.2.1 *Distinction Between Suspension, Repeal, Change, or Enactment of Municipal Laws*. The distinction between the suspension, repeal, change, or enactment of a new law may be of little practical consequence.

Implicit in the authority to enact new laws is the authority to suspend or repeal existing laws. Thus, the repeal or suspension of a law may be accomplished by the Occupying Power subjecting the population of the occupied territory to additional provisions of law that are inconsistent with the earlier law. For example, a law permitting discrimination may be effectively suspended by the Occupying Power requiring officials to apply the law impartially.[163] The power to suspend or repeal laws may also include the power to amend laws. For example, the partial suspension of a tax imposed on the population may also be viewed as a change in the tax rate.[164]

However, new penal provisions may not come into force before they have been published.[165]

11.9.2.2 *Examples of Laws That May Be Suspended, Repealed, Changed, or Enacted by the Occupying Power*. Examples of provisions that may, depending on the circumstances, be suspended, repealed, changed, or enacted by the Occupying Power include legal provisions:

- relating to serious offenses that pose a threat to public order, such as kidnapping, rape, and forcible vehicle larceny;[166]
- relating to the sale or possession of weapons by members of the civilian population;[167]
- relating to measures for the protection of the Occupying Power's forces or administration (*e.g.*, laws relating to the recruitment of persons into the occupied State's armed forces or insurgent groups, relating to curfews or restricted areas);[168]
- relating to political process, such as laws regarding the rights of suffrage and of assembly;[169]
- relating to the supply of food and other items essential to the survival of the civilian population, and the prevention of illicit trade in, or hoarding of, such items;[170]
- relating to the administration of the law, such as repealing laws establishing racial discrimination or promulgating laws requiring the impartial application of the law by local officials;[171] and
- relating to the Occupying Power's obligations under the GC, such as legislation to help provide for child welfare, labor, food, hygiene, and public health of the occupied territory.[172]

11.9.3 Procedural Obligation – Notification to the Population of Changes in Law. The Occupying Power is not required to adhere to the local procedure for amending municipal law. However, the population of the occupied territory must be informed of the alteration, suspension, or repeal of existing laws, and of the enactment of new laws. In particular, penal provisions must be published before entering into force.[173]

In practice, to ensure that all inhabitants are on notice of what is expected, the Occupying Power is to publish a decree or order at, or immediately after, the beginning of the occupation, in order to clarify the applicable laws and orders, and the punishments that can be imposed if the inhabitants fail to comply. Such notification may be provided as part of a proclamation of occupation.[174]

11.10 ORDINARY COURTS IN OCCUPIED TERRITORY

In general, the courts and other tribunals of the occupied territory, like other governmental functions, should continue to operate.[175]For example, ordinary crimes that do not affect the security of the Occupying Power or its personnel should continue to be administered by the courts in the occupied territory.[176]

However, the administration of justice in the occupied territory, like the performance of other governmental functions, is subject to the direction of the Occupying Power.[177]

11.10.1 Authority to Suspend Ordinary Courts. The ordinary courts in occupied territory should be suspended only if: (1) judges and magistrates are unable or unwilling to perform their duties; (2) the courts are corrupt or unfairly constituted, or (3) the local judicial administration has collapsed due to the hostilities preceding the occupation.[178]

In such cases, the Occupying Power may use its own properly constituted, non-political military courts to ensure that offenses against the local laws are properly tried.[179]The Occupying Power may also reform local courts, including by establishing new criminal courts under the law of the occupied State.[180]

11.10.1.1 *Judicial Inability or Unwillingness to Perform Their Duties*. Judges and magistrates may abstain from fulfilling their functions for reasons of conscience.[181]An inability to perform their normal duties may result from a failure of public services, such as public transportation, or threats by insurgents.

11.10.1.2 *Corruption and Unfairness in the Judicial System*. Corruption and unfairness in the judicial system may constitute a serious obstacle to public order and safety, and an obstacle to the Occupying Power's application of the fundamental fair trial protections guaranteed by the GC.[182]The Occupying Power may take appropriate measures to remedy such corruption or unfairness. For example, committees may be established to review the suitability of judges and prosecutors.[183]Similarly, a facility may be established to help resolve conflicting property claims on a voluntary basis.[184]

11.10.2 Prohibition Against Extinguishing Rights and Actions of Enemy Nationals. It is especially forbidden to declare abolished, suspended, or inadmissible in a court of law the rights and actions of the nationals of a hostile party.[185]

This rule has been interpreted to apply solely to enemy areas occupied by a belligerent.[186]

It has been interpreted to prohibit a military commander from arbitrarily annulling the results of civil proceedings between private parties.[187]

11.11 CRIMINAL LAW IN OCCUPIED TERRITORY

This section addresses criminal law in occupied territory. Additional rules apply to judicial proceedings against protected persons in occupied territory, which are also applied by analogy to judicial proceedings against internees in a belligerent's home territory.[188]

11.11.1 General Continuation of Penal Laws of the Occupied Territory. The general principle that the Occupying Power must respect, unless absolutely prevented, the laws in force in the occupied territory also applies to the penal laws of the occupied territory.[189]

In particular, the penal laws of the occupied territory shall remain in force during the occupation, with the exception that they may be repealed or suspended by the Occupying Power where they constitute a threat to its security or an obstacle to the application of the GC.[190]

Subject to the latter consideration and to the necessity of ensuring the effective administration of justice, the courts of the occupied territory shall continue to function in respect of all offenses covered by these laws.[191]

11.11.1.1 *Types of Laws That Are Considered Penal Laws*. Penal laws addressed by this rule may be understood to include all laws, regulations, decrees, orders, and similar measures intended to repress offenses, including the rules of criminal procedure.[192]

11.11.1.2 *Laws Constituting an Obstacle to the Application of the GC*. The Occupying Power may repeal or suspend penal laws constituting an obstacle to the application of the GC. For example, the Occupying Power may abrogate discriminatory measures that are incompatible with humane requirements.[193]

11.11.1.3 *Method of Repeal or Suspension of Penal Laws*. The repeal or suspension of the penal laws of the occupied territory may be achieved directly through an order suspending or repealing the law.[194] The suspension of a penal law may also be done by requiring that a penal law may only be enforced with the approval of the Occupying Power.[195] The repeal or suspension of a penal law may also be done by the Occupying Power's subjecting the population of the occupied territory to additional provisions of law that are inconsistent with application of that penal law.[196]

11.11.1.4 *Continued Administration of Penal Law by Courts in the Occupied Territory*. In general, ordinary crimes that do not affect the security of the Occupying Power or its personnel would be left to the jurisdiction of the courts in the occupied territory.[197]

The continued administration of local and national penal law by the courts in occupied territory is subject to certain modifications by the Occupying Power, such as modifications necessary for ensuring the effective administration of justice.[198]

The Occupying Power may reform or establish new criminal courts in occupied territory as part of its authority to subject the population of the occupied territory to additional provisions of law and its duty to restore and maintain public order.[199] In addition, the Occupying Power may use its properly constituted, non-political military courts in occupied territory for certain types of offenses.[200]

11.11.2 Penal Provisions Enacted by the Occupying Power. As with other types of legal provisions, the Occupying Power may subject the population of the occupied territory to penal provisions: (1) that are essential to enable the Occupying Power to fulfill its obligations under the GC; (2) to maintain the orderly government of the territory; and (3) to ensure the security of the Occupying Power, of the members and property of the occupying forces or administration, and likewise of the establishments and lines of communication used by them.[201]

11.11.2.1 *Publication of Penal Provisions Before Coming Into Force*. The penal provisions enacted by the Occupying Power shall not come into force before they have been published and brought to the knowledge of the inhabitants in their own language.[202] When more than one language is used by the local population, the Occupying Power may adhere to the local practice regarding official publication of legislation.[203]

The effect of these penal provisions shall not be retroactive.[204]

11.11.3 Properly Constituted, Non-Political Military Courts in Occupied Territory. In case of a breach of the penal provisions promulgated by the Occupying Power by virtue of the second paragraph of Article 64 of the GC, the Occupying Power may hand over the accused to its properly constituted, non-political military courts, on condition that these courts sit in the occupied country. Courts of appeals shall preferably sit in the occupied country.[205]

These courts must be regularly constituted courts affording all the judicial guarantees that are recognized as indispensable by civilized peoples.[206] These courts must also comply with requirements for judicial proceedings regarding protected persons in occupied territory.[207]

11.11.3.1 *U.S. Practice for Properly Constituted, Non-Political Military Courts*. The United States has a long history of using provost courts in occupied territory.[208] In U.S. practice, provost courts have been used as properly constituted, non-political military courts. Service regulations have addressed the use of provost courts.[209]

11.11.3.2 *Applicable Law in Properly Constituted, Non-Political Military Courts*. Properly constituted, non-political military courts shall apply only those provisions of law that were applicable prior to the offense, and that are in accordance with general principles of law, in particular the principle that the penalty shall be proportionate to the offense.[210] The requirement that properly constituted, non-political military courts shall apply only those provisions of law that were applicable prior to the offense reflects the general rule that penal laws cannot be retroactive and that there cannot be an offense or penalty unless the act in question was punishable under the law in force at the time the act was committed.[211] The requirement that properly constituted, non-political military courts shall apply only those provisions of law that are in accordance with general principles of law includes the rule that punishments must be based on individual responsibility.[212]

These courts shall also take into consideration the fact that the accused is not a national of the Occupying Power.[213]

These courts also may, at their discretion, convert a sentence of imprisonment to one of internment for the same period.[214]

11.11.4 Limitations on Penalties for Certain, Non-Serious Offenses Solely Intended to Harm the Occupying Power. Protected persons who commit an offense that is solely intended to harm the Occupying Power, but that is not an attempt on the life or limb of members of the occupying forces or administration, nor a grave collective danger, nor seriously damage the property of the occupying forces or administration, or the installations used by them, shall be liable to internment or simple imprisonment, provided the duration of such internment or imprisonment is proportionate to the offense committed.[215]

The minor offense must have been solely intended to harm the Occupying Power in order to trigger these restrictions.[216]For example, offenses such as traveling without a permit or violating exchange control regulations would not fall under this restriction, but, nonetheless, may result in internment or simple imprisonment.

11.11.4.1 *Internment or Imprisonment as the Only Measure Adopted for Depriving Protected Persons of Liberty for Such Offenses*. Furthermore, internment or imprisonment shall, for such offenses, be the only measure adopted for depriving protected persons of liberty.[217]

Penalties that do not deprive the protected person of liberty, such as fines, or less severe measures, such as arrest, are not addressed by this rule.[218]

11.11.4.2 *Discretion of Non-Political, Military Courts to Convert Sentences of Imprisonment to Sentences of Internment*. Properly constituted non-political military courts may, at their discretion, convert a sentence of imprisonment to one of internment for the same period.[219]

11.11.5 Imposing the Death Penalty for Offenses Committed by Protected Persons in Occupied Territory. The GC provides that the penal provisions promulgated by the Occupying Power in accordance with Articles 64 and 65 of the GC may impose the death penalty against a protected person only in cases where the person is guilty of espionage, or serious acts of sabotage against the military installations of the Occupying Power, or of intentional offenses that have caused the death of one or more persons, provided that such offenses were punishable by death under the law of the occupied territory in force before the occupation began.[220]

However, the United States has reserved the right to impose the death penalty in accordance with the provisions of Article 68, paragraph 2, of the GC without regard to whether the offenses referred to in that paragraph are punishable by death under the law of the occupied territory at the time the occupation begins.[221]The United States expressed the position that an Occupying Power would be unable to protect its own forces adequately against the activities of illegal combatants unless it retained the authority to apply the death penalty, and the rule in Article 68 of the GC would be subject to manipulation in that an adversary on the verge of being occupied might repeal the death penalty to encourage subversive activities against an Occupying Power.[222]

The death penalty may not be pronounced against a protected person unless the attention of the court has been particularly called to the fact that since the accused is not a national of the Occupying Power, he or she is not bound to it by any duty of allegiance.[223]

In any case, the death penalty may not be pronounced against a protected person who was under eighteen years of age at the time of the offense.[224]

In practice, if an occupation is undertaken with coalition partners, the death penalty may be suspended in order to address requests of these States.[225]

11.11.6 Deduction From Sentences of Period Under Arrest. In all cases, the duration of the period during which a protected person accused of an offense is under arrest awaiting trial or punishment shall be deducted from any period of imprisonment awarded.[226]

11.11.7 Limitation on Criminal Jurisdiction With Respect to Pre-Occupation Acts. The GC provides for certain limitations on the criminal jurisdiction of the Occupying Power with respect to the pre-occupation acts of protected persons and of nationals of the Occupying Power, who, before the outbreak of hostilities, have sought refuge in the territory of the occupied State.

11.11.7.1 *Pre-Occupation Acts of Protected Persons*. Protected persons shall not be arrested, prosecuted, or convicted by the Occupying Power for acts committed or opinions expressed before the occupation, or during a temporary interruption thereof, with the exception of breaches of the laws and customs of war.[227]

11.11.7.2 *Protection of Nationals of the Occupying Power Who, Before the Outbreak of Hostilities, Have Sought Refuge in the Territory of the Occupied State*. Nationals of the Occupying Power who, before the outbreak of hostilities, have sought refuge in the territory of the occupied State, shall not be arrested, prosecuted, convicted, or deported from the occupied territory, except for offenses committed after the outbreak of hostilities, or for offenses under common law committed before the outbreak of hostilities that, according to the law of the occupied State, would have justified extradition in time of peace.[228] Thus, although protected persons may not be forcibly transferred or deported to another country,[229] nationals of the Occupying Power may be involuntarily removed under certain circumstances.[230]

This rule is an exception to the general principle that a State's own nationals are excluded from the protection of the GC.[231]

11.11.8 Disposition of Accused and Convicted Protected Persons Upon the Close of Occupation. Protected persons who have been accused of offenses or convicted by courts in occupied territory shall be handed over at the close of occupation, with the relevant records, to the authorities of the liberated territory.[232]

Pending their transfer to such authorities, such protected persons continue to be protected by the GC because protected persons whose release, repatriation, or re-establishment may take place after such dates continue to benefit from the protections of the GC.[233]

The practical arrangements for the handing over of such protected persons will depend on the circumstances, including whether the liberation of occupied territory is accompanied by fighting and whether the local administration is able to function.[234]

If occupying forces will remain in occupied territory after the end of occupation, it is permissible, with the consent of the formerly occupied State, for the occupying forces to retain custody, in the formerly occupied territory, of protected persons accused or convicted of offenses.

11.12 MOVEMENT OF PERSONS IN OCCUPIED TERRITORY

11.12.1 Authority of the Occupying Power to Restrict Freedom of Movement. As a general matter, the Occupying Power assumes the authority of the State whose territory is occupied in controlling the movement of persons within the occupied territory, as well as the movement of persons entering and exiting occupied territory. For example, private persons, members of non-governmental organizations, or representatives of foreign States or international organizations seeking to enter occupied territory generally would not be able to do so without authorization from the Occupying Power.

The Occupying Power may withdraw from individuals the right to change their residence, restrict freedom of internal movement, forbid visits to certain districts, prohibit emigration and immigration by protected persons who are nationals of the State whose territory is occupied, and require that all individuals carry identification documents.[235]

11.12.2 Departure of Protected Persons Who Are Not Nationals of the State Whose Territory Is Occupied. Protected persons who are not nationals of the State whose territory is occupied may avail themselves of the right to leave occupied territory subject to the provisions of Article 35 of the GC, and decisions thereon shall be taken according to the procedure that the Occupying Power shall establish in accordance with the Article 35 of the GC.[236]

Article 35 of the GC sets forth rules regarding the departure of protected persons from the home territory of a belligerent State and provides protected persons with a right to depart. But, Article 35 allows a belligerent to prevent such departure if such departure is contrary to the belligerent's national interests, and Article 35 specifies certain procedural requirements.[237]

Thus, for example, an Occupying Power may prevent the departure of protected persons who are not nationals of the State whose territory is occupied if such departure is contrary to the Occupying Power's national interests. The Occupying Power must comply with the requirements of Article 35 of the GC if it does so (*e.g.*, providing for the reconsideration of its refusal of permission to depart by an appropriate court or administrative board).[238]

11.12.3 Prohibition Against Forcible Transfers and Deportations. Individual or mass forcible transfers, as well as deportations of protected persons from occupied territory to the territory of the Occupying Power, or of any other country, occupied or not, are prohibited, regardless of their motive.[239]

The unlawful deportation or transfer of protected persons in violation of this rule constitutes a grave breach of the GC.[240]

11.12.3.1 *Exception for Certain Transfers and Evacuations*. Nevertheless, the Occupying Power may undertake total or partial evacuation of a given area if required for the security of the population or for imperative military reasons. Such evacuations may not involve the displacement of protected persons outside the bounds of the occupied territory except when, for material reasons, it is impossible to avoid such displacement. Persons thus evacuated shall be transferred back to their homes as soon as hostilities in the area in question have ceased.[241] For example, in some cases, the whole of occupied territory, such as an island, may be dangerous, and it may be safer for the protected person to be transferred from occupied territory.[242]

The Occupying Power undertaking such transfers or evacuations shall ensure, to the greatest practicable extent: (1) that proper accommodation is provided to receive the protected persons; (2) that evacuations or transfers are effected with satisfactory conditions of hygiene, health, safety, and nutrition; and (3) that members of the same family are not separated.[243]

11.12.3.2 *Transfer Prohibition Only Applicable to Protected Persons*. Persons in occupied territory who do not receive protected person status under the GC are not subject to this prohibition on forcible transfers.[244] For example, POWs may be transferred from occupied territory to POW camps in the home territory of a belligerent.[245] Similarly, a person who is not entitled to protected person status, such as a national of a neutral or non-belligerent State who travels to an occupied State to fight the Occupying Power, would not be covered by this prohibition.[246]

11.12.3.3 *Notification of Transfers and Evacuations to the Protecting Power*. The Protecting Power shall be informed of any transfers and evacuations as soon as they have taken place.[247]

11.12.3.4 *Voluntary Transfers Not Prohibited*. Only forcible transfers are prohibited; voluntary transfers from occupied territory, such as voluntary transfers of persons belonging to a minority group that had been persecuted by the occupied State, would not be prohibited.[248]

11.12.4 General Prohibition Against Detaining Protected Persons in Dangerous Areas. The Occupying Power shall not detain protected persons in an area particularly exposed to the dangers of war unless the security of the population or imperative military reasons so demand.[249]

11.12.5 Prohibition Against Deporting or Transferring the Occupied State's Civilian Population Into Occupied Territory. The Occupying Power shall not deport or transfer parts of its own civilian population into the territory it occupies.[250]

11.13 PROTECTION OF CHILDREN IN OCCUPIED TERRITORY

Article 50 of the GC provides for certain duties of an Occupying Power with respect to children in occupied territory. "Children" for the purposes of Article 50 of the GC and its obligations with respect to the protection of children in occupied territory may be understood generally to refer to children under fifteen years of age.[251]

Other duties that are not specific to occupied territory exist with regard to children.[252]

11.13.1 Obligation With Respect to the Care and Education of Children. The Occupying Power shall, with the cooperation of the national and local authorities, facilitate the proper working of all institutions devoted to the care and education of children.[253] This obligation goes beyond merely not interfering with such institutions, but also includes the affirmative duty to support them when the responsible authorities of the country fail to do so.[254]

Should the local institutions be inadequate for the purpose, the Occupying Power shall make arrangements for the maintenance and education – if possible, by persons of their own nationality, language, and religion – of children who are orphaned or separated from their parents as a result of the war, and who cannot be adequately cared for by a near relative or friend.[255]

11.13.2 Obligation With Respect to the Identification of Children. The Occupying Power shall take all necessary steps to facilitate the identification of children and the registration of their parentage.[256]

A special section of the National Protected Person Information Bureau set up in accordance with Article 136 of the GC shall be responsible for taking all necessary steps to identify children whose identity is in doubt.[257]

11.13.3 Prohibition Against Changing the Personal Status of Children or Enlisting Them. The Occupying Power may not, in any case, change the personal status of children, nor enlist them in formations or organizations subordinate to it.[258]

11.13.4 Obligation Not to Hinder Preferential Measures for Children and Mothers. The Occupying Power shall not hinder the application of any preferential measures in regard to food, medical care, and protection against the effects of war that may have been adopted prior to the occupation in favor of children under fifteen years, expectant mothers, and mothers of children under seven years.[259]

11.14 FOOD AND MEDICAL SUPPLIES FOR THE CIVILIAN POPULATION

11.14.1 Duty of Ensuring the Food and Medical Supplies for the Population. To the fullest extent of the means available to it, the Occupying Power has the duty of ensuring the sufficiency of food and medical supplies for the population; it should, in particular, bring in the foodstuffs, medical stores, and other articles if the resources of the occupied territory are inadequate.[260] Other articles may be understood to include all urgently required goods that may be essential to the "life of the territory."[261]

11.14.2 Limitation on Requisition of Food and Medical Supplies. The Occupying Power may not requisition foodstuffs, articles, or medical supplies available in the occupied territory, except for use by the occupation forces and administration personnel, and then only if the requirements of the civilian population have been taken into account.[262] Subject to the provisions of other international conventions, the Occupying Power shall make arrangements to ensure that fair value is paid for any requisitioned goods.[263]

11.14.3 Verification of the State of Food and Medical Supplies by the Protecting Power. The Protecting Power shall, at any time, be at liberty to verify the state of the food and medical supplies in occupied territories, except where temporary restrictions are made necessary by imperative military requirements.[264]

11.15 PUBLIC HEALTH AND HYGIENE

11.15.1 General Duty With Respect to Public Health. To the fullest extent of the means available to it, the Occupying Power has the duty of ensuring and maintaining, with the cooperation of national and local authorities, medical and hospital establishments, medical services, and public health and hygiene in the occupied territory, with particular reference to the adoption and application of the prophylactic and preventive measures necessary to combat the spread of contagious diseases and epidemics.[265] The responsibility for providing and maintaining health services falls primarily on the national and local authorities, but where such authorities are unable to provide adequately for the health needs of the civilian population, the Occupying Power then has the duty described above.[266]

11.15.2 Respect for the Work of Civilian Medical Personnel. Medical personnel of all categories must be allowed to carry out their duties.[267] "Medical personnel of all categories" includes a wide range of persons engaged in medical work.[268] It includes even those who are not affiliated with civilian hospitals who qualify for protection under Article 20 of the GC.[269]

11.15.3 Recognition of New Hospitals, Medical Personnel, and Transports. If new hospitals are set up in occupied territory and if the competent organs of the occupied State are not operating there, the occupying authorities must, if necessary, grant them the recognition provided for in Article 18 of the GC.[270] This recognition allows the civilian hospitals to show that they are civilian hospitals and that the buildings that they occupy are not used for any purpose that would deprive these hospitals of protection.[271] In similar circumstances, the occupying authorities must also grant recognition to hospital personnel and transport vehicles under the provisions of Articles 20 and 21 of the GC.[272] This recognition also allows such personnel and vehicles to display their entitlement to protection.[273]

11.15.4 Consideration of Cultural Traditions of the Population i n Adopting Public Health Measures. In adopting measures of health and hygiene, and in their implementation, the Occupying Power must take into consideration the moral and ethical susceptibilities of the population of the occupied territory.[274] The Occupying Power should respect the sentiments and traditions of the population, and should not introduce new methods if they are liable to cause deep disquiet among the population.[275]

11.15.5 Requisition of Civilian Hospitals. The Occupying Power may requisition civilian hospitals only temporarily and only in cases of urgent necessity for the care of military wounded and sick, and then on the condition that suitable arrangements are made in due time for the care and treatment of the hospital's current patients, and for the future needs of the civilian population for hospital accommodation. The material and stores of civilian hospitals cannot be requisitioned so long as they are necessary for the needs of the civilian population.[276]

If the requisition of medical supplies is required for the occupying force, they must take the needs of the civilian population into account and resupply the materials as soon as possible.[277]

11.16 SPIRITUAL ASSISTANCE

The Occupying Power shall permit ministers of religion to give spiritual assistance to the members of their religious communities. The Occupying Power shall also accept consignments of books and articles required for religious needs and shall facilitate their distribution in occupied territory.[278]

Religious assistance addressed by the GC, however, does not include activities that are a pretext for political agitation against the Occupying Power.[279]

11.17 RELIEF EFFORTS AND CONSIGNMENTS

Special rules address relief efforts in occupied territory. Other rules address relief efforts outside the context of occupation.[280]

11.17.1 Collective Relief. If the whole or part of the population of an occupied territory is inadequately supplied, the Occupying Power shall agree to relief schemes on behalf of the affected population, and shall facilitate them by all the means at its disposal. Such schemes, which may be undertaken either by States or by impartial humanitarian organizations such as the ICRC, shall consist, in particular, of the provision of consignments of foodstuffs, medical supplies, and clothing.[281]

All Parties to the GC shall permit the free passage of these consignments and shall guarantee their protection.[282]

A State granting free passage to consignments on their way to territory occupied by an adverse party to the conflict shall, however, have the right to search the consignments, to regulate their passage according to prescribed times and routes, and to be reasonably satisfied (through the Protecting Power) that these consignments are to be used for the relief of the needy population, and are not to be used for the benefit of the Occupying Power.[283]

11.17.1.1 *Diversion of Relief Consignments*. The Occupying Power shall in no way whatsoever divert relief consignments from the purpose for which they are intended, except in cases of urgent necessity, in the interests of the population of the occupied territory, and with the consent of the Protecting Power.[284]

11.17.1.2 *Distribution of Collective Relief Consignments*. The distribution of collective relief consignments referred to in Articles 59 and 60 of the GC shall be carried out with the cooperation and under the supervision of the Protecting Power. This duty may also be delegated, by agreement between the Occupying Power and the Protecting Power, to a neutral State, the ICRC, or any other impartial humanitarian body.[285]

The Occupying Power shall facilitate the rapid distribution of these consignments.[286]For example, it may be necessary for the occupation authorities to remove administrative impediments (*e.g.*, granting permits) to relief organizations distributing consignments.[287]

11.17.1.3 *Charges, Taxes, and Customs Duties on Collective Relief Consignments for Occupied Territories*. Collective relief consignments referred to in Articles 59 and 60 of the GC shall be exempt in occupied territory from all charges, taxes, or customs duties unless such charges, taxes, or customs duties are necessary in the interests of the economy of the territory.[288]All Parties to the GC shall endeavor to permit the transit and transport, free of charge, of such relief consignments on their way to occupied territories.[289]

11.17.2 Individual Relief Consignments. Subject to imperative reasons of security, protected persons in occupied territories shall be permitted to receive individual relief consignments sent to them.[290]

An individual relief consignment is a shipment that is addressed to a particular person.[291]

The occupation authorities may limit or temporarily forbid the entry of individual relief supplies for imperative reasons of security, such as if the volume of relief consignments prevents security screening.[292]

11.17.3 Continuing Responsibilities of the Occupying Power. Relief consignments shall in no way relieve the Occupying Power of any of its responsibilities under Articles 55, 56, and 59 of the GC.[293]The responsibilities of the Occupying Power under Articles 55, 56, and 59 of the GC relate to ensuring the provision of food, medical supplies, and medical services to the population.[294]

11.17.4 National Red Cross or Red Crescent Societies and Other Relief Organizations. Subject to temporary and exceptional measures imposed for urgent reasons of security by the Occupying Power:

• recognized National Red Cross (Red Crescent, Red Lion and Sun) Societies shall be able to pursue their activities in accordance with Red Cross principles, as prescribed by the International Red Cross Conferences. Other relief societies shall be permitted to continue their humanitarian activities under similar conditions;

• the Occupying Power may not require any changes in the personnel or structure of these societies that would prejudice these activities.[295]

The same principles shall apply to the activities and personnel of special organizations of a non-military character that already exist or that may be established for the purpose of ensuring the living conditions of the civilian population by the maintenance of the essential public utility services, by the distribution of relief, and by the organization of rescues.[296]

11.18 ENEMY PROPERTY DURING OCCUPATION

Pillage is prohibited. Enemy property in occupied territory may not be seized or destroyed unless imperatively demanded by the necessities of war. The Occupying Power may take certain control measures with respect to property in occupied territory. Other rules apply to the treatment of public enemy property and private enemy property in occupied territory.

11.18.1Prohibition Against Pillage in Occupied Territory. Pillage is forbidden.[297]In addition to this specific prohibition in the context of occupied territory, pillage is prohibited as a general matter.[298]

11.18.2 Seizure or Destruction of Property During Occupation – Application of the Military Necessity Standard. The general rule that enemy property may not be seized or destroyed unless imperatively demanded by the necessities of war also applies to property during the occupation of enemy territory.[299]In particular, any destruction by the Occupying Power of real (immovable) or personal (movable) property belonging individually or collectively to private persons, to the State of the occupied territory, to other public authorities, or to social or cooperative organizations, is prohibited, except where such destruction is rendered absolutely necessary by military operations.[300]

11.18.2.1 *Assessment of Military Necessity for Seizure or Destruction.* The evaluation of whether destruction of property is militarily necessary in the context of occupation is undertaken by applying the standards that also are applied in the context of combat operations.[301]For example, the evaluation of whether the destruction of property is militarily necessary is made by the responsible commander or other authority of the Occupying Power.[302]

However, such assessment must be made in good faith based on the information available at the time.[303]

11.18.2.2 *Temporary Possession for Direct Military Use.* An Occupying Power may always take temporary possession of enemy property (real or personal, and public or private) where required for direct military use in military operations. In the case of private property, an Occupying Power, where possible, should requisition the property and offer compensation to the owner.[304]

11.18.2.3 *Capture or Seizure and Vesting of Title in the Occupying Power.* In the case of real (immovable) property that is captured or seized, the Occupying Power may use such property for the duration of the occupation but does not take title to the property.[305]Public property captured or seized from the enemy, as well as private property validly captured on the battlefield and abandoned property, is the property of the capturing State.[306]Valid capture or seizure of property requires both an intent to take such action and a physical act of capture or seizure. The mere presence within occupied territory of property that is subject to appropriation under international law does not operate to vest title thereto in the Occupying Power.[307]

11.18.2.4 *Seizure or Destruction of Submarine Cables Connecting an Occupied Territory With a Neutral Territory.* Submarine cables connecting an occupied territory with a neutral territory shall not be seized or destroyed except in the case of absolute necessity. Such cables must likewise be restored and compensation fixed when peace is made.[308]This rule applies only to activities on land and does not deal with seizure or destruction of cables in the open sea.[309]

11.18.3 Property Control Measures. Public and private property within occupied territory may be controlled by the Occupying Power to the degree necessary to prevent its use by or for the benefit of hostile forces, or in a manner harmful to the Occupying Power.[310]For example, in the case of persons who have acted against the security of the Occupying Power, the Occupying Power may freeze their accounts to prevent such funds from being used against the interests of the Occupying Power.[311]

Conservators may be appointed to manage the property of absent persons (including nationals of the United States and of friendly States) and of internees, property managed by such persons, and property of persons whose activities are deemed to be prejudicial to the Occupying Power. However, when the owners or managers of such property are again able to resume control of their property and the risk of its hostile use no longer exists, it must be returned to them.[312]

Measures of property control must not extend to confiscation of private property.[313] For example, if the Occupying Power controls a private business and earns a profit, the owner must be indemnified.[314]

However, the authority of the Occupying Power to impose such controls does not limit its power to seize or requisition property or take such other action with respect to it as may be authorized by other provisions of law.[315]

11.18.4 Determination Whether Property Is Public or Private. The rules for the treatment of enemy property may depend on whether the property is public or private.

For example, private property may not be confiscated, and there is an obligation to pay compensation for requisitions of private property.[316] On the other hand, public property generally may be seized or captured without any obligation to pay compensation.[317]

11.18.4.1 *Beneficial Ownership Test*. Under modern conditions, the distinction between public and private property is not always easy to draw. For the purpose of treatment of property under military occupation, it is often necessary to look beyond strict legal title and to ascertain the character of the property on the basis of the beneficial ownership thereof. Thus, for example, trust funds, pension funds, and bank deposits generated by private persons are not to be regarded as public property simply by reason of their being held by a State-owned bank.[318]

11.18.4.2 *Property of Mixed Ownership*. For the purpose of determining what type of control the Occupying Power may exercise over property (by way of confiscation, seizure, requisition, etc.), the most cogent evidence of public character is such a complete or partial assumption by the State of the economic risk involved in the holding and management of the property in question that the enemy State, rather than private individuals or corporations, would be subjected to a substantial portion of the loss were the property to be appropriated for the use of the Occupying Power.[319] Should property that is ostensibly private be subjected to a large measure of governmental control and management, or perform functions that are essentially public, these facts would tend to indicate that the property should be regarded in practice as public.[320]

If property that is appropriated by the Occupying Power is beneficially owned in part by the enemy State and in part by private interests, the occupation authorities should compensate the private owners to the extent of their interest. Such compensation should bear the same relationship to the compensation that would be paid if the property were entirely privately owned. The Occupying Power may take those measures it deems necessary to ensure that no portion of the compensation paid on account of private interests accrues to the enemy State.[321]

11.18.4.3 *Property of Unknown Ownership*. If it is unknown whether certain property is public or private, it should be treated as public property until its ownership is ascertained.[322]

11.18.5 Treatment of Enemy Public Property. In general, an Occupying Power may capture or seize the real (immovable) and personal (movable) property of the occupied State and use it for military operations or the administration of the occupied territory. No compensation needs to be paid to the occupied State for the use or taking of such property.

11.18.5.1 *Public Real (Immovable) Property Susceptible of Direct Military Use*. Real property of the occupied State that is of direct military use, such as forts, arsenals, dockyards, magazines, barracks, railways, bridges, piers, wharves, airfields, and other military facilities, may remain in the hands of the Occupying Power until the close of the hostilities, and may be destroyed or damaged by the Occupying Power, if it deems it necessary to military operations.[323]

11.18.5.2 *Public Real (Immovable) Property That Is Essentially of a Non-Military Nature*. The Occupying Power shall be regarded only as administrator and usufructuary of public buildings, real estate, forests, and agricultural estates belonging to the hostile State, and situated in the occupied territory. It must safeguard the capital of these properties, and administer them in accordance with the rules of usufruct.[324] This real (immovable) property that is essentially of a non-military nature may not be damaged or destroyed unless such destruction is rendered absolutely necessary by military operations.[325]

The term usufruct means literally "to use the fruit."[326] The Occupying Power may use and enjoy the benefits of public real (immovable) property belonging to an enemy State, but does not have the right of sale or unqualified use of such property. As administrator or usufructuary, the Occupying Power should not exercise its rights in such a wasteful and negligent manner as seriously to impair the property's value.[327]

The Occupying Power may, however, lease or utilize public lands or buildings, sell the crops, cut and sell timber, and work the mines. The term of a lease or contract should not extend beyond the conclusion of the war.[328]

11.18.5.3 *Public Personal (Movable) Property*. An army of occupation may only take possession of cash, funds, and realizable securities that are strictly the property of an enemy State, depots of arms, means of transport, stores and supplies, and, generally, all personal (movable) property belonging to the State that may be used for military operations.[329]

Thus, all personal (movable) property belonging to an enemy State susceptible of military use may be taken possession of, and utilized for, the benefit of the Occupying Power. Under modern conditions of warfare, a large proportion of State property may be regarded as capable of being used for military purposes. However, personal (movable) property that is not susceptible of military use must be respected and may not be appropriated.[330]

11.18.6 Treatment of Enemy Private Property. Private property may not be confiscated.[331]

Private personal (movable) property susceptible to direct military use may be seized, with a receipt to be given to allow for compensation to be paid at the conclusion of the war.[332]

Other private property may be subject to requisition, with payment in cash given as soon as possible.[333]

Private property may be forfeited to the Occupying Power as a result of the valid imposition of penalties.[334]For example, a regulation closing a road for security reasons could specify that, as a penalty, unauthorized private vehicles found on the road would be subject to forfeiture without any compensation.[335]

11.18.6.1 *Prohibition on Confiscation of Private Property in Occupied Territory*. Private property may not be confiscated.[336]

The prohibition against confiscation of private property extends not only to outright taking in violation of the law of war, but also to any acts that, through the use of threats, intimidation, or pressure, or by actual exploitation of the power of the Occupying Power, permanently or temporarily deprive the owner of the use of such property without the owner's consent, or without authority under international law.[337]

The prohibition against confiscation of private property does not extend to takings by way of contribution, requisition, or the valid imposition of penalties.[338]

11.18.6.2 *Seizure of Private Property Susceptible to Direct Military Use*. All appliances, whether on land, at sea, or in the air, adapted for the transmission of news, or for the transport of persons or things, exclusive of cases governed by naval law, depots of arms, and, generally, all kinds of ammunition of war, may be seized, even if they belong to private individuals, but must be restored and compensation fixed when peace is made.[339]

Private property susceptible of direct military use includes cables, telephone and telegraph facilities, radio, television, telecommunications and computer networks and equipment, motor vehicles, railways, railway plants, port facilities, ships in port, barges and other watercraft, airfields, aircraft, depots of arms (whether military or sporting), documents connected with the conflict, all varieties of military equipment (including that in the hands of manufacturers), component parts of, or material suitable only for use in, the foregoing, and, in general, all kinds of war material.[340]

If private property is seized on the grounds that is it susceptible to direct military use, a receipt should be given the owner, or a record made of the nature and quantity of the property and the name of the owner or person in possession of it, in order that restoration and compensation may be made at the conclusion of the war.[341]

11.18.6.3 *Private Real (Immovable) Property*. Private real (immovable) property may under no circumstances be seized. It may, however, be requisitioned.[342]

11.18.6.4 *Municipal, Religious, Charitable, and Cultural Property*. The property of municipalities, that of institutions dedicated to religion, charity, and education, and the arts and sciences, even when State property, shall be treated as private property. All seizure of, destruction of, or willful damage done to institutions of this character, historic monuments, works of art, and science, is forbidden, and should be made the subject of legal proceedings.[343]The reference to "property of municipalities" above has been interpreted as not granting preferential treatment for all types of property owned by municipalities, but as referring to communal property, such as poor houses, asylums, park buildings, museums, and public records, whether owned by a parish, a town, or a province.[344]

Just as private property may be subject to requisitions and contribution and certain other purposes during a military occupation, the types of property referred to in Article 56 of the Hague IV Regulations may also be subject to such demands.[345] For example, such property may be requisitioned in case of necessity for quartering the forces and the sick and wounded, storage of supplies and material, housing of vehicles and equipment, and generally as prescribed for private property.[346] Such property must, however, be secured against all avoidable injury, even when located in fortified places that are subject to seizure or bombardment.[347]

Additional rules apply to cultural property.[348]

In the practice of the United States, religious buildings, shrines, and consecrated places employed for worship are used only for aid stations, for medical installations, or for the housing of wounded personnel awaiting evacuation, provided in each case that a situation of emergency requires such use.[349]

11.18.7 Requisitions of Private Enemy Property. Although private enemy property may not be confiscated, it may be subject to requisition, which is the method of taking private enemy real (immovable) and personal (movable) property for the needs of the army of occupation.[350]

Requisitions in kind and services shall not be demanded from municipalities or inhabitants except for the needs of the army of occupation. They shall be in proportion to the resources of the country, and of such a nature as not to involve the inhabitants in the obligation of taking part in military operations against their own country.[351]

11.18.7.1 *Method of Requisitioning*. Requisitions must be made under the authority of the commander in the locality.[352] No prescribed method is fixed, but if practicable, requisitions should be accomplished through the local authorities by systematic collection in bulk. They may be made direct by detachments if local authorities fail or if circumstances preclude resort to such authorities.[353]

11.18.7.2 *Types of Property That May Be Requisitioned.* Goods and services that are necessary for the maintenance of the occupation army (*e.g.*, fuel, food, clothing, building materials, machinery, tools, vehicles, furnishings for quarters) may be requisitioned. Billeting of forces in occupied areas is also authorized.[354]

Special additional rules apply to the requisition of foodstuffs, articles necessary to support life, or medical supplies available in the occupied territory.[355] Similarly, additional rules apply to the requisition of civilian hospitals and their supplies.[356]

11.18.7.3 *Prices and Compensation for Requisitioned Articles and Services.* Contributions in kind shall as far as possible be paid for in cash; if not, a receipt shall be given, and the payment of the amount due shall be made as soon as possible.[357]

The prices of articles and services requisitioned will be fixed by agreement if possible, otherwise by military authority.[358]

11.18.7.4 *Method of Enforcing Requisition.* Coercive measures may be used to enforce requisitions, but will be limited to the amount and kind necessary to secure the articles requisitioned.[359]

11.19 PROTECTION OF CULTURAL PROPERTY DURING OCCUPATION

The following rules address the protection of cultural property during occupation. The general rules for the protection of cultural property during hostilities also apply during occupation.[360] For example, military commanders have an obligation to take reasonable measures to prevent or stop any form of theft, pillage, or misappropriation of, and any acts of vandalism directed against, cultural property.[361]

11.19.1 Obligation With Respect to the Safeguarding and Preservation of Cultural Property. Any Party to the 1954 Hague Cultural Property Convention in occupation of the whole or part of the territory of another Party to the 1954 Hague Cultural Property Convention shall as far as possible support the competent national authorities of the occupied country in safeguarding and preserving its cultural property.[362]

Should it prove necessary to take measures to preserve cultural property situated in occupied territory and damaged by military operations, and should the competent national authorities be unable to take such measures, the Occupying Power shall, as far as possible, and in close co-operation with such authorities, take the most necessary measures of preservation.[363]

11.19.2 Respect for Cultural Property by Members of a Legitimate Resistance Movement. Any Party to 1954 Hague Cultural Property Convention whose government is considered by members of a resistance movement as their legitimate government shall, if possible, draw the resistance movement members' attention to the obligation to comply with those provisions of the 1954 Hague Cultural Property Convention dealing with respect for cultural property.[364]

11.19.3 Special Representative for Cultural Property in Occupied Territory and Application for Special Protection. An Occupying Power shall appoint a special representative for cultural property situated in that territory.[365]

The Occupying Power is competent to submit an application for immovable cultural property to be entered into the International Register of Cultural Property Under Special Protection.[366]

11.19.4 Transfers of Cultural Property From and Within Occupied Territory. The requisition of movable cultural property situated in the territory of another Party to the 1954 Hague Cultural Property Convention is prohibited.[367]

If necessary under its obligation to take measures to preserve damaged cultural property, the Occupying Power may arrange for the transport of cultural property to a refuge within occupied territory in accordance with other provisions of the 1954 Hague Cultural Property Convention.[368]Whenever a High Contracting Party occupying territory of another High Contracting Party transfers cultural property to a refuge situated elsewhere in that territory, without being able to follow the procedure for the transport of cultural property under special protection provided for in Article 17 of the Regulations, the transfer in question shall not be regarded as misappropriation within the meaning of Article 4 of the Convention, provided that the Commissioner-General for Cultural Property certifies in writing, after having consulted the usual custodians, that such transfer was rendered necessary by circumstances.[369]

11.20 LABOR OF PROTECTED PERSONS IN OCCUPIED TERRITORY

11.20.1 Types of Labor That May Not Be Compelled. The Occupying Power may not compel protected persons to perform certain types of labor.

11.20.1.1 *Prohibition on Compulsory Service in an Occupying Power's Armed Forces*. The Occupying Power may not compel protected persons to serve in its armed or auxiliary forces. No pressure or propaganda that aims at securing voluntary enlistment is permitted.[370]

Compelling protected persons to serve in its armed or auxiliary forces would in most cases also violate the prohibition against compelling inhabitants of occupied territory to swear allegiance to the hostile State.[371]Compelling a protected person to serve in the forces of a hostile Power is a grave breach of the GC.[372]

11.20.1.2 *Prohibition Against Compelling Protected Persons as Guards for Compulsory Work Installations*. The Occupying Power may not compel protected persons to employ forcible means to ensure the security of the installations where protected persons are performing compulsory labor.[373]

11.20.1.3 *Prohibition Against Requisition Labor Such That Workers Are Mobilized in a Military or Semi-Military Organization*. In no case shall requisition of labor lead to a mobilization of workers in an organization of a military or semi-military character.[374]

11.20.1.4 *Prohibition Against Forcing Inhabitants to Provide Information About the Opposing Army or Its Means of Defense*. A belligerent is forbidden to force the inhabitants of territory occupied by it to furnish information about the army of the other belligerent, or about its means of defense.[375]This obligation also results from the general prohibition against use of physical or moral coercion against protected persons.[376]

11.20.1.5 *Prohibition Against Other Work That Would Involve Protected Persons in the Obligation of Taking Part in Military Operations*. Protected persons may not be compelled to undertake any work that would involve them in the obligation of taking part in military operations.[377]This rule is formulated more broadly than the general rule against compelling a national to take part in operations against his or her own country[378]because it prohibits compulsory service in any military operation.[379]

The purpose of this rule, at least in part, is to ensure that civilians in occupied territory are not required to engage in activities that would forfeit their protections as civilians under the law of war.[380]

The prohibition against forcing the inhabitants of an occupied territory to take part in military operations against their own country precludes requisitioning their services in work directly promoting the ends of the war, such as construction of fortifications, entrenchments, and military airfields, or the transportation of supplies or ammunition in the zone of operations.[381]

11.20.2 Types of Work That May Be Compelled. The Occupying Power may not compel protected persons to work unless they are over eighteen years of age, and then only on work that is necessary for: (1) the needs of the army of occupation; (2) the public utility services; or (3) the feeding, sheltering, clothing, transportation, or health of the population of the occupied country.[382]

Thus, the Occupying Power may compel protected persons to engage in these three types of work, provided that such work is consistent with the other prohibitions against employing protected persons.[383] For example, even if it is permitted to compel civilian engineers in occupied territory to provide information about the telecommunications system of an occupied country in order to facilitate restoration of service, they may not be required to provide such information where it will be used to locate and attack the armed forces of the occupied territory.

11.20.2.1 *Examples of Services That May Be Requisitioned From Inhabitants*. The services that may be obtained from inhabitants by requisition include those of professionals, such as engineers, physicians, and nurses; and of artisans and laborers, such as clerks, carpenters, butchers, bakers, and truck drivers. The officials and employees of:

- railways, trucking companies, airlines, canals, and river or coastal steamship companies;
- cable, telegraph, telephone, radio, postal, and similar services;
- gas, electric, and water works; and
- sanitary authorities;

whether employed by the State or private companies, may be requisitioned to perform their duties only so long as the duties required do not directly concern the operations of war against their own country.[384]

The Occupying Power may also requisition labor to restore the general condition of the public works to that of peace, including the repair of roads, bridges, railways, and telecommunication networks, and to perform services on behalf of the local population, such as the care of the wounded and sick, and the burial of the dead.[385]

In addition, inhabitants over eighteen years of age may be compelled to perform work necessary to meet the maintenance needs of the occupation forces (as opposed to its strategic or tactical needs), such as providing telephone, water, or electricity services to the facilities of the occupation army from which the occupied territory is administered, or providing support to prisons, police stations, and other facilities necessary for the maintenance of order among the civilian population.[386]

11.20.2.2 *Requisition of Police and Other Services Essential to Good Order*. In general, police, firefighters, prison guards, and others who provide services essential to good order and security in occupied territory may be compelled by an Occupying Power to continue to provide those services.[387] Such a requirement is consistent with the Occupying Power's obligation to maintain public order in occupied territory.[388]

These officials may not be required to participate in military operations or other measures aimed at countering belligerent acts against the Occupying Power that are performed by privileged combatants under the law of war.[389] For example, civilian police forces in occupied territory may not be compelled to provide security for an occupying force against attacks in compliance with the law of war launched by lawful combatants, including resistance fighters who, if captured, would be entitled to POW status under GPW Article 4. On the other hand, such police forces may be required to continue to perform their normal policing functions with respect to actual or threatened criminal acts, even where the victim of such acts is the Occupying Power.[390] Similarly, civilian firefighters may be required to extinguish fires endangering the Occupying Power's military property that result from attacks by unprivileged combatants.

11.20.3 Working Conditions for Laborers Performing Requisitioned Work. Requisitioned work shall be carried out only in the occupied territory where the persons whose services have been requisitioned are resident, and such persons shall, so far as possible, be kept in their usual places of employment.[391]

Workers shall be paid a fair wage, and the work shall be proportionate to their physical and intellectual capacities.[392]

The legislation in force in the occupied country concerning working conditions, and safeguards as regards, in particular, such matters as wages, hours of work, equipment, preliminary training, and compensation for occupational accidents and diseases, shall be applicable to the protected persons assigned to the work.[393]

11.20.4 Voluntary Work Not Prohibited by the GC. Although the GC prohibits protected persons from being compelled to provide certain types of work related to military operations, there is no prohibition in the law of war to such persons being employed voluntarily and for pay in such work.[394] For example, there is no law of war prohibition against civilian firefighters putting out fires endangering civilian property resulting from attacks by lawful combatants against the Occupying Power's military property.

However, no pressure or propaganda that aims at securing voluntary enlistment in the armed or auxiliary forces of the Occupying Power is permitted.[395]

In addition, certain voluntary humanitarian work is recognized and protected by the GWS.[396]

11.20.5 General Protections Applicable to Workers in Occupied Territory.

11.20.5.1 *Access to Protecting Power.* No contract, agreement, or regulation shall impair the right of any worker, whether voluntary or not and wherever he or she may be, to apply to the representatives of the Protecting Power in order to request the Protecting Power's intervention.[397]

For example, it would be prohibited to forbid workers who are protected persons from applying to the Protecting Power. It would also be prohibited to require any protected persons who are workers, as a condition of work, to renounce their right to apply for assistance to the Protecting Power concerning work conditions or any other matter.[398]

11.20.5.2 *No Use of Measures to Create Unemployment to Induce Work for the Occupying Power.* All measures aiming at creating unemployment or at restricting the opportunities offered to workers in an occupied territory, in order to induce them to work for the Occupying Power, are prohibited.[399]

Measures intended to reduce unemployment would not be prohibited by this rule.[400]

11.21 JUDGES AND OTHER PUBLIC OFFICIALS

11.21.1 Continued Service of Judges and Other Public Officials. The Occupying Power may not alter the status of public officials or judges in the occupied territories, or in any way apply sanctions to or take any measures of coercion or discrimination against them, should they abstain from fulfilling their functions for reasons of conscience.[401]Public officials may be understood to include officials at both the national and local levels who fulfill public duties.[402]

This prohibition does not prejudice the application of the second paragraph of Article 51 of the GC.[403]Thus, a public official may be compelled to work to meet the needs of the army of occupation or for the public utility services, such as water, electricity, or sanitation.[404]Similarly, a public official may be compelled to provide certain police services.[405]

This prohibition does not affect the right of the Occupying Power to remove public officials from their posts.[406]For example, the Occupying Power may remove the political leadership and other political agents from their posts to prevent them from undermining the Occupying Power's administration.[407]

11.21.2 Oath of Public Officials. An Occupying Power may not require the inhabitants of occupied territory, including officials, to swear allegiance to it.[408]However, the Occupying Power may require such officials as are continued in their offices to take an oath to perform their duties conscientiously and not to act to its prejudice.[409]Any official who declines to take the oath may be removed; but, regardless of whether the official takes the oath, the official is required to obey the legitimate orders of the Occupying Power.[410]

11.21.3 Salaries of Public Officials. The salaries of civil officials of the hostile government who remain in the occupied territory and continue the work of their offices, especially those who can properly continue it under the circumstances arising out of the war – such as judges, administrative or police officers, and officers of city or communal governments – are paid from the public revenues of the occupied territory, until the military government has reason wholly or partially to dispense with their services.[411]

Based on consistent practice, salaries or incomes connected with purely honorary titles would be suspended.[412]Similarly, the Occupying Power need not continue to pay salaries that are rewards for loyalty to the prior regime, and may reform the pay system of public servants to increase transparency and fairness, as well as to create incentives for meritorious service.[413]

11.22 PUBLIC FINANCES AND TAXES

As a result of assuming the functions of government of the occupied territory, the financial administration of the occupied territory passes into the hands of the Occupying Power. During the occupation, the fiscal laws of the enemy State remain in effect, but may be changed or suspended by the Occupying Power under certain circumstances, as discussed below.

11.22.1 Taxes. If, in the occupied territory, the Occupying Power collects the taxes, dues, and tolls imposed for the benefit of the State, it shall do so, as far as is possible, in accordance with the rules of assessment and incidence in force, and shall in consequence be bound to defray the expenses of the administration of the occupied territory to the same extent as the legitimate Government was so bound.[414]

11.22.1.1 *Supervision of Taxes Collected by Local Authorities*. The words "for the benefit of the State" were inserted in Article 48 of the Hague IV Regulations to exclude local taxes, dues, and tolls collected by local authorities for local purposes.[415]The Occupying Power of such revenue and prevent its hostile use.[416]

11.22.1.2 *Changes in Taxes or New Taxes*. The Occupying Power shall collect taxes, as far as is possible, in accordance with the rules of assessment and incidence in force. This is an example of the general rule that the Occupying Power shall respect the laws in force in the occupied territory unless absolutely prevented.[417]Modifications to the tax laws may be justified by practical considerations and considerations of public order and safety. If, due to the flight or unwillingness of local officials, it is impracticable to follow the rules of incidence and assessment in force, then the total amount of taxes to be paid may be allotted among the districts, towns, etc., and the local authorities may be required to collect it.[418]The Occupying Power may suspend the tax laws of the occupied territory.[419]Similarly, the Occupying Power may also reduce the rate of taxes under the existing tax laws. Unless required to do so by considerations of public order and safety, the Occupying Power must not create new taxes.[420]Additional revenue may be raised in some other form, such as monetary contributions or customs duties.[421]

U.N. Security Council resolutions may provide additional authority for the Occupying Power to amend the tax laws.[422]

11.22.1.3 *Spending of Tax Revenue*. The first charge upon such taxes is for the cost of the administration of the occupied territory. The balance may be used for the purposes of the Occupying Power.[423]

The Occupying Power may use tax revenue to defray the costs of maintaining order in the occupied territory or for expenditures that benefit the local population (*e.g.*, infrastructure improvements).[424]

The Occupying Power may refrain from spending money for the support of any activity opposed to its military interests or to the restoration of order in the occupied territory, even if the State whose territory is occupied formerly allocated tax revenues to such activity.[425]

11.22.1.4 *Exemption of the Occupying Power From Local Taxation*. As the paramount authority in the occupied territory,[426]the Occupying Power is exempt from indigenous taxation in the occupied territory unless it specifically waives its sovereign immunity and consents to be taxed.[427]Its personnel are also generally exempt from local taxation, as part of their immunity from local law.[428]

In practice, the Occupying Power often issues an order specifying that no direct taxes of any kind may be levied or assessed within the occupied territory on the persons, agencies, property, instrumentalities, or transactions of the Occupying Power.[429]

11.22.1.5 *Scope of the Occupying Power's Power to Collect Taxes*. The power of the Occupying Power to collect taxes extends only to persons or property under its actual control.[430]For example, persons and property wholly outside occupied territory generally may not be taxed,[431]but the property of the absent inhabitants that is within the occupied territory may be taxed.

11.22.1.6 *Social Welfare Taxes*. An Occupying Power is often an employer of local civilian labor. Local law may provide that employers are responsible for the deduction and transfer to indigenous agencies of unemployment, health insurance, pensions, and similar welfare contributions. However, generally the agencies of the Occupying Power do not act as a collector for the local authorities, and will not be responsible for the employer's share of such welfare taxes. On the other hand, the inhabitants retain their obligation to pay their share of such contributions out of their remuneration.[432]

11.22.2 Contributions. If, in addition to continuing to collect taxes under the existing law of the occupied territory as permitted by Article 48 of the Hague IV Regulations, the Occupying Power levies other money contributions in the occupied territory, this shall only be for the needs of the army or the administration of the territory in question.[433]

11.22.2.1 *Prohibited Purposes for Levying Contributions*. Contributions may not be levied for other than the needs of the occupying forces and the administration of the occupied territory. For example, contributions may not be levied:

- for the enrichment of the Occupying Power;
- for the payment of war expenses generally;[434]

• for the purposes of collective punishment;[435] or

• for the purposes of impoverishing the population in order to pressure the enemy to sue for peace.[436]

11.22.2.2 *Methods of Levying Contributions*. No contribution shall be collected except under a written order, and on the responsibility of a "Commander-in-chief."[437] The term "Commander-in-chief" may be understood to refer to the highest military officer charged with the administration of the occupied territory.[438] Commanders of small units or detachments may not order the collection of contributions.[439]

The collection of contributions shall only be effected as far as possible in accordance with the rules of assessment and incidence of the taxes in force.[440]

For every contribution, a receipt shall be given to the contributors.[441] The receipt is intended to secure for the contributors the possibility of being indemnified afterward by their own government, and does not imply a promise of reimbursement by the Occupying Power.[442]

11.22.2.3 *Forced Loans*. The Occupying Power may seek contributions from the inhabitants of an occupied territory in the form of forced loans. The Occupying Power is required to repay such loans. As forced loans are viewed as a form of contribution, they are governed by the rules applicable to contributions.[443]

11.22.2.4 *Non-Cash Contributions*. As contributions are money demands, commodities may not be exacted as contributions. However, if the inhabitants of the occupied territory use certain commodities, rather than money, as a medium of exchange and receivable in payment of tax obligations, contributions in-kind limited to such commodities would be permissible.[444] Additionally, if the Occupying Power finds it difficult to secure prompt money payment, it may accept securities and bills of exchange from contributors in lieu of money.[445]

11.22.3 Customs Duties. The Occupying Power has the right to continue to exact existing duties, as part of its right to collect existing taxes.[446] Such collection must comply with the rules for the collection of existing taxes.[447]

The Occupying Power may also exact new duties as a form of contributions levied against the enemy or its trade.[448] Such new duties must comply with the rules for contributions.[449]

11.22.3.1 *Exemptions for Certain Relief Shipments*. Relief shipments for POWs, relief shipments for internees, and other relief consignments intended for occupied territory are exempt from customs duties.[450]

11.22.3.2 *Power to Suspend Customs Duties and Tariffs or Exempt Certain Goods*. The Occupying Power may suspend customs duties and tariffs for shipment of goods into the occupied territory.[451] It may also exempt certain types of goods (*e.g.*, humanitarian goods, goods for its forces) from customs duties and tariffs.[452]

11.22.4 Enemy Public Debts.

11.22.4.1 *Debts Owed to Occupied Territory*. Many jurists have taken the view that the Occupying Power is generally not permitted to collect pre-occupation debts owed to the sovereign of the occupied territory, as it is not a party to the agreement originating the debt.[453] However, the Occupying Power may collect the debts owed to the sovereign, provided that the debts may be legitimately characterized as realizable securities that are strictly the property of the State (*e.g.*, bearer instruments).[454] In addition, the Occupying Power may seize debts owed to insurgent forces.[455]

11.22.4.2 *Existing Debts Owed by the Occupied Territory*. The Occupying Power is under no obligation to pay the debts owed by the occupied territory, although it may choose to do so as a matter of policy.[456]

The Occupying Power may prevent payments from being made from occupied territory to a hostile belligerent.[457]

11.22.4.3 *Contracting New Debt on Behalf of the Occupied Territory*. In general, the Occupying Power may not contract new debts (including through the issuance of debt securities) on behalf of the occupied territory or collect taxes to pay interest on such debt.[458] However, new debt may be undertaken on behalf of the occupied territory if immediately necessary for the welfare of the inhabitants of occupied territory, and if undertaking such debt constitutes a fair and reasonable transaction.[459]

11.22.4.4 *Refinancing or Consolidating Existing Public Debt*. The Occupying Power may refinance or consolidate already existing public debt of the occupied territory if it is clearly in the interest of sound financial administration of that territory, and therefore of direct benefit to the inhabitants.[460]

11.22.5 Currency and Exchange Rates. The Occupying Power may leave the local currency of the occupied area in circulation. The Occupying Power may also authorize domestic authorities to re-issue currency if necessary for ensuring public order and safety.[461]

The Occupying Power may introduce its own currency into the occupied area or issue special currency for use in the occupied area territory, should the introduction or issuance of such currency be necessary.[462]There is a long history of issuing such war currency.[463]The issuance of occupation currency may be necessary to counteract the enemy State's practice of engaging in economic sabotage.[464]

The Occupying Power may also set exchange rates for currency in occupied territory.[465]

The Occupying Power's powers to regulate currency must not be used to confiscate private property.[466]For example, intentional debasement of currency by the establishment of fictitious valuation or exchange rates, or like devices, as well as failure to take reasonable steps to prevent inflation, with the result of enrichment of the Occupying Power, would violate international law.[467]

11.23 OTHER ECONOMIC REGULATION OF OCCUPIED TERRITORY

11.23.1 General Authority of the Occupying Power to Regulate Commercial Intercourse. The Occupying Power has the right to regulate commercial intercourse within, into, or out of the occupied territory. It may subject such intercourse to such prohibitions or restrictions as are essential to the purposes of the occupation.[468]The Occupying Power also may remove existing commercial restrictions or regulations when essential to the purposes of the occupation.[469]For example, the Occupying Power may suspend existing customs tariffs.[470]

The purposes of the occupation that justify economic regulation may include the military interest of the Occupying Power, the needs of the inhabitants of occupied territory, and applicable law of war obligations.

For example, for its military purposes, the Occupying Power may impose a moratorium on business payments until an effective occupation government is in place in order to safeguard financial institutions, to preserve records, and to ensure that funds and other assets of the enemy government are not transferred without the knowledge and approval of the Occupying Power. Similarly, for purposes of security and restoration of public order, the Occupying Power may also take steps to prevent hoarding of supplies, to curb or prevent black markets, and to regulate labor conditions, including strikes.[471]As another case in point, in order to fulfill its obligations with respect to the preservation of cultural property in occupied territory, the Occupying Power may issue commercial regulations intended to prevent illegal traffic in, and looting of, such property.[472]

Additional authority to regulate the economy of occupied territory may be provided by U.N. Security Council resolutions.[473]

11.23.2 Limit on the Overall Burdens Placed on the Economy of the Occupied Territory. The economy of an occupied territory can only be required to bear the expenses of the occupation, and these should not be greater than the economy of the occupied territory can reasonably be expected to bear.[474]For example, requisitions must be in proportion to the resources of the occupied territory.[475]

11.23.3 Control of Business in Occupied Territory. The Occupying Power may exercise controls over private business for the purpose of addressing the needs of the inhabitants of the occupied territory or for military purposes, including by: (1) compelling the continued operation of a business; (2) granting of business subsidies; (3) closing down a business; or (4) assuming control and management of a business.

The Occupying Power may compel a business to continue operations if necessary to serve the needs of the local population or for military purposes.[476]Compulsory work of protected persons, however, must comply with the GC.[477]

The Occupying Power may take steps to increase production from private business, such as by granting subsidies out of available governmental revenues from the occupied territory.[478]

If the Occupying Power determines that the continued operation of a business is detrimental to the interests of the local population or to the Occupying Power, the Occupying Power may close down the business.[479]

The Occupying Power may assume control and management of such a business if necessary to serve the needs of the local population or for military purposes.[480]Title to the business in such circumstances remains with the legal owner, and if the Occupying Power earns a profit from the operation of the business, the legal owner must be indemnified, to avoid a prohibited confiscation of private property.[481]

11.23.4 Regulation of External Trade. The Occupying Power may also regulate foreign trade, including completely suspending such trade.[482] For example, the Occupying Power may halt the export of precious metals and other valuable items that are readily converted or exchanged on the international market, including metals such as copper, jewels, and securities.[483] Commercial relations between the occupied territory and the remaining territory of the enemy and its allies are also normally suspended.[484] Such suspension may be relaxed through licenses to trade.[485]

11.23.5 Exchange Controls, Including Clearing Arrangements, and Asset Controls. The Occupying Power may also institute exchange controls, including clearing arrangements, and, if necessary, the freezing or blocking of certain assets, in order to conserve the monetary assets of the occupied territory, as well as for security purposes.[486] For example, the Occupying Power may regulate, or even prohibit, the flow of funds between the occupied territory and other countries, under the theory that preventing flight of capital assets is critical to maintaining order and stability.[487] Such measures must not, however, be used to enrich the Occupying Power or otherwise circumvent the restrictions placed on requisitions, contributions, seizures, and other measures dealing with property.[488]

11.23.6 Price Controls in Occupied Territory. The Occupying Power may regulate prices in the occupied territory. For example, shortages of commodities and increased demand for certain commodities in the occupied territory may result in a rise in price fluctuations requiring the Occupying Power to resort to measures designed to maintain prices at a reasonable maximum level.[489] However, the Occupying Power may not use its power over price controls for the purpose of exploiting the occupied territory to its own illegal advantage.[490]

XII – Non-Hostile Relations Between Belligerents

Chapter Contents

12.1 INTRODUCTION

This Chapter addresses the legal principles for non-hostile relations between belligerents and the basic mechanisms for implementing them.

More specifically, this Chapter addresses the rules for the protection of certain personnel engaged in non-hostile relations, such as *parlementaires*, and persons protected by military passports, safe-conducts, or safeguards. In addition, this Chapter addresses certain agreements between opposing belligerents, such as cartels, capitulations, and armistices.

12.1.1 Traditional Rule of Non-Intercourse During War. The traditional rule during international armed conflict is that, even without any special proclamation, all intercourse between the territories occupied by belligerent forces, including communication, transportation, and commerce, would cease.[1]

The traditional rule of non-intercourse reflects a belligerent's authority under the law of war to limit and regulate intercourse between persons and territory controlled by or belonging to that belligerent and persons and territory controlled by or belonging to the enemy.[2]For example, States may authorize their military commanders to limit travel and trade within their areas of operations for security purposes. During armed conflict, States have exercised their authority to limit intercourse with the enemy in order to avoid supporting the enemy's war effort and in order to speak with one voice in communications with the enemy State.

In particular, the traditional rule of non-intercourse notifies a State's nationals and other persons subject to its jurisdiction that they may be liable to punishment by that State under its domestic law and war powers if they communicate with or support the State's enemies in war without proper authorization.[3]

Exceptions to the general rule of non-intercourse (such as communication with enemy forces or permission to trade or travel between territories that are controlled by forces hostile to one another) have been granted on behalf of individuals only with the approval of the Government or the highest military authority.[4]

12.1.1.1 *Limitations on the Belligerent's Authority to Regulate Intercourse.* A belligerent's authority to regulate intercourse between territory it controls and territory controlled by the enemy is subject to certain limitations.

Insofar as restrictions on travel and trade imposed by a belligerent affect neutral rights, such restrictions may be limited by the law of neutrality. An Occupying Power's authority to control travel and trade during belligerent occupation is addressed by the law of occupation.[5]

Under the GC, however, protected persons in a belligerent's home country, or in territory occupied by a belligerent, are entitled to leave unless such departure is contrary to the interests of the State.[6]

12.1.2 Necessity for Non-Hostile Relations. Despite the general rule of non-intercourse during war, the conduct of war and the restoration of peace require certain non-hostile relations between belligerents.[7]

12.1.2.1 *Non-Hostile Relations to Facilitate Humanitarian Activities During Armed Conflict*. Non-hostile relations, including local communication between belligerent forces, may be necessary to facilitate the conclusion and implementation of special agreements contemplated by the 1949 Geneva Conventions. These include:

• special agreements to facilitate the protection of the wounded, sick, shipwrecked, and medical personnel,[8]such as

o armistices or local arrangements to permit the removal, exchange, or transport of the wounded and sick;[9]

• special agreements to facilitate the protection of POWs,[10]such as

o agreements to effect the direct repatriation or accommodation in neutral countries of certain seriously wounded and sick POWs during hostilities;[11]and

• special agreements to facilitate the protection of civilians,[12]such as

o agreements to establish areas where civilians or the wounded and sick are protected.[13]

Relatedly, non-hostile relations, including local communication between belligerent forces, may also be necessary:

• to obtain the necessary consent for the appointment of Protecting Powers;[14]

• to establish safe-conduct protections for certain persons or property under the 1949 Geneva Conventions;[15]and

• to conclude special agreements applicable to a non-international armed conflict to bring into effect treaty provisions that would normally only apply during international armed conflict.[16]

12.1.2.2 *Non-Hostile Relations to Facilitate the Restoration of Peace*. Nonhostile relations, including local communication between belligerent forces, may also be important in facilitating the restoration of peace, including by facilitating:

• ceasefires or local armistices that implement a general armistice;

• armistices that are a prelude to a peace treaty;[17]

• the permanent cessation of hostilities through a peace treaty;

• agreements on the repatriation of POWs or retained personnel after the cessation of hostilities;[18]

• measures to protect civilians from the effects of minefields, mined areas, mines, boobytraps, and other devices, such as information sharing with another party or parties to the conflict;[19]and

• the provision of certain assistance to facilitate the marking and clearance, removal, or destruction of explosive remnants of war, in cases where a user of explosive ordnance that has become explosive remnants of war does not exercise control of the territory.[20]

12.2 PRINCIPLE OF GOOD FAITH IN NON-HOSTILE RELATIONS

Absolute good faith with the enemy must be observed as a rule of conduct, including in non-hostile relations between opposing belligerents.[21]In particular, in the context of non-hostile relations, the principle of good faith requires that:

• compacts between enemies, such as truces and capitulations, shall be faithfully adhered to;[22]

• no advantage not intended to be given by the adversary shall be taken;[23]and

• the means of conducting non-hostile relations (such as flags of truce) must not be misused.[24]

On the other hand, the principle of good faith does not prohibit belligerents from: • continuing their military operations while negotiations are ongoing;[25]or

• declining to respond to offers to negotiate, refusing offers to negotiate, or refusing specific offers from the adversary for reasons of military expediency.[26]

12.3 METHODS FOR COMMUNICATION BETWEEN BELLIGERENTS

Belligerents may communicate with one another through a variety of methods, including:

• directly by telecommunications;[27]

• through the traditional mechanism of a display of a flag of truce and the sending of *parlementaires*;[28]

• in conferences between belligerent representatives in an agreed neutral zone;[29]

• through international organizations, such as the United Nations;

- indirectly through another State, such as a Protecting Power;[30] and
- indirectly through the ICRC or other impartial humanitarian organization.[31]

12.4 THE WHITE FLAG OF TRUCE TO INITIATE NEGOTIATIONS

In the past, the normal means of initiating negotiations between belligerents has been the display of the white flag of truce.[32]

12.4.1 Meaning of the White Flag—a Desire to Communicate. As a legal matter, the white flag, when used by military forces, indicates a desire to communicate with the enemy. The hoisting of a white flag has no other legal meaning in the law of war.[33]

The hoisting of a white flag may indicate that the party hoisting it desires to open communication with a view to an armistice (*e.g.*, to enable forces to collect the wounded) or a surrender.[34] If hoisted during a military action by an individual combatant or a small party of combatants, it may signify merely that those persons or forces wish to surrender.[35] Although the white flag has been used with this intent, the display of the white flag does not necessarily mean that the person or forces displaying it are prepared to surrender. Moreover, enemy forces in the immediate area might not have the same intent as the individual or forces displaying the white flag, especially where the display of the white flag was not authorized by the individual or force's commander.

12.4.2 Rules for the Party Displaying the White Flag. Forces displaying a flag of truce must show clearly that they intend to engage in non-hostile relations. They bear the burden of communicating their intent to the adversary. To indicate that the hoisting of the white flag is authorized by its commander, the appearance of the flag should be accompanied or followed promptly by a complete cessation of fire from that side.[36]

The commander authorizing the hoisting of the flag should also promptly send a *parlementaire* to communicate the commander's intent.[37]

12.4.2.1 *Prohibition on Improper Use of the Flag of Truce*. It is especially forbidden to make improper use of a flag of truce.[38] It would be improper to use a flag of truce to feign an intention to negotiate, surrender, or otherwise suspend hostilities when there is no such intention.[39]

It is an abuse of the flag of truce if the force that sends a *parlementaire* does not halt and cease fire while the *parlementaire* is approaching, or is being received by, the other party.[40]

Improper use of a flag of truce also includes its use while engaging in attacks or in order to shield, favor, or protect one's own military operations, or otherwise to impede enemy military operations.[41] For example, flags of truce may not be used surreptitiously to obtain military information, or with the purpose of obtaining time to effect a withdrawal, secure reinforcements, resupply, or conduct other operations.[42]

12.4.3 Rules for the Party to Whom the White Flag Is Displayed. A party is not required to cease firing or other military operations when a white flag is raised by the other side.[43]

It is essential to determine with reasonable certainty that the flag is shown by actual authority of the enemy commander before basing important action upon that assumption.[44] For example, forces should not assume that all enemy forces in the locality intend to surrender and expose themselves to hostile fire based on the enemy's display of a white flag.[45]

Fire must not be directed intentionally on the person carrying the white flag or on persons near him or her unless there is a clear manifestation of hostile intent by those persons.[46]

12.5 RULES FOR *PARLEMENTAIRES*

After the display of the white flag, a *parlementaire* would be sent to conduct negotiations, traveling under the display and protection of the white flag of truce.

12.5.1 The *Parlementaire* and Party. *Parlementaires* ordinarily are agents employed by commanders of belligerent forces in the field, to go in person within the enemy lines, for the purpose of communicating or negotiating openly and directly with the enemy commander.[47]

12.5.1.1 Parlementaire – *Notes on Terminology*. The term *parlementaire* is derived from "parley" and was adopted by governments at the First Hague Peace Conference in 1899.[48] These provisions were repeated in the 1907 Hague IV Regulations. Article 32 of the Hague IV Regulations provides that "a person is regarded as bearing a flag of truce who has been authorized by one of the belligerents to enter into communication with the other, and who advances bearing a white flag."[49] Although the word *parlementaire* has been translated as a person bearing a flag of truce, the *parlementaire* need not physically carry the white flag.[50]

12.5.1.2 *Authorization and Credentials of the* Parlementaire. A *parlementaire* must be authorized by a belligerent to enter into communications with the other side.[51] To demonstrate the authorization to negotiate, a *parlementaire* must be duly authorized in a written instrument signed by the commander of his or her forces.[52]

12.5.1.3 *Members of a* Parlementaire *Party*. A *parlementaire* may come alone, or he or she may request to have others, such as an interpreter, accompany him or her. In the past, flag-bearers or drummers accompanied *parlementaires* to reduce the risk that the *parlementaires* would be inadvertently attacked.[53] In modern warfare, members of a *parlementaire's* party may include, for example, a driver and communications personnel, and they would take similar measures to make their status and purpose clear to the other side.[54]

12.5.2 Refusal or Reception of a *Parlementaire*. The commander to whom a flag of truce is sent is not in all cases obliged to receive it.[55] A commander may decline to receive a *parlementaire* for reasons of military necessity, such as if the reception of the *parlementaire* would interfere with the commander's operations.[56] However, a belligerent may not declare beforehand, even for a specified period – except in case of reprisal for abuses of the flag of truce – that it will not receive *parlementaires*.[57]

A commander may declare the formalities and conditions upon which he or she will receive a *parlementaire* and fix the hour and place at which the *parlementaire* must appear.[58] The receiving commander may set the details of the meeting, such as the time, place, number of persons accompanying the *parlementaire*, authorized method of transport (*e.g.*, foot or vehicle), and the frequency of meetings if more than one will take place. For example, the receiving commander may limit the size of the *parlementaire*'s party and need not allow an unnecessary repetition of *parlementaire* visits.[59]

Although commanders may refuse to receive *parlementaires* and other enemy envoys seeking to negotiate, commanders may not refuse the unconditional surrender of the adversary or declare that they will refuse unconditional surrender.[60]

12.5.3 Duties and Liabilities of the *Parlementaire*. Persons who display a white flag during combat accept the risk that they might be inadvertently wounded or killed while attempting to communicate with opposing forces.[61] *Parlementaires* bear the burden of communicating their status to the enemy.[62] They should take appropriate measures to help clarify their status to enemy forces. For example, the *parlementaire's* party should move slowly and deliberately so that it is not inadvertently attacked.[63]

While within the lines of the enemy, the *parlementaire* must obey all instructions given to him or her. The *parlementaire* may be required to deliver his or her message to a subordinate of the commander.[64]

12.5.4 Rights of Inviolability of the *Parlementaire*. A *parlementaire* and members of the *parlementaire's* party (*e.g.*, an interpreter) have a right to inviolability.[65] For example, it would be a serious violation of good faith to attack deliberately a *parlementaire's* party that has been received and acknowledged.[66]

12.5.4.1 *Security Measures Are Consistent With Inviolability*. The receiving commander, however, may take all the necessary steps to prevent the envoy from taking advantage of his or her mission to obtain information.[67] For example, the envoy and the envoy's interpreter may be blindfolded or hooded for security purposes.[68]

12.5.4.2 *Loss of Rights of Inviolability of* Parlementaire. A *parlementaire* loses his or her rights of inviolability if it is proved in a clear and uncontestable manner that he or she has taken advantage of his or her privileged position to provoke or commit an act of treachery.[69]

Such acts of treachery include sabotage or the secret gathering of information about the adversary while under the adversary's protection.[70]

A *parlementaire* does not commit an act of treachery if he or she reports what he or she observed in plain sight during his or her mission.[71]

12.5.4.3 *Detention of* Parlementaire. In case of abuse of a flag of truce, a commander to whom a flag of truce is sent has the right to detain the envoy temporarily.[72]

In addition to the right of detention for abuse of his or her position, a *parlementaire* may be detained for other imperative security reasons. For example, a *parlementaire* admitted during an engagement may be detained pending the conclusion of the immediate fighting.[73] In addition, a *parlementaire* may be detained in case he or she has seen anything that may be detrimental to the enemy, or otherwise obtained such knowledge, or if the envoy's departure would reveal information on the movement of forces.[74]

A *parlementaire* should be detained only so long as circumstances imperatively require, and information regarding the *parlementaire*'s detention, as well as any other action against the *parlementaire* or against any member of the *parlementaire*'s party, should be sent to his or her commander.[75]

12.5.5 Neutral Zone and Other Appropriate Measures to Facilitate Negotiations. If it is anticipated that negotiations between belligerents may be prolonged, a neutralized area may by agreement of the parties be set aside for the site of the negotiations.[76]The belligerents may agree that no combat activity will take place within or over the area, and that forces will be detailed to maintain the security of the area.[77]The area designated as neutral on maps interchanged by representatives of the belligerents may be marked with searchlights, balloons, and other devices to ensure that it will not be attacked.[78]

Other appropriate measures may also be agreed upon to facilitate negotiations. For example, special signals that the *parlementaire's* party should give or routes that the *parlementaire's* party must take may be agreed upon to help identify the *parlementaire's* party to the opposing forces.

12.6 MILITARY PASSPORTS, SAFE-CONDUCTS, AND SAFEGUARDS

Persons within an area occupied by a belligerent may be protected from molestation or interference through military passports, safe-conducts, and safeguards. These devices are a matter of international law only when granted or posted by arrangement with the enemy or with a neutral State.[79]

12.6.1 Military Passports, Safe-Conducts, and Safeguards –Notes on Terminology. The terms *pass* or *permit* may be used instead of *passport*. *Pass* has sometimes been used for a general permission to do certain things, while *permit* has sometimes been used like *safe-conduct*, to signify permission to do a particular thing.[80]

The terms *military passport* and *safe-conduct* have to some degree been used interchangeably, but *military passport* generally has been used in a broader sense than *safe-conduct*. [81]

Ultimately, however, the purpose of the document will be more important than the nomenclature in deciding the nature of the document issued.[82]

12.6.2 Military Passport. A military passport is a document issued by order of a commander of belligerent forces, authorizing a person or persons named therein residing or sojourning within territory occupied by such forces to travel unmolested within such territory, with or without permission to pass, or to pass and return by designated routes through the lines, subject to such further conditions and limitations as the commander may prescribe.[83]

Unlike a passport for peacetime travel, such as a passport issued by the U.S. State Department, a military passport would be issued under a commander's authority to regulate commerce and travel within territory controlled by his or her forces.[84]

12.6.3 Safe Conduct. Documents like passports, issued by the same authority and for similar purposes to persons residing or traveling outside of the occupied areas who desire to enter and remain within or pass through such areas, are called safe-conducts.[85]Similar documents issued by the same authority to persons residing within or without the occupied areas to permit them to carry specified goods to or from designated places within those areas, and to engage in trade otherwise forbidden by the general rule of non-intercourse, are also called safe-conducts.[86]

Safe-conducts for goods in which the grantee is given a continuing right for a prescribed period, or until further orders, to engage in the specified trade, are sometimes called licenses to trade.[87]

12.6.3.1 *Safe-Conduct Pass for Ambassadors and Diplomatic Agents of Neutral Governments Accredited to the Enemy*. Ambassadors and other diplomatic agents of neutral governments accredited to the opposing party to the conflict may receive a safe-conduct pass through territory under the opposing force's control, unless there are military or other security reasons to the contrary, including the safety of the personnel in question, and unless they may reach their destination conveniently by another route.[88]There is no international law requirement, however, for issuing such a safe-conduct pass; such passes usually would be granted by national-level authorities and not subordinate officers.[89]

12.6.3.2 *Licenses to Trade*. Licenses to trade must, as a general rule, emanate from the supreme authority of the State.[90]In an international armed conflict, a State controlling territory of the enemy may grant licenses to trade that relax its prohibitions on trading with the enemy.[91]

Licenses to trade issued by military authorities may be either general or special. A *general license* generally or partially relaxes the exercise of the rights of war in regard to trade in relation to any community or individuals liable to be affected by their operation.[92] A *special license* is one given to individuals for a particular voyage or journey for the importation or exportation of particular goods.[93]

12.6.3.3 *Safe-Conducts Contemplated by the 1949 Geneva Conventions*. Certain safe-conduct protection may be granted by special agreements contemplated in the 1949 Geneva Conventions. Such agreements include:

- agreements for the removal and passage of certain personnel, such as vulnerable civilians, from besieged areas;[94]
- agreements for the passage of consignments;[95]
- agreements for the removal or transport of the wounded and sick;[96]
- agreements for the passage of medical aircraft for the transport of the military or civilian wounded and sick;[97]
- agreements for the safe passage of chartered medical transport ships;[98] and
- agreements for the special transport of relief shipments for POWs or civilian internees.[99]

12.6.4 Safeguard. A safeguard may refer to a detachment of forces posted for the protection of, or a written instrument affording protection by a belligerent to, enemy or neutral persons or property.

A safeguard falls within the law of war, however, only when granted and posted by arrangement with the enemy or a neutral. For example, guards or written orders posted by a belligerent for the protection of its own personnel or property would not be governed by the law of war.[100]

The effect of a safeguard is to pledge the honor of the nation that the person or property will be respected by its armed forces.[101] It does not commit the government to its protection or defense against attacks by enemy armed forces or other hostile elements. Safeguards have been used to protect cultural property or other civilian property or personnel meriting special protection.[102]

12.6.4.1 *Personnel Serving as a Safeguard*. A safeguard is a detachment, guard, or detail posted by a commander for the protection of persons, places, or property of the enemy or of a neutral.[103]

Combatants serving as safeguards occupy a protected status.[104] They may not be attacked, and it is customary to send them back, together with their equipment and arms, to their own armed forces when the locality is occupied by the enemy and as soon as military exigencies permit.[105]

12.6.4.2 *Written Safeguard*. The term safeguard also refers to a written order left by a commander with an enemy subject, or posted upon enemy property, for the protection of that person or property.[106] It is usually directed to the succeeding commander and requests a grant of protection.[107]

12.6.4.3 *Uniform Code of Military Justice Offense of Forcing a Safeguard*. The violation of a safeguard is a serious violation of the law of war.[108] The Uniform Code of Military Justice makes forcing a safeguard punishable.[109] "Forcing a safeguard" means to perform an act or acts in violation of the protection of the safeguard.[110]

12.6.5 Revocation of Passports, Safe-Conducts, and Similar Documents. In general, passports, as well as safe-conducts, may be revoked for good reasons of military expediency by the person who issued them or that person's superior officer. Revocation must never be used as a means of detaining the holder, who is always allowed to withdraw in safety.[111]

In addition, if the holder commits any wrongful act, such as using the opportunity given by a passport or safe-conduct to obtain military information or if the holder exceeds its terms, the privilege may be withdrawn.[112]

12.7 CARTELS

In its narrower sense, a cartel is an agreement entered into by opposing belligerents for the exchange of POWs.[113] In its broader sense, it is any convention concluded between belligerents for the purpose of arranging or regulating certain kinds of non-hostile intercourse that would otherwise be prohibited by reason of the existence of the armed conflict (*e.g.*, postal communication or trade in certain goods or commodities).[114]

Both parties to a cartel are honor bound to observe its provisions scrupulously. A cartel is voidable by a party upon definite proof that it has been intentionally violated in an important particular by the other party.[115]

12.8 CAPITULATIONS – NEGOTIATED INSTRUMENTS OF SURRENDER

12.8.1 Definition of Capitulation. A capitulation is an agreement (sometimes with certain conditions) entered into between commanders of belligerent forces for the surrender of a body of forces, a fortress, or other defended locality, or of a district of the theater of operations.[116] For example, a capitulation agreement could involve the surrender of a small unit, such as a squad, platoon, company, or battalion, or the surrender of larger forces, such as a division or corps.[117]The surrender of military forces also may be arranged between belligerents at national levels without the involvement of military commanders, and possibly through third parties. A capitulation agreement may be negotiated between opposing military forces in local implementation of a surrender negotiated at national levels.

12.8.1.1 *Capitulation Compared to Armistice.* A *capitulation* surrenders the capitulating unit, while an *armistice* suspends fighting between opposing forces.

12.8.1.2 *Capitulation Compared to Unconditional Surrender.* A *capitulation* is a surrender pursuant to an agreement. A surrender may be effected without resort to a capitulation agreement.[118]For example, individuals or units may surrender themselves unconditionally to the opposite side without a specific capitulation agreement.[119]

On the other hand, an unconditional surrender may be effected through a capitulation instrument.[120]Such an instrument would lack any conditions on the surrender, but may specify the way in which the surrender is to be implemented or may clarify for all concerned that the surrender is to be conducted on an unconditional basis.

12.8.2 Authority of Commanders to Conclude Capitulation Agreements. Commanders are generally presumed to have the authority to conclude capitulation agreements with respect to forces under their command and areas under their control.[121]For example, if a commander of military forces concludes that continued fighting has become impossible and is unable to communicate with his or her superiors, under the law of war he or she may assume that he or she has authority to surrender his or her forces or position.[122]

However, a commanding officer's powers generally do not extend beyond the forces and territory under his or her command.[123]Thus, a commander may not surrender forces that are not under his or her command. Similarly, unless so authorized by his or her government, a commander lacks the authority to bind his or her government to a permanent cession of the place or places under his or her command, to surrender sovereignty over territory, or to agree to other terms of a political nature or such as will take effect after the termination of hostilities.[124]

However, capitulation agreements that have been concluded by political authorities may include such terms.[125]

A State is not bound by the terms of a capitulation agreement that it has not authorized its commanders to conclude and may repudiate such terms.[126]For example, a State would not be bound to honor political conditions accepted by a capitulating commander or by a commander accepting capitulation, if those commanders lack such authority.[127]

12.8.2.1 *Surrenders Contrary to Domestic Law or Superior Orders.* The fact that a commander surrenders in violation of orders or the law of his or her own State does not impair the validity of the surrender.[128]Commanders who surrender in violation of orders or the law of their own State may be punished by their State.

Under the Code of Conduct for U.S. armed forces, a commander must never surrender the members of his or her command while they still have the means to resist.[129]Under the Uniform Code of Military Justice, shameful surrenders are punishable. In addition, compelling or attempting to compel a commander to surrender or striking colors or flag to an enemy without proper authority is punishable.[130]

12.8.3 Rules of Military Honor. Capitulations agreed upon between belligerents must take into account the rules of military honor.[131]

Conditions involving unnecessary disgrace or ignominy should not be insisted upon.[132]

Capitulations may include the right of the capitulating forces to surrender with colors displayed or other indications of professional respect for the capitulating forces.[133]For example, it may be appropriate to allow surrendering officers to keep their side arms.[134]

Even if the capitulation does not explicitly have such provisions, the capitulation agreement should be executed with honor and respect for the adversary. Treating the adversary honorably diminishes psychological stigma from capitulation and thereby provides an incentive for the adversary to capitulate rather than fight on without any chance of success.

12.8.4 Form of Capitulation Agreement or Instrument. There is no specified form for capitulation instruments. They may be concluded either orally or in writing, but in order to avoid disputes, they should be reduced to writing. The agreement should contain in precise terms every condition to be observed on either side, excepting such conditions as are clearly imposed by the law of war. Details of time and procedure should be prescribed in exact and unequivocal language. Even in case of an unconditional surrender, when the terms are dictated by the victor, they should nevertheless be embodied in a written capitulation as soon as practicable.[135]

12.8.5 General Effect of Capitulation. The general effect of concluding a capitulation agreement is that of an unconditional surrender. In other words, absent specific terms in the capitulation agreement to the contrary, the capitulation agreement should be understood to create the effect of an unconditional surrender by the capitulating party.[136]

The capitulating party must generally cease operations and maintain the military status quo at the time in which the capitulation becomes effective.[137]For example, the capitulating forces must not engage in offensive operations against opposing forces.[138]Similarly, although forces may destroy their own weapons and intelligence information to prevent them from falling into the hands of the enemy before they capitulate, after the capitulation is effective, the capitulating forces must abstain from all destruction and damage to their own facilities and equipment, unless expressly permitted by the capitulation agreement.[139]The capturing side is free to confiscate as war booty or, at its discretion, destroy the weapons, ammunition, and military equipment of the capitulating side.[140]

12.9 CAPITULATIONS – SUBJECTS USUALLY ADDRESSED

In general, a capitulation agreement is understood to have the effect of an unconditional surrender under the law of war. However, specific topics may be addressed in the capitulation agreement that attach conditions to the surrender or specify the way in which the surrender is to be effected.

12.9.1 Scope of Surrendered Forces and Territory and Time of Surrender. The capitulation agreement may specify the force or territory that is surrendered and the exact time when the surrender is to take effect. If a place or area is surrendered, provisions relative to the withdrawal of the defenders from it and the entering into possession of it by the victorious forces may also be specified in the agreement.

A commander's competence to capitulate is limited to forces under his or her command.1[141]To avoid misunderstandings, capitulation should state the forces to be surrendered and to what extent detached forces and personnel are included in the surrender of the main body.[142]

12.9.2 Disposition of Surrendered Forces. The capitulation instrument may specify the movements and administration of the surrendered forces after the surrender.

In general, surrendering military forces and others entitled to POW status who fall into the power of the enemy during international armed conflict become POWs under the GPW. Similarly, the disposition of surrendered medical personnel and the wounded and sick belonging to the surrendered party would be addressed by the GWS and GWS-Sea. Insofar as matters are addressed by those treaties, there is little need for similar provisions in a capitulation instrument.

However, special circumstances, such as inability of the victor to guard, evacuate, and maintain large numbers of POWs or to occupy the area in which enemy military forces are present, may justify the victorious commander in allowing the defeated force to remain in its present positions, to withdraw, or to disperse after having been disarmed and having given their paroles, provided that the giving of paroles is not forbidden by the laws of their own country and that they are willing to give their paroles.[143]

12.9.2.1 *Orders Given by the Victor*. It is normally stipulated that the orders of the victorious commander will be scrupulously carried out by the surrendered forces and that those who fail to comply with such orders or with the terms of the surrender itself may be punished.[144]

12.9.2.2 *Disarmament of Surrendered Forces*. Normally provisions are included to govern the disposition of enemy arms, equipment, and other property in the hands of the force that has surrendered. Officers have sometimes been allowed to retain their side arms.[145]In addition, POWs are allowed to keep certain articles under the GPW.[146]

12.9.2.3 *Prohibitions on Acts of Destruction by the Surrendered Forces*.

Although a capitulation, like an unconditional surrender, generally has the effect of prohibiting acts of destruction by surrendered forces, the capitulation may also specifically prohibit destruction by the surrendered forces of their materiel or installations, or communications, transportation facilities, and other public utilities in the area concerned.

12.9.3 Disposition of Detainees Held by the Surrendered Forces. The disposition of POWs, civilian internees, and other persons held in the custody of the surrendered forces may be addressed in a capitulation.[147]

12.9.4 Minefields and Other Defensive Measures. The provision of facilities and of information on such matters as minefields and other defensive measures may be addressed in a capitulation.[148]

12.9.5 Civil Administration of the Area Concerned. The civil administration of the area concerned, if a place or area is surrendered, may also be addressed in a capitulation.

12.10 CAPITULATIONS – VIOLATIONS AND DENUNCIATION

Once capitulation agreements have been made, they must be scrupulously observed by all parties.[149]

12.10.1 Denunciation of Capitulation Agreements Based on Directed Violations. If violations of a capitulation agreement have been directed by the commander who capitulated or by higher authority, the other belligerent may denounce the capitulation and resume hostilities.[150]

Like action may also be taken if the capitulation was obtained through a breach of faith.[151]

A capitulation may not, however, be denounced because one of the parties has been induced to agree to it by a means consistent with the law of war, such as a ruse, or by that party's own incapacity, such as through mistake of fact.[152]

12.10.2 Violation of Capitulation Agreements by Individuals. Deliberate violations of the terms of a capitulation by individuals are punishable as a war crime.[153]For example, a member of the capitulated force may be punished by his or her State or the enemy State for carrying out attacks on enemy military equipment or personnel in violation of the terms of the capitulation agreement. Similarly, destroying one's own military equipment in violation of the terms of the capitulation agreement would also be punishable.[154]

Unwitting violations of a capitulation agreement, however, are not punishable. For example, individuals who return from a patrol and find their unit to have surrendered and, lacking knowledge of capitulation, continue to fight, would not be committing a war crime.

12.10.2.1 *Violation of Capitulation Agreements and POW Status*. Violation of a capitulation agreement, like other pre-capture law of war violations, is not a basis for denying a person POW status, if that person otherwise qualifies for POW status under the GPW.[155]

12.10.2.2 *Violation of Capitulation Agreements by POWs*. Members of the armed forces who have capitulated may become POWs. Violations of the terms of a capitulation agreement by a POW may also be punishable as misconduct as a POW, either by the Detaining Power or the State to which the POW belongs when that POW has been repatriated.[156]

12.11 ARMISTICES AND OTHER CEASE-FIRE AGREEMENTS

12.11.1 Definition of Armistice. An armistice may also be described as the cessation of active hostilities for a period agreed upon by the belligerents.[157]An armistice suspends military operations by mutual agreement between the belligerent parties.[158]

12.11.1.1 *Armistice as an Agreement*. An armistice is an agreed suspension of hostilities. For example, a unilateral suspension of operations by one party would not be an armistice.

12.11.1.2 *Armistice as a Suspension of Hostilities and Not a Peace Treaty*. An armistice is not a partial or a temporary peace; it is only the suspension of military operations to the extent agreed upon by the parties to the conflict.[159]War as a legal state of hostilities between parties may continue, despite the conclusion of an armistice agreement.[160]In some cases, however, armistice agreements may be intended to be a prelude to peace treaties.[161]In some cases, armistice agreements may persist for a long time.[162]

12.11.1.3 *Armistice – Notes on Terminology*. An armistice also has been referred to as a *truce, local truce, ceasefire*, *cessation of hostilities*, and *suspension of arms*.[163]Although in some cases distinctions may be drawn between these different terms, *armistice* may also be used in a general sense to encompass these terms.[164]

12.11.2 Types of Armistice and the Authority to Conclude Armistice Agreements. Armistice agreements have been categorized based on the degree to which they suspend hostilities.

An armistice may be general or local. The first suspends the military operations of the belligerent States everywhere; the second only between certain fractions of the belligerent armies and within a fixed radius.[165]

The degree to which the parties wish to suspend hostilities affects what authorities are needed to conclude the armistice agreement. An armistice agreement must be concluded by authorities who are competent to agree to and to enforce its terms. Thus, armistices that include more substantive and expansive terms must be approved by more senior authorities. For example, a commander would not have the authority to conclude an armistice agreement that binds units or areas that are not under his or her command. Similarly, if an armistice contains political terms, it must be made under authorization from the governments concerned or subject to approval by them.[166]

12.11.2.1 *General Armistice.* A *general armistice* suspends all military operations between opposing forces.

General armistices are usually of a combined political and military character. They usually precede the negotiations for peace, but may be concluded for other purposes. Due to its political importance and the fact that it would be binding on all of a belligerent's forces, a general armistice is normally negotiated by senior officials and is subject to approval by each State.[167]For example, an armistice agreement may be negotiated by senior military commanders and be subject to approval of each State.[168]

12.11.2.2 *Local or Partial Armistice.* A local armistice (also called a partial armistice) suspends operations between certain portions of the belligerent forces or within a specified area.[169]In some cases, a local armistice has been distinguished from a suspension of arms as implicating a large number of forces, a greater geographic area, and broader interests than the local military requirements that are addressed in a suspension of arms.[170]

12.11.2.3 *Suspension of Arms.* A suspension of arms is a form of local armistice concluded between commanders of military forces for some local military purpose, such as to bury the dead, to collect the wounded, to arrange for exchange of detainees, or to enable a commander to communicate with his or her government or superior officer.[171]

Commanders are presumed to have the authority to conclude suspensions of arms for forces and areas within their control.

12.11.3 Form of Armistice Agreements. No special form for an armistice is prescribed. If possible, armistice agreements should be reduced to writing to avoid misunderstandings and for the purpose of reference should differences of opinion arise. Armistice agreements should be drafted with the greatest possible precision and clarity.[172]

12.11.3.1 *Languages.* Unless agreement is possible for an armistice to be drawn up in one language, an armistice should be drawn up in the language of each belligerent force, with each side retaining a copy in its language. Each belligerent should confirm the text in each version to ensure consistency in each language.

12.11.4 General Effects of an Armistice Agreement. The precise effect of an armistice agreement may depend on the terms of the agreement. In the absence of specific provisions to the contrary, the following effects should be understood to result from an armistice agreement.

12.11.4.1 *Commencement, Notification, and Effective Time of the Armistice.* An armistice must be notified officially and in good time to the competent authorities and to the forces. Hostilities are suspended immediately after the notification, or on the date fixed.[173]

An armistice for the parties commences at the moment it is signed, in the absence of specific provision to the contrary.[174]For example, an armistice agreement may specify that hostilities are suspended for the parties upon receipt of notification by their commanders, rather than upon signing.[175]

Although an armistice is binding upon the belligerents from the time agreed for commencement of the armistice (or the time of signature if the time of commencement is not specified), officers of the armed forces are responsible for adhering to the armistice only from the time when they receive official information of its existence from their State.[176]

12.11.4.2 *Duration of an Armistice and the Resumption of Operations.* If the duration of an armistice is not prescribed, belligerents may resume operations at any time, provided always that the enemy is warned within the time agreed upon, in accordance with the terms of the armistice.[177]

If the duration is for a definite period of time and there is no positive agreement to the contrary, hostilities may be resumed without notice after the prescribed period of time.

12.11.4.3 *General Prohibition on Offensive Military Operations.* An armistice agreement generally would be understood to prohibit offensive military operations, such as conducting attacks or seizing territory.[178]Such acts would be examples of serious violations of an armistice.[179]

12.11.4.4 *Armistice No Excuse for Relaxing Vigilance.* The existence of an armistice does not warrant relaxation of vigilance in the service of security and protection, or in the preparedness of forces for action or exposing positions to the enemy.[180]For example, belligerents may continue to gather intelligence during an armistice, and may rely on that intelligence after the armistice has ended.[181]

12.11.4.5 *Preparations for Resuming Hostilities and Espionage Generally Not Prohibited.* During an armistice, in the absence of stipulations to the contrary, each belligerent is authorized:

- to make movements of forces within its own lines;
- to receive reinforcements;
- to construct new fortifications, installations, and bases;
- to build and repair transportation and communications facilities;
- to seek information about the enemy;
- to bring up supplies and equipment; and
- in general, to take advantage of the time and means at its disposal to prepare for resuming hostilities.[182]

Belligerents are not bound to discontinue espionage during an armistice, but the risks incurred by spies are the same as at other times.[183]For example, spies may be punished by an opposing belligerent if captured.[184]

12.11.4.6 *Armistice and Besieged Localities.* Absent express provision in the armistice agreement to the contrary, an armistice does not give authorities of a besieged place the right to receive food, water, or other supplies.[185]

However, armistice negotiations may be an opportunity for belligerents to attempt to conclude local agreements for the removal of wounded, sick, infirm and aged persons, children, and maternity cases, or for the passage of ministers of all religions, medical personnel, and medical equipment on their way to such areas.[186]In addition, armistice negotiations may be an opportunity for commanders to seek to make arrangements to permit the free passage of certain consignments.[187]

12.12 ARMISTICES – SUBJECTS USUALLY ADDRESSED

The following topics are often addressed in an armistice agreement: (1) the scope of the armistice, including start time and duration; (2) what military activities are prohibited by the armistice; (3) the relations between forces and the local populations during the armistice; (4) humanitarian activities to occur during the armistice; and (5) mechanisms for implementing the armistice.

Armistice provisions may address a range of political and military issues, provided that belligerents agree upon the terms and provided that the approving authorities are competent to address them.[188]

12.12.1 Scope of an Armistice, Including Start Time and Duration. The armistice should specify its scope—when and where it applies. The start time and duration of an armistice should be specified as precisely as possible (*e.g.*, date, time, and applicable time zone) to avoid misunderstanding and an unintentional resumption of hostilities.[189]

It may be appropriate for the armistice to specify that it starts at a later time to allow all forces to receive notification of the armistice before it becomes effective.[190]

If the armistice is of an indefinite duration, the terms and manner of giving notice that that armistice is to be ended (including the length of time between the delivery of the notice and when combat operations may recommence) should be specified in the armistice agreement.[191]

12.12.2 What Military Activities Are Prohibited by an Armistice. Although an armistice is generally understood to prohibit offensive operations, the armistice agreement may specify the extent to which offensive operations are prohibited and other military activities that are also prohibited during the armistice.[192]

12.12.3 Relations Between Forces and the Local Population During an Armistice. It rests with the Contracting Parties to settle, in the terms of the armistice, what intercourse may be held in the theater of war with the inhabitants of one belligerent State and those of the other.[193]

If nothing is stipulated, intercourse (*e.g.*, communication, movement, or commerce) remains suspended, as during actual hostilities.[194]For example, absent explicit agreement, a belligerent retains the right to control all communications or movement of members of the civilian population between its lines and enemy lines (*e.g.*, denying movement to reduce the risk of enemy espionage activities, denying trade to avoid strengthening enemy armed forces).[195]

The armistice agreement may also stipulate responsibilities of each party for civil administration of areas under its respective control (*e.g.*, public safety and public health needs, maintenance of public utilities including communications facilities). The armistice agreement may also specify the responsibilities of each belligerent for civil administration of neutral areas or areas where control is shared.[196]

12.12.4 Humanitarian Activities to Occur During an Armistice. An armistice agreement may arrange for a variety of humanitarian activities, such as:

• search for and collection of the sick and wounded from the battlefield, by, for example, medical personnel or authorized civilians (such as non-governmental organizations); and

• the provision or transport of medical care or food supplies intended for the civilian population, including displaced persons.

12.12.4.1 *Disposition of Detainees*. Because armistice agreements are generally only a temporary suspension of hostilities and not a complete cessation of active hostilities, POWs generally need not be repatriated upon the conclusion of an armistice agreement.[197]If it is desired that detainees, such as POWs, retained personnel, or civilian internees, should be released or exchanged, specific provisions in this regard should be made in the armistice agreement.[198]

12.12.5 Mechanisms for Implementing an Armistice. An armistice agreement may specify mechanisms to help implement it.

12.12.5.1 *Neutral Zone Created by an Armistice*. An armistice may create a "neutral zone," situated in order to minimize risk of unintentional confrontation between opposing forces.[199]It may be necessary for forces to withdraw in order to establish the zone.[200]

It is usually agreed that these lines are not to be crossed or the neutral zone entered except by *parlementaires* or other parties by special agreement for specified purposes, such as to bury the dead and collect the wounded.[201]

12.12.5.2 *Consultative Mechanism*. The armistice agreement may provide for the establishment of a commission composed of representatives of the opposing forces to supervise the implementation of the armistice agreement.[202]The agreement may specify that other representatives, such as representatives from neutral States or representatives from local authorities, are included on the commission.[203]

12.12.5.3 *Signals*. The belligerents may agree to use certain signals to help implement the armistice. For example, signals may be agreed upon to indicate the passage of *parlementaires*, the start or end of the armistice, or for other reasons.

12.12.5.4 *Maps Showing Boundaries or the Location of Forces*. An armistice agreement may also include an exchange of maps or other imagery showing the lines of opposing forces (or the location of forces) at the time of the start of the armistice.[204]Such maps may facilitate understanding of applicable boundaries, such as the location of neutral zones, as well as reduce the risk of unintended confrontation.

12.13 ARMISTICES – VIOLATIONS AND DENUNCIATION

An armistice, like other formal arrangements between belligerents, engages the honor of all parties for the exact and complete fulfillment of every obligation thereby imposed.[205]Serious violations of the armistice by one of the parties gives an opposing party the right of denouncing the armistice, and deliberate violations of the armistice by individuals are punishable.

12.13.1 Serious Violations as a Basis for Denouncing an Armistice. Any serious violation of the armistice by one of the parties gives an opposing belligerent the right of denouncing the armistice, and even, in cases of urgency, of recommencing hostilities immediately.[206]

12.13.1.1 *Examples of Serious Violations of an Armistice*. Serious violations of an armistice include acts in contravention of the terms of the agreement or acts that are wholly inconsistent with the status of the suspension of hostilities.[207]Such acts may include:

• conducting attacks against the enemy;

• a deliberate advance or construction of works beyond the line agreed upon;

• the seizure of any point outside the lines; and

• the utilization of the occasion to withdraw forces from an unfavorable position commanded by the enemy.[208]

12.13.1.2 *Process of Denunciation – Warning Requirement*. Absent urgent necessity, a delay should occur between denunciation of the armistice and the resumption of hostilities.[209]This delay allows the denunciation to act as warning and to ensure that the party recommencing hostilities does not rely on the other party's compliance with the armistice as a way of facilitating its offensive military operations.

It would be dishonorable and perfidious for either party, without warning, to resume hostilities during the period of an armistice, with or without a formal denunciation thereof, except in case of urgency and upon convincing proof of intentional and serious violation of its terms by the other party.[210]

Nevertheless, under Article 40 of the Hague IV Regulations, upon definite proof of such a violation of an armistice, if the delay incident to formal denunciation and warning seems likely to give the violator a substantial advantage of any kind, the other party is free to resume hostilities without warning and with or without a formal denunciation.[211]For example, the law of war does not prohibit armed forces that reasonably believe an adversary is committing armistice violations from taking necessary and proportionate action in self-defense immediately in response.

12.13.1.3 *Process of Denunciation – Approval of Co-Belligerents Not Required.* A belligerent faced with serious violations of an armistice is not required to obtain the permission of its co-belligerents, even though those States may be parties to the armistice agreement, before recommencing hostilities.[212]

12.13.2 Violations of an Armistice by Individuals. A violation of the terms of an armistice by private persons acting on their own initiative only entitles the injured party to demand the punishment of the offenders or, if necessary, compensation for the losses sustained.[213]

12.13.2.1 *"Private Persons" Defined for the Purposes Article 41 of the Hague IV Regulations.* A private person, in the sense of Article 41 of the Hague IV Regulations, refers to any person, including a member of the armed forces, who acts on his or her own responsibility.[214]

12.13.2.2 *Effect of Violations of an Armistice by Individuals.* Deliberate violation of the terms of an armistice by individuals is punishable as a war crime. Such violations by individual members of the armed forces or subordinate officers do not justify denunciation of the armistice unless they are proved to have been committed with the knowledge and actual or tacit consent of their own government or commander. Consent may be inferred in the event of a persistent failure to punish such offenders.[215]

12.13.2.3 *Violation of Armistice Agreements and POW Status.* Violations of an armistice agreement, like other pre-capture law of war violations, are not a basis for denying a person POW status, if that person otherwise qualifies for POW status under the GPW.[216]

12.14 U.N. SECURITY COUNCIL CEASE-FIRES

The U.N. Security Council may call upon or demand that hostilities be ceased.[217]In addition, the U.N. Security Council may call upon or demand that members of the United Nations conclude cease-fire or armistice agreements.[218]The U.N. Security Council may also establish a formal cease-fire.[219]

Cease-fires established by the U.N. Security Council may be interpreted in accordance with the general principles of international law governing armistices.[220]

XIII – Naval Warfare

Chapter Contents

13.1 INTRODUCTION

This Chapter addresses the law of war rules that apply to naval operations, especially those aspects of naval warfare that differ from warfare on land or in the air. This Chapter also briefly addresses the legal boundaries of the ocean, which may be relevant to the application of the law of war.

Navy publications have provided discussion of other public international law relating to naval operations, including discussion of the law of the sea applicable during peacetime.[1]

Some topics that are related to neutrality law and that are relevant to the conduct of both naval and air warfare, such as the conduct of visit and search and neutral commerce, are addressed in Chapter XV, the Law of Neutrality.[2]

13.1.1 The Law of the Sea During Armed Conflict. The law of the sea is a body of treaty and customary international law. Its rules have been developed principally with peacetime situations in mind. Nothing in the law of the sea impairs a State's inherent right of individual or collective self-defense, or rights during armed conflict.[3]

For example, the law of the sea provides that the high seas are reserved for peaceful purposes.[4]However, the use of the high seas for peaceful purposes is understood not to impose restraints upon military operations that would otherwise be consistent with international law, or to impair a State's inherent right of self-defense.[5]The use of outer space for peaceful purposes also has been interpreted in this way.[6]

As another example, certain rights under the law of the sea (*e.g.*, certain rights of navigation of vessels, or rights of States with respect to exclusive economic zones) must be exercised with due regard for the rights and duties of other States.[7]To the extent this obligation applies during armed conflict, what regard would be due would depend on military necessity and other principles and rules of the law of war, which are specially adapted to the circumstances of armed conflict.[8]

The United States and the LOS Convention. The U.13.1.2 N. Convention on the Law of the Sea (LOS Convention) was opened for signature on December 10, 1982.[9]

The United States is not a Party to the LOS Convention. The United States did not sign the LOS Convention when it opened for signature because of several major problems in the Convention's deep seabed mining provisions.[10]However, in 1983, the United States announced that it was prepared to accept and act in accordance with the balance of interests reflected in the LOS Convention relating to traditional uses of the oceans – such as navigation and overflight – and that the United States would exercise and assert its navigation and overflight rights and freedoms on a worldwide basis in a manner that is consistent with the balance of interests reflected in the LOS Convention.[11]For example, consistent with the LOS Convention, the United States has claimed a 12-nautical mile territorial sea.[12]Similarly, the United States has established a contiguous zone extending 24 nautical miles from the baselines of the United States determined in accordance with international law, but in no case within the territorial sea of another nation.[13]

The Agreement on the Implementation of Part XI of the LOS Convention was adopted on August 17, 1994.[14]Because this agreement addressed the objections that the United States previously expressed to Part XI of the LOS Convention, President Clinton recommended that the Senate give its advice and consent to accession to the LOS Convention and to ratification of the Agreement on the Implementation of Part XI.[15]Subsequent administrations have also supported U.S. accession to the LOS Convention and the ratification of the Agreement on the Implementation of Part XI of the LOS Convention.[16]

13.2 LEGAL BOUNDARIES OF THE OCEANS

The legal classifications of ocean areas may be relevant to the application of the law of war and therefore affect military operations during armed conflict by, for example:

- determining the legal boundaries of airspace above those waters;[17]
- determining whether waters are neutral waters;[18]
- determining the authority that a belligerent State has with respect to neutral vessels in an area;[19]or
- determining the authority that a neutral State has with respect to belligerent vessels in an area.[20]

Waters are often divided analytically between *national waters* (*i.e.*, internal waters, territorial seas, and archipelagic waters), which are subject to the sovereignty of a State, and *international waters*, which are not subject to the sovereignty of any State.[21]In addition, special rules apply to international straits and archipelagic sea lanes.[22]

13.2.1 Territory – Notes on Terminology. In some cases, the word "territory" is used to describe the land, waters, and airspace subject to the sovereignty of a State.[23]In other cases, the word "territory" is used to describe only the land that is subject to the sovereignty of a State.[24]

Thus, in describing waters that are not subject to the sovereignty of a State (*e.g.*, the exclusive economic zone and continental shelf, and high seas), the word "territory" should not be used. Coastal States may exercise limited "sovereign rights" over specific functional areas in the exclusive economic zone and on the continental shelf, but these rights do not imply sovereignty over these areas.[25]

13.2.2 National Waters. National waters, which are subject to the sovereignty of a State, include internal waters, territorial seas, and archipelagic waters.

13.2.2.1 *Internal Waters*. A State has sovereignty over its internal waters.[26]

Internal waters are generally understood to be those waters on the landward side of the baseline of the territorial sea.[27]The coastal baseline must be defined in accordance with specific rules of international law as reflected in the LOS Convention.[28]

13.2.2.2 *Territorial Seas*. The sovereignty of a State extends, beyond its land territory and its internal waters, to a belt of sea adjacent to its coast, described as the territorial sea.[29]

The United States has claimed a 12 nautical mile territorial sea and recognizes territorial sea claims of other nations up to a maximum breadth of 12 nautical miles.[30]Previously, the United States claimed a three-nautical mile territorial sea.[31]This claim was made in the context of asserting the rights of the United States as a neutral State.[32]

Coastal States often make specific maritime claims, but the United States does not recognize those maritime claims that are not in conformity with customary international law, as reflected in the LOS Convention.[33]

13.2.2.3 *Archipelagic Waters*. An archipelagic State's sovereignty extends to certain waters enclosed by archipelagic baselines drawn in accordance with the LOS Convention.[34]

An "archipelago" means a group of islands, including parts of islands, interconnecting water, and other natural features that are so closely interrelated that they form an intrinsic geographic, economic, and political entity or that historically have been regarded as such.[35]An "archipelagic State" means a State constituted wholly by one or more archipelagos and may include other islands.[36]The Philippines and Indonesia are examples of archipelagic States.[37]

13.2.2.4 *Innocent Passage of Foreign Vessels Through Territorial Seas and Archipelagic Waters*. During peacetime, all ships enjoy a right of innocent passage through territorial seas and archipelagic waters.[38]A Coastal State, however, has a right of protection, which includes the right to suspend temporarily in specified areas of its territorial sea the innocent passage of foreign ships if such suspension is essential for the protection of its security.[39]

During armed conflict, the right of innocent passage would not apply between belligerent States. During armed conflict, belligerent States may restrict the right of neutral vessels to conduct innocent passage through territorial seas and archipelagic waters belonging to a belligerent State, such as by establishing maritime zones.[40]During armed conflict, neutral States may regulate, and even prohibit, belligerent warships and prizes from entering their territorial seas and archipelagic waters.[41]

13.2.3 International Waters. International waters, which are not subject to the sovereignty of any State, include contiguous zones, exclusive economic zones (EEZs), and the high seas. These waters, which are seaward of the territorial sea, are waters in which States have certain freedoms, such as freedom of navigation and overflight.

13.2.3.1 *High Seas Freedoms*. States may exercise certain freedoms on the high seas, such as freedom of navigation and freedom of overflight.[42]For example, the high seas freedoms that warships may exercise include: task force maneuvering, flight operations, military exercises, surveillance, intelligence gathering activities, and ordnance testing and firing.[43]These rights and freedoms must be exercised with due regard for the interests of other States in their exercise of the freedom of the high seas and other applicable rights.[44]

Within a State's EEZ, other States enjoy high seas freedoms of navigation and overflight, and other rights, but must exercise those rights with due regard for the rights and duties of the coastal State.[45]

13.2.3.2 *Contiguous Zones*. A contiguous zone is an area extending seaward from the territorial sea to a maximum distance of 24 nautical miles from the baseline, in which the coastal State may exercise the control necessary to prevent or punish infringement of its customs, fiscal, immigration, and sanitary laws and regulations that occurs within its territory or territorial sea.[46]

The United States has established a contiguous zone extending 24 nautical miles from the baselines of the United States determined in accordance with international law, but in no case within the territorial sea of another nation.[47]

13.2.3.3 *Exclusive Economic Zones (EEZs)*. An Exclusive Economic Zone (EEZ) is a zone of limited, generally resource-related rights and jurisdiction adjacent to the territorial sea and may not extend beyond 200 nautical miles from the baseline.[48]The United States has established an exclusive economic zone.[49]

Although a coastal State does not have sovereignty over its EEZ, it has certain, generally economic rights over these areas that must be exercised with due regard for the rights and duties of other States, such as the high seas freedoms of other States.[50]

13.2.3.4 *High Seas*. International law regarding the high seas applies to all parts of the sea that are not included in the EEZ, territorial sea, internal waters, or archipelagic waters of an archipelagic State.[51]No State may validly purport to subject any part of the high seas to its sovereignty.[52]States may exercise certain freedoms on the high seas, such as freedom of navigation and freedom of overflight.[53]

13.2.4 Chart Illustrating the Legal Boundaries of the Oceans and Airspace. This chart, reproduced from the 2007 Commander's Handbook on the Law of Naval Operations, seeks to summarize the maximum permissible claims of States with respect to the boundaries of the oceans and airspace, as reflected in the LOS Convention.[54]

13.3 OVERVIEW OF RULES FOR NAVAL ENGAGEMENTS

In general, the rules for conducting attacks, such as bombardments, by naval forces are the same as those for land or air forces.[55]

13.3.1 Areas of Naval Warfare. As a general rule, naval forces may attack military objectives wherever located outside neutral territory.[56]In certain cases (*e.g.*, involving belligerent use of neutral territory as a base of operations), hostilities may be conducted in neutral territory to redress violations of neutrality.[57]

13.3.2 Classification of Vessels. The law of land warfare has divided enemy nationals into different categories in order to facilitate the protection of the civilian population from hostilities.[58]Similarly, the law of naval warfare has sought to classify enemy vessels to protect those that are civilian or non-combatant in character.

In general, all vessels of an enemy State may be understood to fall into one of three classes: (1) warships and naval and military auxiliaries; (2) merchant vessels; and (3) exempt vessels.

Warships and naval and military auxiliaries are generally liable to attack and capture.[59]

Merchant vessels are generally liable to capture, but are liable to attack if they forfeit their protection.[60]Exempt vessels are not liable to capture or attack, unless they forfeit their protection.[61]

13.3.3 Vessels Entitled to Conduct Attacks. During international armed conflict at sea, warships are the only vessels that are entitled to conduct attacks.[62]

Other vessels, such as auxiliary vessels and merchant vessels, are not entitled to conduct attacks in offensive combat operations.[63]All vessels, however, may defend themselves (including resisting attacks by enemy forces).[64]

13.3.3.1 *Entitlement of Vessels to Conduct Attacks During Non-International Armed Conflict.* The United States is not a Party to any treaties that would prohibit the use of warships and auxiliaries in non-international armed conflict, nor has the United States recognized such a prohibition in customary international law. Accordingly, State vessels other than warships may be used to conduct attacks against non-State armed groups during noninternational armed conflict.[65]For example, international law does not prohibit auxiliaries from conducting attacks in a non-international armed conflict. Similarly, a State may use its law enforcement authorities to address insurgent groups, and there would be no objection to using a law enforcement vessel as part of operations against insurgents.[66]

In some cases, the acts of hostility by insurgents on the high seas may be regarded as piracy.[67]

13.3.4 Shipwrecked Persons. Shipwrecked persons are *hors de combat*, and may not be made the object of attack.[68]Shipwrecked persons include personnel involved in forced landings at sea by or from aircraft, but do not include, for example, combatant personnel engaged in attacks.[69]

As far as military exigencies permit, after each naval engagement, all possible measures should be taken without delay to search for and collect the wounded, sick, and shipwrecked, and to recover the dead.[70]

13.3.5 Surrender by Enemy Vessels. The general rules on the protection of persons *hors de combat*, including the rule prohibiting the attack of persons who have surrendered, also apply to enemy vessels.[71]In particular, it is forbidden to make an enemy vessel the object of attack if it has genuinely, clearly, and unconditionally surrendered, in circumstances in which it is feasible to accept such surrender.[72]

Once an enemy vessel has clearly indicated a readiness to surrender, the attack must be discontinued.[73]Indicia of surrender by vessels may include:

- hauling down her flag;
- hoisting a white flag;[74]
- surfacing (in the case of submarines);[75]
- stopping engines and responding to the attacker's signals;7[76]or
- taking to lifeboats.

13.4 ENEMY WARSHIPS

Enemy warships and naval and military auxiliaries are subject to attack, destruction, or capture anywhere beyond neutral territory.

13.4.1 Character of Warships. A warship is generally understood to be a ship belonging to the armed forces of a State bearing the external markings distinguishing the character and nationality of such ships, under the command of an officer duly commissioned by the government of that State and whose name appears in the appropriate service list of officers, and manned by a crew that is under regular armed forces discipline.[77]

Warships are generally understood to possess certain privileges and immunities from the jurisdiction of other States.[78]

13.4.2 Attacks Against Enemy Warships. In general, enemy warships are military objectives.[79]However, warships that have surrendered or that are exempt vessels may not be made the object of attack.[80]

13.4.3 Captured Enemy Warships – No Prize Procedure. Prize procedures are not used for captured enemy warships because their ownership vests immediately in the captor's government by the fact of capture.[81]As public movable property, warships are seizable as war booty.[82]Similarly, prize procedures are not used for captured enemy military aircraft.[83]

13.5 ENEMY MERCHANT VESSELS

13.5.1 Capture of Enemy Merchant Vessels. Enemy merchant vessels may be captured wherever located beyond neutral territory. Prior exercise of visit and search is not required, provided positive determination of enemy status can be made by other means.[84]

13.5.1.1 *Captured Enemy Merchant Vessels – Notes on Terminology*. A captured neutral or enemy merchant vessel is called a *prize*.

13.5.1.2 *Use of Prize Procedures to Complete Transfer of Title*. Prize procedures are usually used to complete the transfer of title of captured property, such as enemy merchant ships.[85]

13.5.1.3 *Destruction of Captured Enemy Merchant Vessels*. When military circumstances preclude sending or taking in such vessel or aircraft for adjudication as an enemy prize, it may be destroyed after all possible measures are taken to provide for the safety of passengers and crew.[86]Documents and papers relating to the prize should be safeguarded and, if practicable, the personal effects of passengers should be saved.[87]Every case of destruction of a captured enemy prize should be reported promptly to higher command.[88]

The destruction of neutral prizes involves similar procedures, but a more serious responsibility.[89]

13.5.2 Attack of Enemy Merchant Vessels. Enemy merchant vessels may be made the object of attack, outside neutral territory, if they constitute a military objective.[90]

In particular, enemy merchant vessels may be attacked and destroyed by warships, either with or without prior warning, in any of the following circumstances:[91]

- persistently refusing to stop upon being duly summoned to do so;
- actively resisting visit and search or capture;
- sailing under convoy of enemy warships or enemy military aircraft;
- armed with systems or weapons beyond that required for self-defense against terrorist, piracy, or like threats;
- if incorporated into, or assisting, the enemy's military intelligence system;
- if acting in any capacity as a naval or military auxiliary to an enemy's armed forces; or
- if otherwise integrated into the enemy's war-fighting/war-sustaining effort such that a belligerent warship's compliance with the rules of the 1936 London Protocol would, under the circumstances of the specific encounter, subject the warship to imminent danger or would otherwise preclude mission accomplishment.

The 1936 London Protocol provides that "except in the case of persistent refusal to stop on being duly summoned, or of active resistance to visit or search, a warship, whether surface vessel or submarine, may not sink or render incapable of navigation a merchant vessel without having first placed passengers, crew and ship's papers in a place of safety."[92]Merchant vessels in the sense of this rule, however, do not include merchant vessels that constitute military objectives.[93]

13.5.2.1 *Attack of Civilian Passenger Vessels*. If a civilian passenger vessel constitutes a military objective and thus is liable to attack, any attack must comply with other applicable rules related to attacks.[94]

In particular, attacks against civilian passenger vessels engaged in passenger service must comply with the requirement that the expected loss of life or injury to civilians, and damage to civilian objects incidental to the attack, must not be excessive in relation to the concrete and direct military advantage expected to be gained.[95]

13.5.3 Detention of Personnel From Enemy Merchant Vessels. Officers and crews of captured enemy merchant ships may be detained.[96]If detained, such persons are POWs.[97]

Other civilian enemy nationals on board such captured ships as private passengers are subject to the discipline of the captor.[98]If detained, such persons must, at a minimum, be afforded certain fundamental guarantees of humane treatment.[99]

Nationals of a neutral State on board captured enemy merchant vessels should not be detained unless they have participated in acts of hostility or resistance against the captor, or are otherwise in the service of the enemy.[100]

13.6 ENEMY VESSELS EXEMPT FROM CAPTURE OR DESTRUCTION

Certain classes of enemy vessels are exempt from capture or destruction, provided they are innocently and solely employed in the activities that enjoy exempted status. Exempt vessels may not be used for purposes outside their innocent role while taking advantage of their harmless appearance. Warships may not be disguised as exempt vessels.[101]

13.6.1 Duties of Exempt Vessels. These specially protected vessels must not take part in hostilities or assist the enemy's military effort in any manner. Such vessels and boats are subject to the regulations of a belligerent naval commander operating in the area.[102] They must not hamper the movement of combatants, must submit to identification and inspection procedures, and may be ordered out of harm's way. Refusal to provide immediate identification upon demand is ordinarily sufficient legal justification for capture or destruction.

13.6.2 Classes of Exempt Vessels. Specifically exempt vessels include: (1) cartel vessels; (2) hospital ships; (3) vessels charged with religious, non-military scientific, or philanthropic missions; (4) vessels granted safe conduct; and (5) small fishing and trade vessels.[103]

13.6.2.1 *Cartel Vessels*. Vessels and aircraft designated for, and engaged in, the exchange of POWs are privileged from capture.[104]

13.6.2.2 *Hospital Ships That Meet the Requirements of the GWS-Sea*. Hospital ships, including military hospital ships, commissioned civilian hospital ships, and authorized neutral hospital ships, that meet the requirements of the GWS-Sea are exempt from capture.[105]

Hospital ships do not forfeit their exempt status if they are armed for self-defense purposes or if they possess or use communications equipment with encryption. Hospital ships may be armed with defensive weapon systems, including when necessary certain crew-served weapons, as a prudent anti-terrorism/force protection (AT/FP) measure to defend against small boat attacks.[106] The 1949 Geneva Conventions do not directly address weapons systems for hospital ships, but they do expressly provide for the arming of crew members with small arms for the maintenance of order and for the self-defense of the crew or the sick and wounded.[107] The arming of the ship with crew-served weapons for self-defense of the ship is not prohibited and is consistent with the ship's humanitarian purpose and the crew's duty to safeguard the wounded and sick.

The GWS-Sea provides that hospital ships may not use or possess "secret codes" as means of communication; however, state practice has evolved to accept that modern communications systems (*e.g.*, satellite communications and video teleconference systems) and navigational technology (*e.g.*, global positioning systems) require encryption.[108] Such systems must not be used for military purposes or in any way that is harmful to an adversary.[109]

13.6.2.3 *Vessels Charged With Religious, Nonmilitary Scientific, or Philanthropic Missions*. Vessels charged with religious, nonmilitary scientific, or philanthropic missions are exempt from capture.[110] Vessels engaged in the collection of scientific data of potential military application, however, would not be included within this exemption.

13.6.2.4 *Vessels Granted Safe Conduct*. Vessels and aircraft guaranteed safe conduct by prior arrangement between the belligerents are exempt from capture.[111]

13.6.2.5 *Small Coastal Fishing Vessels and Small Boats Engaged in Local Coastal Trade*. Small coastal (not deep-sea) fishing vessels and small boats engaged in local coastal trade are exempt from capture.[112]

13.7 SUBMARINE WARFARE

13.7.1 General Principle – Same Rules Applicable to Both Submarine and Surface Warships. Submarine warships must comply with the same law of war rules that apply to surface warships.[113] For example, in their action with regard to merchant ships, submarines must conform to the law of war rules to which surface vessels are subject.[114] In general, submarines must provide for the safety of passengers, crew, and ship's papers before destruction of an enemy merchant vessel.[115] However, the same exceptions to this rule that permit surface ships to attack enemy merchant vessels that are military objectives also permit submarines to conduct such attacks.[116]

13.7.2 Different Application of Law of War Rules in the Context of Submarine Warfare. Although submarines must comply with the same law of war rules as surface ships, a law of war rule may apply differently in the context of submarine warfare because of the different circumstances of submarine warfare as compared to surface warfare.

For example, like surface warships, submarine warships also have an obligation to search for and collect the shipwrecked, wounded, and sick after an engagement.[117] This obligation, however, is subject to certain practical limitations, and the practical limitations faced by submarines may be different than those faced by surface vessels.[118] For example, although a surface warship might be able to take on board survivors after an engagement, a submarine may have limited passenger carrying capabilities. Thus, it may be necessary to rely on other measures (*e.g.*, such as passing the location of possible survivors to a surface ship, aircraft, or shore facility capable of rendering assistance) to comply with the law of war obligation.[119]

13.8 BELLIGERENT CONTROL OF THE IMMEDIATE AREA OF NAVAL OPERATIONS

13.8.1 Belligerent Right to Establish Special Restrictions in the Immediate Area of Naval Operations. Within the immediate area or vicinity of naval operations, to ensure proper battle space management and self-defense objectives, a belligerent State may establish special restrictions upon the activities of neutral vessels and aircraft, and may prohibit altogether such vessels and aircraft from entering the area.[120]

The immediate area or vicinity of naval operations is that area within which hostilities are taking place, or belligerent State forces are actually operating.[121]

Belligerent State control over neutral vessels and aircraft within the immediate area of naval operations is based upon a belligerent State's right to ensure the security of its forces and its right to conduct hostilities without interference from neutrals.[122]

A belligerent State may not purport to deny access to neutral States, or to close an international strait to neutral shipping, pursuant to this authority unless another route of similar convenience remains open to neutral traffic.[123]

13.8.2 Belligerent Right to Control Communication of Neutral Merchant Vessels or Aircraft at Sea. A belligerent warship may exercise control over the communications of any neutral merchant vessel or civil aircraft whose presence in the immediate vicinity of naval operations might otherwise endanger or jeopardize those operations.[124]

A neutral merchant ship or civil aircraft within that area that fails to conform to a belligerent warship's directions concerning communications may thereby assume enemy character and risk being fired upon or captured.[125]Any transmission to an opposing belligerent State of information concerning military operations or military forces is inconsistent with the neutral State's duties of abstention and impartiality, and renders the neutral State's vessel or aircraft making such a communication liable to capture or destruction.[126]

Legitimate distress communications should be permitted to the extent that the success of the operation is not prejudiced thereby.[127]

13.9 MARITIME AND AIRSPACE ZONES: EXCLUSION,WAR, OPERATIONAL,WARNING, AND SAFETY

Belligerent States may establish various maritime and airspace zones during armed conflict. The legal rules that apply to the establishment and enforcement of a zone may depend on the function of the zone and, in particular, whether additional belligerent rights are asserted as a consequence of vessels entering the zone.

Neutral or non-belligerent States have established such zones.[128]

13.9.1Authority to Establish Zones. The authority necessary to establish a zone may depend on the location of the zone and the belligerent rights that are asserted as a consequence of vessels entering the zone.

The establishment of a zone in a State's waters may rely on its sovereignty over those waters.[129]Similarly, the establishment of a zone in a State's airspace may rely on its sovereignty over that airspace.[130]The right of a belligerent State to establish a zone outside its territory may rely on the belligerent State's: (1) right to interdict contraband; (2) right to control the immediate area of operations; or (3) right of blockade.[131]

Exclusion zones may also be established in accordance with an appropriate resolution adopted by the U.N. Security Council.

13.9.2 Use of Zones to Warn Vessels or Aircraft – War, Operational, Warning, and Safety Zones. A zone may be issued to advise vessels or aircraft to remain clear of an area of naval operations. Such a zone may also provide procedures to reduce the risk of neutral vessels being mistakenly attacked.[132]

Such zones have been used: (1) to identify a particularly dangerous operational area;[133]

(2) to assist in the defense of a particular area;[134]or (3) to assist in the defense of particular naval forces (*i.e.*, a defensive bubble).[135]

The establishment of such a zone does not relieve the proclaiming belligerent State of its obligation under the law of war to refrain from attacking vessels and aircraft that do not constitute military objectives. Thus, a vessel or aircraft that is otherwise protected does not forfeit its protection from being made the object of attack simply by entering a zone of the ocean on the high seas established by a belligerent State.[136]

In some cases, however, the fact that a vessel or aircraft enters a zone without authorization may be probative in assessing whether it is entitled to protection (*e.g.*, whether it is an enemy military vessel or aircraft, or whether it has acquired the character of enemy military vessels or aircraft).[137] For example, the notification of the zone in advance may mean that most neutral or protected vessels and aircraft have departed the area. Similarly, the entry into the zone of an unidentified vessel without authorization may be probative of whether the vessel constitutes a military objective.

13.9.3 Use of Zones to Counter Enemy Logistics. Zones may be used to assist in the implementation of the belligerent's strategy to counter the adversary's logistics.[138] For example, zones may be established to facilitate the interdiction of contraband.[139] As a case in point, a zone may be established to warn neutral vessels and aircraft that they will be subject to visit and search if they attempt to enter the zone without authorization.

Zones may not be employed for the purpose of starving the civilian population.[140] Such zones must not impose an unreasonable burden on neutral commerce in free goods.[141]

13.9.4 Use of Zones to Prevent Movement – Exclusion Zones. Belligerents may establish exclusion zones that prohibit the entry of vessels or aircraft without authorization from that belligerent.[142] Such zones may suspend the right of innocent passage through non-neutral waters.[143]

The extent, location, and duration of the exclusion zone and the measures imposed should not exceed what is required by military necessity.[144]

Belligerents establishing such zones must provide safe passage through the zone for neutral vessels and aircraft where the geographical extent of the zone significantly impedes free and safe access to the ports and coasts of a neutral State and, unless military requirements do not permit, in other cases where normal navigation routes are affected.[145]

13.9.5 Procedures for Declaring a Zone. As a matter of practice, maritime and/or aerial warnings (*e.g.*, HYDROLANT, HYDROPAC, Notice to Airmen (NOTAM)) would ordinarily be promulgated as part of the establishment of a maritime or airspace zone.

13.10 BLOCKADE

A blockade is an operation by a belligerent State to prevent vessels and/or aircraft of all States, enemy as well as neutral, from entering or exiting specified ports, airfields, or coastal areas belonging to, occupied by, or under the control of an enemy belligerent State.[146]

13.10.1 Purpose of Blockade and Belligerent Rights Associated With Blockade. The purpose of a blockade is to deprive the adversary of supplies needed to conduct hostilities.[147]

A blockade enables the blockading State to control traffic in the blockaded area. A blockade also enables the blockading State to take measures on the high seas to deny supplies to a blockaded area.[148] For example, a blockading State has the right to visit and search vessels on the high seas to enforce its blockade.[149]

13.10.2 Criteria for Blockades. In order to be binding, a blockade must meet certain criteria.

13.10.2.1 *Authority for the Establishment of a Blockade.* A blockade must be established by proper legal authority. The establishment of a blockade has been accomplished by a declaration of the belligerent State, or by the commander of the blockading force acting on behalf of the belligerent State.[150]

The U.N. Security Council is also expressly given the authority by the Charter of the United Nations to authorize blockades.[151] A blockade of the ports or coasts of a State is generally regarded as a measure involving force against that State for the purposes of the Charter of the United Nations.[152]

13.10.2.2 *Notification of the Blockade.* It is customary for the belligerent State establishing the blockade to notify all affected States of its imposition.[153] The commander of the blockading forces will usually also notify local authorities in the blockaded area.[154]

Notification should include, as a minimum, the date the blockade is to begin, its geographic limits, and the grace period granted neutral vessels and aircraft to leave the area to be blockaded.[155]

The form of the notification is not material so long as it is effective, and a variety of methods may be used.[156]

The notification of the blockade may establish a presumption of knowledge of the blockade that is required in the offense of breach or attempted breach of a blockade.[157]

13.10.2.3 *Effectiveness of the Blockade.* In order to be binding, a blockade must be effective.[158] This requirement of effectiveness is intended to prevent "paper" blockades.[159]

The requirement of effectiveness means that the blockade must be maintained by forces that are sufficient to render ingress or egress of the blockaded area dangerous.[160]The requirement of effectiveness does not preclude temporary absence of the blockading force, if such absence is due to stress of weather or to some other reason connected with the blockade (*e.g.*, pursuit of a blockade runner).[161]

The forces that are necessary to make a blockade effective depend on the specific military circumstances.[162]The blockade may be maintained by forces that are some distance from the shore.[163]

13.10.2.4 *Impartial Application of the Blockade*. A blockade must be applied impartially to the vessels and aircraft of all States.[164]The requirement that a belligerent must apply a blockade impartially to the vessels and aircraft of all States is intended to prevent measures of discrimination by the blockading belligerent in favor of or against the vessels and aircraft of particular States, including its own or allied vessels and aircraft.[165]

13.10.2.5 *Limitations on the Scope of the Blockade*. A blockade must not bar access to, or departure from, neutral ports and coasts.[166]

A blockade may not be used for the purpose of starving the civilian population, and the expected incidental harm to the civilian population may not be excessive in relation to the expected military advantage to be gained from employing the blockade.[167]

13.10.3 Special Entry and Exit Authorization. In certain cases, vessels or aircraft may enter or exit a blockaded area.

13.10.3.1 *Neutral Warships and Military Aircraft*. Although neutral warships and military aircraft enjoy no positive right of access to blockaded areas, the belligerent State imposing the blockade may authorize their entry and exit. Such special authorization may be made subject to such conditions as the blockading force considers necessary and expedient.[168]

13.10.3.2 *Neutral Vessels and Aircraft in Evident Distress*. Neutral vessels and aircraft in evident distress should be authorized entry into a blockaded area, and subsequently authorized to depart, under conditions prescribed by the officer in command of the blockading force or responsible for maintenance of the blockading instrumentality (*e.g.*, mines).[169]

13.10.3.3 *Passage of Relief Consignments Under the GC*. Similarly, neutral vessels and aircraft engaged in the carriage of qualifying relief supplies for the civilian population and the sick and wounded should be authorized to pass through the blockade cordon, subject to the right of the blockading force to prescribe the technical arrangements, including search, under which passage is permitted.[170]

13.10.4 Breach and Attempted Breach of Blockade. Breach or attempted breach of a blockade subjects a neutral vessel or aircraft to capture.[171]

13.10.4.1 *Acts of Breach or Attempted Breach of Blockade*. Breach of blockade is the passage of a vessel or aircraft through a blockade without special entry or exit authorization from the blockading belligerent.[172]

Attempted breach of a blockade occurs from the time a vessel or aircraft leaves a port or airfield with the intention of evading the blockade and, for vessels exiting the blockaded area, continues until the voyage is completed.[173]A temporary anchorage in waters occupied by the blockading vessels does not justify capture, in the absence of other grounds.[174]

A presumption arises that certain merchant vessels or aircraft are attempting to breach a blockade where those vessels or aircraft are bound for a neutral port or airfield serving as a point of transit to the blockaded area.[175]

13.10.4.2 *Knowledge of the Existence of Blockade*. Knowledge of the existence of the blockade is essential to the offenses of breach of blockade and attempted breach of blockade; presumed knowledge is sufficient.[176]

Knowledge may be presumed once a blockade has been declared and appropriate notification provided to affected governments.[177]For example, a vessel in a blockaded port is presumed to have notice of the blockade as soon as it commences.[178]

A vessel sailing ignorantly (*i.e.*, with neither presumptive nor actual knowledge) to a blockaded port is not liable to capture, although it may be turned away from the blockaded area.[179]

13.11 NAVAL MINES

Naval mines are lawful weapons, *i.e.*, they are not illegal *per se*.[180]However, specific rules apply to their use. These rules have developed largely to mitigate the risk these mines pose to neutral vessels.[181]Different rules apply to landmines.[182]

13.11.1 Background on Naval Mines. Naval mines have been used for area denial, coastal and harbor defense, anti-surface and anti-submarine warfare, and blockade. Naval mines are sometimes classified as either armed or controlled mines, with controlled mines having no destructive capability until affirmatively activated by an arming order (thereby becoming armed mines).[183]

13.11.2 Peacetime Mining. Naval mines may not be emplaced in internal waters, the territorial sea, or archipelagic waters of another State in peacetime without that State's consent.[184]

13.11.2.1 *A State's Archipelagic Waters and Territorial Sea*. A State's right to mine its archipelagic waters and territorial sea during peacetime is subject to the right of innocent passage of foreign vessels.[185]

If armed mines are emplaced in a State's own archipelagic waters or territorial sea, appropriate international notification of the existence and location of such mines is required.[186]Because the right of innocent passage may be suspended only temporarily,[187]armed mines must be removed or rendered harmless as soon as the security threat that prompted their emplacement has terminated.[188]

Armed mines may not be emplaced in international straits or archipelagic sea lanes during peacetime.[189]

Emplacement of controlled mines in a State's own archipelagic waters or territorial sea is not subject to notification or removal requirements.[190]

13.11.2.2 *International Waters*. Controlled mines may be emplaced in international waters (*i.e.*, beyond the territorial sea of any State) if they do not unreasonably interfere with other lawful uses of the oceans. What constitutes an "unreasonable interference" may involve a number of factors, including:

- the rationale for their emplacement (*e.g.*, the self-defense requirements of the emplacing State);
- the extent of the area to be mined;
- the hazard (if any) to other lawful ocean uses; and
- the duration of their emplacement.[191]

Because controlled mines do not constitute a hazard to navigation, international notice of their emplacement is not required.[192]

Armed mines may not be emplaced in international waters prior to the outbreak of armed conflict, except under the most demanding requirements of individual or collective self-defense. Should armed mines be emplaced in international waters under such circumstances, prior notification of their location must be provided. A State emplacing armed mines in international waters during peacetime must maintain an on-scene presence in the area sufficient to ensure that appropriate warning is provided to ships approaching the danger area. All armed mines must be expeditiously removed or rendered harmless when the imminent danger that prompted their emplacement has passed.[193]

13.11.3 Naval Mining During Armed Conflict. Belligerent States may lawfully employ naval mines subject to certain restrictions.

Neutral States that lay automatic contact mines off their coasts must observe the same rules and take the same precautions as those that are imposed on belligerents.[194]In addition, a neutral State must inform ship owners, by a notice issued in advance, where automatic contact mines have been laid. This notice must be communicated at once to the Governments through the diplomatic channel.[195]

13.11.3.1 *General Obligation to Take Feasible Precautions When Using Naval Mines*. The general obligation to take feasible precautions for the protection of civilians applies when using naval mines.[196]In particular, when anchored automatic contact mines are employed, every possible precaution must be taken for the security of peaceful shipping.[197]

13.11.3.2 *International Notification When Using Armed Mines*. International notification of the location of emplaced armed mines must be made as soon as military exigencies permit.[198]Such notification may be conducted through official public announcements, communications to the United Nations, bilateral diplomatic notification, or notices to mariners.[199]

In particular, when anchored automatic contact mines are employed, belligerents undertake to do their utmost to render these mines harmless within a limited time and, should they cease to be under surveillance, to notify the danger zones as soon as military exigencies permit, by a notice addressed to ship owners, which must also be communicated to the Governments through the diplomatic channel.[200]

13.11.3.3 *Surveillance of Minefields*. Surveillance or monitoring of minefields may be an appropriate precaution to reduce the risk of harm to peaceful vessels by permitting warning to be given to peaceful vessels that unwittingly venture near the minefield.[201]

13.11.3.4 *Recording of Minefield Locations*. The location of minefields must be carefully recorded to ensure accurate notification and to facilitate subsequent removal and/or deactivation.[202]

13.11.3.5 *Restrictions on Where Naval Mines May Be Placed*. Mines may not be emplaced by belligerents in neutral waters.[203]

Naval mines may be employed to channelize neutral shipping, but not in a manner to deny transit passage of international straits or archipelagic sea lanes passage of archipelagic waters by such shipping.[204]

Mining of areas of indefinite extent in international waters is prohibited. Reasonably limited barred areas may be established by naval mines, provided neutral shipping retains an alternate route around or through such an area with reasonable assurance of safety.[205]

13.11.3.6 *Prohibition Against Mining Off Enemy Coasts and Ports Solely to Intercept Commercial Shipping*. Naval mines may not be emplaced off the coasts and ports of the enemy with the sole objective of intercepting commercial shipping.[206]Mining for some other purpose would not violate this rule.[207]For example, mines may otherwise be used in the strategic blockade of enemy ports, coasts, and waterways, even if commercial shipping is incidentally affected.[208]

13.11.3.7 *Neutralization of Naval Mines*. Anchored mines must become harmless as soon as they have broken their moorings.[209]In particular, it is forbidden to lay anchored automatic contact mines that do not become harmless as soon as they have broken loose from their moorings.[210]

Unanchored mines not otherwise affixed or imbedded on the ocean floor must become harmless within an hour after loss of control over them.[211]In particular, it is forbidden to lay unanchored (*i.e.*, free-floating) automatic contact mines, except when they are so constructed as to become harmless one hour at most after the person who laid them ceases to control them.[212]

13.11.4 Post-Conflict Naval Demining Obligations. The removal of naval mines at the end of the conflict may be the subject of specific agreements among States.[213]

At the close of a war to which Hague VIII is applicable, Parties to Hague VIII have an obligation to remove the naval mines that they had laid, each State removing its own mines.[214]As regards anchored automatic contact mines laid by one of the belligerents off the coast of the other, their position must be notified to the other party by the State that laid them, and each State must proceed with the least possible delay to remove the mines in its own waters.[215]

13.12 TORPEDOES

It is forbidden to use torpedoes that do not become harmless when they miss their mark.[216]Such torpedoes may become a hazard to innocent shipping, and therefore torpedoes must be designed to become harmless when they have missed their mark. For example, torpedoes have been designed to become harmless upon completion of their propulsion run, such as by sinking to the bottom.

13.12.1 Torpedoes – Notes on Terminology. In the 19th century, the term "torpedo" was used to refer to any explosive munition that operated by contact against the hull of a ship, including relatively stationary munitions that, in modern parlance, would be called mines.[217]However, by the time of the 1907 Hague VIII Convention, the term "torpedo" was used in the modern sense to refer to munitions that propelled through the water.[218]

13.13 DECEPTION BY NAVAL FORCES, INCLUDING THE USE OF ENEMY OR NEUTRAL FLAGS

In general, the rules concerning good faith, ruses, and perfidy also apply to naval warfare.[219]For example, it would be prohibited for a belligerent warship to convey falsely the appearance of a hospital ship.[220]On the other hand, a variety of deceptions are not prohibited, including camouflage, deceptive lighting, disguised electronic signatures, and dummy ships, among others.[221]

The rules concerning deception by belligerent warships, especially with respect to the flying of neutral flags or with respect to disguise as merchant vessels, may be considered more permissive than similar situations in land warfare.[222]

13.13.1 Belligerent Warship's Use of False Colors and Other Disguises. In particular, according to custom, it is permissible for a belligerent warship to use false colors and to disguise her outward appearance in other ways in order to deceive an enemy, provided that before going into action such warship shows her true colors.[223]

For example, a belligerent warship may fly enemy or neutral colors and display enemy or neutral markings in order to deceive the enemy into believing the vessel is of enemy or neutral nationality or is not a warship.[224]Such deceptions may be used in order to facilitate the approach of enemy vessels or to escape enemy vessels.[225]

Warships, however, may not seek to claim falsely the status of vessels that are exempt from capture or destruction. These vessels include: (1) cartel vessels; (2) hospital ships; (3) vessels charged with religious, non-military scientific, or philanthropic missions; (4) vessels granted safe conduct; and (5) small coastal fishing vessels and small boats engaged in local coastal trade.[226]

XIV – Air and Space Warfare

Chapter Contents

14.1 INTRODUCTION

This Chapter addresses the international law applicable to U.S. air and space operations during international armed conflict. This Chapter also discusses the application to air operations of some of the law of war rules discussed elsewhere in this manual. Air Force publications have provided discussion of other aspects of public international law relating to air operations, including discussion of rules applicable during peacetime.[1]

14.1.1 Application of the 1944 Chicago Convention With Respect to Military Aircraft and With Respect to States' Rights During Armed Conflict. The 1944 Chicago Convention primarily addresses States' obligations regarding international civil aviation.[2]The 1944 Chicago Convention does not restrict State action during war.[3]In addition, the 1944 Chicago Convention generally is not applicable to State aircraft.[4]However, the 1944 Chicago Convention imposes requirements with respect to entry by State aircraft into foreign airspace and with respect to the issue of due regard for the safety of navigation of civil aircraft.[5]

14.1.1.1 *1944 Chicago Convention and Freedom of Action of States During Armed Conflict.* According to Article 89 of the 1944 Chicago Convention, "[i]n case of war, the provisions of this Convention shall not affect the freedom of action of any of the contracting States affected, whether as belligerents or as neutrals."[6]Article 89 may be understood as an example, reflected in this case in the Convention, of the application of the general principle that the law of war, as a body of law specially adapted to the circumstances of armed conflict, is the controlling body of law with respect to armed conflict.[7]Under Article 89, a State's rights under the law of war (and the law of neutrality) would prevail in the event of conflict with obligations under the 1944 Chicago Convention.

The provisions of the 1944 Chicago Convention may be relevant during armed conflict insofar as they articulate restrictions that continue to apply to civil aircraft. For example, neutral civil aircraft engaged in international navigation would still be required to seek permission from a foreign neutral State before carrying munitions of war or implements of war through that neutral State's airspace.[8]

The 1944 Chicago Convention's requirements with respect to entry by State aircraft into foreign airspace and with respect to the issue of due regard for the safety of navigation of civil aircraft are discussed below.[9]

14.1.1.2 *1944 Chicago Convention and Military Aircraft or Other State Aircraft.* The 1944 Chicago Convention provides generally that the Convention "shall be applicable only to civil aircraft, and shall not be applicable to state aircraft."[10]State aircraft include aircraft used in military service.[11]The United States has made statements interpreting this provision of the 1944 Chicago Convention.[12]

The 1944 Chicago Convention, however, imposes obligations with respect to entry into foreign airspace by State aircraft and an obligation with respect to State aircraft and the safety of navigation of civil aircraft.[13]

14.1.1.3 *Requirement for State Consent Prior to Entry by Foreign State Aircraft.* The 1944 Chicago Convention provides that State aircraft (which include military aircraft) are not permitted to enter the airspace of another State without that State's consent.[14]During armed conflict, the requirement for a State's consent to entry into its airspace would generally continue to apply with respect to neutral military aircraft seeking to enter the airspace of foreign States.

This requirement, however, clearly would not apply to belligerent military aircraft conducting operations in enemy airspace.

The requirement for a State's consent to entry into its airspace may be characterized as applicable with respect to belligerent military aircraft and a neutral State's airspace; however, even if belligerent military aircraft enter neutral airspace with that neutral State's consent, such entry may involve violations of neutrality.[15]

There are exceptions to the requirement for State consent to entry into its airspace by State aircraft in certain cases of violations of neutrality.[16]

There is an exception during peacetime to the requirement for a State's consent to entry into its airspace by foreign State aircraft when such entry is due to distress and there is no reasonable safe alternative.[17]

14.1.1.4 *Due Regard for the Safety of Navigation of Civil Aircraft.* The 1944 Chicago Convention also provides that "[t]he contracting States undertake, when issuing regulations for their state aircraft, that they will have due regard for the safety of navigation of civil aircraft."[18]Assuming that an obligation to exercise due regard would be applicable during armed conflict, what regard would be due in any particular set of circumstances would depend on military necessity, and other principles and rules of the law of war, which are specially adapted to the circumstances of armed conflict. For example, the use of force against enemy civil aircraft that constitute military objectives and thus may be made the object of attack under the law of war would not be prohibited.[19]

As reflected in Article 89 of the 1944 Chicago Convention, the Convention does not restrict the freedom of action of States as belligerents or neutrals.[20]Nonetheless, as a practical matter, modern air warfare is often conducted in complex airspace, and U.S. forces have routinely and intensively coordinated flight operations with national civil aviation authorities. This coordination is important for a variety of purposes, including ensuring mission accomplishment, avoiding fratricide and mid-air collisions, and ensuring the safety of international civil aviation. DoD policy has required that, in the event of combat operations during armed conflict, aircraft commanders, consistent with military necessity, take measures to minimize hazards to civil air or surface traffic.[21]

Similarly, under the law of the sea, military operations must be conducted with due regard for the high seas freedom of overflight in international airspace.[22]

14.1.2 Past Attempts to Conclude Treaties About Air Warfare. Initial attempts to conclude agreements on the law of war governing means and methods in the air context were made during the 1899 and 1907 Hague Peace Conferences, before air power had become a significant factor in warfare. Extensive efforts were made in 1922-23 to adopt a code of laws specifically applicable only to air warfare; however, the proposed rules were not ratified by any State.[23]The United States has not ratified a treaty applicable solely to air operations during armed conflict, although the United States has ratified treaties that have included specific references to aspects of war in the air, such as protection for medical aircraft in the 1949 Geneva Conventions.[24]

14.2 LEGAL BOUNDARIES OF AIRSPACE

The legal classifications of airspace may be relevant to the application of the law of war and therefore may affect military operations during armed conflict by:

- determining whether airspace is neutral airspace;[25]
- determining the authority that a belligerent State has with respect to neutral aircraft in an area; or
- determining the authority that a neutral State has with respect to belligerent aircraft in an area.

14.2.1 Lateral Boundaries of Airspace. The lateral boundaries of the airspace are determined by the status of the land or water directly beneath them.

Airspace is often divided between *national airspace* (*i.e.*, airspace over a State's land territory, internal waters, territorial seas, and archipelagic waters), which is subject to the sovereignty of a State, and *international airspace*, which is not subject to the sovereignty of any State.[26]In addition, special rules apply to international straits and archipelagic sea lanes.[27]

The division between *national airspace* and *international airspace* is similar to the division between *national waters* and *international waters*.[28]One notable distinction, however, is that aircraft do not enjoy the same right of innocent passage over territorial seas that ships enjoy through territorial seas.[29]

14.2.1.1 *National Airspace.* Every State has complete and exclusive sovereignty over the airspace above its territory.[30]In other words, a State has sovereignty over the airspace that is above land or waters that are subject to its sovereignty.[31]Thus, a State's national airspace includes:

• airspace over that State's land territory, including its internal waters;[32]

• airspace over that State's territorial sea;[33]and

• airspace over that State's archipelagic sea.[34]

14.2.1.2 *International Airspace.* International airspace may be understood as all airspace that is not subject to the sovereignty of any State, including:[35]

• airspace over States' contiguous zones;[36]

• airspace over States' Exclusive Economic Zones;[37]and • airspace over the high seas.[38]

14.2.2 Vertical Boundary Between Airspace and Outer Space. A State's sovereignty over its airspace does not extend to outer space, which is not subject to the sovereignty of any State.[39]

It is generally accepted that orbiting objects are in outer space.[40]The United States has expressed the view that there is no legal or practical need to delimit or otherwise define a specific boundary between airspace and outer space.[41]

14.2.3 Flight Information Regions. Flight Information Regions (FIRs) are areas of airspace allocated by the International Civil Aviation Organization (ICAO). Every portion of airspace belongs to a defined flight information region. Within each flight information region, a flight information service and an alerting service are provided by the national authorities responsible for air traffic control. These services are the most basic levels of air traffic service, and they provide information to aviators necessary for the safe and efficient conduct of flights.[42]A flight information region normally encompasses substantial areas of international airspace and does not necessarily reflect international or national airspace borders.[43]Under Article 3 of the 1944 Chicago Convention, flight information region rules do not apply to State aircraft, including military aircraft, and military aircraft are free to operate in international airspace without the consent of or notice to coastal State authorities.[44]Thus, for example, a flight information region could not limit the rights of belligerent military aircraft during armed conflict.

14.2.4 Air Defense Identification Zone (ADIZ). In general terms, an ADIZ is an area of airspace over land or water in which the ready identification, location, and control of aircraft may be required in the interests of national security.[45]The United States has established ADIZs.[46]

14.2.4.1 *Establishment of an ADIZ.* The legal basis for a State to establish an ADIZ is its right to establish conditions and procedures for entry into its national airspace.[47]

The establishment of an ADIZ by a State does not constitute a claim to international airspace as its own. The ADIZ is merely a reference point for the initiation of identification procedures for aircraft on a course to enter national airspace.[48]

Because the establishment of an ADIZ by a State that extends to international airspace does not confer sovereignty over that airspace, the ADIZ, and its implementing regulations and operational practices, must respect the rights of other States to use international airspace. The United States does not recognize the right of a State to apply its ADIZ laws and regulations to foreign aircraft, either civil aircraft or State aircraft, if those aircraft do not intend to enter the national airspace of that State.[49]

14.2.4.2 *Failure to Comply With Identification Requirements.* Failure by an aircraft to identify itself once it enters an ADIZ does not, of itself, entitle the declaring State to use force against that aircraft, although it may subject the aircraft to interception.[50]Depending on the circumstances, a persistent failure to comply with instructions may be evidence that the aircraft, in fact, poses a threat to the declaring State.[51]

Establishment of an ADIZ, like other warning zones, does not relieve the proclaiming State of its obligations under applicable international law, such as its obligation under the law of war to refrain from attacking aircraft that do not constitute lawful military objectives, or its obligation to refrain from attacks against civil aircraft under peacetime international law.[52]

14.3 AIRCRAFT STATUS

14.3.1 State Versus Civil Aircraft. The 1944 Chicago Convention distinguishes between State aircraft and civil aircraft.[53]

Aircraft used in military, customs, and police services are deemed to be State aircraft for the purposes of the 1944 Chicago Convention.[54]More generally, State aircraft may be understood to include other aircraft operated by a government for sovereign, non-commercial purposes.[55]In U.S. practice, DoD contract aircraft have not qualified as State aircraft, unless specifically designated as such by the United States.[56]

The distinction between State aircraft and civil aircraft may be important for a variety of reasons. For example, the 1944 Chicago Convention generally does not apply to State aircraft.[57]

In addition, enemy State aircraft are subject to seizure as war booty.[58]As another example, neutral State aircraft are not subject to visit and search or diversion.[59]And, although not necessary for an aircraft to constitute a military objective, the status of an aircraft as State aircraft may also be an important factor in determining that an aircraft constitutes a military objective.[60]

14.3.2 Nationality of Aircraft. State aircraft possess the nationality of the State that operates them. Civil aircraft possess the nationality of the State in which they are registered.[61]The nationality of an aircraft may be important in assessing whether it constitutes an enemy or neutral aircraft.[62]

14.3.3 Status of Military Aircraft. Military aircraft may be understood as aircraft that are designated as such by a State that operates them. The United States has not ratified a treaty that requires certain qualifications before an aircraft may be designated as military aircraft. In general, military aircraft are operated by commissioned units of the armed forces of a State, bearing the military markings of that State, and commanded by a member of the armed forces of that State.

In addition to combat aircraft such as fighters and bombers, other types of aircraft operated by the armed forces of a State may also be designated as military aircraft, such as transport, reconnaissance, and meteorological aircraft. Unmanned aircraft, *i.e.*, aircraft that are remotely piloted or controlled, may also be designated as military aircraft.

14.3.3.1 *Military Aircraft – Rights and Liabilities*. Of all aircraft, only military aircraft are entitled to engage in attacks in armed conflict.[63]Likewise, during armed conflict, enemy military aircraft generally constitute military objectives and, thus, may be made the object of attack, whether in the air or on the ground.[64]There may be exceptions to the general rule that enemy military aircraft are liable to being made the object of attack, such as in the case of enemy military medical aircraft or enemy military aircraft that have surrendered.[65]

Belligerent military aircraft generally may not enter neutral airspace.[66]

As State aircraft, military aircraft, like warships, are customarily accorded certain privileges and immunities by friendly foreign States.[67]

14.3.3.2 *Military Aircraft – Markings*. Military aircraft are customarily marked to signify both their nationality and military character.[68]Markings may help distinguish friend from foe and help preclude misidentification of aircraft as neutral or as civil.[69]However, circumstances may exist where such markings are superfluous.[70]

A single marking may be used to signify both an aircraft's nationality and its military character.[71]

14.3.3.3 *Military Aircraft – Command and Crew*. Military aircraft are commanded by members of the armed forces of that State.[72]The crew, however, may include civilian members, and such personnel are expressly entitled to POW status under the GPW.[73]

14.4 STATUS OF AIRCREW ON MILITARY AIRCRAFT

14.4.1 Military Aircrew. Aircrew who are members of the armed forces of a State have the rights, duties, and liabilities of combatants.[74]For example, they are entitled to POW status if they fall into the power of the enemy during international armed conflict, and they have legal immunity from domestic law for acts done under military authority and in accordance with the law of war.

14.4.1.1 *Wearing of Uniform by Military Aircrew*. Military aircrew are required to distinguish themselves from the civilian population in the same manner as other combatants.[75]

The wearing of flying clothing distinctive to the armed forces satisfies this requirement. In particular, military aircrew should wear military uniforms (including distinctive flying clothing) in case they become separated from the aircraft.[76]

14.4.2 Civilian Members of Military Aircrew and Other Persons Authorized to Accompany the Armed Forces. Civilian members of military aircrew fall into the category of persons authorized to accompany the armed forces.[77]Other civilians who are present on military aircraft or who support the operations of military aircraft may also fall into the category of persons authorized to accompany the armed forces, provided they have received such authorization.

As persons authorized to accompany the armed forces, civilian members of military aircrews are entitled to POW status if they fall into the power of the enemy during international armed conflict, and they have legal immunity from the enemy's domestic law for providing authorized support services to the armed forces. Civilians who work in or on military objectives, such as military aircraft, assume a certain risk of harm from attacks against military objectives.[78]

14.4.3 Downed Aircrew. There are a number of law of war issues related to downed aircrew.

14.4.3.1 *Protection of Aircrew While Parachuting From an Aircraft in Distress*. In general, aircrew parachuting from an aircraft in distress are treated as though they are *hors de combat*, *i.e.*, they must not be made the object of attack.[79] Aircrew parachuting from an aircraft in distress who engage in hostile acts or attempt to evade capture, either while descending or after reaching the ground, forfeit any protection from being made the object of attack.[80]

14.4.3.2 *Protection of Downed Aircrew at Sea*. Downed aircrew at sea are generally to be regarded as shipwrecked, as the definition of shipwrecked includes forced landings at sea by or from aircraft.[81] As shipwrecked persons, they must not be made the object of attack, and there are obligations to search for and collect them.[82]

14.4.3.3 *Rescue of Downed Aircrew*. The rescue of downed aircrew to prevent their capture by the adversary is generally regarded as a combat activity that enemy military forces may legitimately oppose.[83] Medical personnel, ships, or aircraft engaged in the rescue of downed aircrew must not hamper or interfere with the efforts of opposing military forces to capture downed aircrew or they will forfeit protection from being made the object of attack as a result.[84]

The United States has understood the prohibition against the use of riot control agents as a method of warfare not to prohibit the use of riot control agents to rescue downed aircrew in certain circumstances.[85]

14.4.3.4 *Downed Aircrew – Evasion From, and Capture by, the Enemy*. Downed aircrew may use enemy uniforms to evade capture by the enemy, but, in some cases, such use may risk liability to treatment as a spy.[86] Likewise, downed aircrew may use civilian clothes to evade capture by the enemy.[87]

If downed aircrew fall into the power of the enemy during international armed conflict, they are entitled to POW status.[88] Downed aircrew remain entitled to POW status even if they unsuccessfully attempt to escape capture by enemy forces or if they attack enemy forces upon reaching the ground. As POWs, downed aircrew are entitled to protection from violence from the civilian population and others.[89]

14.4.3.5 *Downed Aircrew in Neutral Territory*. Military aircrew forced to land in neutral territory due to navigational failure, combat damage, mechanical failure, or other emergencies are subject to internment by the neutral State for the duration of the conflict.[90]

14.5 MEASURES SHORT OF ATTACK: INTERCEPTION, DIVERSION, AND CAPTURE

14.5.1 Interception. During armed conflict, a party may choose to intercept an aircraft through a variety of ways, including closing to visual range or to a distance where the target aircraft is within the range of weapons systems. Interception of civil aircraft that are not military objectives should be exercised with due regard for the safety of such aircraft.[91]

14.5.1.1 *Purposes for Conducting Interception*. Interception may have a variety of purposes, including, for example:

• to assist in the obligation to distinguish between military objectives and civilian objects and the obligation to protect the civilian population,[92] such as by

o warning a civil aircraft to refrain from entering an area of active operations; or

o helping identify whether the aircraft constitutes a military objective;

• as part of the belligerent right of visit and search, including the belligerent right to enforce a blockade,[93] such as by

o facilitating identification of an aircraft (*e.g.*, enemy or neutral);

o forcing an aircraft to divert and land at a specific airfield for search or capture; or

• to be in a position to attack the aircraft.

14.5.1.2 *Airspace Where Interception May Be Conducted.* As a general rule, belligerent military aircraft may not intercept aircraft in neutral airspace.[94]

In addition, interception may not be conducted during passage through neutral international straits or neutral archipelagic sea lanes.[95]

14.5.2 Diversion of Aircraft. Diversion and search of civil aircraft may be conducted outside neutral airspace as part of the belligerent right of visit and search (*e.g.*, to help determine whether aircraft are liable to capture for carriage of contraband or for breach of blockade).[96]

Interference with civil aircraft of neutral States must be justified by military necessity. If, upon interception outside of neutral airspace, reasonable grounds exist for suspecting that the intercepted civil aircraft, its cargo, or its personnel are liable to capture,[97]then it may be directed to proceed to a belligerent airfield that is both reasonably accessible and suitable for the type of aircraft involved for visit and search.[98]Should such an airfield not be available, the intercepted civil aircraft may be diverted from its declared destination.[99]

Certain aircraft are exempt from the belligerent right to divert aircraft for purposes of visit and search: (1) neutral military aircraft; and (2) neutral civil aircraft accompanied by neutral military aircraft of the same nationality.[100]

14.5.2.1 *Failure to Comply by Civil Aircraft.* An enemy civil aircraft that persistently fails to comply with military instructions becomes a military objective subject to attack.[101]

Failure to comply with military instructions from intercepting aircraft does not in itself render a neutral or non-belligerent civil aircraft a military objective. However, it may provide strong evidence that the civil aircraft is in fact being used for a military or hostile purpose.

14.5.3 Capture of Aircraft and Goods on Board Aircraft. Enemy civil aircraft and goods on board such aircraft may be captured outside neutral airspace.[102]

Any enemy State aircraft (including military, law enforcement, and customs aircraft), as public movable property, is subject to capture as war booty with ownership passing immediately to the captor government.[103]Similarly, prize procedure is not used for captured enemy warships.[104]

Neutral civil aircraft engaged in certain activity in violation of their neutral status are liable to capture.[105]Even if not liable to capture, neutral civil aircraft are subject to visit and search.[106]

14.6 BELLIGERENT CONTROL OF AVIATION IN THE IMMEDIATE VICINITY OF HOSTILITIES

Belligerents may prohibit or establish special restrictions upon flight activities in the immediate vicinity of hostilities to prevent such activities from jeopardizing military operations. In some cases, this right may be distinct from a belligerent State's right to establish airspace zones during armed conflict.[107]

As with the belligerent right to control the immediate area of naval operations, this right is based on a belligerent State's right to ensure the security of its forces and its right to conduct hostilities without interference from neutrals.[108]However, belligerent control of aviation in the immediate vicinity of hostilities may be applicable in the national airspace of belligerents.[109]

14.7 AIRSPACE ZONES

During armed conflict, States may establish airspace zones and associated procedures intended to prohibit aircraft from entering or flying in designated areas, including areas in international airspace. Such zones may be established for a variety of purposes, including to decrease the risk of inadvertent attack of civil or neutral aircraft, to control the scope of the conflict, or to enhance the predictability and effectiveness of ongoing operations.

The legal rules that apply to the establishment and enforcement of a zone are discussed in connection with the establishment and enforcement of maritime zones.[110]

In some cases, such as with Air Defense Identification Zones, the zone may be established during peacetime.[111]

14.8 ATTACKS AGAINSTMILITARY OBJECTIVES IN THE AIR

14.8.1 The general rules on conducting attacks also apply to attacks against military objectives in the air.[112]In general, enemy military aircraft may be made the object of attack. Medical Aircraft. Specific rules found in the 1949 Geneva Conventions address the protection of aircraft that are engaged exclusively in specified medical functions.[113]

14.8.2 Protection of Persons Who Surrender or Who Are Otherwise *Hors De Combat* on Board Enemy Aircraft. The general rules on the protection of persons who are *hors de combat* (such as those who have been incapacitated and those who have effectively surrendered) also apply to persons on board enemy aircraft.[114]

Although the capture of enemy military aircraft and aircrew may be of significant intelligence value, there are often significant practical obstacles to identifying when persons on board an aircraft are *hors de combat* and to accepting their surrender. Despite these practical difficulties, if surrender is offered in good faith and circumstances do not preclude enforcement, then surrender must be respected.[115]Persons who are conducting attacks against enemy military aircraft must assess in good faith whether surrender is offered in good faith and can feasibly be accepted based on the information that is available to them at the time.[116]

14.8.2.1 *Difficulty in Identifying Persons on Board Aircraft as Hors De Combat*. As a practical matter, it may be quite difficult to establish that persons on board enemy aircraft are *hors de combat* and may no longer be made the object of attack.

First, it is very difficult, if not impossible, to assess whether the enemy aircraft has actually been disabled and does not pose any threat. For example, aircraft may feign symptoms of distress to evade enemy attacks.[117]Moreover, even if an aircraft has been disabled in some respects, the aircraft may not have lost its means of combat, and weapons on board the aircraft may still pose a threat.

As a practical matter, it may also be difficult for the pilot of an attacking aircraft to know when an adversary is attempting to surrender or has surrendered. Broadcasting on the international GUARD frequency (aircraft emergency frequency), rocking the aircraft's wings, lowering the landing gear, and other signals (such as the flashing of navigational lights) are sometimes cited as indications of a pilot or aircrew's desire to surrender, but they are not recognized in law as signals of surrender. Consequently, absent an explicit message of offering surrender, an intention to surrender cannot be presumed from the conduct of the aircraft.

14.8.2.2 *Feasibility of Accepting Surrender by Enemy Aircraft*. In many circumstances, it may not be feasible to accept the surrender of enemy aircraft.[118]For example, it may not be possible to enforce surrender when the engagement takes place over enemy territory.[119]

14.8.3 Attacks Against Civil Aircraft. During armed conflict, civil aircraft are generally considered civilian objects, but may be made the object of attack, outside neutral territory, if they constitute a military objective.

14.8.3.1 *Protection of Civil Aircraft in Peacetime*. Under customary international law applicable during peacetime, States have an obligation to refrain from resorting to the use of weapons against civil aircraft in flight.[120]However, this obligation does not modify in any way the rights and obligations of States set forth in the Charter of the United Nations.[121]Thus, a State's use of force against civil aircraft in the exercise of the inherent right of self-defense would be permitted.

14.8.3.2 *Forfeiture of Protection From Being Made the Object of Attack During Armed Conflict*. Civil aircraft may be made the object of attack, outside neutral territory, if the aircraft constitutes a military objective.[122]

In particular, civil aircraft forfeit any protection from being made the object of attack if they acquire the character of enemy military aircraft by:

• taking a direct part in the hostilities on the side of the enemy; or

• acting in any capacity as a naval or military auxiliary to the enemy's armed forces.[123]

In addition, enemy civil aircraft forfeit protection from being made the object of attack:[124]

• when persistently refusing to comply with directions from intercepting aircraft;[125]

• when flying under convoy of enemy warships or military aircraft;

• when armed with systems or weapons beyond that required for self-defense against terrorism, piracy, or like threats;

• when incorporated into or assisting the enemy's military intelligence system;[126]or

• when otherwise integrated into the enemy's war-fighting or war-sustaining effort.

14.8.3.3 *Attack of Civilian Passenger Aircraft*. If a civilian passenger aircraft constitutes a military objective and thus is liable to attack, any attack must comply with other applicable rules related to attacks.[127]

In particular, attacks against civilian passenger aircraft engaged in passenger service must comply with the requirement that the expected loss of life or injury to civilians, and damage to civilian objects incidental to the attack, must not be excessive in relation to the concrete and direct military advantage expected to be gained.[128]

14.9 AIR ATTACKS AGAINSTMILITARY OBJECTIVES ON THE GROUND

The general rules on conducting attacks also apply to air attacks against military objectives on the ground.[129]

14.9.1 Prohibition Against Bombardment of Undefended Cities, Towns, and Villages. The aerial bombardment of towns, villages, dwellings, or buildings that are undefended is prohibited.[130]The phrase "by whatever means" was inserted in the Hague IV Regulations to clarify that bombing attacks by air were included in this rule.[131]

An "undefended" city, town, or village in this sense is a term of art. For example, a city may only be declared "undefended" if it is open for immediate physical occupation by opposing military ground forces.[132]

14.9.2 Selection of Weapons in Conducting Attacks From the Air Against Ground Military Objectives. It may be the case that commanders will have a variety of weapons with which to conduct a potential aerial bombardment.

Certain weapons are prohibited *per se*, and it is specifically provided that feasible precautions be taken in connection with certain weapons.[133]It must be emphasized, however, that the selection of the appropriate weapon for conducting an aerial bombardment remains primarily a military judgment rather than a legal one. In particular, there is no law of war requirement to use precision-guided weapons when non-precision-guided weapons may be used in compliance with the law of war.[134]

Commanders have a general obligation to take feasible precautions in planning and conducting attacks in order to reduce the risk of harm to the civilian population.[135]The selection of weapons may be among the available precautions that a commander could take in order to reduce the risk of harm to the civilian population.

14.9.3 Protection of Enemy Ground Forces Who Are *Hors de Combat*. When aircrew or aircraft operators assess that an enemy combatant has been placed *hors de combat*, they must refrain from making such persons the object of attack.[136]

In order to place a person *hors de combat*, the person's surrender must be (1) genuine; (2) clear and unconditional; and (3) under circumstances where it is feasible for the opposing party to accept the surrender.[137]

In many cases, it would not be feasible for a party that is conducting air operations to accept the surrender of an enemy person. In some cases, however, it may be appropriate, in order to facilitate such surrender, to communicate steps for enemy units to take to communicate clearly their intention to surrender.[138]

14.9.3.1 *Difficulty in Identifying Persons Placed Hors De Combat*. The identification by aircraft of an enemy combatant on the ground placed *hors de combat* may pose a particular challenge. It may be difficult for aircrew or aircraft operators to determine whether an enemy combatant is dead, injured, merely taking cover, or feigning injury or surrender to avoid attack.[139]On the other hand, it may be difficult for enemy combatants on the ground to communicate clearly to aircraft that they wish genuinely to surrender unconditionally or that they have been incapacitated. Aircrew or aircraft operators must assess in good faith whether persons have been placed *hors de combat* based on the information that is available to them at the time.[140]

14.9.3.2 *Feasibility of Air Units to Accept the Surrender of Ground Forces*. For a communication of surrender to place a person *hors de combat*, the person must make the surrender under circumstances where it is feasible for the opposing party to accept the surrender.[141]For example, it must be possible for the opposing party to take that person into custody.[142]

In many cases, it would not be feasible for the aircraft to land and take the person into custody or for nearby ground forces to take the person into custody.[143]

14.10 INTERNATIONAL LAW ANDWARFARE IN OUTER SPACE

14.10.1 Classification of Outer Space. Outer space is governed by a separate legal regime than that for airspace.[144]

Outer space may be viewed as analogous to the high seas in certain respects.[145]For example, no State may claim sovereignty over outer space.[146]In addition, the space systems of all nations have rights of passage through space without interference.[147]

14.10.2 Application of International Law to Activities in Space.

14.10.2.1 *Treaties Specifically Addressing Space Activities*. The United States is a Party to certain treaties that address space activities. The Treaty on Principles Governing the Activities of States in the Exploration and Use of Outer Space, including the Moon and Other Celestial Bodies (Outer Space Treaty), imposes restrictions on certain military operations in outer space (*i.e.*, it does not exempt military spacecraft or military space activities from its purview).[148]The Outer Space Treaty provides for State responsibility for the activities of nongovernmental entities in outer space, including the moon and other celestial bodies.[149]

Other treaties that specifically address space activities include: • Agreement on the Rescue of Astronauts, the Return of Astronauts, and the Return of Objects Launched into Outer Space;[150]

• Convention on International Liability for Damage Caused by Space Objects;[151]and

• Convention on Registration of Objects Launched into Outer Space.[152]

Certain provisions of these treaties may not be applicable as between belligerents during international armed conflict.[153]

14.10.2.2 *Application of General International Law to Activities and Use of Outer Space*. The Outer Space Treaty reaffirms the duty of States Parties to comply with existing international law in carrying out activities in outer space. Article III of the Outer Space Treaty provides that "States Parties to the Treaty shall carry on activities in the exploration and use of outer space, including the moon and other celestial bodies, in accordance with international law, including the Charter of the United Nations, in the interest of maintaining international peace and security and promoting international cooperation and understanding."[154]

Although existing international law, such as the Charter of the United Nations, generally applies to States Parties' activities in outer space, international law that prescribes certain conditions for national claims of sovereignty does not apply to outer space because outer space is not subject to national appropriation.[155]

Certain treaties apply only in certain geographical locations (such as a State's own territory), and thus might not create obligations applicable to a State's activities in outer space. However, law of war treaties and the customary law of war are understood to regulate the conduct of hostilities, regardless of where they are conducted, which would include the conduct of hostilities in outer space. In this way, the application of the law of war to activities in outer space is the same as its application to activities in other environments, such as the land, sea, air, or cyber domains.

14.10.3 Outer Space Treaty Restrictions on Military Activities. The Outer Space Treaty imposes restrictions on certain military operations in outer space. Other treaties may also impose restrictions on military activities in outer space. For example, the Treaty Banning Nuclear Testing in the Atmosphere, Oceans, and Outer Space (Limited Test Ban Treaty) prohibits nuclear weapon test explosions in outer space.[156]

14.10.3.1 *Restriction on Nuclear Weapons and Other Kinds of Weapons of Mass Destruction in Outer Space*. Article IV of the Outer Space Treaty provides that "States Parties to the Treaty undertake not to place in orbit around the Earth any objects carrying nuclear weapons or any other kinds of weapons of mass destruction, install such weapons on celestial bodies, or station such weapons in outer space in any other manner."[157]

The prohibition on placing weapons of mass destruction "in orbit around the earth" refers only to their placement in full orbit around the Earth; thus, the Outer Space Treaty does not ban the use of nuclear or other weapons of mass destruction that go into a fractional orbit or engage in suborbital flight.[158]For example, intercontinental ballistic missiles (ICBMs) will travel a portion of their trajectory in outer space; but because ICBMs would enter outer space only temporarily, their entry into outer space with nuclear warheads would not violate this prohibition.[159]By contrast, some arms control treaties have prohibited the production, testing, or deployment of systems, including missiles, that place nuclear weapons or other weapons of mass destruction into either full earth orbit or a *fraction* of an earth orbit.[160]

In addition, this rule in Article IV of the Outer Space Treaty does not establish any prohibitions with respect to weapons that are not weapons of mass destruction (*e.g.*, anti-satellite laser weapons or other conventional weapons).[161]

14.10.3.2 *Restrictions on Military Activities on the Moon and Other Celestial Bodies*. Article IV of the Outer Space Treaty places certain prohibitions on military activities on the moon and other celestial bodies: (1) the establishment of military bases, installations, and fortifications; and (2) the testing of any type of weapons; and (3) the conduct of military maneuvers.[162]

These activities are prohibited only on the moon and other celestial bodies, not in outer space itself.

Article IV also recognizes the unimpeded right to: (1) the use of military personnel for scientific research or other peaceful purposes on outer space missions; and (2) the use of any equipment or facility necessary for the peaceful exploration of the moon and other celestial bodies.[163]

14.10.4 General Use of Outer Space for Peaceful Purposes. The United States has expressed the view that outer space should be used only for peaceful purposes.[164]This view is consistent with the Preamble to the Outer Space Treaty.[165]

The United States has interpreted use of outer space for "peaceful purposes" to mean "non-aggressive and beneficial" purposes consistent with the Charter of the United Nations and other international law.[166]This interpretation of "peaceful purposes" is similar to the interpretation given to the reservation of the high seas for "peaceful purposes" in the LOS Convention.[167]

For example, observation or information-gathering from satellites in space is not an act of aggression under the Charter of the United Nations and, thus, would be a use of space for peaceful purposes.[168]Similarly, lawful military activities in self-defense (*e.g.*, missile early warning, use of weapon systems) would be consistent with the use of space for peaceful purposes, but aggressive activities that violate the Charter of the United Nations would not be permissible.[169]

Article IV of the Outer Space Treaty provides that "[t]he moon and other celestial bodies shall be used by all States Parties to the Treaty exclusively for peaceful purposes."[170]Article IV specifies restrictions on military operations on the moon and other celestial bodies.[171]

14.10.5 Outer Space Treaty Provisions on Cooperation, Mutual Assistance, and Potentially Harmful Interference. Article IX of the Outer Space Treaty provides that in the exploration and use of outer space, States Parties shall be guided by the principle of cooperation and mutual assistance and shall conduct all their activities in outer space with due regard to the corresponding interests of all other States Parties.[172]For example, States should conduct their activities in space with due regard for the rights of other States to have their space systems pass through, and conduct operations in, space without interference.

Article IX of the Outer Space Treaty also requires States Parties to undertake "appropriate international consultations" before proceeding with any activity or experiment planned by it or its nationals in outer space if that State Party has reason to believe that its activity or experiment would cause potentially harmful interference with the activities of other States Parties in the peaceful exploration and use of outer space.[173]Conversely, a State Party that has reason to believe that an activity or experiment planned by another State Party in outer space would cause potentially harmful interference with its activities in the peaceful exploration and use of outer space may request consultation concerning the activity or experiment.[174]

XV – The Law of Neutrality

Chapter Contents

15.1 INTRODUCTION

This Chapter addresses the law of neutrality.

Issues of neutrality law can raise questions of national policy regarding an armed conflict.[1] In U.S. practice, such national policies would be developed through the National Security Council (NSC) process by the Department of State and other departments and agencies represented on the National Security Council.[2]

Some of the rules described in this Chapter were formulated long ago. Moreover, treaties concerning the law of neutrality might, by their terms, apply only to a limited set of international armed conflicts, and the rules prescribed in those treaties might not reflect customary international law.[3] In addition, it may be important to consider the implications of more recent treaties that might be applicable to a specific legal issue. In particular, the Charter of the United Nations may, in certain respects, be understood to be consistent with, and, in other respects, to modify, rules reflected in the law of neutrality.[4]

15.1.1 Matters Addressed by the Law of Neutrality. The law of neutrality prescribes the legal relationship between belligerent States and neutral States. The law of neutrality regulates relations between: (1) belligerent States, vessels, aircraft, and persons; and (2) neutral States, vessels, aircraft, and persons. Under the law of neutrality, these categories of belligerents and neutrals have corresponding rights, duties, and liabilities. Special rules have been developed to address situations on land, at sea, and in the air. Certain rules found in the law of neutrality have also been applied in other contexts that are closely analogous, such as a State's duties to prevent a non-State armed group's use of its territory for hostile expeditions against another State.[5]

15.1.2 Classification of States as Belligerent, Neutral, or Non-Belligerent.

15.1.2.1 *Belligerent State*. Belligerent State refers to a State that is engaged in an international armed conflict, whether or not a formal declaration of war has been made.[6]

15.1.2.2 *Neutral State*. Neutral State refers to a State that is not taking part in the armed conflict.[7] In some cases, States formally state their neutral status in relation to an armed conflict.[8]

15.1.2.3 *Non-Belligerent State*. The term "non-belligerent" or "non-belligerent State" has been used to refer to a State that is not participating in the armed conflict. The term "non-belligerent Power" is used in the 1949 Geneva Conventions.[9]

In addition, the term "non-belligerent" has been used to refer to States that sought to refrain from active participation in hostilities, but that did not adhere to the duties of strict impartiality to which neutrals have traditionally been required to adhere.[10]Such departure from the traditional duties of impartiality has, at times, been controversial.[11]

15.1.3 Purpose of the Law of Neutrality. The law of neutrality seeks to preserve friendly relations between belligerent and neutral States by permitting States to avoid taking sides in an armed conflict.[12]The law of neutrality also seeks to prevent additional States from being drawn into an armed conflict by establishing a clear distinction between belligerent and neutral States.[13]

In particular, the law of neutrality seeks to minimize the effects of armed conflict on States that are not party to the conflict, including by lessening the effect of war on neutral commerce.

15.1.4 Application of Treaties on Neutrality and Customary International Law. The treaties that address neutrality may be limited in their application as a matter of treaty law, but provisions of these treaties may reflect customary international law. For example, many of the treaties that address the law of neutrality were concluded before World War II, and have not been universally ratified by States.[14]Moreover, certain treaties only apply between the Parties to the treaty and, in some cases, only if all the belligerent States are also Parties to the treaty.[15]The principles reflected in those treaties, however, may still be applicable insofar as they reflect customary international law, even if they do not apply as a matter of treaty law.[16]However, if the factual circumstances of the current context are quite different from those underlying the formulation of the original treaty rule, it may be incorrect to assume that the particular treaty rule reflects (or should reflect) a rule of customary international law applicable to the current factual circumstances.

15.1.5 Domestic Neutrality Laws. States may have domestic legislation relating to the law of neutrality. In some cases, these statutes implement a State's obligations under the law of neutrality, such as its obligations as a neutral to prevent the arming of belligerent warships in neutral ports.[17]These statutes may also serve to implement a State's international obligations outside the context of international armed conflict, such as its peacetime obligations to prevent its territory from being used as a base of operations for hostile expeditions against friendly States.[18]Domestic statutes may also implement the rights of States under the law of neutrality, such as the right of belligerents to conduct captures.[19]

Domestic neutrality statutes may also help implement a State's national policy with respect to neutrality. The law of neutrality permits neutral States a degree of policy discretion with respect to an armed conflict.[20]Some neutrality treaties recognize such domestic legislation.[21]Such legislation, however, must be applied impartially among belligerents.[22]In addition, a State may have an obligation to notify other States of its domestic neutrality statutes.[23]

15.2 APPLICATION OF THE LAW OF NEUTRALITY

The application of the law of neutrality may depend on a State's national policy towards an armed conflict. A State's obligations under *jus ad bellum*, including its obligations under the Charter of the United Nations, may alter significantly its rights and duties under the law of neutrality.

15.2.1 Armed Conflict and the Application of the Law of Neutrality.

15.2.1.1 *Application of the Law of Neutrality and the National Policies of States Towards an Armed Conflict*. Whether the law of neutrality governs a State's relations with belligerents in an armed conflict may depend on that State's national policy towards that armed conflict.[24]For example, if an armed conflict occurs, a State may choose to join the armed conflict, in which case it would no longer be neutral and the law of war rather than the law of neutrality would govern its relations with opposing belligerents.[25]Similarly, if two States conduct hostilities against one another, but refuse to recognize a state of armed conflict, third States may reject this position and invoke the law of neutrality to protect their rights in relation to the armed conflict.[26]In addition, outside States that recognize the belligerency of a rebel faction in a civil war may choose to apply the law of neutrality in their relations with the rebel faction and with the government.[27]

15.2.1.2 *Application of Certain Duties of Neutral States Only in Certain International Armed Conflicts*. The duties of neutral States to refrain from certain types of support to belligerent States do not apply to all armed conflicts to which *jus in bello* rules apply; rather, such duties are only triggered in armed conflicts of a certain duration and intensity.[28]

However, belligerent States have fundamental duties to respect the sovereignty of neutral States in all international armed conflicts.[29]

15.2.1.3 *Application of Certain Rules in the Law of Neutrality Outside the Context of International Armed Conflict*. Certain parts of the law of neutrality may be viewed as applicable outside the context of international armed conflict because certain duties that the law of neutrality imposes are also found in international law applicable in peacetime. For example, under the Charter of the United Nations, States must respect the sovereignty of other States.[30]

Thus, States have duties of non-intervention and neutrality in relation to a non-international armed conflict against a friendly State.[31]

15.2.1.4 *Proclamations of Neutrality and Other Notifications of Neutral Status*. A formal declaration by nonparticipating States of their intention to adopt a neutral status generally would not be required for a State to retain its neutral status.[32]However, in light of the importance of national policy in determining whether the law of neutrality is applicable, States traditionally issued proclamations of neutrality in order to make known their neutral status in relation to a conflict.[33]These proclamations of neutrality would state the determination of the State to observe the duties of neutrality and warn its nationals of the penalties they would incur for joining or assisting a belligerent State.[34]

Although the practice of issuing formal proclamations of neutrality has declined, States have continued to make public statements of neutrality to indicate their national policy and legal status in relation to an armed conflict.[35]States may also communicate their neutral status through diplomatic channels or use other means they deem appropriate.

15.2.2 Qualified Neutrality. The United States has taken the position that certain duties of neutral States may be inapplicable under the doctrine of qualified neutrality. The law of neutrality has traditionally required neutral States to observe a strict impartiality between parties to a conflict, regardless of which State was viewed as the aggressor in the armed conflict.[36]However, after treaties outlawed war as a matter of national policy, it was argued that neutral States could discriminate in favor of States that were victims of wars of aggression.[37]Thus, before its entry into World War II, the United States adopted a position of "qualified neutrality" in which neutral States had the right to support belligerent States that had been the victim of flagrant and illegal wars of aggression.[38]This position was controversial.[39]

15.2.3 The Law of Neutrality Under the Charter of the United Nations. The Charter of the United Nations may, in certain respects, be understood to be consistent with, and, in other respects, to modify, rules reflected in the law of neutrality.

15.2.3.1 *Consistency Between the Rules in the Law of Neutrality and the Charter of the United Nations*. In certain cases, the rules on the use of force prescribed in the Charter of the United Nations would be understood to be consistent with rules in the law of neutrality.[40]

For example, a neutral State's acts of participation in a war of aggression against another member of the United Nations would likely violate not only its duties under the law of neutrality,[41]but also the Charter's prohibition on the unlawful use of force.[42]

In the past, the law of neutrality was often viewed as not regulating a State's decision as to whether to adopt a position of neutrality or hostility towards another State.[43]Thus, under this view, the requirements of the law of neutrality could have been avoided by neutral States if they decided to enter the war or by belligerent States if they chose to declare war on a neutral.[44]The Charter of the United Nations clarifies that such decisions to resort to force may not be made without a proper legal basis.[45]Thus, insofar as provisions of the Charter are coincident with certain rules in the law of neutrality, such requirements of the Charter may not be evaded in the way that arguably certain requirements of the law of neutrality previously could have been evaded.

15.2.3.2 *Modification of Rules in the Law of Neutrality*. The Charter of the United Nations and decisions by the U.N. Security Council may, in certain circumstances, qualify rights and obligations under the law of neutrality.[46]

For example, under the Charter of the United Nations, States may have obligations to support military operations that have been authorized by the U.N. Security Council.[47]In such circumstances, the obligations under the Charter would prevail over the obligations under the law of neutrality.[48]For example, a State that is not participating in the conflict that received in its territory military forces operating pursuant to a U.N. Security Council resolution might be required to return them, rather than to intern them under the law of neutrality.[49]Moreover, U.N. Security Council resolutions may obligate States to take measures that would not be required by neutrality law. For example, the U.N. Security Council may require States to impose restrictions on private conduct within their jurisdiction, such as to prevent private individuals from selling weapons to certain States or non-State armed groups.[50]

15.2.4 Neutrality Under Regional and Collective Self-Defense Arrangements. Regional and collective self-defense agreements may affect States' rights to maintain a neutral status because the States may be required to provide assistance or take military action in collective self-defense until the U.N. Security Council has taken measures necessary to maintain international peace and security.[51]The obligations by a State under such an agreement may prevail over the State's right under the customary law of neutrality to be impartial among the belligerents and to abstain from the armed conflict.

15.3 OVERVIEW OF THE NEUTRALITY LAW'S FRAMEWORK OF RECIPROCAL RIGHTS AND DUTIES

The law of neutrality establishes a framework of reciprocal rights and duties for neutrals and belligerents.

15.3.1 Neutral Rights. Belligerent States are bound to respect the sovereign rights of neutral States (*e.g.*, the inviolability of a neutral State's territory).[52]The exercise by a neutral State of its rights under the law of neutrality may not be considered an unfriendly act by either of the belligerents.[53]

15.3.1.1 *Inviolability of Neutral Territory - Prohibition on Unauthorized Entry*. The territory of neutral States is inviolable.[54]The inviolability of neutral territory prohibits any unauthorized entry into the territory of a neutral State, its territorial waters, or the airspace over such areas by armed forces or instrumentalities of war.[55]

15.3.1.2 *Inviolability of Neutral Territory - Prohibition on Hostile Acts or Other Violations of Neutrality*. The inviolability of neutral territory also requires belligerent forces to refrain from committing hostile acts or other acts that, if knowingly permitted by a neutral State, would constitute a violation of neutrality in neutral territory (including neutral lands, neutral waters, and neutral airspace).[56]For example, belligerent warships may not exercise the belligerent right of visit and search, and may not capture vessels in neutral waters.[57]

However, belligerent forces may use force in self-defense or as part of self-help enforcement actions against enemy forces that have committed violations of neutrality when the neutral State is unwilling or unable to address such violations.[58]

15.3.2 Neutral Duties -- Abstention From Participation in Hostilities and Impartial Conduct Toward Contending Parties. The principal duties of a neutral State are to abstain from any participation in the conflict and to be impartial in conduct towards contending parties.[59]

Certain other duties may be viewed as a consequence of these principal duties, but different publicists have categorized these duties differently. For example, the duty of a neutral State to refrain from supporting one side in the conflict may be viewed as a function of its duty to abstain from participation in hostilities.[60]On the other hand, the duty of a neutral State to refrain from supporting one side in the conflict may also be viewed as a result of its duty of impartiality.[61]

The duties of a neutral State may also be classified in terms of: (1) abstention (obligations to refrain from taking certain actions); (2) prevention (obligations to take certain actions); and (3) acquiescence (obligations to accept certain actions by belligerents).[62]

15.3.2.1 *Duty to Refrain From Providing War-Related Goods and Services to Belligerents*. Neutral States have an obligation to refrain from providing war-related goods and services to belligerents.[63]A neutral State is prohibited from supplying a belligerent State in any manner, either directly or indirectly, with warships, ammunition, or war material of any kind.[64]The neutral State is also prohibited from providing money or loans to a belligerent State.[65]

A neutral State also has a duty to refrain from placing its various governmental agencies at the disposal of a belligerent in such a way as to aid it directly or indirectly in the prosecution of the war.[66]

Although a neutral State may not supply war-related goods and services to belligerents, a neutral State is not called upon to prevent the export or transport, on behalf of one or other of the belligerents, of arms, munitions, or anything that can be of use to an armed force.[67]Commercial transactions between belligerent States and neutral corporations, companies, citizens, or persons resident in a neutral State are not prohibited.[68]A belligerent State may purchase from such corporations, companies, citizens, or persons supplies, munitions, or anything that may be of use to an armed force, which can be exported or transported without involving the neutral State.[69]

The supply of services and goods to a belligerent State, however, must not convert neutral territory into a base of operations.[70]

If a neutral State prohibits the export or transport of arms, munitions, or anything that can be of use to an armed force, every measure of restriction or prohibition taken by a neutral State in regard to the provision of such supplies or munitions must be impartially applied by it to all belligerent States.[71]

15.3.2.2 *Duty to Prevent Violations of Neutrality Within Its Jurisdiction*. A neutral State has certain obligations to prevent violations of its neutrality within its territory. A neutral State has an obligation to prevent violations of its neutrality committed by belligerent forces.[72] A neutral State also has an obligation to prevent violations of its neutrality committed by persons within its jurisdiction.[73]

15.3.2.3 *Public Expressions of Sympathy Are Not Violations of Neutrality*. Public expressions of sympathy for one belligerent or disapproval of another belligerent are not violations of duties of neutrality.[74]

15.3.2.4 *Authorized Humanitarian Assistance Is Not a Violation of Neutrality*. Humanitarian assistance given in accordance with international law is also not a violation of duties of neutrality.[75] For example, a neutral State's performing of the role of the Protecting Power in accordance with the 1949 Geneva Conventions is not regarded as unneutral conduct.[76]

15.3.3 Correlative or Reciprocal Nature of Rights and Duties Under the Law of Neutrality. The rights and duties of belligerents and neutrals under the law of neutrality may be understood as correlative or reciprocal.

The duties of neutrals often correspond to rights of belligerents, and rights of neutrals often correspond to the duties of belligerents.[77] For example, the duty of a neutral to abstain from hostilities is also a right of a belligerent not to have its adversary aided in hostilities by the neutral State. The right of a neutral State to have its territory be inviolable is also a duty of belligerent States to avoid conducting hostilities in neutral territory. Similarly, the ability of a neutral to assert its rights may depend on whether it has fulfilled its corresponding neutral duties.[78] For example, the right of a neutral State to have its territory be inviolable may not be asserted to the extent it has failed in its duty to ensure that its territory is not used as a base of operations by a belligerent.

In light of the correlative nature of the rights and duties of belligerents and neutrals under the law of neutrality, a single rule in the law of neutrality may reflect multiple rights and duties of both belligerents and neutrals. For example, hostile acts are not to be committed on neutral territory. This rule involves the right of the neutral State for its territory to be inviolable, and involves a duty on the part of the neutral State to prevent such acts. This rule also reflects a duty of the belligerent State to refrain from such acts, and a right of the belligerent State not to have such acts conducted by an opposing belligerent.

15.4 REMEDIES FOR VIOLATIONS OF NEUTRALITY LAW

15.4.1 Distinction Between Violations of Neutral Duties and the End of Neutral Status. Violations of neutrality by belligerent or neutral States should be distinguished from the end of a State's neutral status. Rather, whether violations of neutrality result in the end of the neutral status of a State may depend on the national policies of that State and the belligerent States.[79]

Acts that are incompatible with the relationship between the neutral State and a belligerent State under the law of neutrality need not end the neutral State's neutrality and bring that State into the conflict as a belligerent. For example, despite Vichy France's violations of its neutral duties during World War II and despite the Allied invasion of French North Africa, diplomatic relations persisted between the United States and France during World War II.[80]

15.4.2 Belligerent Use of Self-Help When Neutral States Are Unable or Unwilling to Prevent Violations of Neutrality. Should the neutral State be unable, or fail for any reason, to prevent violations of its neutrality by the forces of one belligerent entering or passing through its territory (including its lands, waters, and airspace), the other belligerent State may be justified in attacking the enemy forces on the neutral State's territory.[81] This view has been reflected in the military manuals of other States.[82] For example, consistent with the *jus ad bellum* requirements for self-defense, belligerent forces may act in self-defense when attacked or threatened with attack from enemy forces unlawfully present in neutral territory,[83] including by taking appropriate action to counter the use of neutral territory as a base of enemy operations when the neutral State is unwilling or unable to prevent such violations.[84]

15.4.3 Neutral State's Use of Force to Defend Its Neutrality. A neutral State may also engage in self-help to prevent violations of neutrality on its territory. Such self-help is also an obligation of the neutral State.[85]

The fact that a neutral State resists, even by force, attempts to violate its neutrality cannot be regarded as a hostile act.[86] For example, a State whose territory is adjacent to a theater of war normally mobilizes a portion of its forces to prevent the forces of either belligerent from entering its territory, to intern such persons as may be permitted to enter, and generally to carry out its duties of neutrality.[87]Such military operations are not regarded as hostile acts against the belligerent State.

15.4.4 Redress of Certain Violations of Neutrality. In addition to the obligations to prevent violations of neutrality, a State may also have obligations to redress violations of neutrality that have been committed.

For example, violations of the law of neutrality may result in liability for payment of damages in certain cases.[88]As a case in point, if harm is caused in a neutral State by the unauthorized entry of belligerent forces, the offending State may be required, according to the circumstances, to respond in damages.[89]

Similarly, when a ship has been captured in the territorial waters of a neutral State, the neutral State must employ, if the prize is still within its jurisdiction, the means at its disposal to release the prize with its officers and crew, and to intern the prize crew.[90]If the prize is no longer in the jurisdiction of the neutral power, the captor State, on the demand of the neutral State, must liberate the prize with its officers and crew.[91]The neutral State has a duty to make this demand.[92]

15.5 PROHIBITION ON THE USE OF NEUTRAL TERRITORY AS A BASE OF OPERATIONS

Neutral territory (including neutral lands, waters, and airspace) may not be used as a base of operations against belligerent forces.[93]For example, belligerent States are forbidden to use a neutral State's ports and waters as a base of naval operations against their adversaries,[94]and a neutral State has a corresponding obligation to prevent such use.[95]

The prohibition against the use of neutral territory as a base of operations extends to any hostile expeditions against a belligerent State, and thus includes such expeditions by non-State actors.[96]For example, a neutral State has an obligation to prevent the arming of any vessel within its jurisdiction that is intended to engage in hostile operations against a State with which it is at peace and to prevent the departure from its jurisdiction of any vessel intended to engage in hostile operations that has been adapted entirely or partly within such jurisdiction for use in war.[97]

The prohibition on the use of neutral territory as a base of operations includes prohibitions against (1) outfitting hostile expeditions with supplies and services; (2) recruiting forces in neutral territory; (3) establishing military communications facilities in neutral territory; and (4) moving belligerent forces or convoys of military supplies on land.

15.5.1 Prohibition on Outfitting Hostile Expeditions With Supplies and Services for a Belligerent From Neutral Territory. It is forbidden to permit the use of the neutral State's territory for the fitting out of hostile expeditions.[98]As a case in point, a belligerent State's warships may not make use of a neutral State's ports, roadsteads, and territorial waters to replenish or increase their supplies of war materials or their armaments, or for completing their crews.[99]

15.5.2 Prohibition Against Recruiting and Forming Corps of Combatants. Units of combatants may not be formed nor may recruiting agencies be established on the territory of a neutral State to assist a belligerent State.[100]This means that the establishment of recruiting agencies, the enlistment of personnel, the formation and organization of hostile expeditions on a neutral State's territory, and the passage across its frontiers of organized bodies of personnel intending to enlist are prohibited.[101]

15.5.2.1 *Exception for Authorized Medical Personnel.* This prohibition does not extend to medical personnel and units of voluntary aid societies of a neutral State who are duly authorized to provide medical assistance to a belligerent State.[102]In no circumstance shall such assistance by medical personnel and units of voluntary aid societies of the neutral State be considered as interference in the conflict.[103]

15.5.2.2 *Exception for Persons Crossing a Neutral State's Frontiers Separately to Assist a Belligerent State.* The prohibition against organizing units of combatants is directed against organized bodies, which only require being armed to become an immediate fighting force.[104]Thus, the responsibility of a neutral State is not engaged by the fact of persons crossing the frontier separately to offer their services to one of the belligerent States.[105]Individuals crossing the frontier singly or in small bands that are unorganized create no obligation on the neutral State.[106]

Neutral States shall not oppose the voluntary departure of nationals of belligerent States even though they leave simultaneously in great numbers; but they may oppose the voluntary departure of their own nationals going to enlist in the armed forces of belligerent States.[107]

Neutral States are not required to enact legislation forbidding their nationals to join the armed forces of belligerent States.[108]

Neutral States, however, may not send regularly constituted military units across the frontier in the guise of "volunteers" or small unorganized bands.[109]

15.5.3 Prohibition Against Establishment or Use of Belligerent Communications Facilities in Neutral Territory. Belligerent States are prohibited from erecting on the territory of a neutral State any apparatus for the purpose of communicating with belligerent forces on land or sea.[110] Belligerent States are prohibited from using any installation of this kind established by them before the armed conflict on the territory of a neutral State for purely military purposes, and which has not been opened for the service of public messages.[111]

15.5.3.1 *Use of Neutral Facilities by Belligerents Not Prohibited.* A neutral State does not, however, have to forbid or restrict the use on behalf of belligerent States of communications equipment and facilities belonging to the neutral State, or belonging to companies or private individuals.[112]

That a neutral State, if it so desires, may transmit messages by means of its communications facilities does not imply that the neutral State may use such facilities or permit their use to lend assistance to the belligerents on one side only.[113] Every measure of restriction or prohibition taken by a neutral State in regard to the use of its communications facilities and equipment must be impartially applied by it to all belligerent States.[114] It also must see to the same obligation being observed by companies or private individuals owning such communication equipment and facilities.[115]

15.5.4 Movement of Belligerent Forces and Convoys of Supplies Through Neutral Land Territory. Belligerent States are forbidden to move forces or convoys of either munitions of war or supplies across the land territory of a neutral State.[116] This rule only prohibits the official acts of a belligerent State in convoying or shipping munitions and supplies through a neutral State as part of an expedition; it does not prohibit the shipment of such supplies by private persons.[117] However, a neutral State may authorize passage through its territory of the wounded and sick, including the overflight of its territory by certain medical aircraft.[118]

15.6 NEUTRAL PERSONS

15.6.1 Definition of Neutral Person. The nationals of a State that is not taking part in the war are considered neutral persons.[119]

15.6.2 Forfeiture of Protections of Neutral Status by a Neutral Person. A national of a neutral State would not retain the protections that a person would derive by virtue of the neutral status of his or her State of nationality if that person commits:

• hostile acts against a belligerent State; or

• acts in favor of a belligerent State, particularly if that person voluntarily enlists in the ranks of the armed forces of one of the parties to the armed conflict.[120]

For example, neutral persons who materially support one belligerent generally forfeit their neutral character or assume certain liabilities with respect to the other belligerent.[121]

15.6.2.1 *No More Severe Treatment Than Nationals of an Opposing Belligerent State.* A neutral person, however, shall not be more severely treated by the belligerent State as against whom that person has abandoned his or her neutrality than a national of the other belligerent State could be for the same act.[122]

Although a neutral person who takes certain actions may be liable to treatment as an unprivileged belligerent, the forfeiture of neutral protections does not necessarily result in that person being regarded as an unprivileged belligerent.

For example, if a national of a neutral State is a member of the armed forces of a belligerent State, that person would be entitled to all the protections that a member of the armed forces of the belligerent State would enjoy under the law of war, such as POW status if that person falls into the power of the enemy during international armed conflict.[123]

15.6.2.2 *Acts Not Considered in Favor of a Belligerent That Would Forfeit Protections of Neutral Status.* The followings acts would not be considered as committed in favor of a belligerent State that would forfeit a neutral person's protections of neutral status:

- furnishing supplies or making loans to a belligerent State, provided that the person who furnishes the supplies or who makes the loans lives neither in the territory of the belligerent State nor in the territory occupied by it, and that the supplies do not come from these territories; or
- rendering services in matters of police or civil administration.[124]

In addition, merely expressing sympathy for a belligerent's cause would not cause a person to forfeit one's neutral character, although incitement or recruitment would be treated differently.

Certain types of humanitarian assistance under the 1949 Geneva Conventions would also not be regarded as a violation of neutrality by the persons engaging in those activities.[125]

15.6.3 Neutral Persons in the Home Territory of a Belligerent. Neutral persons in the home territory of a belligerent State are not "protected persons" under the GC while their State of nationality has normal diplomatic relations with the belligerent State.[126]

In general, neutral persons residing permanently in the territory of a belligerent State, whether protected persons or not, are to a great extent regarded, both by that belligerent and the opposing side, like other residents of that territory, so long as they remain there.[127] By contrast, the neutral person who is a transient visitor in the territory of a belligerent is provided particular respect, so far as the hazards of war permit, and provided that the person's actions are consistent with that person's neutral character.[128]

15.6.4 Neutral Persons Resident in Occupied Territory. Neutral persons resident in occupied territory are not entitled to claim different treatment, in general, from that accorded the other inhabitants.[129] They must refrain from all participation in the war and from all hostile acts, and must observe strictly the rules of the Occupying Power.[130]

All nationals of neutral States, whether resident in or temporarily visiting an occupied territory, may be punished for offenses committed by them to the same extent and in the same manner as enemy nationals.[131] In addition, it may be possible to extradite nationals of neutral States who have committed offenses to their home States for prosecution. If nationals of neutral States are not "protected persons," they may be deported or expelled for just cause.[132]

In the event that such a person is arrested, suspicions must be verified by a serious inquiry, and the arrested neutral person must be given an opportunity to present a defense, and to communicate with his or her national consul if requested.[133] In addition, the minimum standards for humane treatment and for detention procedures would be applicable.[134]

15.6.4.1 *Protected Person Status of Neutral Persons in Occupied Territory*. Neutral persons, among other persons, who find themselves in the hands of an Occupying Power generally would be regarded as "protected persons" under the GC.[135] Neutral persons who travel to occupied territory for the purpose of fighting the Occupying Power are not regarded as "protected persons" under the GC.[136]

15.6.4.2 *Diplomatic Agents and Consular Personnel in Occupied Territory*. Diplomatic agents of neutral States must be treated with all courtesy and must be permitted such freedom of action as it is possible to allow, with due regard to the necessities of the war. The same is true of consular personnel of neutral States, except those who are enemy nationals.[137]

15.7 NEUTRAL WATERS

Belligerent States are forbidden to use a neutral State's ports and waters as a base of naval operations against their adversaries, and in particular to erect or employ any apparatus for communicating with belligerent forces on land or sea.[138] For example, a belligerent State's warships may not make use of a neutral State's ports, roadsteads, and territorial waters to replenish or increase their supplies of war materials or their armaments, or for completing their crews.[139]

Belligerent State forces must refrain from acts of hostility in neutral waters, including the exercise of visit and search.[140]

The neutral State has an affirmative duty to police its waters to prevent violations of neutrality in its waters.[141] If a neutral State is unable or unwilling to detect and expel belligerent forces unlawfully present in its waters, the opposing belligerent State may undertake such self-help enforcement actions as may be necessary to terminate the violation of neutrality.

15.7.1 Waters That Are Considered Neutral. The waters that are subject to the sovereignty of a neutral State are considered neutral. Neutral waters may be understood to include the following waters belonging to a neutral State:

- the territorial sea, including up to 12 nautical miles;[142]
- archipelagic waters;[143] and
- ports, roadsteads, and internal waters.[144]

For the purpose of applying the law of neutrality, all ocean areas *not* subject to the territorial sovereignty of any State (*i.e.*, all waters seaward of neutral States' territorial seas) are not considered neutral waters. The following waters are not considered neutral waters:

- a neutral State's contiguous zone;[145]
- a neutral State's exclusive economic zone;[146] and
- the high seas.[147]

15.7.2 A Neutral State's Regulations Concerning Belligerent Warships and Prizes in Its Waters. A neutral State may adopt laws or regulations governing the presence of belligerent warships and their prizes in its waters.

A neutral State must apply impartially to opposing belligerents the conditions, restrictions, or prohibitions made by the neutral State in regard to the admission into its ports, roadsteads, or territorial waters, of belligerent warships or of their prizes.[148] Nevertheless, a neutral Power may forbid a belligerent vessel that has failed to conform to the orders and regulations made by it, or that has violated neutrality, to enter its ports or roadsteads.[149]

15.7.3 24-Hour Limit on Stay of Belligerent Warships in Neutral Waters. In the absence of special provisions to the contrary in the legislation of a neutral State, a belligerent State's warships are generally prohibited from remaining in that neutral State's ports, roadsteads, or territorial waters for more than 24 hours.[150]

A belligerent warship may not prolong its stay in a neutral port beyond the permissible time except on account of damage or stress of weather. It must depart as soon as the cause of the delay is at an end.[151]

The regulations as to the question of the length of time that these vessels may remain in neutral ports, roadsteads, or waters, do not apply to warships devoted exclusively to religious, scientific, or philanthropic purposes.[152] Warships engaged in the collection of scientific data of potential military application would not be considered to be devoted exclusively to scientific purposes and would not be exempt.[153]

15.7.3.1 *Notification to Belligerent Warships to Depart a Neutral State's Waters Upon Outbreak of Hostilities*. If a neutral State, which has been informed of the outbreak of hostilities, learns that a belligerent State's warship is in one of the neutral State's ports or roadsteads, or in its territorial waters, it must notify such warship to depart within 24 hours or within the time prescribed by local regulations.[154]

15.7.4 Passage of Belligerent Warships and Prizes Through a Neutral State's Waters. A neutral State may allow the passage of belligerent warships and prizes through its waters. The neutrality of a State is not affected by the mere passage through its territorial waters of warships or prizes belonging to belligerents.[155]

A neutral State may, on a nondiscriminatory basis, suspend passage of belligerent warships and prizes through its waters (with the exception of international straits and archipelagic sea lanes).[156]

Although the general practice has been to close neutral territorial seas to belligerent submarines, a neutral State may elect to allow passage of submarines provided they do not engage in hostile acts while in territorial waters.[157]

Neutral States customarily authorize passage through their territorial sea of ships carrying the wounded, sick, and shipwrecked, whether or not those waters are otherwise closed to belligerent vessels.[158]

15.7.4.1 *Right of Entry in Distress for Belligerent Vessels*. Belligerent vessels, including warships, have a right of entry in distress whether caused by *force majeure* or damage resulting from enemy action.[159] The right of entry in distress does not prejudice the measures that a neutral State may take after entry has been granted, such as measures to intern the ship if it remains when it is not entitled to remain.[160]

15.8 PASSAGE OF BELLIGERENT VESSELS AND AIRCRAFT THROUGH INTERNATIONAL STRAITS AND ARCHIPELAGIC SEA LANES

Although a neutral State may suspend the passage of belligerent warships through its waters, a neutral State may not suspend, hamper, or otherwise impede the access of belligerent vessels and aircraft through international straits overlapped by neutral waters or archipelagic sea lanes of a neutral State. Belligerent vessels and aircraft transiting through international straits overlapped by neutral waters or archipelagic sea lanes of a neutral State must adhere to certain requirements.

15.8.1 Passage of Belligerent Vessels and Aircraft Through International Straits Overlapped by Neutral Waters. All ships and aircraft (including those belonging to belligerent and neutral States) have a right of transit passage through, over, and under all straits used for international navigation.[161]Such passage by vessels carrying contraband of war or by belligerent warships is not regarded as compromising the neutrality of the State with water overlapping such straits.[162]Neutral States bordering international straits may not suspend, hamper, or otherwise impede the right of transit passage through international straits.[163]

Belligerent forces transiting through international straits overlapped by neutral waters must proceed without delay, must refrain from the threat or use of force against the neutral State, and must otherwise refrain from acts of hostility and other activities not incident to their transit.[164]For example, belligerent forces must refrain from exercising the right of visit and search while transiting through international straits that are overlapped by neutral waters.[165]Belligerent forces in transit may, however, take defensive measures consistent with their security, including the launching and recovery of military devices, screen formation steaming, and acoustic and electronic surveillance, and may respond in self-defense to a hostile act or a demonstration of hostile intent.[166]

15.8.2 Passage of Belligerent Vessels and Aircraft Through Archipelagic Sea Lanes of a Neutral State. Belligerent ships or aircraft, including surface warships, submarines, and military aircraft, retain the right of unimpeded archipelagic sea[167]lanes passage through, under, and over neutral archipelagic sea lanes.[168]Neutral archipelagic States shall not suspend or hamper the right of transit passage through their archipelagic sea lanes.[169]

Belligerent State forces exercising the right of archipelagic sea lanes passage must refrain from acts of hostility and other activities not incident to their transit. [170]For example, belligerent State forces must refrain from exercising the right of visit and search while exercising the right of archipelagic sea lanes passage.[171]Belligerent State forces exercising the right of archipelagic sea lanes passage may, however, engage in those activities that are incident to their normal mode of continuous and expeditious passage, and that are consistent with their security, including formation steaming, acoustic and electronic surveillance, and the launching and recovery of military devices.[172]

15.9 ADDITIONAL RULES APPLICABLE TO NEUTRAL PORTS, ROADSTEADS, AND INTERNAL WATERS

In addition to the general rules applicable to neutral waters,[173]certain other rules apply to neutral ports, roadsteads, and internal waters.

15.9.1 Maximum Number of Belligerent Warships in One Neutral Port or Roadstead. In the absence of special provisions to the contrary in the legislation of a neutral State, the maximum number of warships belonging to a belligerent that may be in one of the ports or roadsteads of that neutral State simultaneously is three.[174]

15.9.2 Detention of Belligerent Ships That, and Personnel Who, Are Not Entitled to Remain in a Neutral Port. If, notwithstanding the notification of the neutral State, a belligerent ship of war does not leave a port where it is not entitled to remain, the neutral State is entitled to take such measures as it considers necessary to render the ship incapable of taking to sea during the war, and the commanding officer of the ship must facilitate the execution of such measures.[175]

When a belligerent ship is detained by a neutral Power, the officers and crew are likewise detained.[176]

The officers and crew thus detained may be left in the ship or kept either on another vessel or on land, and may be subjected to the measures of restriction that may appear necessary to impose upon them. A sufficient number of persons for looking after the vessel must, however, always be left on board. The officers may be left at liberty on giving their word not to quit the neutral territory without permission.[177]

15.9.3 Departure of Belligerent Warships in Relation to the Departure of Ships of the Opposing Belligerent. When warships of opposing belligerent States are present simultaneously in a neutral State's port or roadstead, a period of not less than 24 hours must elapse between the departure of the respective enemy warships.[178]The order of departure is determined by the order of arrival, unless the ship that arrived first is in such circumstances that an extension of its stay is permissible.[179]

A belligerent warship may not leave a neutral port or roadstead until 24 hours after the departure of a merchant ship flying the flag of its adversary.[180]

15.9.4 Supplies and Repairs of Belligerent Warships in Neutral Ports and Roadsteads.

15.9.4.1 *Food and Fuel for Belligerent Warships in Neutral Ports and Roadsteads.* A neutral State has discretion to issue and implement regulations regarding the supply of food and supplies for belligerent warships in its territory.[181]However, such regulations must be applied impartially among belligerent States.[182]A belligerent State's warships may take on food supplies and provisions in a neutral State's ports and roadsteads, but only to bring up their supplies to the peacetime standard.[183]

Similarly, a belligerent State's warship may only take on sufficient fuel to enable it to reach the nearest port in its own country.[184]The belligerent State's warship may, however, fill up its bunkers built to carry fuel when the neutral State has adopted this method of determining the amount of fuel to be supplied.[185]

15.9.4.2 *Repairs of Belligerent Warships in a Neutral State's Ports and Roadsteads.* A belligerent State's warships may only carry out such repairs in a neutral State's ports and roadsteads as are absolutely necessary to render the warships seaworthy; they may not add to or repair weapons systems, or enhance any other aspect of their war fighting capability.[186]The neutral State shall decide what repairs are necessary to restore seaworthiness; repairs must be accomplished with the least possible delay.[187]

If the 1928 Pan American Maritime Neutrality Convention is applicable, then damage found to have been produced by the enemy's fire must not be repaired.[188]However, whether such repairs are prohibited by customary international law is less clear.[189]Some States have allowed such repairs provided they are limited to rendering the ship sufficiently seaworthy to continue its voyage safely.[190]

A belligerent warship damaged by enemy fire that will not (or cannot) be put to sea once the lawful period of stay has expired must be interned.[191]

15.9.5 Prizes in Neutral Ports or Roadsteads. A prize (*i.e.*, a captured neutral or enemy merchant ship) may only be brought into a neutral port or roadstead because of unseaworthiness, stress of weather, or want of fuel or provisions.[192]It must leave as soon as such circumstances that justified its entry cease.[193]Neutral ports may not be used as harbors of safety in which prizes may be kept indefinitely.[194]

If the prize does not leave as soon as such circumstances that justified it entry cease, the neutral State must order the prize to leave at once, and should it fail to do so, the neutral State must employ the means at its disposal to release it with its previous officers and crew, and to intern the prize crew.[195]

If a prize is brought into a neutral State's port under circumstances other than because of unseaworthiness, stress of weather, or want of fuel or provisions, the neutral State must release the prize.[196]The prize crew should be interned and the vessel restored to its former crew.[197]

15.9.5.1 *Article 23 of Hague XIII – Prizes Sequestrated Pending Prize Court Decision.* Article 23 of Hague XIII permits a neutral State to allow prizes to enter its ports and roadsteads when they are brought there to be sequestrated pending the decision of a prize court.[198]However, the United States ratified Hague XIII subject to a reservation to Article 23.[199]

15.10 NEUTRAL AIRSPACE

In general, belligerent military aircraft may not enter neutral airspace except to address violations of neutrality by enemy forces when the neutral State is unwilling or unable to address such violations.

15.10.1 Airspace That Is Considered Neutral. The airspace over a neutral State's land territory and the neutral State's territorial and archipelagic waters is subject to the sovereignty of the neutral State and, thus, is considered neutral.[200]

The airspace over international waters, including the airspace over a neutral State's contiguous zone and exclusive economic zone, is not considered neutral airspace.[201]

15.10.2 Prohibition on Entry by Belligerent Military Aircraft Into Neutral Airspace. Belligerent military aircraft are forbidden to enter neutral airspace, subject to certain exceptions.[202]

Belligerent military aircraft have the right to pass through international straits overlapped by neutral waters and archipelagic sea lanes of a neutral State.[203]

Belligerent States' medical aircraft may enter neutral airspace subject to certain conditions.[204]

Belligerent aircraft in evident distress may be permitted to enter neutral airspace and to land in neutral territory under such safeguards as the neutral State may wish to impose. The neutral State must require such aircraft to land and must intern both aircraft and crew.[205] This situation would fall under the general duty of a neutral State with respect to the treatment of belligerent State forces that enter, or seek to enter, its territory.[206]

15.10.3 Duties of Neutral States With Respect to Their Airspace. As with other neutral territory, neutral States have an affirmative duty to prevent the violation of their airspace by belligerent military aircraft.[207] For example, if a belligerent military aircraft enters neutral airspace, the neutral State is obliged to use the means at its disposal to require the belligerent military aircraft to land within its territory.[208] After the aircraft lands, the neutral State must intern the aircraft and its crew for the duration of the armed conflict.

If a neutral State is unable or unwilling to prevent the unlawful entry or use of its airspace by a belligerent State, the opposing belligerent State's forces may undertake such self-help enforcement measures as the circumstances may require.[209] For example, belligerents with missile defense capabilities may be justified in intercepting enemy missiles transiting neutral airspace if the neutral state cannot, or will not, prevent such airspace incursions.

15.11 BELLIGERENT RIGHT OF ANGARY

The right of angary recognizes the right of belligerents to requisition (upon payment of just compensation) neutral property transiently within their territory, or in territory that they have occupied, where the property is urgently required for the conduct of the war.[210]

15.11.1Neutral Property Transiently Within Belligerent Jurisdiction. The belligerent right of angary applies to property that is transiently within belligerent jurisdiction. For example, neutral ships have been subject of the exercise of the right of angary.[211]

The right of angary does not apply to property of neutral ownership that has acquired enemy character.[212]

15.11.2 Railway Material Coming From the Territory of a Neutral State. Railway material coming from the territory of a neutral State, whether it be the property of that State or of companies or private persons, and recognizable as such, shall not be requisitioned or utilized by a belligerent State except where and to the extent that it is absolutely necessary.[213] It shall be sent back as soon as possible to the country of origin.[214] A neutral State may likewise, in case of necessity, retain and utilize to an equal extent railway material coming from the territory of a belligerent State.[215] Compensation shall be paid by one party or the other in proportion to the railway material used, and to the period of usage.[216]

Railway materials would include, for example, railroad cars and locomotives. This rule on railway material has sought to balance between the necessities of war (*i.e.*, that such material might be of great military utility) and the interests and rights of neutrals.[217]

This rule also reflects a compromise between the different views of States.[218]

15.12 NEUTRAL COMMERCE AND CARRIAGE OF CONTRABAND

Although neutral States must not provide war-related goods and services to belligerents, neutral persons are not prohibited from such activity by the law of neutrality.[219] The law of neutrality's rules on neutral commerce and the carriage of contraband have sought to balance the right of neutral persons to conduct commerce free from unreasonable interference against the right of belligerent States to interdict the passage of war materials to the enemy.[220]

Neutral merchant vessels and civil aircraft engaged in legitimate neutral commerce are subject to visit and search, but generally may not be captured or destroyed by belligerent forces.[221] On the other hand, neutral merchant vessels and civil aircraft are subject to capture and other penalties if they engage in certain conduct.[222]

Contraband goods are liable to capture at any place beyond neutral territory, if their destination is the territory belonging to, or occupied by, an opposing belligerent State.[223]

15.12.1 Classes of Goods That May Be Considered Contraband. *Contraband* consists of goods that are destined for an enemy of a belligerent and that may be susceptible to use in armed conflict.[224] Items susceptible to use in armed conflict may be understood to include war-sustaining commerce, *i.e.*, commerce that indirectly but effectively supports and sustains the belligerent State's war fighting capability (*e.g.*, imports of raw materials used for the production of armaments and exports of products whose proceeds are used by the belligerent State to purchase arms and armaments).[225]

15.12.1.1 *Distinction Between Absolute and Conditional Contraband.* Whether an item is susceptible to use in armed conflict may depend on the character of the item. Traditionally, contraband has been divided into two categories, *absolute* and *conditional.*

Absolute contraband consisted of goods whose character is such that they are obviously destined for use in armed conflict, such as munitions, weapons, uniforms, and the like. *Conditional contraband* consisted of goods equally susceptible to either peaceful or warlike purposes, such as foodstuffs, construction materials, and fuel.[226]

Whether it is possible to distinguish between absolute contraband and conditional contraband may depend on the extent to which the enemy government controls imports and the scale of the conflict. For example, during World War II, belligerent States largely did not distinguish between absolute and conditional contraband because of the involvement of virtually the entire population in support of the war effort and because the belligerent States exercised governmental control over all imports.[227]

15.12.1.2 *Free Goods That Are Exempt From Capture as Contraband.* Certain goods qualify as "free goods," meaning that they are exempt from capture by belligerent States as contraband even though they are destined for enemy territory. Free goods include the following:

• equipment exclusively intended for the treatment of wounded and sick members of armed forces or for the prevention of disease, provided that the particulars regarding the voyage of such equipment have been notified to and approved by the opposing belligerent State;[228]

• consignments of certain types of relief goods (*e.g.*, medical supplies and religious materials for civilians; clothing and medicine for children under fifteen, expectant mothers, and maternity cases), under certain conditions;[229]

• certain types of relief consignments intended for the benefit of the population of occupied territory;[230]

• items destined for POWs, including individual parcels and collective relief shipments containing food, clothing, medical supplies, religious objects, and educational, cultural, and athletic articles;[231]and

• other goods that are specifically exempted from capture by an applicable treaty or by a special arrangement between belligerent States.[232]

In practice, neutral States have provided belligerent States of both sides with information regarding the nature, timing, and route of shipments of goods constituting exceptions to contraband, and obtained approval for their safe conduct and entry into belligerent owned or occupied territory.[233]

15.12.1.3 *Contraband Lists.* Belligerent States may declare contraband lists at the initiation of hostilities to notify neutral States of the type of goods considered to be absolute or conditional contraband, as well as those not considered to be contraband at all (*i.e.*, exempt or "free goods"). The precise nature of a belligerent State's contraband list may vary according to the circumstances of the conflict.[234]

Although there has been no conflict of a similar scale and magnitude to World War II, in the years following the conclusion of that conflict State practice indicates that, to the extent international law may continue to require publication of contraband lists, the requirement may be satisfied by a listing of exempt goods.[235]

15.12.2 Requirement of Enemy Destination. Contraband goods are liable to capture at any place beyond neutral territory if their destination is the territory belonging to, or occupied by, the enemy.

15.12.2.1 *Ultimate Destination – Doctrine of Continuous Voyage.* It is immaterial whether the carriage of contraband is direct, involves transshipment, or requires overland transport.[236]Under the doctrine of continuous voyage, the ultimate destination is determinative, and contraband goods may be captured, even if there are neutral ports that are intended to be visited between the point of capture and the ultimate destination.[237]

15.12.2.2 *Presumption of Destination of Enemy Territory in Certain Cases.* When contraband is involved, a destination of enemy-owned or enemy-occupied territory may be presumed when:

• the neutral vessel is to call at an enemy port before arriving at a neutral port for which the goods are documented;

• the goods are documented to a neutral port serving as a port of transit to an enemy, even though they are consigned to a neutral; or

• the goods are consigned "to order" or to an unnamed consignee, but are destined for a neutral State in the vicinity of enemy territory.[238]

These presumptions of enemy destination constitute sufficient cause for naval commanders to order a capture.[239]However, these presumptions may be rebuttable during subsequent prize court proceedings by an affirmative showing of innocent destination.[240]These presumptions of the destination of enemy territory are not sufficient to establish that the property is destined for use by an enemy government or its armed forces, which would be necessary to establish a basis for the forfeiture of the property if the property is classified as conditional contraband.[241]

15.12.3 Certificate of Noncontraband Carriage. A certificate of noncontraband carriage is a document issued by a belligerent consular or other designated official to a neutral vessel (navicert) or neutral aircraft (aircert) certifying that the cargo being carried has been examined, usually at the initial place of departure, and has been found to be free of contraband. The purpose of such a navicert or aircert is to facilitate belligerent control of contraband goods with minimal interference and delay of neutral commerce.[242]The navicert or aircert may be viewed as a type of commercial passport.[243]

The certificate is not a guarantee that the vessel or aircraft will not be subject to visit and search or that cargo will not be seized. (Changed circumstances, such as a change in status of the neutral vessel, between the time of issuance of the certificate and the time of interception at sea may cause it to be invalidated.) Conversely, the absence of a navicert or aircert is not, in itself, a valid ground for seizure of cargo.[244]

Navicerts and aircerts issued by one belligerent State do not limit the visit and search rights of an opposing belligerent State. When a neutral ship or aircraft accepts a navicert or aircert from one belligerent State, this may affect how the other belligerent State views the neutrality of that aircraft or vessel.[245]

A similar procedure was used during the Cuban Missile Crisis, when the United States issued clearance certificates (clearcerts).[246]

15.13 BELLIGERENT RIGHT OF VISIT AND SEARCH OF MERCHANT VESSELS AND CIVIL AIRCRAFT

Belligerent warships and military aircraft have a right of visit and search of merchant vessels and civil aircraft outside of neutral territory.

15.13.1 Purposes of Visit and Search. The belligerent right of visit and search may be viewed as a necessary part of the belligerent's right to capture enemy merchant vessels and civil aircraft, and to capture neutral merchant vessels and civil aircraft that have engaged in violations of neutrality.[247]Thus, for example, visit and search has been conducted with the object of:[248]

• ascertaining the character of the vessel or aircraft and nationality (including assessing whether a vessel or aircraft that is flagged to a neutral State has acquired enemy character by engaging in service to the enemy);[249]

• verifying whether it conveys contraband cargo;[250]

• verifying whether it has committed a breach of blockade;[251]or

• verifying whether the vessel or aircraft has committed another violation of neutrality making it liable to capture.[252]

15.13.2 Types of Neutral Vessels and Aircraft That Are Exempt From Visit and Search. Certain neutral vessels and aircraft are exempt from the belligerent right of visit and search: (1) neutral warships; (2) neutral State aircraft (including military aircraft); (3) ships of neutral States used only on government non-commercial service; and (4) neutral merchant vessels under convoy of neutral warships of the same nationality, and neutral aircraft accompanied by neutral military aircraft of the same nationality.

15.13.2.1 *Neutral Warships*. Neutral warships may not be subjected to visit and search because they have complete immunity from the jurisdiction of any State other than the flag State.[253]

15.13.2.2 *Neutral State Aircraft*. Neutral State aircraft (including military aircraft) are immune from visit and search by foreign States.[254]

15.13.2.3 *Ships of a Neutral State Used Only on Government Non-Commercial Service*. Ships owned or operated by a neutral State and used only on government noncommercial service may not be subjected to visit and search because they also have complete immunity from the jurisdiction of any State other than the flag State.[255]

15.13.2.4 *Neutral Merchant Vessels Under Convoy of Neutral Warships of the Same Nationality and Neutral Aircraft Accompanied by Neutral Military Aircraft of the Same Nationality*. Neutral merchant vessels under convoy of neutral warships of the same nationality are also exempt from visit and search.[256]However, the convoy commander may be required to provide in writing to the commanding officer of an intercepting belligerent State's warship information as to the character of the vessels and their cargoes that would otherwise be obtained by a visit and search.[257]Neutral merchant vessels under convoy of neutral warships of the same nationality are exempt from visit and search because the neutral State has provided an assurance that the neutral vessel is not engaged in violations of neutrality.[258]

If a convoy commander determines that a vessel under his or her charge possesses enemy character or carries contraband cargo, the commander is obliged to withdraw his or her protection from the offending vessel, making it liable to visit and search, and possible capture, by opposing belligerent warships.[259]

Neutral civil aircraft accompanied by neutral military aircraft of the same flag may also be exempt from visit and search if the following two elements are met:

• the flag State of a neutral military aircraft warrants that the neutral civil aircraft is not carrying contraband cargo; and

• the commander of the neutral military aircraft provides to the intercepting belligerent military aircraft upon request all information as to the character and cargo of the neutral civil aircraft that would otherwise be obtained by a visit and search.[260]

15.13.3 Where Belligerents May Not Exercise the Right of Visit and Search. As an act of hostility, the belligerent right of visit and search may not be conducted within neutral territory (*e.g.*, a neutral State's territorial seas and neutral airspace).[261]This prohibition on the exercise of the belligerent right of visit and search extends to international straits overlapped by neutral territorial seas and to archipelagic sea lanes.[262]

15.13.4 Procedure for Visit and Search of Merchant Vessels and Aircraft. Belligerent States have discretion in formulating their procedures for conducting the visit and search of neutral merchant vessels and aircraft.[263]Generally, the belligerent warship or aircraft visits or intercepts the vessel or aircraft, shows its true colors, and provides a clear signal to the merchant vessel or civil aircraft that it is expected to submit to visit and search.[264]

During armed conflict, specific rules of engagement or other special instructions may be issued by the operational chain of command to provide guidance on the visit and search procedure to be carried out by U.S. warships and military aircraft when exercising the belligerent State's right of visit and search of merchant vessels and civil aircraft. For example, the issuance of certificates of non-contraband carriage may be part of such procedures.[265]In the absence of specific guidance from the operational chain of command, Military Department or Service regulations or guidance may provide the applicable procedures.[266]

15.13.4.1 *Use of Force During Visit and Search*. Merchant vessels or civil aircraft that comply with instructions given to them may not be made the object of attack; merchant ships or civil aircraft that refuse to comply may be stopped by force.[267]Merchant ships or civil aircraft that resist visit and search assume the risk of resulting damage.[268]Such vessels or aircraft also may be deemed to acquire the character of enemy merchant ships or civil aircraft.[269]

15.13.4.2 *Visit and Search of Merchant Vessels by Military Aircraft*. Belligerent military aircraft have a right to conduct visit and search of merchant vessels.[270]Ordinarily, visit and search of a vessel by a belligerent State's aircraft is accomplished by directing and escorting the merchant vessel to the vicinity of a belligerent warship, which would carry out the visit and search at sea, or to a belligerent port.[271]

15.13.4.3 *Visit and Search of Civil Aircraft by Military Aircraft*. Belligerent military aircraft have a right to conduct visit and search of civil aircraft.[272]Such visit and search is conducted through interception and, if necessary, diversion.[273]

15.14 ACQUISITION OF ENEMY CHARACTER BY NEUTRAL-FLAGGED MERCHANT VESSELS AND NEUTRAL-MARKED CIVIL AIRCRAFT

All vessels operating under an enemy flag, and all aircraft bearing enemy markings, possess enemy character. However, the fact that a ship flies a neutral flag, or that an aircraft bears neutral markings, does not necessarily establish the neutral character of the vessel or aircraft. Vessels or aircraft may acquire enemy character from (1) the ownership or control of the vessel or aircraft, or (2) their conduct.

15.14.1 Acquisition of Enemy Character by Vessels or Aircraft Through Belligerent Ownership or Control. Any merchant vessel or civil aircraft that is owned or controlled by an enemy State or person has enemy character, regardless of whether it is operating under a neutral flag or bears neutral markings.[274] A neutral flag or neutral markings cannot serve as a device to protect vessels or aircraft from seizure whose actual status indicates either continued ownership or control by individuals who themselves possess enemy character.[275] Such vessels may be subject to treatment as enemy merchant vessels or civil aircraft, including being subject to capture.[276]

15.14.2 Acquisition of Enemy Character by Vessels or Aircraft Through Conduct. Certain conduct by vessels or aircraft may lead them to acquire enemy character and make them liable to treatment as either (1) enemy warships or military aircraft, or (2) enemy merchant ships or civil aircraft. Humanitarian assistance given to the wounded, sick, or shipwrecked of belligerent forces by neutral vessels does not lead to such vessels acquiring enemy character if such assistance is provided in accordance with the GWS-Sea.[277]

15.14.2.1 *Acquiring the Character of an Enemy Warship or Military Aircraft.* Neutral merchant vessels and civil aircraft acquire enemy character, and may be treated by a belligerent as enemy warships and military aircraft, when engaged in either of the following acts:[278]

• taking a direct part in the hostilities on the side of the enemy; or

• acting in any capacity as a naval or military auxiliary to the enemy's armed forces.[279]

For example, such vessels and aircraft may be made the object of attack outside of neutral territory.[280]

15.14.2.2 *Acquiring the Character of an Enemy Merchant Vessel or Civil Aircraft.* Neutral merchant vessels and civil aircraft acquire enemy character and may be treated by a belligerent as enemy merchant vessels or civil aircraft when engaged in either of the following acts:[281]

• operating directly under enemy control, orders, charter, employment, or direction; or

• resisting an attempt to establish identity, including resisting visit and search.[282]

For example, such vessels and aircraft may be captured and, under certain circumstances, destroyed.[283]

15.15 CAPTURE OF NEUTRAL VESSELS AND AIRCRAFT

Certain neutral merchant vessels and civil aircraft may be captured outside neutral territory.

15.15.1 Grounds for the Capture of Neutral Vessels and Aircraft. Neutral-flagged merchant vessels or neutral-marked civil aircraft that have acquired enemy character are liable to capture.[284] In addition, neutral merchant vessels and civil aircraft are, in general, liable to capture by a belligerent State's warships and military aircraft if performing any of the following acts:[285]

• carrying contraband;[286]

• carrying personnel in the military or public service of the enemy;[287]

• communicating information in the interest of the enemy;[288]

• breaching or attempting to breach a blockade;[289]

• violating regulations established by a belligerent within the immediate area of naval operations;[290]

• avoiding an attempt to establish identity, including visit and search;[291] or

• presenting irregular or fraudulent papers; lacking necessary papers; or destroying, defacing, or concealing papers.[292]

15.15.2 Procedure for Capture and Condemnation. Belligerent States have discretion in formulating their procedures for conducting the capture and condemnation of neutral merchant vessels and aircraft.[293] Prior exercise of the right of visit and search is not required for the capture of neutral-flagged vessels or neutral-marked aircraft that have acquired enemy status, if positive determination of status can be obtained by other means.[294]

Captured neutral merchant vessels and civil aircraft are sent to a port or airfield under a belligerent State's jurisdiction as a prize for adjudication by a prize court. Ordinarily, a belligerent State's warship will place a prize master and prize crew on board a captured vessel for this purpose. Should that be impracticable, the prize may be escorted into port by a belligerent State's warship or military aircraft. In the latter circumstances, the prize must obey the instructions of its escort or risk forcible measures.[295]

A prize may not be brought into a neutral port, except under emergency circumstances.[296] A prize court cannot be set up by a belligerent on neutral territory or on a vessel in neutral waters.[297]

15.15.2.1 *Use of Force During Capture.* Neutral vessels or aircraft attempting to resist proper capture lay themselves open to forcible measures by a belligerent State's warships and military aircraft, and assume the risk of resulting damage.[298]The same rule applies to resistance during visit and search.[299]

15.15.3 Destruction of Neutral Prizes. Although the destruction o f a neutral prize is not absolutely forbidden, it involves a much more serious responsibility than the destruction of an enemy prize.[300]Thus, a higher standard is applicable than for the destruction of enemy prizes.[301]

Every reasonable effort should be made to avoid destruction of captured neutral vessels and aircraft.[302]A capturing officer, therefore, should not order such destruction without being entirely satisfied that the prize can neither be sent to a belligerent State port or airfield nor, in his or her opinion, properly be released.[303]

Should it become necessary that the prize be destroyed, the capturing officer must provide for the safety of the passengers and crew.[304]In that event, all documents and papers relating to the prize should be preserved. If practicable, the personal effects of passengers should also be safeguarded.[305]

15.15.4 Personnel on Board Captured Neutral Vessels and Aircraft.

15.15.4.1 *Officers and Crews of Captured Neutral Merchant Vessels and Civil Aircraft.* The officers and crews of captured neutral merchant vessels and civil aircraft who are nationals of a neutral State should not be made POWs,[306]even if the vessel or aircraft has acquired the character of enemy merchant vessels or aircraft.[307]However, if the vessel or aircraft has acquired the character of an enemy warship or military aircraft,[308]the officers and crew may be held as POWs.[309]

15.15.4.2 *Enemy Nationals Found Onboard Neutral Merchant Vessels and Civil Aircraft.* Belligerents have a right to remove certain enemy persons from neutral vessels or aircraft, even if there are no grounds for the capture of the vessel or aircraft as prize.[310]Enemy nationals found onboard a neutral State's merchant vessels or civil aircraft as passengers who are:

• a current member of an enemy military force;[311]

• en route to join, or be incorporated into, an enemy's armed forces;

• employed in the public service of the enemy State; or

• engaged in, or suspected of engagement in, service in the interests of the enemy State;[312]

also may be interned until a determination of their status has been made.[313]

15.16 BELLIGERENT FORCES TAKING REFUGE IN NEUTRAL TERRITORY

Neutral States have rights and duties with respect to the treatment of belligerent State forces that enter, or seek to enter, their territory. Generally, a neutral State is required to intern a belligerent State's forces to ensure that they do not return to the armed conflict.[314]

15.16.1 Discretion of the Neutral State in Offering Asylum to Belligerent Forces Seeking Refuge. A neutral State has discretion in whether to permit belligerent forces seeking refuge to enter its territory.

15.16.1.1 *Neutral Right to Deny Access or to Impose Conditions on Belligerent Forces.* A neutral State is not bound to permit a belligerent State's forces to enter its territory.[315]The neutral State may impose conditions on belligerent forces seeking refuge in its territory.[316]

In cases of large bodies of belligerent forces seeking refuge in neutral territory, these conditions will usually be stipulated in an agreement concluded by the representatives of the neutral State and the senior officer of the forces.[317]

15.16.1.2 *Neutral Duty to Disarm and Intern the Belligerent Forces.* If the belligerent forces are permitted to seek refuge in neutral territory, the neutral State must take appropriate measures to ensure that these forces will no longer participate in the armed conflict. If such measures were not taken, then the neutral State's territory might serve as an unlawful sanctuary or base of operations for the belligerent.[318]

Thus, belligerent forces received in neutral territory must be disarmed, and appropriate measures must be taken to prevent their leaving the neutral State.[319]In particular, a neutral State that receives on its territory personnel belonging to the belligerent forces shall intern them, as far as possible, at a distance from the theater of war.[320]For example, if belligerent military aircraft land in a neutral State, that State must intern the aircraft, aircrew, and accompanying military personnel for the duration of the war.[321]

15.16.2 Neutral Reception of the Wounded, Sick, and Shipwrecked. Like other belligerent personnel received in neutral land territory, the wounded, sick, and shipwrecked who are received within a neutral jurisdiction are also generally to be guarded so that they can no longer participate in hostilities. However, ground transports of belligerent wounded and sick personnel may be authorized to pass through neutral land territory by special agreement.[322]

15.16.2.1 *Wounded, Sick, and Shipwrecked Landed in Neutral Ports*. Wounded, sick, or shipwrecked persons who are landed in neutral ports with the consent of the local authorities shall, failing arrangements to the contrary between the neutral State and the belligerent States, be so guarded by the neutral State, where so required by international law, that such persons cannot again take part in operations of war.[323]

The costs of hospital accommodation and internment shall be borne by the belligerent State on whom the wounded, sick, or shipwrecked persons depend.[324]

15.16.2.2 *Wounded, Sick, and Shipwrecked Taken Aboard a Neutral State's Warship or Aircraft*. If wounded, sick, or shipwrecked persons are taken on board a neutral State's warship or a neutral State's military aircraft, it shall be ensured, where so required by international law, that they can take no further part in operations of war.[325]

15.16.2.3 *Wounded and Sick Disembarked From Belligerent Medical Aircraft*. Unless otherwise agreed between the neutral State and the belligerent States, the wounded and sick who are disembarked, with the consent of the local authorities, on neutral territory by medical aircraft, shall be detained by the neutral State, where so required by international law, in such a manner that they cannot again take part in military operations.[326]

The cost of their accommodation and internment shall be borne by the State on which they depend.[327]

15.16.3 Conditions of Internment in a Neutral State. The neutral State may keep belligerent forces in camps and even confine them in military compounds or in places set apart for the purpose of internment.[328]Belligerent forces should be interned, as far as possible, at a distance from the theater of war.[329]

15.16.3.1 *Provision of POW Treatment and Application of the GWS and GWSSea by Analogy*. Under Article 4B(2) of the GPW, persons who are entitled to POW status if they fall into the power of the enemy during international armed conflict are generally entitled to POW treatment, as a minimum, if they are interned by a neutral State under its duties under international law.[330]

Provisions of the GWS and GWS-Sea apply by analogy to wounded and sick, to members of the medical personnel, and to chaplains of the armed forces of the parties to the conflict, who are received or interned in neutral territory, as well as to deceased persons of a State that is a party to a conflict found there.[331]

15.16.3.2 *Belligerent State Acting as the Protecting Power*. Where diplomatic relations exist between the parties to the conflict and the neutral or non-belligerent Power concerned, the parties to a conflict on whom these persons depend shall be allowed to perform towards them the functions of a Protecting Power as provided in the GPW, without prejudice to the functions that these parties normally exercise in conformity with diplomatic and consular usage and treaties.[332]

15.16.3.3 *Reimbursement of Expenses at the Conclusion of Peace*. In the absence of a special convention to the contrary, the neutral Power shall supply the interned with the food, clothing, and relief required by humanity.[333]At the conclusion of peace, the neutral Power's expenses incurred supporting the internment shall be made good.[334]

The provisions of the GPW regarding the costs of POW internment and the financial resources of POWs are not applicable to the situation of persons treated as POWs who have been interned by a neutral State.[335]

15.16.4 Parole of Belligerent Personnel Interned in Neutral Territory. The neutral State shall decide whether interned officers can be left at liberty on giving their parole not to leave the neutral territory without permission.[336]

Neutral States may release enlisted personnel on parole, prescribe penalties for violations of parole, and authorize personnel to depart neutral territory temporarily so long as these actions remain consistent with its obligations as a neutral State.[337]The granting of leave to an interned officer to return to his or her own country, however, would be considered an exceptional measure, and a neutral State inclined to grant such permission would be prudent in the first instance to obtain the consent of the opposing belligerent State.

A neutral State may demand the return of persons who have been released on parole but have left the neutral State in violation of their parole.[338]If such persons return to the State in whose armed forces they serve, that State is obliged to return them to the neutral State at its request.[339]

15.16.5 Military Equipment and Supplies of Belligerent Forces Taking Refuge. The munitions, arms, vehicles, equipment, and other supplies that the interned forces are allowed to bring with them into a neutral State are likewise detained by the neutral State. A belligerent State's military equipment and supplies, whether its own or captured, which are brought on to neutral territory must be returned at the end of the armed conflict to the State to which the items belong. Similarly, captured war material found in the possession of the belligerent forces that take refuge on neutral territory is the property of their State, regardless of its origin.[340]

15.16.6 Retention of Medical Personnel and Chaplains in a Neutral State. Medical and religious personnel serving with belligerent forces that are interned in neutral territory may be retained only in so far as the health and the numbers of such forces so require.[341]Subject to these requirements, they must be returned as soon as possible to the State to which they belong.[342]Chaplains are in the same position, their retention being dependent upon the spiritual needs of the interned forces.[343]Medical personnel and chaplains who are retained must be accorded similar treatment to those retained under the GPW.[344]

15.17 POWS OR INTERNEES BROUGHT TO, OR RECEIVED BY, A NEUTRAL STATE

15.17.1 Escaped POWs Received by a Neutral State. POWs who have escaped to neutral territory are deemed to have successfully escaped from the Detaining Power.[345]

A neutral State may deny the admission of escaped POWs or receive them. A neutral State that receives escaped POWs shall leave them at liberty. If it allows them to remain in its territory, it may assign them a place of residence.[346]

15.17.2 POWs Brought Into a Neutral State by Belligerent Forces Taking Refuge. POWs brought into a neutral State by belligerent forces that take refuge in the neutral State are treated like POWs who have escaped to neutral territory.[347]This means that the POWs regain their liberty from the belligerent forces that previously held them captive, and that if the neutral State allows them to remain on its territory, it may assign them a place of residence. POWs brought into a neutral State by belligerent forces taking refuge are released from their captivity by the belligerent forces that previously held them because of a concern that the detention operations by the belligerent forces would constitute a form of continuation of hostilities on neutral territory and because the POWs probably would have been freed had the belligerent forces not taken refuge in neutral territory.[348]

15.17.3 POWs Brought to a Neutral State by Special Agreement Among the Neutral State and the Parties to a Conflict. POWs may be brought to a neutral State for internment by a special agreement among the neutral State and the parties to the conflict.[349]Wounded and sick POWs may also be brought to neutral countries for accommodation by a special agreement among the neutral State and the parties to the conflict.[350]

Such special agreements must not adversely affect the situation of POWs, nor restrict the rights that the GPW confers upon them.[351]

15.17.4 Certain Classes of Protected Persons Brought to a Neutral State by Special Agreement Among the Neutral State and the Parties to a Conflict. Certain classes of civilian internees may be brought to a neutral State for internment by a special agreement among the neutral State and the parties to the conflict.[352]Agreements may also be concluded for children under the age of fifteen who are orphaned or separated from their families to stay in neutral countries.[353]

Such special agreements must not adversely affect the situation of protected persons nor restrict the rights that the GC confers upon them.[354]

15.18 AUTHORIZED PASSAGE OF WOUNDED AND SICK COMBATANTS THROUGH NEUTRAL TERRITORY

Ground transports of wounded and sick combatants may pass through a neutral State's land territory with the permission of the neutral State. Subject to rules established by the neutral State, medical aircraft of belligerent States may fly over the territory of neutral States, land on it in case of necessity, or use it as a port of call.

15.18.1 Authorized Ground Transports of Wounded and Sick Combatants Through Neutral Territory. A neutral State may authorize the passage into its territory of wounded or sick persons belonging to belligerent forces, on condition that such transports shall carry neither personnel nor material of war.[355]In such a case, the neutral Power is bound to take whatever measures of safety and control are necessary for the purpose.[356]

However, if any such persons are committed to the care of the neutral State or remain in the neutral State's territory, they must be guarded so as to ensure they do not take part again in the war.[357]

15.18.1.1 *Discretion of the Neutral State to Authorize Such Passage*. The neutral State has the right, but not the obligation, to permit such passage; if provided, such passage should be provided on an impartial basis to all belligerent States.[358]

It is not necessary to obtain the consent of the other belligerent States before permitting the passage of sick and wounded personnel, but it would be advisable to do so if considerable numbers are involved.[359]

15.18.1.2 *Personnel or Material of War Accompanying the Wounded or Sick*. A neutral State permitting the passage of sick and wounded into and through its territory must take measures of safety and control to ensure that combatants or military supplies do not accompany them. If combatants accompany the passage of the wounded and sick, they should be interned.[360]Similarly, any military supplies must be seized and placed in safe custody until the end of the conflict.[361]

Medical personnel and materials necessary for the care of the wounded and sick of a convoy of evacuation may be permitted to accompany the convoy.[362]

15.18.1.3 *Wounded and Sick POWs Brought Into Neutral Territory Under Authorized Ground Transports*. Wounded and sick POWs brought by one of the belligerents as part of a convoy of evacuation that is granted passage through neutral territory, and belonging to the hostile party, must be guarded by the neutral State so as to ensure they do not take part again in the operations of the war.[363]

Thus, wounded and sick and POWs brought into neutral territory by the Detaining Power as part of a convoy of evacuation granted right of passage through neutral territory are not treated like escaping POWs or POWs brought by belligerent forces seeking asylum in neutral territory (*i.e.*, transported to their own country or liberated).[364]Instead, they are treated like other wounded, sick, and shipwrecked combatants who are received within neutral territory, and they must be guarded by the neutral State.[365]

15.18.1.4 *Wounded and Sick Combatants Committed to the Care of the Neutral State*. Wounded and sick combatants brought under these conditions into neutral territory by one of the belligerents, and who do not simply pass through neutral territory but are committed to the care of the neutral State, must be guarded by the neutral State so as to ensure they do not take part again in the operations of the war.[366]

15.18.2 Medical Aircraft and Neutral Territory. Subject to conditions and restrictions established by the neutral State, medical aircraft of belligerent States may fly over the territory of neutral States, land on it in case of necessity, or use it as a port of call.[367]Belligerent States shall give the neutral States previous notice of their passage over the neutral State and obey all summons to alight, on land or water.[368]

The neutral State may place conditions or restrictions on the passage or landing of medical aircraft on its territory.[369]Such conditions or restrictions shall be applied equally to all belligerent States.[370]Belligerent States' medical aircraft would be immune from attack only when flying on routes, at heights, and at times specifically agreed upon between the belligerent States and the neutral State concerned.[371]

As a general matter, the wounded and sick who are disembarked, with the consent of the local authorities, on neutral territory by medical aircraft, shall be detained by the neutral State, where so required by international law, in such a manner that they cannot again take part in military operations.[372]

XVI – Cyber Operations

Chapter Contents

16.1 INTRODUCTION

This Chapter addresses the law of war and cyber operations. It addresses how law of war principles and rules apply to relatively novel cyber capabilities and the cyber domain. As a matter of U.S. policy, the United States has sought to work internationally to clarify how existing international law and norms, including law of war principles, apply to cyber operations.[1]

Precisely how the law of war applies to cyber operations is not well-settled, and aspects of the law in this area are likely to continue to develop, especially as new cyber capabilities are developed and States determine their views in response to such developments.[2]

16.1.1 Cyberspace as a Domain. As a doctrinal matter, DoD has recognized cyberspace as an operational domain in which the armed forces must be able to defend and operate, just like the land, sea, air, and space domains.[3]

Cyberspace may be defined as "[a] global domain within the information environment consisting of interdependent networks of information technology infrastructures and resident data, including the Internet, telecommunications networks, computer systems, and embedded processors and controllers."[4]

16.1.2 Description of Cyber Operations. Cyberspace operations may be understood to be those operations that involve "[t]he employment of cyberspace capabilities where the primary purpose is to achieve objectives in or through cyberspace."[5]Cyber operations: (1) use cyber capabilities, such as computers, software tools, or networks; and (2) have a primary purpose of achieving objectives or effects in or through cyberspace.

16.1.2.1 *Examples of Cyber Operations*. Cyber operations include those operations that use computers to disrupt, deny, degrade, or destroy information resident in computers and computer networks, or the computers and networks themselves. Cyber operations can be a form of advance force operations, which precede the main effort in an objective area in order to prepare the objective for the main assault. For example, cyber operations may include reconnaissance (*e.g.*, mapping a network), seizure of supporting positions (*e.g.*, securing access to key network systems or nodes), and pre-emplacement of capabilities or weapons (*e.g.*, implanting cyber access tools or malicious code). In addition, cyber operations may be a method of acquiring foreign intelligence unrelated to specific military objectives, such as understanding technological developments or gaining information about an adversary's military capabilities and intent.

16.1.2.2 *Examples of Operations That Would Not Be Regarded as Cyber Operations*. Cyber operations generally would not include activities that merely use computers or cyberspace without a primary purpose of achieving objectives or effects in or through cyberspace. For example, operations that use computer networks to facilitate command and control, operations that use air traffic control systems, and operations to distribute information broadly using computers would generally not be considered cyber operations. Operations that target an adversary's cyberspace capabilities, but that are not achieved in or through cyberspace, would not be considered cyber operations. For example, the bombardment of a network hub, or the jamming of wireless communications, would not be considered cyber operations, even though they may achieve military objectives in cyberspace.

16.1.3 Cyber Operations – Notes on Terminology. DoD doctrine and terminology for cyber operations continue to develop.

16.1.3.1 *"Cyber" Versus "Cyberspace" as an Adjective*. The terms "cyber" and "cyberspace" when used as an adjective (*e.g.*, cyber attack, cyber defense, cyber operation) are generally used interchangeably.

16.1.3.2 *Cyber Attacks or Computer Network Attacks*. The term "attack" often has been used in a colloquial sense in discussing cyber operations to refer to many different types of hostile or malicious cyber activities, such as the defacement of websites, network intrusions, the theft of private information, or the disruption of the provision of internet services. Operations described as "cyber attacks" or "computer network attacks," therefore, are not necessarily "attacks" for the purposes of applying rules on conducting attacks during the conduct of hostilities.[6]Similarly, operations described as "cyber attacks" or "computer network attacks" are not necessarily "armed attacks" for the purposes of triggering a State's inherent right of self-defense under *jus ad bellum*.[7]

16.2 APPLICATION OF THE LAW OF WAR TO CYBER OPERATIONS

Specific law of war rules may apply to cyber operations, even though those rules were developed before cyber operations were possible. When no more specific law of war rule or other applicable rule applies, law of war principles provide a general guide for conduct during cyber operations in armed conflict.

16.2.1 Application of Specific Law of War Rules to Cyber Operations. Specific law of war rules may be applicable to cyber operations, even though these rules were developed long before cyber operations were possible.

The law of war affirmatively anticipates technological innovation and contemplates that its existing rules will apply to such innovation, including cyber operations.[8]Law of war rules may apply to new technologies because the rules often are not framed in terms of specific technological means. For example, the rules on conducting attacks do not depend on what type of weapon is used to conduct the attack. Thus, cyber operations may be subject to a variety of law of war rules depending on the rule and the nature of the cyber operation. For example, if the physical consequences of a cyber attack constitute the kind of physical damage that would be caused by dropping a bomb or firing a missile, that cyber attack would equally be subject to the same rules that apply to attacks using bombs or missiles.[9]

Cyber operations may pose challenging legal questions because of the variety of effects they can produce. For example, cyber operations could be a non-forcible means or method of conducting hostilities (such as information gathering), and would be regulated as such under rules applicable to non-forcible means and methods of warfare.[10]Other cyber operations could be used to create effects that amount to an attack and would be regulated under the rules on conducting attacks.[11]Moreover, another set of challenging issues may arise when considering whether a particular cyber operation might be regarded as a seizure or destruction of enemy property and should be assessed as such.[12]

16.2.2 Application of Law of War Principles as a General Guide to Cyber Operations. When no specific rule applies, the principles of the law of war form the general guide for conduct during war, including conduct during cyber operations.[13]For example, under the principle of humanity, suffering, injury, or destruction unnecessary to accomplish a legitimate military purpose must be avoided in cyber operations.[14]

Certain cyber operations may not have a clear kinetic parallel in terms of their capabilities and the effects they create.[15]Such operations may have implications that are quite different from those presented by attacks using traditional weapons, and those different implications may well yield different conclusions.[16]

16.3 CYBER OPERATIONS AND *JUS AD BELLUM*

Cyber operations may present issues under the law of war governing the resort to force (*i.e.*, *jus ad bellum*).[17]

16.3.1 Prohibition on Cyber Operations That Constitute Illegal Uses of Force Under Article 2(4) of the Charter of the United Nations. Article 2(4) of the Charter of the United Nations states that "[a]ll Members shall refrain in their international relations from the threat or use of force against the territorial integrity or political independence of any state, or in any other manner inconsistent with the Purposes of the United Nations."[18]

Cyber operations may in certain circumstances constitute uses of force within the meaning of Article 2(4) of the Charter of the United Nations and customary international law.[19]

For example, if cyber operations cause effects that, if caused by traditional physical means, would be regarded as a use of force under *jus ad bellum*, then such cyber operations would likely also be regarded as a use of force. Such operations may include cyber operations that: (1) trigger a nuclear plant meltdown; (2) open a dam above a populated area, causing destruction; or (3) disable air traffic control services, resulting in airplane crashes.[20]Similarly, cyber operations that cripple a military's logistics systems, and thus its ability to conduct and sustain military operations, might also be considered a use of force under *jus ad bellum*.[21] Other factors, besides the effects of the cyber operation, may also be relevant to whether the cyber operation constitutes a use of force under *jus ad bellum*.[22]

Cyber operations that constitute uses of force within the meaning of Article 2(4) of the Charter of the United Nations and customary international law must have a proper legal basis in order not to violate *jus ad bellum* prohibitions on the resort to force.[23]

16.3.2 Peacetime Intelligence and Counterintelligence Activities. International law and long-standing international norms are applicable to State behavior in cyberspace,[24] and the question of the legality of peacetime intelligence and counterintelligence activities must be considered on a case-by-case basis. Generally, to the extent that cyber operations resemble traditional intelligence and counter-intelligence activities, such as unauthorized intrusions into computer networks solely to acquire information, then such cyber operations would likely be treated similarly under international law.[25] The United States conducts such activities via cyberspace, and such operations are governed by long-standing and well-established considerations, including the possibility that those operations could be interpreted as a hostile act.[26]

16.3.3 Responding to Hostile or Malicious Cyber Operations. A State's inherent right of self-defense, recognized in Article 51 of the Charter of the United Nations, may be triggered by cyber operations that amount to an armed attack or imminent threat thereof.[27] As a matter of national policy, the United States has expressed the view that when warranted, it will respond to hostile acts in cyberspace as it would to any other threat to the country.[28]

Measures taken in the exercise of the right of national self-defense in response to an armed attack must be reported immediately to the U.N. Security Council in accordance with Article 51 of the Charter of the United Nations.[29]

16.3.3.1 *Use of Force Versus Armed Attack*. The United States has long taken the position that the inherent right of self-defense potentially applies against any illegal use of force.[30] Thus, any cyber operation that constitutes an illegal use of force against a State potentially gives rise to a right to take necessary and proportionate action in self-defense.[31]

16.3.3.2 *No Legal Requirement for a Cyber Response to a Cyber Attack*. There is no legal requirement that the response in self-defense to a cyber armed attack take the form of a cyber action, as long as the response meets the requirements of necessity and proportionality.[32]

16.3.3.3 *Responses to Hostile or Malicious Cyber Acts That Do Not Constitute Uses of Force*. Although cyber operations that do not constitute uses of force under *jus ad bellum* would not permit injured States to use force in self-defense, those injured States may be justified in taking necessary and appropriate actions in response that do not constitute a use of force.[33] Such actions might include, for example, a diplomatic protest, an economic embargo, or other acts of retorsion.[34]

16.3.3.4 *Attribution and Self-Defense Against Cyber Operations*. Attribution may pose a difficult factual question in responding to hostile or malicious cyber operations because adversaries may be able to hide or disguise their activities or identities in cyberspace more easily than in the case of other types of operations.[35]

A State's right to take necessary and proportionate action in self-defense in response to an armed attack originating through cyberspace applies whether the attack is attributed to another State or to a non-State actor.[36]

16.3.3.5 *Authorities Under U.S. Law to Respond to Hostile Cyber Acts*. Decisions about whether to invoke a State's inherent right of self-defense would be made at the national level because they involve the State's rights and responsibilities under international law. For example, in the United States, such decisions would generally be made by the President. The Standing Rules of Engagement for U.S. forces have addressed the authority of the U.S. armed forces to take action in self-defense in response to hostile acts or hostile intent, including such acts perpetrated in or through cyberspace.[37]

16.4 CYBER OPERATIONS AND THE LAW OF NEUTRALITY

The law of neutrality may be important in certain cyber operations. For example, under the law of neutrality, belligerent States are bound to respect the sovereign rights of neutral States.[38] Because of the interconnected nature of cyberspace, cyber operations targeting networked information infrastructures in one State may create effects in another State that is not a party to the armed conflict.[39]

16.4.1 Cyber Operations That Use Communications Infrastructure in Neutral States. The law of neutrality has addressed the use of communications infrastructure in neutral States, and in certain circumstances, these rules would apply to cyber operations. The use of communications infrastructure in neutral States may be implicated under the general rule that neutral territory may not serve as a base of operations for one belligerent against another.[40]In particular, belligerent States are prohibited from erecting on the territory of a neutral State any apparatus for the purpose of communicating with belligerent forces on land or sea, or from using any installation of this kind established by them before the armed conflict on the territory of a neutral State for purely military purposes, and which has not been opened for the service of public messages.[41]

However, merely relaying information through neutral communications infrastructure (provided that the facilities are made available impartially) generally would not constitute a violation of the law of neutrality that belligerent States would have an obligation to refrain from and that a neutral State would have an obligation to prevent.[42]This rule was developed because it was viewed as impractical for neutral States to censor or screen their publicly available communications infrastructure for belligerent traffic.[43]Thus, for example, it would not be prohibited for a belligerent State to route information through cyber infrastructure in a neutral State that is open for the service of public messages, and that neutral State would have no obligation to forbid such traffic. This rule would appear to be applicable even if the information that is being routed through neutral communications infrastructure may be characterized as a cyber weapon or otherwise could cause destructive effects in a belligerent State (but no destructive effects within the neutral State or States).[44]

16.5 CYBER OPERATIONS AND *JUS IN BELLO*

This section addresses *jus in bello* rules and cyber operations.

16.5.1 Cyber Operations That Constitute "Attacks" for the Purpose of Applying Rules on Conducting Attacks. If a cyber operation constitutes an attack, then the law of war rules on conducting attacks must be applied to those cyber operations.[45]For example, such operations must comport with the requirements of distinction and proportionality.[46]

For example, a cyber attack that would destroy enemy computer systems could not be directed against ostensibly civilian infrastructure, such as computer systems belonging to stock exchanges, banking systems, and universities, unless those computer systems met the test for being a military objective under the circumstances.[47]A cyber operation that would not constitute an attack, but would nonetheless seize or destroy enemy property, would have to be imperatively demanded by the necessities of war.[48]

16.5.1.1 *Assessing Incidental Injury or Damage During Cyber Operations*. The principle of proportionality prohibits attacks in which the expected loss of life or injury to civilians, and damage to civilian objects incidental to the attack, would be excessive in relation to the concrete and direct military advantage expected to be gained.[49]For example, in applying this prohibition to cyber operations, it might be important to assess the potential effects of a cyber attack on computers that are not military objectives, such as private, civilian computers that hold no military significance, but that may be networked to computers that are valid military objectives.[50]

In assessing incidental injury or damage during cyber operations, it may be important to consider that remote harms and lesser forms of harm, such as mere inconveniences or temporary disruptions, need not be considered in assessing whether an attack is prohibited by the principle of proportionality.[51]For example, a minor, brief disruption of internet services to civilians that results incidentally from a cyber attack against a military objective generally would not need to be considered in a proportionality analysis.[52]In addition, the economic harms in the belligerent State resulting from such disruptions, such as civilian businesses in the belligerent State being unable to conduct e-commerce, generally would not need to be considered in a proportionality analysis.[53]

Even if cyber operations that constitute attacks are not expected to result in excessive incidental loss of life or injury or damage such that the operation would be prohibited by the principle of proportionality, the party to the conflict nonetheless would be required to take feasible precautions to limit such loss of life or injury and damage in conducting those cyber operations.[54]

16.5.2 Cyber Operations That Do Not Amount to an "Attack" Under the Law of War. A cyber operation that does not constitute an attack is not restricted by the rules that apply to attacks.[55]Factors that would suggest that a cyber operation is not an "attack" include whether the operation causes only reversible effects or only temporary effects. Cyber operations that generally would not constitute attacks include:

• defacing a government webpage;

• a minor, brief disruption of internet services;

• briefly disrupting, disabling, or interfering with communications; and

• disseminating propaganda.

Since such operations generally would not be considered attacks under the law of war, they generally would not need to be directed at military objectives, and may be directed at civilians or civilian objects. Nonetheless, such operations must not be directed against enemy civilians or civilian objects unless the operations are militarily necessary.[56]Moreover, such operations should comport with the general principles of the law of war.[57]For example, even if a cyber operation is not an "attack" or does not cause any injury or damage that would need to be considered under the principle of proportionality in conducting attacks, that cyber operation still should not be conducted in a way that unnecessarily causes inconvenience to civilians or neutral persons.

16.5.3 Duty to Take Feasible Precautions and Cyber Operations. Parties to a conflict must take feasible precautions to reduce the risk of incidental harm to the civilian population and other protected persons and objects.[58]Parties to the conflict that employ cyber operations should take precautions to minimize the harm of their cyber activities on civilian infrastructure and users.[59]

The obligation to take feasible precautions may be of greater relevance in cyber operations than other law of war rules because this obligation applies to a broader set of activities than those to which other law of war rules apply. For example, the obligation to take feasible precautions to reduce the risk of incidental harm would apply to a party conducting an attack even if the attack would not be prohibited by the principle of proportionality.[60]In addition, the obligation to take feasible precautions applies even if a party is not conducting an attack because the obligation also applies to a party that is subject to attack.[61]

16.5.3.1 *Cyber Tools as Potential Measures to Reduce the Risk of Harm to Civilians or Civilian Objects.* In some cases, cyber operations that result in non-kinetic or reversible effects can offer options that help minimize unnecessary harm to civilians.[62]In this regard, cyber capabilities may in some circumstances be preferable, as a matter of policy, to kinetic weapons because their effects may be reversible, and they may hold the potential to accomplish military goals without any destructive kinetic effect at all.[63]

As with other precautions, the decision of which weapon to use will be subject to many practical considerations, including effectiveness, cost, and "fragility," *i.e.*, the possibility that once used an adversary may be able to devise defenses that will render a cyber tool ineffective in the future.[64]Thus, as with special kinetic weapons, such as precision-guided munitions that have the potential to produce less incidental damage than other kinetic weapons, cyber capabilities usually will not be the only type of weapon that is legally permitted.

16.5.4 Prohibition on Improper Use of Signs During Cyber Operations. Under the law of war, certain signs may not be used improperly.[65]These prohibitions may also be applicable during cyber operations. For example, it would not be permissible to conduct a cyber attack or to attempt to disable enemy internal communications by making use of communications that initiate non-hostile relations, such as prisoner exchanges or ceasefires.[66]Similarly, it would be prohibited to fabricate messages from an enemy's Head of State falsely informing that State's forces that an armistice or cease-fire had been signed.[67]

On the other hand, the restriction on the use of enemy flags, insignia, and uniforms only applies to concrete visual objects; it does not restrict the use of enemy codes, passwords, and countersigns.[68]Thus, for example, it would not be prohibited to disguise network traffic as though it came from enemy computers or to use enemy codes during cyber operations.

16.5.5 Use of Civilian Personnel to Support Cyber Operations. As with non-cyber operations, the law of war does not prohibit States from using civilian personnel to support their cyber operations, including support actions that may constitute taking a direct part in hostilities.[69]

Under the GPW, persons who are not members of the armed forces, but who are authorized to accompany them, are entitled to POW status.[70]This category was intended to include, *inter alia*, civilian personnel with special skills in operating military equipment who support and participate in military operations, such as civilian members of military aircrews.[71]It would include civilian cyber specialists who have been authorized to accompany the armed forces.

Civilians who take a direct part in hostilities forfeit protection from being made the object of attack.[72]

16.6 LEGAL REVIEW OF WEAPONS THAT EMPLOY CYBER CAPABILITIES

DoD policy requires the legal review of the acquisition of weapons or weapon systems.[73] This policy would include the review of weapons that employ cyber capabilities to ensure that they are not *per se* prohibited by the law of war.[74] Not all cyber capabilities, however, constitute a weapon or weapons system. Military Department regulations address what cyber capabilities require legal review.[75]

The law of war does not prohibit the development of novel cyber weapons. The customary law of war prohibitions on specific types of weapons result from State practice and *opinio juris* demonstrating that a type of weapon is illegal; the mere fact that a weapon is novel or employs new technology does not mean that the weapon is illegal.[76]

Although which issues may warrant legal analysis would depend on the characteristics of the weapon being assessed, a legal review of the acquisition or procurement of a weapon that employs cyber capabilities likely would assess whether the weapon is inherently indiscriminate.[77] For example, a destructive computer virus that was programmed to spread and destroy uncontrollably within civilian internet systems would be prohibited as an inherently indiscriminate weapon.[78]

XVII – Non-International Armed Conflict (NIAC)

Chapter Contents

17.1 INTRODUCTION

This Chapter addresses the law of war rules applicable to armed conflict not of an international character, or non-international armed conflict (NIAC). Non-international armed conflicts are those armed conflicts that are not between States.[1]In particular, this Chapter addresses the rules applicable to State armed forces conducting military operations against non- State armed groups. The application of the law of war to non-international armed conflict may be complex.[2]

In U.S. practice, in certain cases, the rules applicable in international armed conflict have been applied as a matter of policy to military operations in non-international armed conflict.

17.1.1 Non-International Armed Conflict – Notes on Terminology. Non-international armed conflict is commonly referred to by the acronym "NIAC." Although there has been a range of views on what constitutes a non-international armed conflict, the intensity of the conflict and the organization of the parties are criteria that have been assessed to distinguish between non-international armed conflict and internal disturbances and tensions, such as riots, isolated and sporadic acts of violence, and other acts of a similar nature.[3]A variety of terms have been used to describe factual situations that often may be characterized as non-international armed conflict.

17.1.1.1 *NIAC and Civil War*. Civil war is a classic example of a noninternational armed conflict. For example, a non-international armed conflict could involve the open rebellion of segments of a nation's armed forces (sometimes called dissident armed forces) against the incumbent regime, each claiming to be the legitimate government.[4]In some cases of civil war, the insurgent party has been recognized as a belligerent, and, at least in some respects, the law of international armed conflict would be applied by the States choosing to recognize the insurgent party as a belligerent.[5]

17.1.1.2 *NIAC and Internal Armed Conflict*. In some cases, the term internal armed conflict is used as a synonym for non-international armed conflict. Such usage may reflect a traditional definition of non-international armed conflict as only those armed conflicts occurring within the borders of a single State.[6]Non-international armed conflicts, however, are classified as such simply based on the status of the parties to the conflict, and sometimes occur in more than one State.[7]The mere fact that an armed conflict occurs in more than one State and thus may be characterized as international "in scope" does not render it "international in character."[8]

17.1.1.3 *Transnational or Internationalized NIACs*. Sometimes the terms of "transnational" or "internationalized" are used to describe certain non-international armed conflicts.

"Transnational" has been used to indicate that the non-international armed conflict takes place in more than one State.

"Internationalized" has been used to indicate that multiple States may be involved in a non-international armed conflict.

17.1.1.4 *NIAC and Guerilla or Unconventional Warfare*. Guerrilla warfare may be understood to be military and paramilitary operations conducted in enemy-held or hostile territory by irregular, predominantly indigenous forces.[9]Guerrilla operations or unconventional warfare are common during non-international armed conflict. Such operations, however, are a method of warfare that has been employed in international armed conflicts and occupation as well.

17.1.1.5 *NIAC and Rebellion or Insurrection*. Rebellion, insurrection, or insurgency may also be types of non-international armed conflict.[10]Counter-insurgency operations generally occur in the context of non-international armed conflict, but could occur in the context of an international armed conflict and occupation as well.

17.1.1.6 *NIAC and Terrorism*. Issues surrounding terrorism and the activities of terrorist groups can arise in the contexts of non-international armed conflict, international armed conflict, and, of course, can arise outside the context of armed conflict altogether. Acts of terrorism are prohibited during international armed conflict and during non-international armed conflict.[11]

17.1.1.7 *NIAC and Small Wars or Low-Intensity Conflict*. Non-international armed conflict has sometimes been discussed using the term "low-intensity" conflict.[12]The term "small wars" has also been used in military doctrine to describe situations that may be characterized as non-international armed conflicts.[13]"Low-intensity conflict" and "small wars" are not synonymous with non-international armed conflict, but there is a high degree of overlap between those categories and non-international armed conflict.

17.1.2 Important Commonalities Between the Law Applicable to International Armed Conflict and the Law Applicable to Non-International Armed Conflict.

17.1.2.1 *Common Baseline Rules*. Certain baseline rules, in particular relating to the humane treatment of detainees, must be observed regardless of the character of the armed conflict.[14]The fact that certain baseline rules are common to international armed conflict and non-international armed conflict means that it may be unnecessary to determine the character of the armed conflict in order to assess whether the law has been violated.

17.1.2.2 *Foundational Principles of the Law of War*. The foundational principles of the law of war are common to both international armed conflict and non-international armed conflict.[15]Thus, reference to first principles in the law of war may be most useful in assessing the rules applicable during non-international armed conflict.[16]

17.1.2.3 *Rules for Conducting Operations Against Unprivileged Belligerents*. Rules for conducting operations against unprivileged belligerents are found in both the law applicable to international armed conflict and the law applicable to non-international armed conflict.[17]The rules for States conducting military operations against unprivileged belligerents in international armed conflict are not significantly different from the rules for States conducting military operations against non-State armed groups during non-international armed conflict.

17.1.3 Important Differences Between the Law Applicable to International Armed Conflict and the Law Applicable to Non-International Armed Conflict.

17.1.3.1 *Nationality and Territoriality Exclusions in the Law of International Armed Conflict*. Certain rules applicable to international armed conflict reflect concepts (*e.g.*, nationality and territory) that preclude the application of those rules to internal armed conflicts. For example, nationals who are in the power of their State of nationality would not be provided POW status under the GPW or protected person status under the GC.[18]Thus, even if the GPW and GC could otherwise be deemed applicable to a civil war, these exclusions based on nationality would limit the application of many of the provisions of the GPW and GC (as a matter of treaty law) to internal armed conflicts. In any case, it remains true that fundamental principles guaranteeing humane treatment (*e.g.*, Common Article 3 of the 1949 Geneva Conventions) would apply in any such circumstances.

Similarly, it is the essence of belligerent occupation that it should be exercised over foreign, enemy territory;[19]thus, occupation law rules would not apply to internal armed conflict.[20]

Certain non-international armed conflicts, however, are not internal armed conflicts.[21]

17.1.3.2 *Prevalence of Customary Law Applicable to NIAC as Opposed to Treaty Law.* There are fewer treaty provisions that address non-international armed conflict than that address international armed conflict.[22]Put another way, practitioners are generally more likely to encounter situations regulated by customary law in non-international armed conflict than in international armed conflict. Certain guidelines may be helpful in assessing customary international law applicable to non-international armed conflict.[23]

17.1.3.3 *Important Substantive Differences Between the Law Applicable to International Armed Conflict and the Law Applicable to Non-International Armed Conflict.* There are important substantive differences between the law applicable to international armed conflict and the law applicable to non-international armed conflict, including the following three examples.

First, the different circumstances that typically arise in non-international armed conflicts as compared to international armed conflicts may need to be considered in applying the principle of distinction.[24]

Second, States have greater latitude to compel enemy persons to switch allegiance or to serve the State in hostilities during non-international armed conflict than States have to compel enemy nationals during international armed conflict.[25]

Third, States have greater latitude to use their domestic law against enemy armed groups in non-international armed conflict than States have to use their domestic law against enemy forces or enemy nationals in international armed conflict.[26]

17.2 APPLICATION OF INTERNATIONAL LAW TO NIACS

In some cases, there may be important substantive differences between the rules applicable in international armed conflict and the rules applicable in non-international armed conflict.[27]In some cases, only the general essence of a rule that applies during international armed conflict applies during non-international armed conflict, as opposed to the detailed provisions in some treaties relating to many aspects of international armed conflict.[28]

The extent to which the law of war rules that apply during international armed conflict must or should apply during non-international armed conflict has not been clearly defined as the law of war has developed.[29]The discretion afforded States in applying law of war rules to non- international armed conflicts results, in part, because treaty provisions applicable to international armed conflict have been presumed not to apply to non-international armed conflict unless explicitly made applicable. For example, in the 1949 Geneva Conventions, only Common Article 3 applies to non-international armed conflict.[30]The discretion afforded States in this regard may also be understood to result from the wide range of circumstances that constitute non-international armed conflict. The United States has objected to efforts to make the applicability of the rules of international armed conflict turn on subjective and politicized criteria that would eliminate the distinction between international and non-international conflicts.[31]

In the sections that follow, which reflect the practice of the U.S. armed forces in applying the law of war to non-international armed conflict, the rules articulated may exceed the requirements of applicable customary international law and treaty law.

17.2.1 Treaties That Apply to NIAC. Relatively few treaties have provisions that expressly apply to non-international armed conflicts. Some treaties, however, may apply implicitly to non-international armed conflict.

17.2.1.1 *Treaties That Have Provisions That Explicitly Apply to NIAC.* Certain treaties to which the United States is a Party have provisions that explicitly apply to noninternational armed conflict. These treaties include:

- the 1949 Geneva Conventions (*i.e.*, Common Article 3);[32]
- the 1954 Hague Cultural Property Convention;[33]
- the CCW Amended Mines Protocol;[34]
- the Amended CCW, including Protocols I, III, and IV;[35]
- the CCW Protocol V on Explosive Remnants of War;[36]and
- AP III.[37]

Treaties to which the United States is not a Party that have provisions applicable to "armed conflict not of an international character" include:

- AP II;[38]and
- the Rome Statute of the International Criminal Court.[39]

17.2.1.2 *Implicit Application of Treaty Provisions to Situations in NIAC.* Some treaties may apply implicitly to certain situations in non-international armed conflict. For example, the Genocide Convention does not expressly refer to non-international armed conflict, but recognizes that acts of genocide are criminal "whether committed in time of peace or time of war" and whether they are committed by "constitutionally responsible rulers, public officials or private individuals."[40]Similarly, the Convention Against Torture recognizes that "No exceptional circumstances whatsoever, whether a state of war or a threat of war, internal political instability or any other public emergency, may be invoked as a justification of torture"; thus, the state of non-international armed conflict could not be justification for torture.[41]The prohibitions in Article 1 of the Chemical Weapons Convention have been interpreted to apply to non-international armed conflict.[42]

In addition, the prohibitions and restrictions on acquisition and development of biological weapons in the Biological Weapons Convention effectively prevent the use of biological weapons by States in non-international armed conflict.[43]

Further, the obligations in the Child Soldiers Protocol relate implicitly to noninternational armed conflict.[44]

17.2.1.3 *Human Rights Treaties and NIAC.* During an internal non-international armed conflict, a State would continue to be bound by applicable human rights treaty obligations.[45]

The applicability of human rights treaty obligations during non-international armed conflict may depend on a variety of factors. Such applicability depends on the terms of the particular treaty in question, and whether the State has exercised an authorized derogation from its provisions due to an emergency situation. The applicability of a human rights treaty obligation with respect to an individual, such as an obligation under the International Covenant on Civil and Political Rights, for example, may depend on whether the person is located outside the territory of the State Party.[46]In addition, law of war rules constitute the *lex specialis* during situations of armed conflict, and as such, serve as the controlling body of law with regard to the conduct of hostilities and the protection of war victims.[47]

17.2.2 Assessing Customary International Law Applicable to NIAC. As a consequence of the fewer treaty provisions applicable to non-international armed conflict, many of the rules applicable to non-international armed conflict are found in customary international law. The following guidelines may be helpful in assessing the customary international law applicable to non-international armed conflict.

17.2.2.1 *Use of Law of War Principles to Discern Rules Applicable to NIAC.* The fundamental principles of the law of war also provide the foundation for the rules applicable during non-international armed conflict.[48]As during international armed conflict, the principles of the law of war form the general guide for conduct during non-international armed conflict, when no specific rule applies.[49]

However, the application of law of war principles may differ insofar as the circumstances in international armed conflicts may often be quite different from the circumstances in noninternational armed conflicts.[50]

17.2.2.2 *Considered Absence of a Restriction in NIAC.* Under general principles of legal interpretation, when a rule mentions specific circumstances or conditions in which it applies, it may give rise to a presumption that the rule was not intended to apply in other related circumstances or conditions that are not specifically mentioned.[51]

Thus, if a treaty addresses both international armed conflict and non-international armed conflict, and provides for a restriction in international armed conflict but does not provide for that restriction in non-international armed conflict, then this omission may, to some extent, reflect States' views that such restrictions were not applicable in non-international armed conflict.[52]Similarly, States negotiated and adopted AP I and AP II at the same diplomatic conference, and the omission from AP II of restrictions present in AP I may, to some extent, reflect States' views that such restrictions were not applicable in non-international armed conflict.[53]

17.2.2.3 *Application of IAC Rules by Analogy.* If an action is not prohibited by the law of war applicable to international armed conflict, it generally would not be prohibited by the law of war applicable to non-international armed conflict.

For example, analogous provisions of the GPW and GC may be helpful for understanding the baseline standards in international law for detention because the baseline standards applicable to all detainees during armed conflict (*e.g.*, Common Article 3 of the 1949 Geneva Conventions) are not more favorable than the treatment and protections applicable to POWs and civilian internees under the GPW and GC, respectively.[54]Thus, in some instances it may be appropriate to implement measures during detention of persons during non-international armed conflict by analogy to the internment of POWs during international armed conflict or by analogy to the internment of protected persons in occupied territory.[55]

17.2.2.4 *Application of Law Enforcement Rules*. The law applicable to noninternational armed conflict generally has been crafted to reflect baseline rules that States respect even in addressing common criminals.[56] Thus, if an action would be permissible under the domestic law enforcement rules of many States, it likely would be permissible as a matter of customary international law during non-international armed conflict.

17.2.3 Application of Humanitarian Rules and the Legal Status of the Parties to the Conflict. The application of humanitarian rules to enemy non-State armed groups does not affect the legal status of such groups.[57] For example, a State's decision to apply humanitarian rules in military operations against a non-State armed group would not constitute an implicit recognition of a government that such an armed group has purported to establish nor an implicit recognition of the legitimacy of the armed group's cause. Such application also would not implicitly provide the members of the armed group with any legal immunity from prosecution.

The principle that the application of humanitarian rules to an armed group does not affect the legal status of that armed group has been recognized in a number of treaties. For example, by its express terms, the application of Common Article 3 of the 1949 Geneva Conventions shall not affect the legal status of the parties to the conflict.[58] Also by express treaty terms, the application of the provisions of the CCW and its annexed Protocols to parties to a conflict that are not High Contracting Parties that have accepted the CCW or its annexed Protocols shall not change their legal status or the legal status of a disputed territory, either explicitly or implicitly.[59] Further, by the express terms of the 1954 Hague Cultural Property Convention, its provisions that relate to non-international armed conflict shall not affect the legal status of the parties to the conflict.[60]

17.2.4 Binding Force of the Law of War on Insurgents and Other Non-State Armed Groups. The law of war applicable in a non-international armed conflict is binding upon all parties to the armed conflict, including State armed forces and non-State armed groups. A variety of explanations have been offered for this principle.

Customary law of war rules are binding on a State, even if it is not a Party to a treaty containing the rule.[61] Similarly, customary law of war rules are binding on those parties to the armed conflict that intend to make war and to claim the rights of a belligerent, even if they are not States.[62]

Treaty provisions that address non-international armed conflict provide that they apply not only to the State, but to each party to the conflict.[63] In many cases, these treaty provisions would also be binding on non State armed groups as a matter of customary international law.[64]

As a practical matter, non-State armed groups would often be bound by their State's treaty obligations due to the very fact that the leaders of those non-State armed groups would claim to be the State's legitimate representatives.[65] Other practical considerations, such as the desire to be seen as legitimate, may also contribute to their compliance with the law of war.[66]

17.3 SPECIAL AGREEMENTS BETWEEN PARTIES TO THE CONFLICT

Parties to a conflict may enter into agreements to bring into force law of war rules.[67] For example, pursuant to Common Article 3 of the 1949 Geneva Conventions, the parties to the conflict should further endeavor to bring into force, by means of special agreements, all or part of the other provisions of the 1949 Geneva Conventions.[68] Similarly, pursuant to the 1954 Hague Cultural Property Convention, the parties to the conflict shall endeavor to bring into force, by means of special agreements, all or part of the other provisions of that Convention.[69] Parties to a non-international armed conflict may wish to conclude agreements on these and a variety of other subjects, such as:

- temporary ceasefire agreements to collect the wounded;
- agreements to permit passage of medical or other relief supplies for the civilian population;
- agreements for the accommodation of detainees in other States;[70]
- agreements to establish hospital or safety zones;
- agreements for the cessation of hostilities; and
- agreements for post-conflict clearance of remnants of war.[71]

17.3.1 Communications Between Parties to the Conflict. The procedures that are used for non-hostile relations between belligerents during international armed conflict may also be applied by parties to a non-international armed conflict.[72]

As is the case during international armed conflict, parties to a non-international armed conflict may agree to specific means of communication between them.

17.4 A STATE'S USE OF ITS DOMESTIC LAW AND NIAC

17.4.1 Ability of a State to Use Its Domestic Law Against Non-State Armed Groups. A fundamental principle of the international legal order is the sovereign equality of States, which generally prohibits States from exercising sovereignty over one another.[73] However, the principle of the sovereign equality of States is not applicable in armed conflicts between a State and a non-State armed group. A State may exercise both sovereign and belligerent rights over non-State armed groups.[74] This means that a State may use not only its war powers to combat non-State armed groups, but it may also use its domestic law, including its ordinary criminal law, to combat non-State armed groups. The limits imposed by international law on a State's action against non-State armed groups do not alter the basic principle that the State may exercise its sovereign powers against the non-State armed group.[75]

17.4.1.1 *A State's Power to Prosecute Hostile Activities*. An important consequence of the fact that States may exercise sovereignty over persons belonging to a non- State armed group is that a State may prosecute individuals for participating in hostilities against it. Such conduct frequently constitutes crimes under ordinary criminal law (*e.g.*, murder, assault, illegal destruction of property).

Although, during international armed conflict, lawful combatants are afforded certain immunities from the enemy State's jurisdiction,[76] persons belonging to non-State armed groups lack any legal privilege or immunity from prosecution by a State that is engaged in hostilities against that group. On the other hand, the non-State armed group lacks authority to prosecute members of the State armed forces. In addition, the non-State status of the armed group would not render inapplicable the privileges and immunities afforded lawful combatants and other State officials. Thus, for example, members of the armed forces of a State would continue to benefit from any privileges or immunities from the jurisdiction of foreign States that sought to exercise jurisdiction with respect to the actions of such State armed forces in a non-international armed conflict.[77]

17.4.1.2 *Range of Activities Subject to Prosecution.* A State may also use its domestic law to make punishable a wide range of activity that extend beyond the activities that constitute actual fighting against the State. For example, joining the non-State armed group, providing material support to the armed group, failing to report the treasonous activities of the armed group, and other conduct may be punishable under a State's domestic law.[78]

17.4.1.3 *Other Sovereign Authorities*. In addition to the power to criminalize certain conduct, a State may use its law and other regulatory powers in its effort to address the threat of non-State armed groups. For example, the State could use its authority to tax, regulate, seize, or destroy property (*e.g.*, weapons, vehicles, food, and medical equipment) within its jurisdiction as part of its effort against the non-State armed group.[79] The use of these sovereign powers would be subject to domestic law restrictions, and might not depend on whether such action would be imperatively required by the necessities of war – the standard for the seizure of destruction of enemy property during international armed conflict.[80] In any event, however, it would not be permissible for the State to seek to starve civilians as a method of combat.[81]

17.4.2 Emergency Laws and Regulations. Many States have laws permitting the government to alter or suspend laws (such as a declaration of martial law, and the establishment of curfews and other controls on the movement of persons and traffic), to enact emergency regulations (such as the establishment of monetary or trade regulations, or the rationing of food, fuel, and other critical materials), and to take other steps to protect the public (such as the issuance of identification cards, the development of detention rules for members of non-State armed groups, and the establishment of special emergency courts).[82]

The full range of actions that a State may take under its domestic law during noninternational armed conflict would depend on the content of that law, including applicable constitutional restrictions.[83]

17.4.3 Special Courts. As part of its emergency regulations, a State may establish special or emergency courts for cases involving unprivileged belligerents or other persons suspected of committing offenses related to the non-international armed conflict. Such courts must be regularly constituted and afford all the judicial guarantees that are recognized as indispensable by civilized peoples.[84] Such courts may distinguish based on nationality.[85] The procedures of such courts may deviate from those applicable during ordinary proceedings, but deviations should be warranted by practical need.[86]

17.4.4 Reintegration Programs and Amnesty. States have used reconciliation and reintegration programs during hostilities as alternatives to prosecution to seek to de-radicalize and rehabilitate violent extremists.[87]

Although amnesty is normally left to the discretion of the State, AP II provides that, at the end of hostilities, the authorities in power shall endeavor to grant the broadest possible amnesty to persons who have participated in the armed conflict, or those deprived of their liberty for reasons related to the armed conflict, whether they are interned or detained.[88]

17.5 PRINCIPLE OF DISTINCTION IN NIAC

As discussed below, the principle of distinction applies during non-international armed conflict. It may be important to note certain differences between the situations that typically arise in non-international armed conflict as compared to those typically arising in international armed conflict.

17.5.1 Discrimination in Conducting Attacks Against the Enemy in NIAC. Parties to a conflict must conduct attacks in accordance with the principle of distinction.[89]As during international armed conflict, an adversary's failure to distinguish its forces from the civilian population does not relieve the attacking party of its obligations to discriminate in conducting attacks.[90]On the other hand – also as during international armed conflict – such conduct by the adversary does not increase the legal obligations on the attacking party to discriminate in conducting attacks against the enemy. For example, even though tactics used by non-State armed groups may make discriminating more difficult, State armed forces though obligated to be discriminate – are not required to take additional protective measures to compensate for such tactics.

17.5.1.1 *Increased Difficulty in Identifying Enemy Forces and Other Military Objectives*. During international armed conflict, State armed forces generally are readily distinguishable from the civilian population. Traditionally, conventional armed forces would often confront one another, with the civilian population of each opposing State remaining to the rear of the lines separating their respective military forces.[91]During non-international armed conflict, however, discriminating in conducting attacks against the enemy may be more difficult because non-State armed groups often seek to blend in with the civilian population [92]

17.5.1.2 *Different Support Structures for Non-State Armed Groups*. In addition to non-State armed groups, other military objectives may also be more difficult to identify because non-State armed groups often do not use military infrastructure (*e.g.*, military bases, logistics facilities) to conduct and sustain their operations. Rather, non-State armed groups may seek to use ostensibly civilian buildings and resources to conduct and sustain their operations. Denying non-State armed groups such support may be particularly important to the success of military operations and justifiable under the law of war.[93]

17.5.1.3 *Increased Strategic Importance of Minimizing Incidental Civilian Casualties*. For various reasons, there may be an increased emphasis by State armed forces on minimizing the risk of incidental civilian casualties, even beyond the requirements of the law of war. The sympathy and support of the civilian population are frequently important objectives in non-international armed conflict.[94]In order to ensure such support, commanders and their forces may operate under rules of engagement that are more restrictive than what the law of war requires.[95]

17.5.2 Distinguishing State Forces From the Civilian Population in NIAC. During noninternational armed conflict, as during international armed conflict, the principle of distinction prohibits the use of protected persons or objects to shield, favor, or impede military operations.[96] However, it may be important to consider certain differences in the circumstances arising in international armed conflict.

17.5.2.1 *Positioning Military Forces Near the Civilian Population to Win Their Support and to Protect Them*. During non-international armed conflict, insurgents or terrorists may seek to attack the civilian population, and the use of the State's forces to protect the civilian population from such attacks may be a key objective of State operations in non-international armed conflict.[97]Thus, positioning military forces near the civilian population may be essential to the protection of the civilian population, and States have not interpreted such practices to be inconsistent with the principle of distinction.[98]

17.5.2.2 *Role of Civilian Personnel, Including Law Enforcement Personnel, in Addressing Non-State Armed Groups*. Members of States' civilian agencies, such as judges, prosecutors, and police and other members of its law enforcement apparatus, often play a critical role in addressing non-State armed groups.[99]Although such personnel might be viewed by the adversary as "military objectives" or as taking a direct part in hostilities, States have not interpreted the principle of distinction to require the separation of such personnel from the civilian population in non-international armed conflict.

17.6 RESPECT AND HUMANE TREATMENT OF PERSONS TAKING NO ACTIVE PART IN HOSTILITIES IN NIAC

All persons who do not take a direct part or who have ceased to take part in hostilities, whether or not their liberty has been restricted, are entitled to respect for their person, honor and convictions, and religious practices. They shall in all circumstances be treated humanely, without any adverse distinction.[100]

In addition, the following prohibitions apply:

17.6.1 Prohibition on Declaring That No Quarter Be Given. It is prohibited to order that there shall be no survivors.[101]

17.6.2 Prohibition on the Taking of Hostages. The taking of hostages is prohibited.[102]

17.6.3 Prohibition on Using Protected Persons and Objects to Shield, Favor, or Impede Military Operations. It is prohibited to use civilians, persons placed *hors de combat*, or other protected persons to shield, favor, or impede military operations.[103]

17.6.4 Prohibition on Pillage. Pillage is prohibited.[104]There is an affirmative obligation to take measures to protect the wounded, sick, and shipwrecked from pillage, and the dead from being despoiled.[105]

17.6.5 Prohibition on Acts of Terrorism. It is prohibited to commit, or threaten to commit, acts of terrorism against persons who do not take a direct part or who have ceased to take part in hostilities.[106]

17.6.6 Prohibition on Offering of Rewards for Persons Dead or Alive. As during international armed conflict, it is prohibited to offer a reward for enemy persons to be turned over dead or alive.[107]On the other hand, as in international armed conflict, there is no prohibition against offering rewards for the apprehension of insurgents or for giving information leading to the apprehension or killing of insurgents in combat.[108]

17.6.7 Prohibition on Collective Punishment. Collective punishments, whether administered by a court or through administrative measures, are prohibited.[109]Collective penalties are prohibited as a general matter.[110]

17.7 RULES ON CONDUCTING ATTACKS IN NIAC

Parties to a conflict must conduct attacks in accordance with the principles of distinction and proportionality. In particular, the following rules must be observed: • Combatants may not direct attacks against civilians, civilian objects, or other protected persons and objects.[111]

• The distinctive emblem must not be used while engaging in attacks.[112]

• Combatants must refrain from attacks in which the expected loss of life or injury to civilians, and damage to civilian objects incidental to the attack, would be excessive in relation to the concrete and direct military advantage expected to be gained.[113]

• Combatants must take feasible precautions in planning and conducting attacks to avoid incidental harm to civilians and civilian objects.[114]

• In conducting attacks, combatants must assess in good faith the information that is available to them.[115]

• Specific rules apply to the use of certain types of weapons.[116]

These rules apply to all parties to a non-international armed conflict, including persons belonging to non-State armed groups, and persons who decide to participate in hostilities of their own initiative. However, persons who belong to non-State armed groups, or who decide to participate in hostilities of their own initiative, may also be subject to the State's domestic law.[117]

17.7.1 AP II Rule on Works and Installations Containing Dangerous Forces. AP II provides that works or installations containing dangerous forces, namely dams, dykes, and nuclear electrical generating stations, shall not be made the object of attack, even where these objects are military objectives, if such attack may cause the release of dangerous forces and consequent severe losses among the civilian population.[118]The United States has objected to this rule in the context of international armed conflict and does not view it as reflecting customary international law in either international or non-international armed conflict insofar as the rule deviates from the regular application of the principles of distinction and proportionality.[119]

However, the Executive, in submitting AP II to the Senate for its advice and consent to ratification in 1987, did not recommend a reservation from or an understanding applicable to this provision of AP II based on an assessment that preserving the option to attack works or installations containing dangerous forces would not be as important in internal conflicts as preserving that option would be in international armed conflicts.[120]

17.8 IMPARTIAL HUMANITARIAN ORGANIZATIONS AND HUMANITARIAN ACTIVITIES DURING NIAC

An impartial humanitarian body, such as the International Committee of the Red Cross, may offer its services to the parties to the conflict.[121] The civilian population may, even on its own initiative, offer to collect and care for the wounded, sick, and shipwrecked.[122] Similarly, the United Nations Educational, Scientific and Cultural Organization may offer its services to the parties to the conflict.[123]

17.8.1 State Consent for Humanitarian Organizations. The activities of relief organizations are subject to the consent of the State concerned.[124]

States may withhold consent for, *inter alia*, legitimate military reasons, but should not arbitrarily withhold consent.[125] The safety of personnel of humanitarian organizations is a legitimate consideration for a government in consenting to their operations. Where a State has accepted the services of an impartial humanitarian organization, the State must not regard such service, such as the provision of medical care, as unlawful and subject to punishment.[126]

17.9 PROTECTION OF THE CIVILIAN POPULATION IN NIAC

17.9.1 Displacement of the Civilian Population. The displacement of the civilian population shall not be ordered for reasons related to the conflict unless the security of the civilians involved or imperative military reasons so demand.[127] Should such displacements have to be carried out, all possible measures shall be taken in order that the civilian population may be received under satisfactory conditions of shelter, hygiene, health, safety, and nutrition.[128] Civilians shall not be compelled to leave their own territory for reasons connected with the conflict.[129] A State must not compel civilians to leave its territory for reasons connected to the conflict, while insurgents that control territory must not compel civilians to leave the area under their authority.[130]

17.9.1.1 *Security of the Civilians Involved or Imperative Military Reasons*. Legitimate reasons to order the movement of the civilian population may include, for example:

- affording the civilian population greater protection from insurgents; and
- reducing the support provided to insurgents from elements of the civilian population.[131]

Legitimate reasons to order the movement of civilians do not include the use of individuals or groups of civilians around military objectives as involuntary human shields.[132]

17.9.2 Prohibition on Starvation of Civilians as a Method of Combat. Starvation of civilians as a method of combat is prohibited. It is therefore prohibited to attack, destroy, remove, or render useless, for that purpose, objects indispensable to the survival of the civilian population, such as foodstuffs, agricultural areas for the production of foodstuffs, crops, livestock, drinking water installations and supplies, and irrigation works.[133]

17.9.2.1 *Starvation of Enemy Forces Not Prohibited*. It is only actions that are for the purpose of starving civilians as a method of combat that are prohibited under this rule; measures to starve enemy forces are not prohibited.[134] For example, States may institute general food control programs that involve the destruction of crops and the adequate provision of the civilian population with food.[135]

Military action intended to starve enemy forces, however, must not be taken where it is expected to result in incidental harm to the civilian population that is excessive in relation to the military advantage anticipated to be gained.[136]

Feasible precautions to reduce the risk of harm to the civilian population or other reasonable measures to mitigate the burden to the civilian population may also be warranted when seeking to starve enemy forces.[137]

17.10 PROTECTION OF CHILDREN IN NIAC

17.10.1 General Protection and Care of Children. Children s hall be provided with the care and aid they require.[138]

Children shall receive an education, including religious and moral education, in keeping with the wishes of their parents, or in the absence of parents, of those responsible for their care.[139]

All appropriate steps shall be taken to facilitate the reunion of families temporarily separated.[140]

Measures shall be taken, if necessary, and whenever possible with the consent of their parents or persons who by law or custom are primarily responsible for their care, to remove children temporarily from the area in which hostilities are taking place to a safer area within the country and ensure that they are accompanied by persons responsible for their safety and wellbeing.[141]

The death penalty shall not be pronounced on persons who were under the age of eighteen years at the time of the offense and shall not be carried out on pregnant women or mothers of young children.[142]

17.10.2 Children and Participation in Non-International Armed Conflict. Children who have not attained the age of fifteen years shall neither be recruited in the armed forces or groups nor allowed to take part in hostilities.[143]Children who are captured and who have taken a direct part in hostilities remain entitled to the special protections afforded to children.[144]

The use or recruitment of child soldiers is an offense in U.S. law.[145]The United States has additional obligations as a Party to the Child Soldiers Protocol.[146]

17.11 PROTECTION OF CULTURAL PROPERTY IN NIAC

17.11.1 Application of Certain Provisions of the 1954 Hague Cultural Property Convention. In the event of an armed conflict not of an international character occurring within the territory of one of the Parties to the 1954 Hague Cultural Property Convention, each party to the conflict shall be bound to apply, as a minimum, the provisions of the 1954 Hague Cultural Property Convention that relate to respect for cultural property.[147]

17.11.2 Obligations to Respect Cultural Property. The obligation to respect cultural property includes essentially negative duties, *i.e.*, duties to refrain from acts of hostility directed against cultural property and duties to refrain from the use of cultural property in support of military operations where such use is not imperatively necessary.[148]

For example, parties to a conflict must not direct acts of hostility against cultural property, its immediate surroundings, or appliances in use for its protection, unless such action is required by imperative military necessity.[149]

In addition, no use should be made of cultural property, its immediate surroundings, or appliances in use for its protection, for purposes that are likely to expose it to destruction or damage in the event of armed conflict, unless such action is required by imperative military necessity.[150]

17.12 USE OF CAPTURED OR SURRENDERED ENEMY PERSONNEL IN NIAC

In contrast to the rules during international armed conflict, State forces may use captured or surrendered enemy personnel in operations against enemy non-State armed groups. The cooperation of enemy personnel may not, however, be procured through illegal methods.

17.12.1 Compelling Captured or Surrendered Enemy Personnel to Take Part in the Conflict. During international armed conflict, it is prohibited to compel the nationals of the hostile party to take part in the operations of war directed against their own country.[151]Additional restrictions apply with respect to POWs, retained personnel, protected persons in the home territory of a belligerent, and protected persons in occupied territory.[152]

These rules, however, do not apply during non-international armed conflict. Under international law, a State may compel its nationals to serve in its armed forces and to fight against non-State armed groups.[153]For example, in contrast to POWs, captured insurgents who are nationals of that State could be required to serve in that State's armed forces or to take part in operations directed against their former comrades.[154]

17.12.2 Methods of Gaining Cooperation. Inhumane treatment or other illegal methods (such as threats to commit unlawful acts) may not be used to gain the cooperation of captured enemy persons.[155]

However, a combination of threats of criminal punishment and inducements (*e.g.*, amnesty, monetary rewards) may be used to seek to gain the cooperation of captured enemy persons.[156]For example, law enforcement authorities often use the cooperation of members of criminal gangs or conspiracies to thwart and to prosecute the other participants or leaders in that criminal enterprise.[157]

17.13 WEAPONS IN NIAC

17.13.1 Prohibited Weapons in NIAC. The use of the following types of weapons is prohibited during non-international armed conflict:

- weapons calculated to cause superfluous injury;[158]
- inherently indiscriminate weapons;[159]
- poison, poisoned weapons, poisonous gases, and other chemical weapons;[160]
- biological weapons;[161]
- weapons that injure by fragments that are non-detectable by X-rays;[162]
- certain types of mines, booby-traps, and other devices;[163]and

• blinding lasers.[164]

17.13.2 Certain Types of Weapons With Specific Rules on Use in NIAC. Certain types of weapons are subject to specific rules that apply to their use by the U.S. armed forces in noninternational armed conflict. These weapons include:

• mines, booby-traps, and other devices (except certain specific classes of prohibited mines, booby-traps, and other devices);[165]

• incendiary weapons;[166]

• laser weapons (except blinding lasers);[167]and

• explosive ordnance.[168]

17.14 PROTECTION OF THE WOUNDED, SICK, SHIPWRECKED, AND DEAD IN NIAC

17.14.1 Respect, Protection, Humane Treatment, and Medical Care of the Wounded, Sick, and Shipwrecked. All the wounded, sick, and shipwrecked, whether or not they have taken part in the armed conflict, shall be respected and protected.[169]

17.14.1.1 *Types of Persons Who Are Considered Wounded, Sick, and Shipwrecked in NIAC.* The wounded, sick, and shipwrecked in non-international armed conflict may be understood to include:

• persons who have been rendered unconscious or otherwise have been incapacitated

because of their wounds, sickness, or shipwreck;[170]

• persons who have surrendered as a consequence of their health; and[171]

• persons who have been shipwrecked (*i.e.*, helpless persons in distress at sea or stranded on the coast) from any cause, including forced landings at sea by or from aircraft.[172]

Of course, any person who commits hostile acts or attempts to evade capture forfeits protection as someone who is placed *hors de combat.*[173]

17.14.1.2 *Meaning of "Respect and Protection" of the Wounded, Sick, and Shipwrecked.* The wounded, sick, and shipwrecked must be respected and protected at all times. This means that they should not be knowingly attacked, fired upon, or unnecessarily interfered with.[174]

Persons who are part of armed forces or groups, however, are deemed to have accepted the risk of harm due to deliberate proximity to military objectives.[175]Although the presence of such persons does not serve to exempt nearby military objectives from attack due to the risk that such persons would be incidentally harmed, feasible precautions to reduce the risk of harm to such persons who are wounded, sick, or shipwrecked must be taken.[176]

The respect and protection afforded the wounded, sick, and shipwrecked do not immunize them from search, or other necessary security measures, or capture and detention, even if they are receiving medical care.[177]

Booby-trapping the wounded or sick is expressly prohibited.[178]

17.14.2 Humane Treatment and Practicable Medical Care and Attention. In all circumstances, the wounded, sick, and shipwrecked shall be treated humanely and shall receive, to the fullest extent practicable and with the least possible delay, the medical care and attention required by their condition.[179]There shall be no distinction among them founded on any grounds other than medical ones.[180]

17.14.3 Search, Collection, and Protection of the Wounded, Sick, Shipwrecked, and Dead. Whenever circumstances permit, and particularly after an engagement, all possible measures shall be taken, without delay, to search for and collect the wounded, sick, and shipwrecked, to protect them against pillage and ill-treatment, to ensure their adequate care, and to search for the dead, prevent their being despoiled, and decently dispose of them.[181]

The obligations to search for, collect, and take affirmative steps to protect the wounded, sick, and shipwrecked are subject to practical limitations; military commanders are to judge what is possible, and to what extent they can commit their personnel to these duties.[182]

17.14.3.1 *Mistreatment and Booby-Trapping of the Dead Prohibited.* The mistreatment of the dead is prohibited.[183]In addition, the dead may not be booby-trapped.[184]

17.15 PROTECTION OF MEDICAL AND RELIGIOUS PERSONNEL AND MEDICAL TRANSPORTS IN NIAC

17.15.1 Protection of Medical and Religious Personnel. Medical and religious personnel shall be respected and protected and shall be granted all available help for the performance of their duties. They shall not be compelled to carry out tasks that are not compatible with their humanitarian mission.[185]

17.15.1.1 *Types of Persons Who Are Considered Medical and Religious Personnel in NIAC*. Medical and religious personnel include those persons who are exclusively (*e.g.*, permanently) engaged in those duties.[186] For example, persons who intermittently take a direct part in hostilities are not considered medical and religious personnel.[187]

17.15.1.2 *Meaning of "Respect and Protection" of Medical and Religious Personnel*. The respect and protection accorded to medical and religious personnel mean that they must not knowingly be attacked, fired upon, or unnecessarily prevented from discharging their proper functions.[188]

Medical and religious personnel who are part of armed forces or groups, however, are deemed to have accepted the risk of harm due to their deliberate proximity to military objectives.[189] Although the presence of such personnel would not serve to exempt nearby military objectives from attack due to the risk that such personnel would be incidentally harmed, feasible precautions must be taken to reduce the risk of harm to such personnel.[190]

The respect and protection afforded medical and religious personnel do not immunize them from search, or from other necessary security measures, or from capture and detention.[191] AP II and applicable treaties to which the United States is a Party (such as the 1949 Geneva Conventions) do not afford medical and religious personnel belonging to non-State armed groups retained personnel status if captured.

17.15.2 Protection of Medical Units and Transports. Medical units and transports shall be respected and protected at all times and shall not be the object of attack.[192] The protection to which medical units and transports are entitled shall not cease unless they are used to commit hostile acts, outside their humanitarian function. Protection may, however, cease only after a warning has been given setting, whenever appropriate, a reasonable time-limit, and after such warning has remained unheeded.[193]

17.15.2.1 *Types of Units and Vehicles That Are Considered Medical Units and Transports*. Medical units and transports include those units and vehicles that are exclusively (*e.g.*, permanently) engaged in those activities.[194] For example, units and transports that intermittently are used in hostilities are not considered medical units and transports.

17.15.2.2 *Meaning of "Respect and Protection" of Medical Units and Transports*. The respect and protection accorded to medical units and transports mean that they must not knowingly be attacked, fired upon, or unnecessarily prevented from discharging their proper functions.[195]

Medical units and transports that are part of armed forces or groups, however, are deemed to have accepted the risk of harm due to their deliberate proximity to military objectives.[196]

Although the presence of such medical units and transports would not serve to exempt nearby military objectives from attack due to the risk that such medical units and transports would be incidentally harmed, feasible precautions must be taken to reduce the risk of harm to such medical units and transports.[197]

The respect and protection afforded medical units and facilities does not immunize them from search or capture.[198]

17.16 DISPLAY OF THE DISTINCTIVE EMBLEM IN NIAC

Under the direction of the competent authority concerned, the distinctive emblem of the red cross, red crescent, or red lion and sun on a white ground shall be displayed by medical and religious personnel and medical units, and on medical transports. It shall be respected in all circumstances. It shall not be used improperly.[199]

These rules also apply to the distinctive emblem of the red crystal.[200]

17.16.1 Display of the Emblem Under the Direction of the Competent Authority Concerned. The display of the distinctive emblem is under the direction of the competent authority concerned, which authority may authorize the removal or obscuring of the distinctive emblem for tactical purposes, such as camouflage.[201] Similarly, it would be appropriate for the distinctive emblem to be removed if it is assessed that enemy forces will fail to respect the emblem and seek to attack medical personnel; display of the emblem in such circumstances would be deemed not to be feasible.[202]

17.16.2 Improper Use of the Distinctive Emblem. Improper use of the distinctive emblem includes use: (1) while engaging in attacks; (2) in order to shield, favor, or protect one's own military operations; or (3) to impede enemy military operations.[203]

17.17 DETENTION IN NIAC

17.17.1 State Authority to Detain. Law of war treaties have not limited the scope of whom a State may detain for reasons related to a non-international armed conflict, but have prescribed humane treatment for such persons.[204]A State's authority to conduct detention operations has often been understood as incident to the legal basis of the State to engage in operations against the non-State armed group.[205]

The precise legal requirements for a detention regime established by a State in noninternational armed conflict would likely depend a great deal on its domestic law.

17.17.1.1 *Non-Punitive Detention in Non-International Armed Conflict.* No punitive detention may be conducted on a variety of legal theories under international law.[206]For example, although enemy non-State armed groups would not be entitled to POW status, a State may detain persons belonging to enemy armed groups, by analogy to the detention of POWs in international armed conflict. Similarly, although persons would not be protected persons under the GC, a State may detain persons for security reasons, by analogy to the detention of protected persons in international armed conflict or occupation. Other legal rationales for the detention of persons in non-international armed conflict also may be available.

17.17.1.2 *Punitive Detention in Non-International Armed Conflict.* During noninternational armed conflict, a State could also detain persons pursuant to its criminal law. For example, persons who support enemy non-State armed groups could be sentenced to a variety of offenses (*e.g.*, murder, criminal conspiracy, treason or other offenses against the State, material support to terrorism or terrorist organizations).

17.17.2 Detention by Non-State Armed Groups. Non-State armed groups typically would lack domestic legal authority to conduct detention operations. Their actions may be subject to prosecution under domestic statutes making punishable kidnapping, hostage taking, false imprisonment, interference with State officials, etc. Nonetheless, detention during noninternational armed conflict by non-State armed groups is not prohibited by international law. In any case, non-State armed groups that conduct detention operations must provide humane treatment to detainees.

17.17.3 Humane Treatment and Other Applicable Requirements. All persons (including those belonging to the State or those belonging to non-State armed groups) who are detained by the adverse party are entitled to the protections of Common Article 3 of the 1949 Geneva Conventions, including humane treatment.[207]Although detainees are afforded humane treatment, they do not receive POW status.[208]

Chapter VIII addresses the baseline rules for the humane treatment of detainees that apply to all U.S. military operations, including those in non-international armed conflict.[209]

17.18 NON-INTERVENTION AND NEUTRAL DUTIES IN NIAC

17.18.1 Duty of Non-Belligerent States to Refrain From Supporting Hostilities by Non- State Armed Groups Against Other States. Under international law, principles of friendly relations and non-intervention require States to refrain from supporting non-State armed groups in hostilities against other States.[210]

These are closely analogous to the duties that neutral States have under the law of neutrality to refrain from supporting military operations against belligerents.[211]

In particular, States are required to take all reasonable steps to ensure that their territory is not used by non-State armed groups for purposes of armed activities—including planning, threatening, perpetrating, or providing material support for armed attacks—against other States and their interests.[212]The failure to fulfill this duty may have consequences in regard to whether other States that are threatened by these armed activities must seek the consent of that State before taking action in self-defense in that State's territory.[213]

Violations of law of war treaties applicable to non-international armed conflict by a State generally have not been understood to provide an independent basis in international law for a non-belligerent State to intervene against the violating State in that conflict.[214]

17.18.2 Duty of Belligerent States to Respect the Sovereignty of Other States. States that are engaged in hostilities against non-State armed groups must respect the sovereignty of other States.[215]In general, States must obtain the consent of a territorial State before conducting military operations against a non-State armed group in that State's territory.[216]

The consent of the territorial State, however, is not necessary when the U.N. Security Council has authorized the military operations.[217]

In addition, the United States has expressed the view that consent is not required when the territorial State is unwilling or unable to prevent its territory from being used by non-State armed groups as a base for launching attacks.[218]Other States have also expressed this view.[219]

It may be unnecessary for a belligerent State to seek consent when there is a strong, reasonable, and objective basis for concluding that the seeking of consent would be likely to undermine materially the effectiveness of the action against the non-State armed group (*e.g.*, reasons of disclosure, delay, incapacity to act) or would increase the risk of armed attack, vulnerability to future attacks, or other development that would give rise to an independent imperative to act in self-defense.[220]

17.18.3 States' Support to Other States in Hostilities Against N on-State Armed Groups. International law does not prohibit States from assisting other States in their armed conflicts against non-State armed groups. To the extent those States intend to conduct hostilities or actually do so, they may incur obligations under the law of war.[221]For example, a State that is conducting detention operations would have obligations to treat detainees (*e.g.*, persons belonging to non-State armed groups) humanely regardless of whether it considers itself a party to the non-international armed conflict.

17.18.4 Liability of Private Individuals for Supporting Non-State Armed Groups. Private individuals who support non-State armed groups that are preparing for, or engaged in, hostilities against a State may be subject to prosecution. Such conduct may be criminalized under a variety of domestic laws relating to treason, hostile expeditions against other States, material support to terrorism, or piracy.[222]

The State that is threatened by such activities may prosecute such individuals for treason or other offenses against the State. States in which such conduct occurs may seek to prosecute such conduct for a variety of reasons, including its duties under international law to refrain from materially supporting hostilities against another State, and to repress terrorism or piracy.

XVIII – Implementation and Enforcement of the Law of War

Chapter Contents

18.1 INTRODUCTION

This Chapter addresses the implementation and enforcement of the law of war. It discusses activities that are undertaken in order to prevent violations of the law of war (such as training and the promulgation of policies, regulations, and orders). It discusses activities to respond to alleged violations, such as reporting, investigation, and corrective or punitive measures.

18.1.1 DoD Policy on Implementing and Enforcing the Law of War. DoD policy has addressed the policies and responsibilities for ensuring DoD compliance with the law of war obligations of the United States.[1] It has been DoD policy that:

• Members of the DoD Components comply with the law of war during all armed conflicts, however such conflicts are characterized, and in all other military operations;[2]

• The law of war obligations of the United States are observed and enforced by the DoD Components and DoD contractors assigned to, or accompanying, deployed armed forces;[3]

• An effective program to prevent violations of the law of war is implemented by the DoD Components.[4]

These policies follow a longer tradition of compliance with the law of war by U.S. armed forces.[5]

18.1.2 National Obligations to Implement and Enforce the Law of War. States, as Parties to treaties, have certain obligations to implement and enforce those treaties. These obligations may be written as a general obligation to undertake to respect and to ensure respect for the treaty. In addition, treaties may provide for more specific obligations to help implement and enforce their provisions, such as obligations with respect to violations of the treaty or dissemination of the text of the treaty.

Different treaties may have different mechanisms to implement and to ensure compliance with that treaty.

18.1.2.1 *General Treaty Obligations to Take Appropriate Actions to Implement and Enforce the Treaty – 1949 Geneva Conventions*. Parties to the 1949 Geneva Conventions undertake to respect and ensure respect for the conventions in all circumstances.[6]This is a general obligation to take the measures that the State deems appropriate in order to fulfill its obligations under the conventions. Although this provision does not reflect an obligation to ensure implementation of the conventions by other States or parties to a conflict, the United States, as a matter of policy, often seeks to promote adherence to the law of war by others.[7]

18.1.2.2 *General Treaty Obligations to Take Appropriate Actions to Implement and Enforce the Treaty – AP III*. Parties to AP III undertake to respect and to ensure respect for AP III in all circumstances.[8]

18.1.2.3 *General Treaty Obligations to Take Appropriate Actions to Implement and Enforce the Treaty – CCW Amended Mines Protocol*. Parties to the CCW Amended Mines Protocol shall take all appropriate steps, including legislative and other measures, to prevent and suppress violations of that Protocol by persons or on territory under its jurisdiction or control.[9]

18.1.2.4 *General Treaty Obligations to Take Appropriate Actions to Implement and Enforce the Treaty – Child Soldiers Protocol*. Each Party to the Child Soldiers Protocol shall take all necessary legal, administrative, and other measures to ensure the effective implementation and enforcement of the provisions of this Protocol within its jurisdiction.[10]

18.1.3 International or Multi-National Actions to Implement and Enforce the Law of War. States sometimes take actions on the international or multinational level to implement and enforce the law of war. For example, NATO Standardization Agreements might specify common procedures to implement obligations under the 1949 Geneva Conventions for the treatment of POWs. States may investigate alleged violations of the law of war through international commissions of inquiry.[11]As another example, States may establish international criminal tribunals to try enemy belligerents.[12]

18.1.4 Law of War Obligations and International Organizations. International organizations, such as the United Nations or NATO, are not Parties to law of war treaties, such as the 1949 Geneva Conventions. Thus, these organizations do not have obligations under those treaty instruments. However, the member States of international organizations are Parties to these treaties, and as a general matter, a State's obligations under law of war treaties would not be rendered inapplicable simply because its forces are part of an international or multinational force or are acting through an international organization.

18.1.5 Application of Implementation and Enforcement Measures Outside the Context of International Armed Conflict. Certain treaty obligations with respect to the implementation and enforcement of the law of war may not apply outside the context of international armed conflict. Nonetheless, many of the domestic law, policies, and regulations that are used to implement and enforce the law of war are applicable outside the context of international armed conflict. Thus, for example, many of the provisions of the Uniform Code of Military Justice may be used to punish crimes that have been committed in the context of a non-international armed conflict or a military operation other than war.

18.2 PRUDENTIAL REASONS SUPPORTING THE IMPLEMENTATION AND ENFORCEMENT OF THE LAW OF WAR

The implementation and enforcement of the law of war have been supported by strong practical considerations. Compliance with the law of war is not only legally required but also in the strong self-interest of everyone subject to the law of war.[13]

18.2.1 Reinforcing Military Effectiveness. Conducting military operations in accordance with the law of war is not fighting with one hand tied behind one's back. Rather, law of war principles and rules are consistent with military doctrines for a profession of arms that are the basis for effective combat operations.[14]

For example, various military doctrines, such as accuracy of targeting, concentration of effort, maximization of military advantage, conservation of resources, avoidance of excessive collateral damage, and economy of force are not only fully consistent with compliance with the law of war, but also reinforce its observance. Use of indiscriminate and excessive force is costly, highly inefficient, and a waste of scarce resources.[15]

Similarly, the necessity of discipline for an effective armed force reinforces the implementation and enforcement of the law of war. An undisciplined force is more likely to commit law of war violations, such as pillaging, detainee abuse, or atrocities against the civilian population.[16]

18.2.2 Encouraging Reciprocal Adherence by the Adversary. The requirement to comply with many law of war rules (such as the obligation to treat detainees humanely) does not depend on whether the enemy complies with that rule.[17]

Nevertheless, reciprocity may be a critical factor in the actual observance of the law of war. Adherence to law of war rules in conducting military operations can encourage an adversary also to comply with those law of war rules.[18]For example, humane treatment of enemy persons detained by U.S. forces can encourage enemy forces to treat detained U.S. persons humanely.[19]Conversely, the maltreatment of detained personnel by U.S. forces may have a dramatic and negative effect on how U.S. personnel in the hands of the enemy are treated and the degree to which the law of war is respected generally.[20]

18.2.3 Maintaining Public Support and Political Legitimacy. The implementation and enforcement of the law of war are also supported by the fact that violations of the law of war are counterproductive to the political goals sought to be achieved by military operations.[21]For example, violations of the law of war in counter-insurgency operations may diminish the support of the local population. Violations of the law of war may also diminish the support of the populace in democratic States, including the United States and other States that would otherwise support or participate in coalition operations. Violations of the law of war committed by one side may encourage third parties to support the opposing side.

18.3 DUTIES OF INDIVIDUAL MEMBERS OF THE ARMED FORCES

Each member of the armed services has a duty to: (1) comply with the law of war in good faith; and (2) refuse to comply with clearly illegal orders to commit violations of the law of war.

18.3.1 Comply With the Law of War in Good Faith. Each member of the armed forces has a duty to comply with the law of war in good faith.[22]In practice, the obligation of individual service members to comply with the law of war in good faith is met when service members: (1) perform their duties as they have been trained and directed; and (2) apply the training on the law of war that they have received.[23]

The duty of each individual service member to comply with the law of war in good faith rests within a broader framework of law of war implementation by the U.S. armed forces.

18.3.1.1 *Implementation of International Obligations by Persons With the Authority to Make the Necessary Decisions and Judgments Required by International Law.* As a general matter, the persons within a party to a conflict who are responsible for implementing an obligation are those persons with the authority to make the necessary decisions and judgments required by that international obligation. It is not the case that every individual who belongs to a party to the conflict is responsible for implementing every obligation held by that party. For example, certain law of war obligations only apply to commanders or to specialized units, such as units providing medical care, conducting detention operations, or engaging in the protection of cultural property. As a case in point, although the specific responsibilities that individual combatants have to implement the principle of proportionality will depend on their role and assigned military duties in planning or conducting an attack, commanders are responsible for making the decisions and judgments required by the principle of proportionality in conducting attacks.[24]

18.3.1.2 *Practical Ways in Which the Requirement to Comply With the Law of War in Good Faith is Met by Individual Servicemembers.* Individual servicemembers are not expected to be experts in the law of war. The burden on individual members of the armed forces to know the relevant requirements of the law of war is addressed in practice through the following ways.

First, servicemembers should receive training commensurate with their duties on the relevant requirements of the law of war.

Second, law of war requirements have also been incorporated into domestic law, policy, regulations, and orders, which often are reviewed by counsel for consistency with the law of war.[25]Moreover, in most cases, the requirements and standards in applicable policies, regulations, and orders will often impose higher standards than the requirements of the law of war.[26]

Third, when appropriate, servicemembers should ask questions through appropriate channels and consult with the command legal adviser on issues relating to the law of war.

18.3.2 Refuse to Comply With Clearly Illegal Orders to Commit Law of War Violations. Members of the armed forces must refuse to comply with clearly illegal orders to commit law of war violations. In addition, orders should not be construed to authorize implicitly violations of law of war.

18.3.2.1 *Clearly Illegal Orders to Commit Law of War Violations*. The requirement to refuse to comply with orders to commit law of war violations applies to orders to perform conduct that is clearly illegal or orders that the subordinate knows, in fact, are illegal. For example, orders to fire upon the shipwrecked would be clearly illegal.[27]Similarly, orders to kill defenseless persons who have submitted to and are under effective physical control would also be clearly illegal.[28]On the other hand, the duty not to comply with orders that are clearly illegal would be limited in its application when the subordinate is not competent to evaluate whether the rule has been violated.[29]

Subordinates are not required to screen the orders of superiors for questionable points of legality, and may, absent specific knowledge to the contrary, presume that orders have been lawfully issued.[30]

18.3.2.2 *Commands and Orders Should Not Be Understood as Implicitly Authorizing Violations of the Law of War*. Commands and orders should not be understood as implicitly authorizing violations of the law of war where other interpretations are reasonably available.[31]

For example, if a commander issues an order to attack a town, one should assume that the order directs attacks on military objectives located in that area.[32]Similarly, speeches by commanders before combat operations to rally members of their command should not be understood to authorize implicitly law of war violations against the enemy.[33]

18.4 COMMANDERS' DUTY TO IMPLEMENT AND ENFORCE THE LAW OF WAR

Military commanders have a duty to take appropriate measures as are within their power to control the forces under their command for the prevention of violations of the law of war.[34]

For example, commanders have obligations to take appropriate measures to prevent pillage and to protect the wounded, sick, and shipwrecked within their control.[35]

18.4.1 Background on Commander's Duties to Implement and Enforce the Law of War. The law of war presupposes that its violation is to be avoided through the control of the operations of war by commanders who are to some extent responsible for their subordinates.[36]

One of the requirements for armed forces to receive the privileges of combatant status is that they operate under a responsible command.[37]In addition, law of war treaties have specified that commanders must take appropriate measures to ensure that the provisions of those treaties are observed.[38]

18.4.2 Discipline of Subordinates. In carrying out their duties to implement and enforce the law of war, commanders may use disciplinary or penal measures.[39]Under international law, commanders have discretion about how to implement and enforce their law of war obligations; there is no absolute or automatic requirement under international law to punish particular offenders within their armed forces in a specific way.[40]

18.4.3 Duty to Investigate Reports of Alleged Law of War Violations. Commanders have a duty to investigate reports of alleged law of war violations committed by persons under their command or against persons to whom they have a legal duty to protect.[41]

18.4.4 Issuance of Guidance, Training of Subordinates, and Other Preventive or Corrective Measures. Apart from disciplinary measures, a variety of other measures may be appropriate to prevent or address violations of the law of war by subordinates. For example, commanders should ensure that members of the armed forces under their command are, commensurate with their duties, aware of their duties under the law of war.[42]After violations have occurred, retraining personnel or revising procedures may be appropriate, particularly where the underlying facts or nature of the violations do not warrant punitive measures.

18.5 ROLE OF JUDGE ADVOCATES AND LEGAL ADVISERS

18.5.1 Legal Advisers. The United States has provided for legal advisers to advise military commanders on the law of war.[43]For example, DoD policy has required that each head of a DoD component make qualified legal advisers available at all levels of command to provide advice about law of war compliance during planning and execution of exercises and operations.[44]

18.5.1.1 *Review of Plans, Policies, Directives, and Rules of Engagement by Legal Advisers*. DoD policy has required that commanders of the combatant commands ensure that all plans, policies, directives, and rules of engagement, and those of subordinate commands and components, are reviewed by legal advisers to ensure their consistency with the law of war and DoD policy on the law of war.[45]

18.5.1.2 *Supervision of Administration of Programs to Address Enemy Violations of the Law of War*. DoD policy has required that commanders of the combatant commands designate a command legal adviser to supervise the administration of those aspects of the command's program dealing with possible, suspected, or alleged enemy violations of the law of war.[46]

18.5.1.3 *Review of the Acquisition of Weapons*. DoD policy has required the legal review of the intended acquisition of weapons.[47]

18.5.2 Law of War Questions During Military Operations. During military operations, questions on the law of war from U.S. forces or coalition partners related to a specific issue should be referred through the operational chain of command for resolution. It may also be appropriate to refer questions to either the office of the Judge Advocate General of a Military Department, the Staff Judge Advocate to the Commandant of the Marines Corps, the General Counsel of a Military Department, the Legal Counsel to the Chairman of the Joint Chiefs of Staff, or the DoD General Counsel.

18.5.3 Role of the DoD Law of War Working Group. The DoD Law of War Working Group is a DoD internal mechanism for coordination on law of war issues. It consists of representatives of the General Counsel, Department of Defense; representatives, at the election of each, of the General Counsel of each Military Department, the Judge Advocate General of each Military Department, the Staff Judge Advocate to the Commandant of the Marine Corps, and the Legal Counsel to the Chairman of the Joint Chiefs of Staff.[48]The DoD Law of War Working Group develops and coordinates law of war initiatives and issues, manages other law of war matters as they arise, and provides advice to the DoD General Counsel on legal matters covered by DoD Directive 2311.01E.[49]This includes the preparation, review, and updating of this manual.

18.6 DISSEMINATION, STUDY, AND OTHER MEASURES TO FACILITATE UNDERSTANDING OF DUTIES UNDER THE LAW OF WAR

A basic step in implementing and enforcing the law of war is to ensure that people understand its requirements. Certain treaties require that Parties disseminate the text of that treaty and promote its study, especially by those personnel who are assigned to implement its obligations.

18.6.1 General Dissemination and Study of Treaties. Certain treaties require that Parties disseminate the treaty and promote study of that treaty by the armed forces and the civilian population.

18.6.1.1 *Dissemination and Study - 1949 Geneva Conventions*. Parties to 1949 Geneva Conventions undertake, in time of peace, as in time of war, to disseminate the text of those Conventions widely as possible in their respective countries, and in particular, to include the study of the Conventions in their programs of military and, if possible, civil instruction.[50]In addition to the entire population of the Party to the GPW, the GPW emphasizes that Parties' armed forces should know the principles of the GPW.[51]Similarly, the GWS and GWS-Sea emphasize that the armed fighting forces, medical personnel, and chaplains should know the principles of those treaties.[52]

18.6.1.2 *Dissemination and Study - CCW and Protocols*. Parties to the CCW also undertake, in time of peace as in time of armed conflict, to disseminate the CCW and those of its annexed Protocols by which they are bound as widely as possible in their respective countries and, in particular, to include the study of these instruments in their program of military instruction, so that those instruments may become known to their armed forces.[53]

18.6.1.3 *Dissemination and Study – AP III*. Parties to AP III undertake, in time of peace as in time of armed conflict, to disseminate this Protocol as widely as possible in their respective countries and, in particular, to include the study of it in their program of military instruction and to encourage the study of it by the civilian population, so that this instrument may become known to the armed forces and to the civilian population.[54]

18.6.1.4 *Dissemination and Study – 1954 Hague Cultural Property Convention*. Parties to the 1954 Hague Cultural Property Convention undertake, in time of peace as in time of armed conflict, to disseminate the text of the Convention and the Regulations for its execution as widely as possible in their respective countries. They undertake, in particular, to include the study thereof in their programs of military and, if possible, civilian training, so that its principles are made known to the whole population, especially the armed forces and personnel engaged in the protection of cultural property.[55]

18.6.1.5 *Dissemination and Study – Child Soldiers Protocol*. Parties to the Child Soldiers Protocol undertake to make the principles and provisions of the Child Soldiers Protocol widely known and promoted by appropriate means, to adults and children alike.[56]

18.6.2 Special Instruction or Training. In addition to requirements to disseminate and to promote the study of treaties, treaties also require States to ensure that members of the armed forces who have duties under those treaties are trained commensurate with those duties. DoD policy has required, as a general matter, that personnel are trained in the law of war commensurate with their duties.[57]

Training may involve not only classroom instruction or individualized study, but also, for example, unit training exercises.[58]

In many cases, training on law of war requirements may not be classified as "law of war" training, or may be conducted without acknowledgment that the requirements are law of war requirements.[59] Rather, it may be the case that military forces would be trained according to military doctrines or regulations, which have incorporated law of war requirements and have been reviewed for consistency with the law of war.[60]

The 1949 Geneva Conventions, the CCW Amended Mines Protocol, and the CCW Protocol V on Explosive Remnants of War each have specific provisions relating to special instruction or training.

18.6.2.1 *Special Instructions or Training - 1949 Geneva Conventions*. The GPW and GC further provide that the military or other authorities who assume responsibilities for POWs or protected persons must possess the text of the GPW or the GC and be specially instructed as to its provisions.[61]

18.6.2.2 *Special Instructions or Training - CCW Amended Mines Protocol*. Each Party to the CCW Amended Mines Protocol shall require that armed forces personnel receive training commensurate with their duties and responsibilities to comply with the provisions of the CCW Amended Mines Protocol.[62]

18.6.2.3 *Special Instructions or Training - CCW Protocol V On Explosive Remnants of War*. Each Party to the CCW Protocol V on Explosive Remnants of War shall require that its personnel receive training consistent with the relevant provisions of this Protocol.[63]

18.7 INSTRUCTIONS, REGULATIONS, AND PROCEDURES TO IMPLEMENT AND ENFORCE THE LAW OF WAR

The law of war has traditionally been implemented through military instructions, regulations, and procedures. For example, the Lieber Code, one of the first codifications of the law of war, was called "Instructions for the Government of Armies of the United States in the Field," and was issued as a General Order.[64] Similarly, directives and regulations have been issued to implement law of war obligations relating to detainees and to establish higher standards as a matter of policy.[65]

18.7.1 Treaty Provisions Specifically Contemplating or Requiring Military Instructions, Regulations, and Procedures. Law of war treaties contemplate or in some cases require that such instructions will be issued. In some cases, the implementation of a treaty through military instructions or regulations may be understood as a part of the general requirements for States to take appropriate actions to implement and enforce their obligations under that treaty.[66]

18.7.1.1 *Instructions, Regulations, and Procedures – Hague Conventions on Land Warfare*. Parties to Hague IV shall issue instructions to their armed land forces, which shall be in conformity with the "Regulations respecting the Laws and Customs of War on Land" annexed to Hague IV.[67]

18.7.1.2 *Instructions, Regulations, and Procedures – 1949 Geneva Conventions*. The 1949 Geneva Conventions contemplate that Parties will adopt laws and regulations to ensure the proper application of the Conventions.[68] The requirement to issue implementing instructions may be understood as part of the general obligation of Parties to undertake to respect and to ensure respect for the Conventions.[69]

18.7.1.3 *Instructions, Regulations, and Procedures – CCW Amended Mines Protocol*. Each Party to the CCW Amended Mines Protocol shall require that its armed forces issue relevant military instructions and operating procedures.[70]

18.7.1.4 *Instructions, Regulations, and Procedures – 1954 Hague Cultural Property Convention*. Parties to the 1954 Hague Cultural Property Convention undertake to introduce in time of peace into their military regulations or instructions such provisions as may ensure observance of that Convention, and to foster in the members of their armed forces a spirit of respect for the culture and cultural property of all peoples.[71]

18.7.1.5 *Instructions, Regulations, and Procedures – CCW Protocol V on Explosive Remnants of War*. Each Party to the CCW Protocol V on Explosive Remnants of War shall require that its armed forces and relevant agencies or departments issue appropriate instructions and operating procedures.[72]

18.7.2 Reasons for Implementation Through Instructions, Regulations, and Procedures. In addition to cases in which instructions or regulations are required by a treaty, there are a number of reasons for implementing law of war obligations through instructions, regulations, and other procedures.

18.7.2.1 *Re-characterizing National Obligations Into Rules for Individuals*. First, under the traditional view of international law, obligations are owed between States, and not as between States and individuals.[73] Thus, treaty provisions may need to be re-characterized from obligations that Parties to the treaty promise to one another into rules for individual conduct.

18.7.2.2 *Implementing the Rules in Non-Self-Executing Treaties*. The adoption of instructions, regulations, or procedures may assist in the implementation of non-self-executing treaties. When treaties or provisions of treaties are regarded as "non-self-executing," such provisions do not constitute enforceable domestic law even though the State has ratified them, but require further implementing action by the ratifying State.[74]

However, a State may adopt instructions, regulations, or procedures that are enforceable under its domestic law and that implement the rules reflected in a treaty. Thus, because such instructions, regulations, or procedures are enforceable domestically, the State can enforce the rules in the treaty, even though the treaty remains non-self-executing.

18.7.2.3 *Setting Higher Standards as a Matter of Policy*. In addition, States may choose to implement law of war obligations through instructions, regulations, and procedures because they wish to set a higher standard for their armed forces as a matter of policy. For example, rules of engagement are often more restrictive than relevant law of war requirements.[75] As another example, U.S. military operations are routinely subject to more restrictions and apply standards that are more protective of civilians than required by the law of war.[76]

Although failures to adhere to such more restrictive standards may be punishable under the UCMJ, such failures would not necessarily be violations of the law of war.

18.7.2.4 *Clarifying Ambiguities in the Law*. States may choose to implement law of war obligations through instructions, regulations, and procedures because they wish to clarify their interpretation of the obligations, which otherwise may not be clear because a treaty provision is ambiguous or because the obligation is reflected in customary international law.

18.7.2.5 *Ensuring Detailed Requirements Are Not Neglected, and Assistance in Training*. In cases in which treaty requirements are detailed and extensive, the promulgation of military instructions may also help ensure that treaty requirements are not omitted or neglected. Similarly, the promulgation of military instructions also assists in conducting training or special instruction.[77]

18.7.3 Enforcement of Law of War Obligations Through Military Instructions, Regulations, and Procedures. The implementation of law of war treaties and obligations through military instructions, regulations, and procedures has the effect of making such rules enforceable because military personnel are required to comply with duly issued instructions, regulations, and procedures. Military personnel may be punished for failure to comply with such orders and instructions.[78] Conduct that violates the law of war has often been punished through this mechanism.[79]

18.8 CONSIDERING LAW OF WAR OBLIGATIONS IN THE PLANNING OF MILITARY OPERATIONS

Military commanders and planners should consider law of war obligations in the planning of military operations. DoD policy has required the review of plans by legal advisers to ensure their consistency with the law of war.[80]

It may be especially important to consider affirmative obligations imposed by the law of war. For example, military commanders and planners should plan for their medical facilities to consider needing to treat enemy wounded and sick. Similarly, the requirements to care for detainees should also be considered. Potential responsibilities as an Occupying Power should also be considered.

18.9 STATES' OBLIGATIONS WITH RESPECT TO VIOLATIONS OF THE LAW OF WAR

States have certain obligations with respect to alleged violations of the law of war within their jurisdiction. In some cases, these obligations to address alleged violations of particular treaties may be understood as part of the general obligations to implement and enforce those treaties. In other cases, law of war treaties impose specific requirements on States with respect to alleged violations of obligations in those treaties.

18.9.1 State Responsibility for Violations of the Law of War by Its Armed Forces. A State may be responsible for violations of the law of war committed by persons forming part of its armed forces.[81] In particular, States are responsible for the treatment accorded protected persons under the GC by their agents.[82] State responsibility for violations of the law of war committed by its armed forces or other agents results from principles of State responsibility in international law that are not specific to the law of war.[83]

State responsibility for violations of the law of war results in obligations to compensate other States for violations.[84]

18.9.2 Breaches of the 1954 Hague Cultural Property Convention. Parties to the 1954 Hague Cultural Property Convention undertake to take, within the framework of their ordinary criminal jurisdiction, all necessary steps to prosecute and impose penal or disciplinary sanctions upon those persons, of whatever nationality, who commit or order to be committed a breach of the 1954 Hague Cultural Property Convention.[85]

18.9.3 Grave Breaches of the 1949 Geneva Conventions. Parties to the 1949 Geneva Conventions have certain obligations relating to grave breaches of the 1949 Geneva Conventions. These obligations have been interpreted as declaratory of the obligations of belligerents under customary international law to take measures for the punishment of war crimes committed by all persons, including members of a belligerent's armed forces.[86]These obligations do not affect the right of a belligerent, under customary international law, to try enemy personnel for war crimes other than grave breaches of the 1949 Geneva Conventions.[87]

Parties to the 1949 Geneva Conventions undertake to enact any legislation necessary to provide effective penal sanctions for persons committing, or ordering to be committed, any of the grave breaches of the 1949 Geneva Conventions.[88]

Each Party to the 1949 Geneva Conventions shall be under the obligation to search for persons alleged to have committed, or to have ordered to be committed, such grave breaches, and shall bring such persons, regardless of their nationality, before its own courts.[89]It may also, if it prefers, and in accordance with the provisions of its own legislation, hand such persons over for trial to another High Contracting Party concerned, provided such High Contracting Party has made out a *prima facie* case.[90]

No Party to the 1949 Geneva Conventions shall be allowed to absolve itself or any other Party of any liability incurred by itself or by another Party in respect of grave breaches of the 1949 Geneva Conventions.[91]

18.9.3.1 *Acts Constituting Grave Breaches*. The acts constituting "grave breaches" for the purpose of triggering these obligations are defined differently depending on the particular Geneva Convention. However, all of the Conventions include as grave breaches the following acts against persons protected by the respective Convention:

• willful killing;

• torture or inhuman treatment, including biological experiments; and

• willfully causing great suffering or serious injury to body or health.

The term "grave breaches" was deliberately chosen so as not to indicate that violations of those provisions of the 1949 Geneva Conventions were themselves crimes or that the 1949 Geneva Conventions created an international penal code.[92]

Grave breaches of the GWS and GWS-Sea are those involving any of the following acts, if committed against persons or property protected by the GWS or GWS-Sea:

• willful killing; torture or inhuman treatment, including biological experiments;

• willfully causing great suffering or serious injury to body or health; and

• extensive destruction and appropriation of property, not justified by military necessity and carried out unlawfully and wantonly.[93]

Grave breaches of the GPW are those involving any of the following acts, if committed against persons or property protected by the GPW:

• willful killing;

• torture or inhuman treatment, including biological experiments;

• willfully causing great suffering or serious injury to body or health;

• compelling a POW to serve in the forces of the hostile Power; and

• willfully depriving a POW of the rights of fair and regular trial prescribed in the GPW.[94]

Grave breaches of the GC are those involving any of the following acts, if committed against persons or property protected by the GC:

• willful killing;

• torture or inhuman treatment, including biological experiments;

• willfully causing great suffering or serious injury to body or health;

• unlawful deportation or transfer or unlawful confinement of a protected person;

• compelling a protected person to serve in the forces of a hostile Power;

• willfully depriving a protected person of the rights of fair and regular trial prescribed in the GC;

• taking of hostages; and

• extensive destruction and appropriation of property, not justified by military necessity and carried out unlawfully and wantonly.[95]

18.9.3.2 *Applicability of Grave Breaches Obligations to Non-International Armed Conflict.* The text of the 1949 Geneva Conventions provides that grave breaches relate to violations "against persons or property protected by the Convention." Since Common Article 3 of the 1949 Convention protects persons against some of the acts described as grave breaches, the United States took the position that the obligations created by the grave breaches provisions of the 1949 Geneva Conventions could apply also to violations of Common Article 3.[96]An Appeals Chamber of the ICTY did not accept this view, and understood the grave breaches provisions of the 1949 Geneva Conventions only to create obligations applicable in international armed conflicts.[97]U.S. law makes punishable certain conduct that constitutes a "grave breach of common article 3."[98]

Regardless of whether the obligations in the grave breaches provisions apply with respect to violations of Common Article 3, serious violations of Common Article 3 may nonetheless be punishable.

18.9.3.3 *Suppression of All Acts Contrary to the Provisions of the 1949 Geneva Conventions.* Each High Contracting Party shall take measures necessary for the suppression of all acts contrary to the provisions of the 1949 Geneva Conventions other than the grave breaches.[99]

Such measures could include punishment of offenders. For example, if a U.S. Soldier steals the money that has been impounded from POWs and appropriates it for his or her own use, the theft would not constitute a grave breach of the 1949 Geneva Conventions, but would be an offense under the Uniform Code of Military Justice.[100]

Such measures may also be understood to include a wide range of measures, such as the promulgation or revision of policies and regulations, administrative or corrective measures, or retraining of personnel.[101]

18.9.4 CCW Amended Mines Protocol. Each Party to the CCW Amended Mines Protocol shall take all appropriate steps, including legislative and other measures, to prevent and suppress violations of the CCW Amended Mines Protocol by persons or on territory under its jurisdiction or control.[102]Such measures include appropriate measures to ensure the imposition of penal sanctions against persons who, in relation to an armed conflict and contrary to the provisions of the CCW Amended Mines Protocol, willfully kill or cause serious injury to civilians and to bring such persons to justice.[103]

Penal sanctions under Article 14 of the CCW Amended Mines Protocol only apply in a situation in which an individual: (1) knew, or should have known, that his or her action was prohibited under the CCW Amended Mines Protocol; (2) intended to kill or cause serious injury to a civilian; and (3) knew, or should have known, that the person he or she intended to kill or cause serious injury was a civilian.[104]

U.S. Federal law authorizes prosecutions for such conduct.[105]

18.9.5 War Crimes – Notes on Terminology. The term "war crime" has been used in different ways in different contexts. In contemporary parlance, the term "war crime" is most often used to mean serious violations of the law of war.

18.9.5.1 *War Crimes – Any Violation of the Law of War.* In some cases, the term "war crime" has been used as a technical expression for a violation of the law of war by any person; *i.e.*, under this usage, any violation of the law of war is a war crime.[106]This has been longstanding U.S. military doctrine.[107]

Any violation of the law of war by members of the armed forces of a State could be made punishable by that State through its domestic law, including military law applicable to its armed forces. For example, under military law, violations of the law of war may be made punishable through military orders to comply with law of war requirements.[108]Commanders may also impose non-judicial punishment or take adverse administrative action to address minor violations.[109]

18.9.5.2 *War Crimes – Serious Violations of the Law of War.* Sometimes the term "war crime" is used to refer to particularly serious violations of the law of war. For example, this is generally the usage when the term "war crime" is defined for the purposes of a particular criminal statute.[110]

This usage of "war crime" is understood to exclude minor violations of the law of war. For example, if during an international armed conflict, military medical personnel perform their duties while wearing an armlet displaying the distinctive emblem affixed to their right arm – rather than to their left arm, as specified by Article 40 of the GWS, these personnel may be said to be violating the law of war. However, under this usage of "war crime," such violations generally would not be regarded as a "war crime."

18.9.5.3 *War Crimes – Serious Violations of Domestic Law Applicable During Armed Conflict.* The term "war crime" has also been used to describe offenses, such as espionage and unprivileged belligerency, that although not prohibited by international law, are properly liable to punishment by the belligerent against which they are directed.[111] Generally, this usage does not prevail today, although practitioners may find this usage in older sources.

18.10 METHODS FOR RESPONDING TO VIOLATIONS OF THE LAW OF WAR BY THE ENEMY

In the event of violation of the law of war, it may be possible for the injured State to resort to remedial action of the following types:

- publication of the facts, with a view to influencing public opinion against the offending belligerent;
- protests and demands to the offending party,[112] including demands for compensation or the punishment of enemy persons who have violated the law;[113]
- solicitation of the good offices, mediation, or intervention of neutral States for the purpose of making the enemy observe the law of war;[114]
- petition to the U.N. Security Council;[115]
- punishment of captured offenders as war criminals, either through national or international tribunals;[116]
- retorsion;[117] and
- reprisals.[118]

18.11 PROTESTS AND DEMANDS TO THE OFFENDING PARTY

An initial step in responding to law of war violations by the enemy is to issue a formal or informal complaint to the offending party.

18.11.1 Method of Transmitting Protests and Demands. In view of the lack of diplomatic relations between States involved in an armed conflict, complaints cannot normally be made through the usual diplomatic channels. However, other methods are available, depending on the degree of publicity required. The traditional method of complaining by *parlementaire* directly to the commander of the offending forces remains, although modern communications have meant that the message is more likely now to be transmitted through electronic means.[119] Complaints also may be made through the Protecting Power, an impartial humanitarian organization performing the duties of a Protecting Power, or a Neutral State.[120]

18.11.2 Demands for Redress. In addition to demands to cease committing violations, protests may also include demands for redress by the offending party. For example, a State may demand that a national investigation or international investigation be conducted.[121] A State may demand financial compensation or other reparations.[122] Similarly, a State may demand that offending individuals be held accountable for their offenses through judicial or other processes.[123]

18.12 U.N. SECURITY COUNCIL AND ENFORCEMENT OF THE LAW OF WAR

In some cases, the U.N. Security Council has determined that situations involving violations of the law of war constitute a threat to international peace and security.[124] The U.N. Security Council may take a variety actions to respond to such threats.

18.12.1 Investigation by the U.N. Security Council Under the Charter of the United Nations. Under the Charter of the United Nations, the U.N. Security Council may investigate any dispute, or any situation that might lead to international friction or give rise to a dispute, in order to determine whether the continuance of the dispute or situation is likely to endanger the maintenance of international peace and security.[125] This authority of the U.N. Security Council to investigate situations includes any such situations that involve violations of the law of war. The U.N. Security Council has provided for the establishment of commissions of inquiry or commissions of experts to report on violations of the law of war.[126]

18.12.2 Determining Liability and Determining That Compensation Should Be Paid. The U.N. Security Council may, in appropriate cases, affirm that States are liable for violations of international law and establish mechanisms to facilitate the payment of compensation for such violations.[127]

18.12.3 Authorizing the Use of Force. The U.N. Security Council has authorized the use of force in order to protect civilians who are being attacked in violation of the law of war.[128]

18.12.4 Authorizing International Criminal Tribunals. The U.N. Security Council has established international criminal tribunals for the purpose of prosecuting serious violations of international humanitarian law. The U.N. Security Council has exercised this authority to create the International Criminal Tribunal for the Former Yugoslavia and the International Criminal Tribunal for Rwanda.[129]

The Rome Statute of the International Criminal Court provides that the International Criminal Court may exercise its jurisdiction with respect to crimes when a situation is referred to the Prosecutor by the Security Council acting under Chapter VII of the Charter of the United Nations.[130]The U.N. Security Council has referred situations in Darfur and Libya to the Prosecutor of the International Criminal Court.[131]The United States has objected to certain aspects of the jurisdiction of the International Criminal Court, and there are certain restrictions in U.S. domestic law on support to the International Criminal Court.[132]

18.12.5 U.N. Security Council Role in Certain Treaties.

18.12.5.1 *U.N. Security Council Role in the Biological Weapons Convention and ENMOD Convention*. The Biological Weapons Convention and the ENMOD Convention provide that Parties to the treaties may lodge a complaint with the Security Council, when a Party believes that another Party is in breach of its obligations deriving from the provisions of that treaty.[133]These treaties provide that each Party undertakes to cooperate with any investigation that the Security Council may initiate.[134]In addition, these treaties impose an additional obligation on Parties to provide or support assistance to Parties that request assistance if the Security Council has made an appropriate decision.[135]

18.12.5.2 *U.N. Security Council and the Chemical Weapons Convention*. The Chemical Weapons Convention provides that the Conference of the States Parties shall, in cases of particular gravity, bring the issue, including relevant information and conclusions, to the attention of the U.N. General Assembly and the U.N. Security Council.[136]

18.12.5.3 *U.N. Security Council Role in the Statute of the International Court of Justice*. Under the Statute of the International Court of Justice, a party to a case may have recourse to the Security Council if an opposing party fails to perform the obligations incumbent upon it under a judgment rendered by the Court.[137]

18.13 NATIONAL INVESTIGATIONS OF ALLEGED VIOLATIONS OF THE LAW OF WAR The duties to implement and enforce the law of war also imply duties to investigate reports of alleged violations of the law of war.[138]In addition to taking measures to meet the requirements of DoD policy, commanders may also take other measures they deem appropriate to ensure appropriate investigation and reporting of alleged violations of the law of war within their command.[139]

18.13.1 DoD Policy on Reporting Law of War Violations. DoD policy has required the reporting of possible, suspected, or alleged violations of the law of war for which there is credible information, or conduct during military operations other than war that would constitute a violation of the law of war if it occurred during armed conflict ("reportable incidents").[140]Such policy has been in addition to other DoD policies that address the reporting of criminal incidents.[141]Individual Military Services have implemented procedures to ensure that incidents, including war crimes, are promptly reported.[142]

18.13.1.1 *Requirement for All Military and U.S. Civilian Employees, Contractor Personnel, and Subcontractors Assigned to or Accompanying a DoD Component*. DoD policy has required that all military and U.S. civilian employees, contractor personnel, and subcontractors assigned to or accompanying a DoD Component shall report reportable incidents through their chain of command. Contracts shall require contractor employees to report reportable incidents to the commander of the unit they are accompanying or the installation to which they are assigned, or to the Combatant Commander.[143]

18.13.1.2 *Requirement for Unit Commanders*. DoD policy has required that the commander of any unit that obtains information about a reportable incident shall immediately report the incident through the applicable operational command and Military Department.[144]

18.13.2 DoD Policy on Investigating Law of War Violations. DoD policy has required that all reportable incidents be thoroughly investigated.[145] All the Military Departments and Services have regulations for the conduct of both formal and informal administrative investigations.[146]

In addition, DoD policy has required that higher authorities receiving an initial report request a formal investigation by the cognizant military criminal investigative organization.[147]

18.14 INTERNATIONAL MECHANISMS TO INVESTIGATE ALLEGED LAW OF WAR VIOLATIONS

In some cases, international mechanisms, such as commissions of inquiry, may be used to investigate alleged law of war violations. Commissions of inquiry might be established by treaty.1[148] Commissions of inquiry might also be established by the U.N. Security Council.[149]

18.14.1 Inquiry Procedure in the 1949 Geneva Conventions. At the request of a party to the conflict, an inquiry shall be instituted, in a manner to be decided between the interested Parties, concerning any alleged violation of the 1949 Geneva Conventions.[150] If agreement has not been reached concerning the procedure for the inquiry, the Parties should agree on the choice of an "umpire" who will decide upon the procedure to be followed.[151] Once the violation has been established, the parties to the conflict "shall put an end to it and shall repress it with the least possible delay."[152]

18.14.1.1 *AP I – International Humanitarian Fact-Finding Commission (IHFFC)*. AP I provides for the establishment of an international fact-finding commission.[153] This commission is competent: (1) to inquire into any facts alleged to be a grave breach as defined in the 1949 Geneva Conventions and AP I or other serious violation of the 1949 Geneva Conventions or of AP I; and (2) to facilitate, through its good offices, the restoration of an attitude of respect for the Conventions and AP I.[154]

The commission operates on the basis of mutual consent. Any party to a conflict may ask the commission to conduct an inquiry; but, unless the States involved previously declared that they recognize *ipso facto* and without special agreement, in relation to any other Party to AP I accepting the same obligation, the competence of the Commission, the Commission will only investigate with the consent of the States involved.[155]

Although the IHFFC was officially constituted in 1991, it has not been used in its first two decades.[156]

The United States has not ratified AP I, and has not recognized the competence of the IHFFC.

18.15 PROTECTING POWER AND OTHER NEUTRAL INTERMEDIARIES

The Protecting Power is an organ for the implementation of the 1949 Geneva Conventions and the 1954 Hague Cultural Property Convention.

18.15.1 Background on the Protecting Power. Under international law, a State that lacks normal diplomatic relations with a second State may, with the consent of the second State, designate a third State act to protect the first State's interests as they relate to the second State.[157] This third State is often called a Protecting Power. A State could be a Protecting Power for the purposes of more than one treaty.

18.15.1.1 *Protecting Power Under the 1949 Geneva Conventions*. Under the 1949 Geneva Conventions, belligerents may designate neutral States as "Protecting Powers" to help implement the Geneva Conventions. Specifically, the 1949 Geneva Conventions provide that the Conventions shall be applied with the cooperation and under the scrutiny of the Protecting Powers whose duty it is to safeguard the interests of the parties to the conflict.[158] The Detaining Power has an obligation to seek a Protecting Power if the wounded and sick, shipwrecked, medical personnel and chaplains, POWs, or protected persons under the GC in its custody do not benefit from one.[159] If such protection cannot be arranged, the 1949 Geneva Conventions contemplate that States will use the ICRC or another impartial humanitarian organization to assume the humanitarian functions performed by Protecting Powers.[160]

18.15.1.2 *Protecting Power Under the 1954 Hague Cultural Property Convention*. The 1954 Hague Cultural Property Convention provides that it and the Regulations for its execution shall be applied with the co-operation of the Protecting Powers responsible for safeguarding the interests of the parties to the conflict.[161]

18.15.2 Appointment of a Protecting Power. Within a State, the appointment of a Protecting Power is a decision made by authorities at the national level. Among States, the appointment of a Protecting Power requires the consent of the States whose relations are to be transacted through the Protecting Power. For example, during an international armed conflict, the U.S. designation of a neutral State as its Protecting Power would require agreement of the enemy State, but the consent of States allied with the United States would not be required. The State appointed to be a Protecting Power must be able to fulfill the Protecting Power's responsibilities.[162] This

State should be neutral or non-belligerent with respect to the conflict in question.[163] However, the State need not be a traditionally neutral power, such as Sweden or Switzerland.[164] In cases where a State serving as a Protecting Power enters the armed conflict, another State may assume the responsibilities of the Protecting Power.[165] A State may serve as the Protecting Power for more than one side in an armed conflict.[166]

18.15.2.1 *Agreement for an Impartial and Effective Organization to Perform Protecting Power Duties Under the 1949 Geneva Conventions*. The Parties to the 1949 Geneva Conventions may at any time agree to entrust to an organization that offers all guarantees of impartiality and efficacy the duties incumbent on Protecting Powers by virtue of the 1949 Geneva Conventions.[167]

18.15.2.2 *Obligation on the Detaining Power to Seek a Protecting Power if Persons Protected by the 1949 Geneva Conventions Do Not Benefit From One*. When the wounded and sick, shipwrecked, medical personnel and chaplains, POWs, or protected persons under the GC do not benefit or cease to benefit from the activities of a Protecting Power or an impartial and effective organization that performs the duties of the Protecting Power, the Detaining Power shall request a neutral State or such an organization to undertake the functions performed under the 1949 Geneva Conventions by a Protecting Power designated by the parties to a conflict.[168]

18.15.2.3 *Impartial Humanitarian Organizations Assuming Humanitarian Functions Performed by Protecting Powers Under the 1949 Geneva Conventions*. If the protection of a Protecting Power or an impartial and effective organization that the performs the duties of the Protecting Power cannot be arranged, the 1949 Geneva Conventions contemplate that States will use the ICRC or another impartial humanitarian organization to assume the humanitarian functions performed by Protecting Powers.[169]

The ICRC has often performed such a role during armed conflict.[170]

18.15.3 Activities of the Protecting Power. A Protecting Power assists and verifies compliance with the 1949 Geneva Conventions. For example, under the GPW and GC, the Protecting Power transmits information between belligerents, monitors compliance by the Detaining Power, and takes an active role in promoting the welfare of POWs and protected persons.[171]

The Protecting Power's activities are conducted with the consent of the State on whose territory it serves and the State whose facilities it visits. For example, the delegates of the Protecting Power are subject to the approval of the Power with which they are to carry out their duties.[172] In addition, a Protecting Power must ensure that its delegation does not exceed its humanitarian responsibilities and takes into account the imperative necessities of security of the State wherein they carry out their duties.[173]

18.15.4 Lending of Good Offices to Assist in Dispute Resolution. The 1949 Geneva Conventions contemplate that Protecting Powers shall lend their good offices with a view to settling disagreements between the parties to the conflict regarding protected persons, particularly in cases regarding the application or interpretation of the 1949 Geneva Conventions.[174] Protecting Powers may propose to the parties to the conflict a representative of a neutral Power or the ICRC to participate in the meeting.[175] This mechanism developed from State practice during World War I.[176]

The 1954 Hague Cultural Property Convention also contemplates that Protecting Powers shall lend their good offices in all cases where they may deem it useful in the interests of cultural property, particularly if there is disagreement between the parties to the conflict as to the application or interpretation of the provisions of the 1954 Hague Cultural Property Convention or the Regulations for its execution.[177] The Director-General of the United Nations Educational, Scientific, and Cultural Organization can also have a role in this process.[178]

18.16 COMPENSATION FOR VIOLATIONS OF THE LAW OF WAR

A State that is responsible for an internationally wrongful act is under an obligation to make full reparation for the injury caused by that act.[179] This principle also applies to the law of war in that a State that violates the law of war shall, if the case demands, be liable to pay compensation.[180]

18.16.1 Agreements Between States Settling Claims. Claims may be addressed by international agreements between States.[181]

Peace treaties or other agreements concluded in connection with the cessation of hostilities have sometimes addressed liability for violations of the law of war committed during that conflict.[182] Parties to the 1949 Geneva Conventions, however, cannot absolve one another of liability for grave breaches of the Geneva Conventions.[183]

18.16.2 International Claims Commissions. Compensation claims have also been addressed by international claims commissions established by agreement by the Parties.[184] A claims commission may also be established by the U.N. Security Council.[185]

18.16.3 Payment or Other Compensation *Ex Gratia*. In some cases, States may provide money or other compensation to other States where not legally required (or without an admission of legal liability or fault).[186] For example, although indemnification is not required for injuries or damage incidental to the lawful use of armed force, compensation may be provided as a humanitarian gesture.[187] Similarly, a State may institute mechanisms to make payments to foreign persons who have suffered loss from combat operations, even when no violation of law of war has occurred.

18.16.4 No Private Right to Compensation Under Customary International Law or the 1949 Geneva Conventions. The responsibility of States for violations of the law of war committed by their agents is owed to other States.[188] The fact that such responsibility is owed to other States reflects the predominately inter-State nature of international obligations.[189] Customary international law and the 1949 Geneva Conventions do not provide a private right for individuals to claim compensation directly from a State; rather, such claims are made by other States.[190]

18.17 RETORSION

Retorsion is one of the measures that an injured party may use to seek to persuade an adversary to cease violations of the law of war.

Retorsion may be understood to mean unfriendly conduct, (1) which is not inconsistent with any international obligation of the State engaging in it, and (2) which is done in response to an internationally wrongful act.[191] Retorsion is frequently contrasted with reprisal, which involves measures that would otherwise be unlawful.[192]

Because retorsion, by definition, does not involve the resort to actions that would ordinarily be characterized as illegal, the stringent conditions that apply to reprisal do not apply to retorsion.[193]

Retorsion is a general remedy available to States under international law and is not specific to the law of war. Retorsion might be used by States before, or in connection with, military operations (*e.g.*, an economic embargo). Retorsion, however, could also be used to seek to compel an adversary to adhere to the law of war.

18.18 REPRISALS

Reprisals are extreme measures of coercion used to help enforce the law of war by seeking to persuade an adversary to cease violations. States may resort to reprisals only when certain conditions are met. In addition, there are certain treaty prohibitions on reprisal, and practical considerations may counsel against their use.

18.18.1 Definition of Reprisal. Reprisals are acts taken against a party: (1) that would otherwise be unlawful; (2) in order to persuade that party to cease violating the law.[194]

For example, during the Civil War, the United States authorized reprisals against Confederate forces for murdering and enslaving captured Union soldiers.[195] Reprisals against POWs are now prohibited.[196]

18.18.1.1 *Reprisal – Notes on Terminology*. Some older sources used "reprisal" in a narrower sense only to refer to taking possession of property of the enemy in response to violations of the law of war.[197] Some older sources used the term "retaliation" to describe what is now commonly understood to be "reprisal."[198]

The term "countermeasures" is sometimes used to cover that part of the subject of reprisals not associated with armed conflict, with the term "reprisals" or "belligerent reprisals" sometimes reserved for action taken during international armed conflict.[199]

18.18.1.2 *Acts That Would Otherwise Be Unlawful*. Reprisals involve acts that would otherwise be unlawful. For example, responding to illegal enemy action by withdrawing benefits extended to the enemy where such benefits are not legally required would not be characterized as a reprisal. Rather, such action would be characterized as retorsion, *i.e.*, unfriendly conduct that is not inconsistent with any international obligation of the State engaging in it even though it may be a response to an internationally wrongful act.[200]

18.18.1.3 *In Order to Persuade a Party to Cease Violating the Law*. Reprisals are intended to influence a party to cease committing violations at present and in the future. Reprisals are not revenge or collective punishment.

18.18.2 Conditions for Lawful Reprisals. Customary international law permits reprisals, subject to certain conditions.

18.18.2.1 *Careful Inquiry That Reprisals Are Justified.* Reprisals shall be resorted to only after a careful inquiry into the facts to determine that the enemy has, in fact, violated the law.[201] In many cases, whether a law of war rule has been violated will not be apparent to the opposing side or outside observers. For example, a bombardment that results in the death of civilians may be the result of good faith, reasonable, mistake or may have been justified by the importance of destroying the military objective against which the bombardment was directed.

18.18.2.2 *Exhaustion of Other Means of Securing the Adversary's Compliance With the Law of War.* Other means of securing compliance with the law of war should be exhausted before resorting to reprisals.[202] For example, consideration should be given to using protests and demands, retorsion, or reasonable notice of the threat to use reprisals before resorting to reprisals.[203]

18.18.2.3 *National-Level Authorization for Reprisal.* Reprisals require a variety of conditions and implicate the rights and duties of a State under international law.[204] Thus, the authority to conduct reprisal is generally held at the national level.[205] Service members and units are not to take reprisal action on their own initiative.

18.18.2.4 *Proportionality in Reprisal.* To be legal, reprisals must respond in a proportionate manner to the preceding illegal act by the party against which they are taken.[206] Identical reprisals are the easiest to justify as proportionate, because subjective comparisons are not involved.[207]

However, the acts resorted to by way of reprisal need not be identical nor of the same type as the violations committed by the enemy. A reprisal should not be unreasonable or excessive compared to the adversary's violation.[208]

18.18.2.5 *Public Announcement of Reprisals.* In order to fulfill their purpose of dissuading the adversary from further illegal conduct, reprisals must be made public and announced as such.[209]

18.18.3 Treaty Prohibitions on Reprisals. Although reprisals are generally permissible under customary international law, certain treaties have prohibited certain types of reprisals.

18.18.3.1 *Reprisals Prohibited by the CCW Amended Mines Protocol.* It is prohibited in all circumstances to direct mines, booby-traps, and other devices, either in offense, defense, or by way of reprisals, against the civilian population as such or against individual civilians or civilian objects.[210]

18.18.3.2 *Reprisals Prohibited by the 1949 Geneva Conventions.* Reprisals against the wounded, sick, or shipwrecked persons, personnel, buildings, vessels, or equipment protected by the GWS or GWS-Sea are prohibited.[211] Such persons or property would include:

- combatant personnel who are wounded, sick, or shipwrecked;[212]
- medical personnel and chaplains;[213]
- medical units and facilities;[214] and
- hospital ships.[215]

Measures of reprisal against POWs are prohibited.[216]

Reprisals against protected persons under the GC and their property are prohibited.[217]

18.18.3.3 *Reprisals Prohibited by the 1954 Hague Cultural Property Convention.* Parties to the 1954 Hague Cultural Property Convention shall refrain from any act directed by way of reprisals against cultural property.[218] Cultural property is defined in the 1954 Hague Cultural Property Convention.[219]

18.18.3.4 *AP I Provisions on Reprisals.* AP I prohibits attacks by way of reprisal against:

- the civilian population or civilians;[220]
- civilian objects;[221]
- objects indispensable to the survival of the civilian population such as foodstuffs, crops, livestock, drinking water installations and supplies, and irrigation works;[222]
- the natural environment;[223] and
- works and installations containing dangerous forces, namely dams, dykes, and nuclear electrical generating systems.[224]

The United States has expressed the view that AP I's provisions on reprisal are counter- productive and that they remove a significant deterrent that protects civilians and war victims on all sides of a conflict.[225] The United Kingdom has taken a reservation to AP I's prohibition on certain attacks by way of reprisal.[226] Egypt, Germany, and Italy also reserved the right to react to serious violations of AP I with any means permitted by international law to prevent further violations.[227] France has declared that it will apply the provisions of paragraph 8 of Article 51 insofar as their interpretation does not constitute an obstacle to the use, according to international law, of the means which it considers indispensable for the protection of its civilian population against grave, clear and deliberate violations of the 1949 Geneva Conventions and of AP I by the enemy.[228]

18.18.4 Practical Considerations in the Use of Reprisals. Apart from the strict legal requirements for the resort to reprisal, it will be important to consider the potential practical consequences of the use of reprisals that may counsel strongly against taking such measures. Practical considerations, including longer-term military or political consequences, may include the following factors:

- Taking reprisals may divert valuable and scarce military resources from the military struggle and may not be as effective militarily as steady adherence to the law.
- Reprisals will usually have an adverse impact on the attitudes of governments not participating in the conflict.
- Reprisals may only strengthen enemy morale and will to resist.
- Reprisals frequently lead only to further unwanted escalation of the conflict by an adversary or a vicious cycle of counter-reprisals.
- Reprisals may render resources of an adversary less able to contribute to the rehabilitation of an area after the cessation of hostilities.[229]

18.19 DISCIPLINE IN NATIONAL JURISDICTIONS OF INDIVIDUALS FOR VIOLATIONS OF THE LAW OF WAR

In most cases, individuals are disciplined in national jurisdictions for violations of the law of war. Corrective action may take the form of adverse or corrective administrative actions. Punishment may take the form of non-judicial punishment or judicial actions in military or civilian courts, depending on the circumstances. In some cases, prosecutions in national courts are carried out by charging violations of domestic law, but in other cases, prosecutions are carried out by charging violations of international law.

18.19.1 Non-Judicial Punishment and Adverse or Corrective Administrative Actions. In some cases, it may be appropriate to administer non-judicial punishment in order to punish and repress violations of the law of war. Non-judicial punishment of members of the armed forces is authorized under the Uniform Code of Military Justice.[230]

In some cases, it may be appropriate to take adverse or corrective administrative action to repress violations of the law of war. Adverse and corrective administrative action for members of the armed forces is authorized by military regulations.[231]

18.19.2 Charging of Domestic Prosecutions as Violations of International Law or Domestic Law. In some cases, prosecutions in national courts for conduct constituting violations of the law of war are carried out by charging violations of ordinary domestic law or military law, and in other cases, such prosecutions are carried out by charging violations of international law. For example, the murder of a POW may be prosecuted under the ordinary criminal statutes prohibiting murder. On the other hand, the same conduct may be prosecuted under special criminal statutes that are framed in terms of "war crimes."[232]

The United States has no international law obligation to prosecute an offense as a "war crime" as opposed to an ordinary criminal offense. Prosecutions in national courts remain domestic prosecutions for violations of domestic statutes, even when those domestic statutes were enacted pursuant to treaty obligations or are framed in terms of violations of international law (*e.g.*, war crimes). In general, prosecutions of war crimes as such (*i.e.*, characterized as international law violations as opposed to violations of domestic law) have only been undertaken when a State seeks to punish enemy nationals or persons serving the interests of the enemy State. When members of a State's armed forces or other personnel violate the law of war, that State generally prosecutes those persons for offenses under ordinary domestic law or military law.[233]

18.19.3 U.S. Law That Can Be Used to Punish Violations of the Law of War. A number of U.S. statutes may be used to punish acts that constitute violations of the law of war. When violations of the law of war are committed within the United States, such acts usually constitute violations of Federal and State law and generally may be prosecuted under ordinary criminal statutes. However, a number of other Federal statutes may be used to prosecute conduct that violates the law of war, even though that conduct does not occur within the United

States. The application of any particular statute depends on the terms of that statute and the facts of a particular case.

18.19.3.1 *Uniform Code of Military Justice Offenses.* The principal way for the United States to punish members of the U.S. armed forces for violations of the law of war is through the Uniform Code of Military Justice. Certain persons may be tried for violations of the Uniform Code of Military Justice, including, among others:

- members of a regular component of the U.S. armed forces;
- POWs in the custody of the U.S. armed forces;
- in time of declared war or contingency operations, persons serving with or accompanying an armed force in the field; and
- individuals belonging to one of the eight categories enumerated in Article 4 of the GPW who violate the law of war.[234]

Offenses under the Uniform Code of Military Justice that may be used to punish conduct that constitutes a violation of the law of war include, among others:

- cruelty and maltreatment;[235]
- murder;[236]
- rape and sexual assault;[237]
- failure to obey order or regulation;[238] and
- conduct prejudicial to good order and discipline.[239]

18.19.3.2 *Incorporation of Other Non-Capital Crimes and Offenses in the Uniform Code of Military Justice.* Under the Uniform Code of Military Justice, "all disorders and neglects to the prejudice of good order and discipline in the armed forces, all conduct of a nature to bring discredit upon the armed forces, and crimes and offenses not capital," of which persons subject to the Uniform Code of Military Justice may be guilty, shall be taken cognizance of by a general, special, or summary court-martial, according to the nature and degree of the offense, and shall be punished at the discretion of that court.[240]

This provision may also be used to prosecute conduct that violates the law of war. However, the preemption doctrine prohibits application of this provision to conduct already covered by specific offenses in the Uniform Code of Military Justice.[241]

18.19.3.3 *U.S. General Court-Martial Jurisdiction to Try and Punish Persons Under the Law of War.* Under the Uniform Code of Military Justice, general courts-martial also have jurisdiction to try any person who by the law of war is subject to trial by a military tribunal and may adjudge any punishment permitted by the law of war.[242] For example, this authority could be used to try enemy combatants for violations of the law of war, because such trials are permitted by the law of war.

18.19.3.4 *Title 18 Federal Crimes Punishing Conduct Outside the United States.* Certain Federal statutes specify that they apply to conduct outside the United States and could be used to prosecute conduct that, in some circumstances, would constitute a violation of the law of war. These statutes may specify in detail the circumstances in which they authorize prosecution, which may vary from statute to statute.

These statutes include provisions prohibiting:

- torture;[243]
- genocide;[244]
- murder or manslaughter of foreign officials, official guests, or internationally protected persons;[245]
- piracy;[246]
- acts of terrorism and material support to terrorists;[247]
- certain acts involving biological weapons;[248]
- certain acts involving chemical weapons;[249] and
- certain acts involving nuclear weapons.[250]

18.19.3.5 *Extraterritorial Application of Certain Federal Offenses Through MEJA.* U.S. law makes special provision for certain conduct that is committed outside the United States to be prosecuted as though it were committed inside the United States. U.S. law, under the Military Extraterritorial Jurisdiction Act (MEJA), permits the United States to prosecute individuals who committed certain offenses outside the United States (i) while employed by or accompanying the U.S. armed forces overseas; or (ii) while a member of the U.S. armed forces subject to the Uniform Code of Military Justice.[251] MEJA cannot be used against a member of the Armed Forces who is subject to the Uniform Code of Military Justice unless (i) such member ceases to be subject to the Uniform Code of Military Justice; or (ii) an indictment or information charges that the member committed the offense with one or more other defendants, at least one of whom is not subject to the Uniform Code of Military Justice.[252]

Provisions of MEJA have been implemented through DoD regulations.[253]

Among other things, MEJA provides a way for U.S. domestic law to be used to punish conduct that violates the law of war. For example, under MEJA, a former service member could be prosecuted for a murder committed while a member of the armed forces during armed conflict abroad, even though Uniform Code of Military Justice jurisdiction over that person has ceased.[254]

18.19.3.6 *Special Maritime and Territorial Jurisdiction of the United States Under Title 18*. U.S. law provides for certain offenses to be prosecuted if they are committed in the special maritime and territorial jurisdiction of the United States.[255] Such jurisdiction may include, for example, offenses committed by U.S. nationals on the premises of U.S. diplomatic, consular, military, or other U.S. Government missions or entities in foreign States.[256] This jurisdiction provides another method by which U.S. law may be used to punish conduct that violates the law of war.[257]

18.19.3.7 *U.S. Military Commissions*. U.S. statutes authorize the prosecution of alien unprivileged enemy belligerents for certain offenses.[258] The Secretary of Defense has promulgated additional rules of procedure and rules of evidence applicable to such proceedings.[259]

In the past, military commissions have been used by the United States and other States to prosecute enemy belligerents for violations of the law of war and for acts of unprivileged belligerency. Military commissions have also been used for the trial of offenses under U.S. law where local courts were not open and acting (*i.e.*, where martial law applies), and for the trial of violations of occupation ordinances. These military commissions have been regarded as instrumentalities for the more efficient execution of the war powers vested in Congress and the power vested in the President as Commander-in-chief in war.[260] Military commissions have been used instead of courts-martial because U.S. courts-martial have been adapted to the circumstances of disciplining members of the armed forces and have not been crafted with a view towards certain other offenses that are also committed during armed conflict.[261]

Similarly, the use of military commission proceedings continues to be appropriate in certain circumstances because the rules for such proceedings are specially adapted to reflect the realities of the battlefield and conducting investigations in a war zone.[262]

18.19.3.8 *War Crimes Act*. The War Crimes Act authorizes the prosecution of individuals for certain war crimes if the victim or the perpetrator is either a U.S. national or a member of the U.S. Armed Forces, whether inside or outside the United States.[263] Under this statute, an individual may be prosecuted for conduct:

- defined as a grave breach in any of the 1949 Geneva Conventions, or any protocol to such convention to which the United States is a Party;
- prohibited by Article 23, 25, 27, or 28 of the Annex to the Hague Convention IV, Respecting the Laws and Customs of War on Land, signed 18 October 1907;
- that constitutes a grave breach of common Article 3 (as defined in the statute) when committed in the context of and in association with an armed conflict not of an international character; or
- of a person who, in relation to an armed conflict and contrary to the provisions of the CCW Amended Mines Protocol, when the United States is a Party to such Protocol, willfully kills or causes serious injury to civilians.[264]

18.19.4 U.S. Domestic Law and Policy on Military Jurisdiction.

18.19.4.1 *Limits on Military Jurisdiction Over U.S. Citizens Who Are Not Members of the Armed Forces*. As a matter of U.S. domestic law, the U.S. Constitution places certain limits on the use of military tribunals to punish U.S. citizens who are not members of the armed forces.[265] DoD policy addresses the use of the Uniform Code of Military Justice to try civilians serving with or accompanying the U.S. armed forces.[266] Such law and policy may dictate the forum for prosecution. For example, a person who has been discharged from the U.S. armed forces and is no longer subject to the Uniform Code of Military Justice would be prosecuted under the MEJA or the War Crimes Act rather than the Uniform Code of Military Justice.

18.19.4.2 *Efforts to Maximize Court-Martial Jurisdiction Over Persons Who Are Members of the Armed Forces*. Where the United States and a foreign State both claim jurisdiction over a service member's conduct, DoD policy has been to make efforts to maximize the exercise of court-martial jurisdiction over persons subject to the Uniform Code of Military Justice to the extent possible under applicable agreements.[267]

18.20 PROSECUTION IN INTERNATIONAL AND HYBRID COURTS

In some cases, the prosecution of individuals for violations of the law of war has been undertaken in international tribunals or hybrid tribunals that mix elements of national and international law. The jurisdiction and procedures of these tribunals vary from tribunal to tribunal, and may depend on applicable treaties and customary international law. In general, the decisions of these tribunals are not binding as precedent on the United States, including U.S. courts.[268]However, in some cases, it may be appropriate to consider their decisions as persuasive authority.

18.20.1 Post-World War II International Military Tribunals. After World War II, the United Kingdom, France, the United States, and the Union of Soviet Socialist Republics sought to try the major European Axis war criminals. Established by the London Agreement of August 8, 1945, the International Military Tribunal at Nuremberg conducted the landmark Trial of Major War Criminals, with 21 defendants, in Nuremberg, Germany from November 1945 - October 1946.[269]A similar tribunal was established in Tokyo by U.S. General MacArthur in his role as Supreme Allied Commander to try major Japanese war criminals in the Far East.[270]

18.20.2 International Criminal Tribunal for the Former Yugoslavia (ICTY) and International Criminal Tribunal for Rwanda (ICTR). The U.N. Security Council has used its authority under Chapter VII of the Charter of the United Nations to establish international criminal tribunals to prosecute persons responsible for serious violations of international humanitarian law committed in the territory of the former Yugoslavia since 1991, and in the territory of Rwanda, as well as Rwandan citizens responsible for genocide and other such violations committed in the territory of neighboring States, between January 1, 1994, and December 31, 1994.[271]

The United States has, through its membership in the U.N. Security Council, supported both the efforts of the ICTR and (also acting through NATO) the efforts of the ICTY.[272]

18.20.3 The International Criminal Court (ICC) Created by the Rome Statute. The Rome Statute of the ICC established the ICC, which sits in The Hague in the Netherlands.[273]The United States is not a Party to the Rome Statute.

The Rome Statute provides that the ICC "shall have the power to exercise its jurisdiction over persons for the most serious crimes of international concern" and "shall be complementary to national criminal jurisdictions."[274]The Rome Statute provides that the ICC has jurisdiction in accordance with the Statute with respect to:[275]

- the crime of genocide;
- crimes against humanity;
- war crimes; and
- the crime of aggression.

18.20.3.1 *Attempt to Assert Jurisdiction With Respect to Nationals of Non-Party States*. The ICC may exercise jurisdiction if a situation has been referred to it by the U.N. Security Council acting under Chapter VII of the Charter of the United Nations.[276]The Rome Statute also provides that the ICC may exercise its jurisdiction if one or more of the following States are Parties to the Rome Statute or have accepted the jurisdiction of the ICC on an *ad hoc* basis:

- the State on the territory of which the conduct in question occurred or, if the crime was committed on board a vessel or aircraft, the State of registration of that vessel or aircraft; or
- the State of which the person accused of the crime is a national.[277]

The first of these provisions creates the possibility that the ICC would seek to exercise jurisdiction with respect to the actions of nationals of States that have not ratified the Rome Statute, even if the U.N. Security Council has not referred the situation. The United States has a longstanding and continuing objection to any assertion of jurisdiction by the ICC with respect to nationals of States not Party to the Rome Statute in the absence of consent from such States or a referral by the Security Council.[278]

18.20.3.2 *Article 98 Agreements*. The Rome Statute provides that the ICC may not proceed with a request for surrender or assistance that would require the requested State to act inconsistently with its obligations under international law with respect to the State or diplomatic immunity of a person or property of a third State, unless the Court can first obtain the cooperation of that third State for the waiver of the immunity.[279]The Rome Statute also provides that the ICC may not proceed with a request for surrender that would require the requested State to act inconsistently with its obligations under international agreements pursuant to which the consent of a sending State is required to surrender a person of that State to the ICC, unless the ICC can first obtain the cooperation of the sending State for the giving of consent for the surrender.[280]

The United States has entered into agreements with numerous States, including Parties to the Rome Statute, that require U.S. consent before U.S. personnel may be surrendered by that State to the ICC.[281]

18.20.3.3 *U.S. Law and Policy on Support to the ICC*. U.S. law contains certain restrictions on support to the ICC or ICC activities within the United States.[282]U.S. policy is to work with the international community to prevent and call to account those responsible for the worst human rights abuses, including through support to the International Criminal Court, consistent with U.S. law and our commitment to protecting our personnel.[283]

18.20.3.4 *ICC and the Crime of Aggression*. As adopted in 1998, the Rome Statute does not define the crime of aggression, but provides that the ICC shall exercise jurisdiction over the crime of aggression once a provision is adopted in accordance with articles 121 and 123 of the Rome Statute defining the crime and setting out the conditions under which the ICC shall exercise jurisdiction with respect to this crime.[284]A 2010 Review Conference in Kampala, Uganda adopted amendments concerning the crime of aggression, but the amendments are subject to ratification or acceptance,[285]and certain conditions must be fulfilled before the ICC will be able to exercise jurisdiction with respect to the crime of aggression.[286]The United States has expressed the view that the definitions of "act of aggression" and "crime of aggression" in the Kampala amendments do not reflect customary international law.[287]The United States has expressed a broad range of other concerns about the Kampala amendments, including concerns regarding the possibility of the ICC exercising jurisdiction over the crime of aggression without a prior determination by the Security Council that a State has committed an act of aggression.[288]

18.20.4 Hybrid or Mixed Tribunals. In some cases, an international tribunal may mix elements of domestic and international law. For example, war crimes trials by national tribunals were conducted after World War II under the military government formed by the Allies to govern occupied Germany.[289]

More recently, hybrid tribunals have been established in Sierra Leone, Cambodia, and Lebanon. The Special Court for Sierra Leone was established by a treaty between Sierra Leone and the United Nations.[290]The Special Court applies both international and Sierra Leonean law.[291]Similarly, in Cambodia, the Extraordinary Chambers in the Courts of Cambodia (ECCC) was established under Cambodian domestic law, but regulated by an agreement between Cambodia and the United Nations.[292]The Special Tribunal for Lebanon was established following an agreement between the United Nations and Lebanon to prosecute, *inter alia*, persons responsible for the February 2005 attack resulting in the death of former Lebanese Prime Minister Rafiq Hariri.[293]

18.21 LIMITS ON THE PUNISHMENT OF INDIVIDUALS UNDER THE LAW OF WAR

International law places certain limits on the authority of States, whether acting individually, or together with other States, to punish individuals for violations committed during armed conflict.

18.21.1 Jurisdiction Over War Crimes. States must have jurisdiction over an alleged offense to prosecute a person for committing war crimes or other punishable offenses committed during armed conflict.[294]

Jurisdiction over war crimes has traditionally been exercised by belligerents with respect to offenses committed by or against their nationals.[295]In addition, belligerents have also prosecuted war crimes committed by enemy nationals against nationals of allies and of co- belligerents, and stateless persons.[296]Because jurisdiction over war crimes has sometimes been exercised by belligerents against enemy nationals without regard to the territorial location of the offense, it has sometimes been characterized as "universal" in character.[297]

In the past, neutral States generally did not exercise jurisdiction with respect to alleged law of war violations between belligerents.[298]For example, States have declined to exercise jurisdiction with respect to offenses committed by enemy nationals before those States became involved in an armed conflict with that State.[299]

Some have argued that States may exercise purely universal jurisdiction over war crimes, *i.e.*, jurisdiction to define and prescribe punishment based simply on the character of the offense as a war crime.[300]Until the 1990s, no such attempts were made by States to exercise jurisdiction on this basis.[301]Congress declined to authorize prosecutions for war crimes based on this principle.[302]Efforts by certain States to undertake prosecutions on the basis of this principle have been controversial.[303]Such prosecutions have generally not been successful unless the State concerned has consented.[304]

18.21.2 Tu Quoque. The international law doctrine *tu quoque* may be understood as an argument that a State does not have standing to complain about a practice in which it itself engages.[305]

For example, it would seem unfair for a State to punish members of opposing military forces for committing acts that it considered lawful for members of its armed forces to perform.[306]On the other hand, as a general matter, the fact that criminal acts have been committed by opposing armed forces does not constitute a defense to criminal liability. In addition, as a general matter, the fact that members of its armed forces have committed a violation does not preclude a State from punishing captured members of opposing armed forces for such violations.[307]Similarly, the authority of a State to punish its own citizens, in accordance with municipal criminal law, for violations of international law is not affected by this rule.[308]

18.21.3 Fundamental Fairness Requirements.

18.21.3.1 *Fair Trial Requirements*. Any person charged with a crime under international law has the right to a fair trial on the facts and law.[309]

In all circumstances, persons who are accused of grave breaches of the 1949 Geneva Conventions shall benefit from safeguards of proper trial and defense, which shall not be less favorable than those provided by Article 105 and those following of the GPW.[310]In addition, other fundamental fair trial guarantees should be afforded.[311]

18.21.3.2 *Punishment of Violations of the Law of War*. The punishment of war crimes may vary according to the particular forum in which the prosecution is brought. For example, what punishments are authorized may depend on a State's domestic law or the statute of an international criminal tribunal. In addition, special rules address the punishment of POWs and protected persons. However, the following principles apply, as a matter of international law, to the punishment of violations of the law of war:

• The punishment imposed for a violation of the law of war must be proportionate to the gravity of the offense.

• The death penalty may be imposed for grave breaches of the law.

• Corporal punishment is excluded.

• Punishments should be deterrent, and in imposing a sentence of imprisonment it is not necessary to take into consideration the end of the war, which does not of itself limit the imprisonment to be imposed.[312]

18.21.4 Limitations on the Trial and Punishment of Certain Classes of Persons.

18.21.4.1 *Limitations on the Trial and Punishment of POWs*. The trial and punishment of POWs must comport with the rules prescribed by the GPW.[313]

18.21.4.2 *Limitations on the Trial and Punishment of Protected Persons*. The trial and punishment of protected persons must comport with the rules prescribed by the GC.[314]

18.21.4.3 *Limitations on the Trial and Punishment of Persons Entitled to the Privileges of Combatant Status*. As a matter of international law, persons who are entitled to the privileges of combatant status have a certain legal immunity from the domestic law of foreign States.[315]

18.22 PRINCIPLES OF INDIVIDUAL CRIMINAL RESPONSIBILITY FOR CRIMES UNDER INTERNATIONAL LAW

Individual criminal responsibility exists for certain violations of the international law.

18.22.1 Individual Criminal Responsibility for Acts Constituting Crimes Under International Law. Any person who commits an act that constitutes a crime under international law is responsible therefor and liable to punishment.[316]International law imposes duties and liabilities on individuals as well as States, and individuals may be punished for violations of international law.[317]

18.22.2 Absence of Penalty Under Domestic Law Does Not Relieve a Person of Responsibility. The fact that internal law does not impose a penalty for an act that constitutes a crime under international law does not relieve the person who committed the act from responsibility under international law.[318]

For example, a State may lack domestic legislation that imposes any penalty for an act that violates international law. Nonetheless, just as a State's municipal law cannot excuse its failure to comply with its international legal obligations, a person is not relieved of responsibility for an act constituting a crime under international law because his or her State's municipal law does not impose any penalty for the offense.[319]

18.22.3 Official Position Does Not Relieve a Person of Responsibility. The fact that a person who committed an act which constitutes a crime under international law acted as Head of State or responsible Government official does not relieve him or her of responsibility under international law.[320]

This principle has been reflected in the statutes of international criminal tribunals.[321]

18.22.4 Acting Pursuant to Orders Does Not Relieve a Person of Responsibility. The fact that a person acted pursuant to orders of his or her Government or of a superior does not relieve that person from responsibility under international law, provided it was possible in fact for that person to make a moral choice.[322]This principle has been reflected in the statutes of international criminal tribunals.[323]It may also be understood as part of a broader principle that military personnel cannot justify committing unlawful acts by producing the order of their superior.[324]

Although it is clear that merely the fact that the act at issue was committed pursuant to superior orders does not constitute a defense to criminal responsibility under international law, the precise extent to which superior orders may constitute a defense or excuse may vary according to the forum in which a violation is tried.[325]

In cases in which the illegality of the order is not apparent, the subordinate might lack the wrongful intent necessary to the commission of the crime.[326]Subordinates, absent specific knowledge to the contrary, may presume orders to be lawful.[327]The acts of a subordinate done in compliance with an unlawful order given by a superior are generally excused unless the superior's order is one that a person of ordinary sense and understanding would, under the circumstances, know to be unlawful (*e.g.*, to torture or murder a detainee), or if the order in question is actually known to the accused to be unlawful.[328]

On the other hand, subordinates must refuse to comply with clearly illegal orders to commit violations of the law of war.[329]

That an offense was committed pursuant to superior orders may be considered in mitigation of punishment.[330]The degree to which superior orders should mitigate punishment would depend on the specific circumstances of the case.[331]

18.23 THEORIES OF INDIVIDUAL CRIMINAL LIABILITY

Individuals may be held liable for violations of the law of war whether they have committed them directly or are complicit in the commission of such crimes.[332]

The theories of liability that apply to a law of war violation may vary depending on the particular forum (*e.g.*, U.S. Federal court, U.S. military commission, International Criminal Tribunal for the Former Yugoslavia) in which the violation is being adjudicated. Modes of liability for law of war offenses may include ordering, instigating or directly inciting, command responsibility, aiding and abetting, conspiracy, and joint criminal enterprise. In some cases, these theories of liability may be viewed as ways of attributing an offense that is committed by one person to another person. In other cases, these theories of liability may be viewed as distinct offenses; for example, a first offense is committed by one person and a second offense is committed by another person that is somehow related to the first offense.

18.23.1Ordering. A person who orders another person to commit an offense is generally punishable as though that person had committed the offense directly. This principle is reflected in the Uniform Code of Military Justice.[333]Statutes of international criminal tribunals have also reflected this mode of liability.[334]

18.23.2 Instigating or Directly Inciting. Instigating or direct incitement of an offense is punishable.[335]Statutes of international criminal tribunals have also reflected this mode of liability.[336]

18.23.3 Command Responsibility. Commanders have duties to take necessary and reasonable measures to ensure that their subordinates do not commit violations of the law of war.[337]Failures by commanders of their duties to take necessary and reasonable measures to ensure that their subordinates do not commit violations of the law of war can result in criminal responsibility.[338]

18.23.3.1 *Command Responsibility as a Distinct Offense*. Commanders may be punished directly for their failure to take necessary and reasonable measures to ensure that their subordinates do not commit violations of the law of war. For example, such failures may be punished under the Uniform Code of Military Justice as dereliction of duty or violation of orders to take such measures.[339]

18.23.3.2 *Command Responsibility as a Mode of Liability for an Offense*. In some cases, the failure by commanders to fulfill their duties to take necessary and reasonable measures to ensure that their subordinates do not commit violations of the law of war are not punished directly as breaches of those duties, but instead by imputing responsibility for the offense committed by the subordinates to the commander.

Command responsibility as a mode of liability for an offense is reflected in U.S. statutes governing military commissions.[340]The statutes of international tribunals have also reflected command responsibility as a mode of liability.[341]In some cases, this theory of superior responsibility has been applied to civilian superiors.[342]

Command responsibility, as a mode of liability, is not a form of strict liability.[343]The commander's personal dereliction must have contributed to or failed to prevent the offense; there must be a personal neglect amounting to a wanton, immoral disregard of the action of his or her subordinates amounting to acquiescence in the crimes.[344]

18.23.4Aiding and Abetting. The theory of aiding and abetting holds an individual liable for an offense committed by another based on certain assistance that the individual gave in relation to the crime. Aiding and abetting liability for a crime can be usefully analyzed as consisting of three elements: (1) knowledge of the illegal activity that is being aided and abetted; (2) a desire to help the activity succeed; and (3) some act of helping.[345]

This theory of liability is applicable in prosecutions in Federal court under title 18,[346]prosecutions under the Uniform Code of Military Justice,[347]and prosecutions by U.S. military commission.[348]This theory of liability has also been reflected in the statutes of international criminal tribunals.[349]

18.23.4.1 *Aiding and Abetting as a Principle of State Responsibility*. The principle of aiding and abetting is also reflected in principles of State responsibility.[350]As with the principle of aiding and abetting when applied to individuals, the aid or assistance must be given with a view to facilitating the commission of the wrongful act, and must actually do so, in order for the State giving the aid to incur liability.[351]

18.23.4.2 *Aiding and Abetting Certain Conduct as a Distinct Offense Instead of a Mode of Liability for an Offense*. In some cases, aiding certain conduct is a distinct offense as opposed to a mode of liability for an offense. These cases reflect circumstances in which (or a judgment that) there is a duty to refrain from aiding that conduct.[352]

18.23.5 Conspiracy. The essence of conspiracy is the combination of minds in an unlawful purpose.[353]

The United States has taken the position that conspiracy to violate the law of war is punishable.[354]The United States has used military tribunals to punish unprivileged belligerents for the offense of conspiracy to violate the law of war.[355]

Conspiracy is an offense under the Uniform Code of Military Justice.[356]Conspiracy is an offense under the Military Commissions Act of 2009.[357]Conspiracy to commit an offense is an offense under Title 18.[358]

The use of conspiracy may vary according the forum in which charges are brought. However, as a general matter, charges of conspiracy for violations of the law of war by persons belonging to the enemy should be restricted to cases of offenses in which an overt act has been committed because non-punitive measures, such as security detention, may be available to address threats.[359]

18.23.5.1 *Conspiracy to Commit Genocide*. The Genocide Convention provides that conspiracy to commit genocide shall be punishable.[360]Thus, mere agreement by perpetrators to commit genocide may be punishable even if no preparatory act has taken place.[361]

18.23.5.2 *Conspiracy to Commit Aggressive War*. The International Military Tribunal at Nuremberg tried defendants for participation in a common plan or conspiracy to wage a war of aggression.[362]However, the International Military Tribunal declined to interpret its Charter to authorize prosecutions for conspiracy to commit war crimes and crimes against humanity.[363]

18.23.6 Other Group Criminality Theories – Joint Criminal Enterprise. In addition to conspiracy, analogous theories of group criminality have been applied to violations of the law of war and other violations of international law.[364]Under recognized principles common to the major legal systems of the world, persons who are connected with plans or enterprises involved in the commission of a crime, or who belong to an organization or group engaged in the commission of crime, may also be guilty of that crime.[365] Joining a criminal group may be punishable as a distinct offense separate from the atrocities committed by that criminal group.[366]

Three categories of group criminality or "joint criminal enterprise" may be identified, which vary according to the mental element (*mens rea*) of the defendant with respect to the crime that is committed:[367]

• cases of co-perpetration where all participants in the common design possess the same criminal intent to commit a crime (and one or more of them actually perpetrate the crime, with intent);[368]

• cases where the requisite *mens rea* comprises knowledge of the nature of the system of ill-treatment (*e.g.*, a German concentration camp) and intent to further the common design of ill-treatment (such intent may be proved either directly or as a matter of inference from the nature of the accused's authority within the camp or organizational hierarchy);[369] and

• cases in which a "common purpose" exists because the following requirements

concerning *mens rea* are fulfilled: (i) the intention to take part in a joint criminal enterprise and to further – individually and jointly – the criminal purposes of that enterprise; and (ii) the foreseeability of the possible commission by other members of the group of offenses that do not constitute the object of the common criminal purpose.[370]

XIX Documentary Appendix – Notes on Treaties and Other Relevant Documents

Chapter Contents

19.1 INTRODUCTION

This appendix provides background information about certain treaties and other documents. This appendix is intended to describe DoD views and practice relating to those documents as of the date of publication of this manual.

19.2 LISTS OF TREATIES AND OTHER DOCUMENTS

This section lists: (1) law of war treaties to which the United States is a Party; (2) arms control agreements to which the United States is a Party that are of direct relevance to the law of war; (3) examples of treaties signed but not ratified by the United States; (4) examples of treaties that the United States has neither signed nor ratified; and (5) examples of treaties or documents of mainly historical value.

The categorization of a treaty as a "law of war" treaty or an "arms control" treaty is intended to help practitioners understand the context in which the treaty was concluded and its purposes. For example, a law of war treaty generally focuses on restrictions on the use of weapons during armed conflict, while an arms control treaty generally focuses on restricting development and acquisition of weapons.[1] Some treaties that are characterized below as law of war treaties have elements of arms control, and some treaties that are characterized below as arms control treaties have elements of the law of war.

Bold type within this section indicates an abbreviation used in this manual; a full list of abbreviations is provided at the beginning of the manual.[2]

19.2.1 Law of War Treaties to Which the United States Is a Party. Law of war treaties to which the United States is a Party include:

• Washington Convention Regarding the Rights of Neutrals at Sea of October 31, 1854.[3]

• Hague Convention for the Exemption of Hospital Ships, in Time of War, from the Payment of all Dues and Taxes Imposed for the Benefit of the State of December 21, 1904.[4]

• Hague Convention III of October 18, 1907, Relative to the Opening of Hostilities.[5]

• Hague Convention IV of October 18, 1907, Respecting the Laws and Customs of War on Land (**Hague IV**), and the Annex thereto, entitled Regulations Respecting the Laws and Customs of War on Land (**Hague IV Regulations**).[6]

• Hague Convention V of October 18, 1907, Respecting the Rights and Duties of Neutral Powers and Persons in Case of War on Land (**Hague V**).[7]

• Hague Convention VIII of October 18, 1907, Relative to the Laying of Automatic Submarine Contact Mines (**Hague VIII**).[8]

• Hague Convention IX of October 18, 1907, Concerning Bombardment by Naval Forces in Time of War (**Hague IX**).[9]

• Hague Convention XI of October 18, 1907, Relative to Certain Restrictions with Regard to the Exercise of the Right of Capture in Naval War (**Hague XI**).[10]

• Hague Convention XIII of October 18, 1907, Concerning the Rights and Duties of Neutral Powers in Naval War (**Hague XIII**).[11]

• 1928 Pan American Maritime Neutrality Convention.[12]

• 1930 London Treaty for the Limitation and Reduction of Naval Armament.[13]

• Treaty on the Protection of Artistic and Scientific Institutions and Historic Monuments of April 15, 1935 (**Roerich Pact**).[14]

• Charter of the United Nations (**U.N. Charter**).[15]

• 1949 Geneva Conventions,[16]including the

o Geneva Convention for the Amelioration of the Condition of the Wounded and Sick in Armed Forces in the Field of August 12, 1949 (**GWS**);[17]

o Geneva Convention for the Amelioration of the Condition of Wounded, Sick and Shipwrecked Members of Armed Forces at Sea of August 12, 1949 (**GWS-Sea**);[18]

o Geneva Convention Relative to the Treatment of Prisoners of War of August 12, 1949 (**GPW**);[19]and

o Geneva Convention Relative to the Protection of Civilian Persons in Time of War of August 12, 1949 (**GC**).[20]

• Hague Convention for the Protection of Cultural Property in the Event of Armed Conflict of May 14, 1954 (**1954 Hague Cultural Property Convention**).[21]

• Convention on Prohibitions or Restrictions on the Use of Certain Conventional Weapons Which May Be Deemed to Be Excessively Injurious or to Have Indiscriminate Effects of October 10, 1980, its Protocols I, II, III, IV, and V, its Amended Protocol II, and its Amended Article 1 (**CCW**);[22]

o CCW Amended Article 1;[23]

o Protocol on Non-Detectable Fragments (Protocol I). Geneva, October 10, 1980 (**CCW Protocol I**);[24]

o Protocol on Prohibitions or Restrictions on the Use of Mines, Booby-Traps and Other Devices as amended on May 3, 1996 (Protocol II to the 1980 CCW Convention as amended on May 3, 1996) (**CCW Amended Mines Protocol**);[25]

o Protocol on Prohibitions or Restrictions on the Use of Incendiary Weapons (Protocol III). Geneva, October 10, 1980 (**CCW Protocol III on Incendiary Weapons**);[26]

o Protocol on Blinding Laser Weapons (Protocol IV to the 1980 Convention), October 13, 1995 (**CCW Protocol IV on Blinding Laser Weapons**);[27]and

o Protocol on Explosives Remnants of War (Protocol V to the 1980 Convention), November 28, 2004 (**CCW Protocol V on Explosive Remnants of War**).[28]

• Optional Protocol to the Convention on the Rights of the Child on the involvement of Children in Armed Conflict, May 25, 2000 (**Child Soldiers Protocol**).[29]

• Protocol Additional to the Geneva Conventions of 12 August 1949, and Relating to the Adoption of an Additional Distinctive Emblem (Protocol III), December 8, 2005 (**AP III**).[30]

19.2.2 Arms Control Agreements to Which the United States Is a Party That Are of Direct Relevance to the Law of War. The United States is a Party to the following treaties that contain restrictions on the use of weapons during armed conflict:

• Geneva Protocol for the Prohibition of the Use in War of Asphyxiating, Poisonous or Other Gases, and of Bacteriological Methods of Warfare of June 17, 1925 (**1925 Geneva Gas and Bacteriological Protocol**).[31]

• Convention on the Prohibition of Development, Production and Stockpiling of Bacteriological (Biological) and Toxin Weapons and on Their Destruction of April 10, 1972 (**Biological Weapons Convention**).[32]

• Convention on the Prohibition of Military or Any Other Hostile Use of Environmental Modification Techniques of May 18, 1977 (**ENMOD Convention**).[33]

• Convention on the Prohibition of the Development, Production, Stockpiling and Use of Chemical Weapons and on Their Destruction of January 13, 1993 (**Chemical Weapons Convention**).[34]

19.2.3 Examples of Treaties Signed but Not Ratified by the United States. This section is not comprehensive. It lists examples of treaties that the United States has signed, but not ratified.

A State that has signed a treaty is obliged to refrain from acts that would defeat the object and purpose of a treaty, until it shall have made its intention clear not to become a Party to the treaty.[35]

• *Procès-Verbal* Relating to the Rules of Submarine Warfare set forth in Part IV of the Treaty of London of April 22, 1930 (**1936 London Protocol**).[36]

• Vienna Convention on the Law of Treaties (**VCLT**).[37]

• Protocol (I) Additional to the Geneva Conventions of August 12, 1949, and Relating to the Protection of Victims of International Armed Conflicts of June 8, 1977 (**AP I**).[38]

• Protocol (II) Additional to the Geneva Conventions of August 12, 1949, and Relating to the Protection of Victims of Non-International Armed Conflicts of June 8, 1977 (**AP II**).[39]

• U.N. Convention on the Law of the Sea (**LOS Convention**).[40]

• Rome Statute of the International Criminal Court of July 17, 1998 (**Rome Statute**).[41]

19.2.4 Examples of Treaties That the United States Has Neither Signed Nor Ratified. This section is not comprehensive. It lists examples of law of war and arms control treaties that the United States has neither signed nor ratified:

• Declaration respecting maritime law signed by the Plenipotentiaries of Great Britain, Austria, France, Prussia, Russia, Sardinia, and Turkey, assembled in Congress at Paris, April 16, 1856.[42]

• Hague Declaration on Expanding Bullets of July 29, 1899.[43]

• Hague Convention VI Relating to the Status of Enemy Merchant Ships at the Outbreak of Hostilities of October 18, 1907.

• Hague Convention VII Relating to the Conversion of Merchant Ships into Warships of October 18, 1907.

• First Protocol to the Hague Convention of 1954 for the Protection of Cultural Property in the Event of Armed Conflict of May 14, 1954.

• Ottawa Convention on the Prohibition of the Use, Stockpiling, Production and Transfer of Anti-Personnel Mines and on Their Destruction of September 18, 1997.[44]

• Second Protocol to the Hague Convention of 1954 for the Protection of Cultural Property in the Event of Armed Conflict of March 26, 1999.

• Convention on Cluster Munitions of May 30, 2008.[45]

19.2.5 Examples of Treaties or Documents of Mainly Historical Value. This section is not comprehensive. It lists examples of treaties and documents that are regarded as having mainly historical value:

• General Order No. 100, Instructions for the Government of Armies of the United States in the Field, 1863 (**Lieber Code**).[46]

• Geneva Convention for the Amelioration of the Wounded in Armies in the Field of August 22, 1864 (**1864 GWS**).[47]

• St. Petersburg Declaration Renouncing the Use, in Time of War, of Explosive Projectiles under 400 Grams Weight of December 11, 1868 (**1868 St. Petersburg Declaration**).[48]

• Hague Declaration (IV, 1) to Prohibit for the Term of Five Years the Launching of Projectiles and Explosives from Balloons, and Other Methods of a Similar Nature of July 29, 1899.[49]

• Hague Declaration on Asphyxiating Gases of July 29, 1899.[50]

• Hague Convention II with Respect to the Laws and Customs of War on Land, with Annex of Regulations of July 29, 1899 (**1899 Hague II**).[51]

• Hague Convention X for the Adaptation to Maritime Warfare of the Principles of the Geneva Convention of October 18, 1907 (**Hague X**).[52]

• Hague Declaration XIV Prohibiting the Discharge of Projectiles and Explosives from Balloons of October 18, 1907.[53]

• Washington Treaty Relating to the Use of Submarines and Noxious Gases in Warfare of February 6, 1922.[54]

• General Report of the Commission of Jurists to Consider and Report upon the Revision of the Rules of Warfare, Feb. 19, 1923 (**1923 Hague Air and Radio Rules**).[55]

• Geneva Convention Amelioration of the Condition of the Wounded and Sick of Armies in the Field of July 27, 1929 (**1929 GWS**).[56]

• Geneva Convention Relative to the Treatment of Prisoners of War of July 27, 1929 (**1929 GPW**).[57]

19.3 LIEBER CODE

General Order No. 100, Instructions for the Government of Armies of the United States in the Field, 1863, issued on April 24, 1863, is often called the "Lieber Code" because it was prepared by Francis Lieber.[58]It is an early example of the law of war being implemented through military instructions or regulations.[59]

The Lieber Code was the first comprehensive publication on the law of war for U.S. armed forces and is regarded as an important work of historical significance in the law of war.[60]

The Lieber Code established rules governing martial law, military jurisdiction, the treatment of spies and deserters, and the treatment of POWs. Many key law of war principles, such as the principle of military necessity, were codified in the Lieber Code.[61]However, parts of the Lieber Code reflect 19th century understandings of the law of war that have been modified by treaties that the United States has ratified or by subsequent customary international law. For example, the Lieber Code permitted the denial of quarter in certain circumstances.[62]However, denying quarter in those circumstances is no longer acceptable.[63]

The Lieber Code was prepared during the Civil War. The Confederate forces agreed with some provisions of the Lieber Code, but disagreed with others.[64]

The Lieber Code reflected rules for "regular war" or what today would be classified as international armed conflict. Such rules were applied to the Confederate forces for humanitarian reasons, even though the United States did not recognize the Confederacy as a legitimate government or State.[65]In this way, the Lieber Code is an example of the application of the doctrine of recognition of belligerency.[66]

19.4 1856 PARIS DECLARATION RESPECTING MARITIME LAW

The 1856 Paris Declaration respecting Maritime Law is an early multilateral law of war treaty that was intended to be open to accession by all States, including States that did not participate in its negotiation.[67]This treaty illustrates how law of war treaties may be written with a view towards being able to be accepted and applied by all States. The United States is not a Party to the 1856 Paris Declaration. The 1856 Paris Declaration's provision that blockades must be effective in order to be binding reflects customary international law.[68]

19.5 1864 GWS

The Geneva Convention for the Amelioration of the Wounded in Armies in the Field of August 22, 1864 (1864 GWS) was one of the earliest multilateral law of war treaties.[69]The 1864 GWS provided for the use of the red cross as a distinctive emblem to help identify medical personnel.[70]It also provided for wounded and sick combatants to be collected and cared for regardless of the nation of their armed forces.[71]

The United States acceded to the 1864 GWS on March 1, 1882.[72]The 1864 GWS was replaced by the GWS in relations between the Parties to the GWS.[73]

19.6 1868 ST. PETERSBURG DECLARATION

The Declaration Renouncing the Use, in Time of War, of Explosive Projectiles Weighing Under 400 Grams Weight (1868 St. Petersburg Declaration) was promulgated by an international conference held in St. Petersburg, Russia on December 11, 1868. The 1868 St. Petersburg Declaration prohibits Parties from, "in case of war among themselves, the employment, by their military or naval forces, of any projectile of less weight than four hundred grammes, which is explosive, or is charged with fulminating or inflammable substances."[74]

The United States is not a Party to the 1868 St. Petersburg Declaration. The language in the Preamble of the 1868 St. Petersburg Declaration that notes that the "employment of arms which uselessly aggravate the sufferings of disabled men" would "be contrary to the laws of humanity" is an early statement of the prohibition against weapons calculated to cause unnecessary suffering, which is a principle that is found in treaties to which the United States is a Party and in customary international law.[75]

The prohibition in the Declaration against "any projectile of less weight than four hundred grammes, which is explosive, or is charged with fulminating or inflammable substances" does not reflect customary international law.[76]For example, for many decades without legal controversy, States have used, and continue to use, tracer ammunition, grenades, explosive bullets, or other projectiles of less weight than four hundred grams with a burning or explosive capability.[77]

19.7 1899 AND 1907 HAGUE DECLARATIONS ONWEAPONS

19.7.1 1899 Declaration on Expanding Bullets. The 1899 Declaration on Expanding Bullets prohibits Parties from using "bullets which expand or flatten easily in the human body, such as bullets with a hard envelope which does not entirely cover the core or is pierced with incisions."[78]The Declaration only creates obligations for Parties to the Declaration in international armed conflicts in which all the parties to the conflict are also Parties to the Declaration.[79]

The United States is not a Party to the 1899 Declaration on Expanding Bullets and does not regard the 1899 Declaration on Expanding Bullets as customary international law applicable in either international or non-international armed conflicts.[80]

19.7.2 1899 Declaration on Asphyxiating Gases. The 1899 Declaration on Asphyxiating Gases was concluded at The Hague on July 29, 1899.[81]This 1899 Declaration prohibits the use of projectiles the object of which is the diffusion of asphyxiating or deleterious gases.[82]The United States did not ratify this declaration.

This declaration was followed by the 1922 Washington Treaty on Submarines and Noxious Gases, and the 1925 Geneva Gas and Bacteriological Protocol.[83]The United States is a Party to the 1925 Geneva Gas and Bacteriological Protocol and applies the broader prohibition in it on the use of asphyxiating, poisonous, or other gases, and all analogous liquids, materials, or devices.[84]

19.7.3 1899 and 1907 Declarations on the Discharge of Projectiles and Explosives From Balloons. In the 1899 Hague (IV, 1), Parties agreed to prohibit, for a term of five years, the launching of projectiles and explosives from balloons, or by other new methods of similar nature.[85]The United States deposited its instrument of ratification to the 1899 Hague Declaration (IV, 1) on September 4, 1900.[86]

The 1907 Hague Declaration (XIV) Prohibiting the Discharge of Projectiles and Explosives from Balloons sought to renew the expired 1899 Hague Declaration (IV, 1).[87]The United States deposited its instrument of ratification to the 1907 Hague Declaration (XIV) on November 27, 1909.[88]This treaty was intended to have effect until the end of the Third Hague Peace Conference.[89]The Third Hague Peace Conference never met due to the outbreak of World War I.[90]The application of the 1907 Hague Declaration (XIV) is limited, *inter alia*, due to its general participation clause providing that it only applies if all the parties to the conflict are also Parties to the Declaration.[91]During World War II, the War Department took the position that the 1907 Hague Declaration (XIV) is not binding and will not be observed.[92]

19.8 1899 HAGUE II AND 1907 HAGUE IV CONVENTIONS AND ANNEXED REGULATIONS REGARDING LAND WARFARE

19.8.1 1899 Hague II. The Hague Convention II with Respect to the Laws and Customs of War on Land of July 29, 1899 (1899 Hague II), with its annexed Regulations, was an early multilateral law of war treaty that contains provisions that were incorporated into later law of war treaties. Article 1 of the 1899 Hague II Regulations addressed the criteria for militia and volunteer corps to qualify as lawful belligerents and for their members to be entitled to POW status if captured; these criteria are repeated in Article 4 of the GPW.[93]In addition, Articles 4-20 of the 1899 Hague II Regulations provide basic rules for the care and protection of POWs. The United States deposited its instrument of ratification to the 1899 Hague II on April 5, 1902.[94]

The 1899 Hague II was followed by Hague IV, which replaces the 1899 Hague II as between Parties to Hague IV.[95]Not all the States that ratified the 1899 Hague II have also ratified Hague IV.[96]

19.8.2 Hague IV. States sought to expand upon and clarify provisions of 1899 Hague II and the 1899 Hague II Regulations through the Hague Convention IV Respecting the Laws and Customs of War on Land of October 18, 1907 (Hague IV), and annexed Regulations (Hague IV Regulations).[97]

The United States deposited its instrument of ratification to Hague IV on November 27, 1909.[98]

19.8.2.1 *Hague IV and Customary International Law*. Provisions of Hague IV and Hague IV Regulations have been found to reflect customary international law.[99]For example, Article 42 of the Hague IV Regulations, which provides a standard for when the law of belligerent occupation applies, is regarded as customary international law.[100]

19.8.2.2 *Relationship Between the 1907 Hague IV and the 1899 Hague II*. The Hague IV, duly ratified, shall as between its Parties be substituted for the 1899 Hague II. The 1899 Hague II remains in force as between its Parties that do not also ratify Hague IV.[101]Martens Clause. The Preamble to the 1899 19.8.3 Hague II contains a clause known as the Martens Clause because of its association with a delegate to the Hague Peace Conference, F.F. de Martens:[102]

Until a more complete code of the laws of war is issued, the High Contracting Parties think it right to declare that in cases not included in the Regulations adopted by them, populations and belligerents remain under the protection and empire of the principles of international law, as they result from the usages established between civilized nations, from the laws of humanity, and the requirements of the public conscience.[103]

A similar formulation was included in subsequent treaties, such as the Hague IV, the 1949 Geneva Conventions, and the CCW.[104]

The Martens Clause was a compromise following difficult and unresolved debates at the 1899 Hague Peace Conference about the status of resistance fighters in occupied territory.[105]

However, the language of the clause is not limited to that specific context, and the Martens Clause has been cited in many other contexts.[106]

The Martens clause reflects the idea that when no specific rule applies, the principles of the law of war form the general guide for conduct during war.[107]

19.9 1907 HAGUE X

On November 27, 1909, the United States deposited its instrument of ratification to the 1907 Hague Convention (X) for the Adaption to Maritime Warfare of the Principles of the Geneva Convention of October 18, 1907.[108]

This treaty was followed by the GWS-Sea, which replaced it in relations between Parties to the GWS-Sea.[109]

19.10 1922WASHINGTON TREATY ON SUBMARINES AND NOXIOUS GASES

The United States signed the Treaty Relating to the Use of Submarines and Noxious Gases in Warfare on February 6, 1922.[110]Article 6 of the treaty provided that the treaty shall take effect on the deposit of all the ratifications.[111]France did not ratify the treaty, and it did not enter into force.[112]

The language of Article 5 of the 1922 Washington Treaty dealing with gas warfare corresponds to language in the 1925 Protocol for the Prohibition of the Use in War of Asphyxiating, Poisonous or Other Gases, and of Bacteriological Methods of Warfare signed at Geneva, June 17, 1925.[113]The 1925 Geneva Gas and Bacteriological Protocol, however, also prohibits bacteriological methods of warfare.[114]

19.11 1923 HAGUE AIR AND RADIO RULES

In 1922, the Conference on the Limitation of Armament at Washington adopted a resolution for the appointment of a Commission representing the United States of America, the British Empire, France, Italy, and Japan, and later, the Netherlands, to consider the application of the existing law of war rules to new agencies of warfare and whether changes should be adopted.[115]It was agreed that the program of the Commission would be limited to the preparation of rules relating to aerial warfare, and to rules relating to the use of radio in time of war.[116]With John Bassett Moore as a U.S. Delegate and the President of the Commission, the Commission prepared a set of rules for the control of radio in time of war, as well as a set of rules for aerial warfare.[117]

The 1923 Hague Air Rules were not, however, subsequently adopted as a treaty by the United States. Some provisions in the 1923 Hague Air Rules may reflect customary international law. For example, the 1923 Hague

Air Rules recognize the permissibility of using tracer, incendiary, or explosive projectiles by or against aircraft, including by States that are Parties to the 1868 Declaration of St. Petersburg.[118]

Many of the other provisions in the 1923 Hague Air Rules, however, do not reflect customary international law. For example, the 1923 Hague Air Rules provide that the crew of military aircraft must be exclusively military.[119]The 1949 Geneva Conventions, however, contemplate that crews of military aircraft may include civilian members.[120]As another example, certain efforts in the 1923 Hague Air Rules to limit the effects of attacks also do not reflect customary international law.[121]

19.12 1925 GENEVA GAS AND BACTERIOLOGICAL PROTOCOL

The Geneva Protocol for the Prohibition of the Use in War of Asphyxiating, Poisonous or Other Gases, and of Bacteriological Methods of Warfare, of June 17, 1925, prohibits the use of asphyxiating, poisonous, or other gases, and all analogous liquids, materials, or devices, and the use of bacteriological methods of warfare.[122]

This treaty followed the 1899 Declaration on Asphyxiating Gases and the 1922 Washington Treaty on Submarines and Noxious Gases.[123]This treaty also followed widespread use of chemical weapons in World War I.

On November 25, 1969, President Nixon announced the intention of the administration to seek Senate advice and consent to ratification of the 1925 Geneva Gas and Bacteriological Protocol as part of U.S. policies relating to chemical and biological weapons.[124]The United States deposited its instrument of ratification to the 1925 Geneva Gas and Bacteriological Protocol on April 10, 1975.[125]

The United States took a reservation that the "Protocol shall cease to be binding on the government of the United States with respect to the use in war of asphyxiating, poisonous or other gases, and of all analogous liquids, materials, or devices, in regard to an enemy state if such state or any of its allies fails to respect the prohibitions laid down in this Protocol."[126]This reservation would permit use by the United States of chemical weapons and agents in response, but would not limit in any way the Protocol's prohibition with respect to biological weapons.[127]

Other States, including France, Belgium, Canada, the USSR (now Russia), and the United Kingdom issued similar statements upon ratification.[128]

This treaty was followed by the Chemical Weapons Convention and Biological Weapons Convention, which, *inter alia*, place restrictions on the use, development, production, and possession of weapons addressed by the 1925 Geneva Gas and Bacteriological Protocol.[129]

19.13 1929 GENEVA CONVENTIONS

19.13.1 1929 GWS. The United States deposited its instrument of ratification to the 1929 GWS on February 4, 1932.[130]

The 1929 GWS was replaced by the GWS in relations between Parties to the GWS.[131]

All the Parties to the 1929 GWS have become Parties to the GWS.

19.13.2 1929 GPW. The United States deposited its instrument of ratification to the 1929 GPW on February 4, 1932.[132]

The 1929 GPW was replaced by the GPW in relations between Parties to the GPW.[133]

All the Parties to the 1929 GPW have become Parties to the GPW. Provisions of the 1929 GPW were found to reflect customary international law by war crimes tribunals after World War II.[134]The 1929 GPW may be relevant to understanding provisions of the 1949 GPW because some provisions of the 1949 GPW were drawn from the 1929 GPW or reflect an effort to improve upon the 1929 GPW.

19.14 1930 LONDON TREATY FOR THE LIMITATION AND REDUCTION OF NAVAL ARMAMENT AND 1936 LONDON PROTOCOL

Article 22 of the 1930 London Treaty for the Limitation and Reduction of Naval Armament (1930 London Treaty) and the 1936 London Protocol set forth identical rules regarding submarine warfare, and the obligations of surface warships and submarines with respect to the sinking of merchant vessels, including the obligation to put merchant vessels' passengers and crew in a place of safety.[135]

The United States deposited its instrument of ratification to the 1930 London Treaty on October 27, 1930.[136]Although other aspects of the treaty expired on December 31, 1936, Article 22 remains in force.[137]Signatories to the 1930 London Treaty desired that as many States as possible accept the rules in Article

22; thus, they concluded the 1936 London Protocol with the same language as that in Article 22 and invited other States to accede to the 1936 London Protocol.[138] The United States signed the 1936 Protocol on November 6, 1936.[139]

19.15 1935 ROERICH PACT

The 1935 Roerich Pact was concluded in Washington on April 15, 1935.[140] It provides for the respect and protection of "historic monuments, museums, scientific, artistic, educational and cultural institutions" and their personnel in time of peace as well as in war.[141] Such institutions and personnel receive protection as cultural property under the 1954 Hague Cultural Property or the general protection afforded civilian objects and persons.[142]

The United States deposited its instrument of ratification to the 1935 Roerich Pact on July 13, 1935.[143]

This treaty was followed by the 1954 Hague Cultural Property Convention.[144] All Parties to the Roerich Pact are Parties to the 1954 Hague Cultural Property Convention. The 1954 Hague Cultural Property Convention is supplementary to the Roerich Pact, but the distinctive emblem of the 1954 Hague Cultural Property Convention is to be used instead of the distinguishing flag described in Article III of the 1935 Roerich Pact, in situations where the 1954 Hague Cultural Property Convention and the Regulations for its execution provide for the use of this distinctive emblem.[145]

19.16 1949 GENEVA CONVENTIONS

The four 1949 Geneva Conventions were adopted at a Diplomatic Conference at Geneva on August 12, 1949. More than 193 States have ratified the 1949 Geneva Conventions.[146] The United States signed each of the 1949 Geneva Conventions on August 12, 1949. The United States deposited its instrument of ratification to each convention on August 2, 1955.[147] The United States has viewed many of the protections embodied in the 1949 Geneva Conventions as reflecting appropriate U.S. practice in armed conflict regardless of formal treaty obligations.[148]

The 1949 Geneva Conventions followed earlier multilateral treaties that addressed the same subjects, including the 1864 GWS, the 1907 Hague X, the 1899 and 1907 Hague Conventions on the Law of Land Warfare, and the 1929 Geneva Conventions.[149]

The Commentaries to the 1949 Geneva Conventions, published by the International Committee of the Red Cross, under the general editorship of Jean S. Pictet, have often been helpful in understanding the provisions of the 1949 Geneva Conventions and have often been cited in this manual. As noted by the International Committee of the Red Cross, however, these Commentaries are not an official interpretation of the 1949 Geneva Conventions, which only participant States would be qualified to give.[150]

19.16.1 Common Provisions in the 1949 Geneva Conventions. The four 1949 Geneva Conventions contain a number of common provisions, *i.e.*, provisions that are substantively the same (if not identical in text) among the conventions. The four Geneva Conventions have this duplication, in part, because each convention is designed to be effective, even if a State only ratifies that particular convention.[151] For example, the list of persons entitled to receive POW status in Article 4 of the GPW is repeated in the GWS and GWS-Sea.[152]

19.16.1.1 *Common Terms in the 1949 Geneva Conventions – Notes on Terminology*. In the 1949 Geneva Conventions, "Power" generally refers to a State.[153]

In the 1949 Geneva Conventions, the *Detaining Power* refers to the State that holds the POW or internee.[154]

In the 1949 Geneva Conventions, the *Protecting Power* refers to a neutral State that helps implement the Conventions.[155]

19.16.1.2 *Common Article 2 of the 1949 Geneva Conventions*. Common Article 2 of the four 1949 Geneva Conventions declares that the provisions of each convention apply "to all cases of declared war or of any other armed conflict which may arise between two or more of the High Contracting Parties, even if the state of war is not recognized by one of them."[156] This language is used to help explain when the law of war applies.[157]

Common Article 2 also provides that each convention "shall also apply to all cases of partial or total occupation of the territory of a High Contracting Party, even if the said occupation meets with no armed resistance."[158] Common Article 2, thus, helps explain that the law of belligerent occupation is applicable, even if the occupying forces are not opposed by force.[159]

The 1954 Hague Cultural Property Convention repeats language used in Common Article 2, and the CCW incorporates language from Common Article 2 by reference.

19.16.1.3 *Common Article 3 of the 1949 Geneva Conventions*. Common Article 3 to the 1949 Geneva Conventions has been described as a "Convention in miniature" that addresses non-international armed conflict.[160]It was the first provision in a multilateral law of war treaty that addressed humane treatment during non-international armed conflict to gain widespread ratification by States.[161]

Although Common Article 3 only applies by its terms to non-international armed conflicts, Common Article 3 reflects a minimum yardstick of humane treatment protections that apply more broadly, including during international armed conflict.[162]

19.16.1.4 *POW Status and the 1949 Geneva Conventions*. Article 4 of the GPW explains which persons are entitled to receive POW status under the GPW.[163]This provision is mirrored in the GWS and GWS-Sea, as those conventions are intended to protect persons who are entitled to POW status under the GPW.[164]

The GC excludes from its protection those persons who are entitled to POW status under the GPW.[165]

19.16.1.5 *Retained Personnel and the 1949 Geneva Conventions*. Retained personnel are addressed by both the GPW and the GWS.[166]

19.16.2 GWS. The GWS is often called the first Geneva Convention.[167]The GWS addresses the following subjects on land: (1) the protection of the wounded, sick, and dead; (2) the rights, duties, and liabilities of military medical and religious personnel; and (3) the protection of military medical units, facilities, and transports.[168]

19.16.2.1 *Relationship Between the GWS and Earlier Conventions*. The GWS replaces the Conventions of August 22, 1864, July 6, 1906, and July 27, 1929, in relations between the Parties to the GWS.[169]

19.16.3 GWS-Sea. The GWS-Sea is often called the second Geneva Convention.[170]The GWS-Sea addresses the following subjects at sea: (1) the protection of the wounded, sick, shipwrecked, and dead; (2) the rights, duties, and liabilities of military medical and religious personnel; and (3) the protection of military medical units, facilities, and transports.[171]

19.16.3.1 *Relationship Between the 1907 Hague X and the GWS-Sea*. The GWSSea replaces the Hague X in relations between the Parties to the GWS-Sea.[172]

19.16.4 GPW. The GPW is often called the third Geneva Convention.[173]The GPW addresses the protection of POWs.[174]

19.16.4.1 *Relationship Between the GPW and the 1929 GPW*. The GPW replaces the 1929 GPW in relations between the Parties to the GPW.[175]

19.16.4.2 *Relationship Between the GPW and the 1899 Hague II or 1907 Hague IV*. In the relations between States that are bound by the 1899 Hague II or the 1907 Hague IV, and that are Parties to the GPW, the GPW shall be complementary to Chapter II of the Regulations annexed to the 1899 Hague II or the 1907 Hague IV.[176]

19.16.5 GC. The GC is often called the fourth Geneva Convention.[177]The GC addresses the protection of civilians in the hands of a party a conflict, including civilian internees.[178]The GC also addresses belligerent occupation.

19.16.5.1 *Application of Different Parts of the GC*. Different parts of the GC apply to different situations.

In general, the GC uses the concept of protected person to define the individuals who are entitled to receive its protections.[179]The provisions of Part II of the GC (articles 13-26) are wider in application; they do not only apply to those persons who are protected persons under the GC.[180]These provisions cover the whole of the populations of the countries in conflict.[181]Section I (articles 27-34) of Part III of the GC includes provisions that are common to the home territories of the parties to the conflict and to occupied territories.

Section II (articles 35-46) of Part III of the GC addresses aliens in the home territory of a party to the conflict.

Section III (articles 47-78) of Part III of the GC addresses occupied territories.

Section IV (articles 79-135) of Part III of the GC provides regulations for the treatment of internees.

19.16.5.2 *Relationship Between the GC and the 1899 Hague II and the 1907 Hague IV Conventions*. In the relations between States that are bound by the 1899 Hague II, or the 1907 Hague IV, and that are Parties to the GC, the GC shall be supplementary to Sections II (Articles 22-41 – Hostilities) and III (Articles 42-56 – Military Authority Over the Territory of the Hostile State) of the Regulations annexed to the 1899 Hague II or the 1907 Hague IV.[182]

19.17 1954 HAGUE CULTURAL PROPERTY CONVENTION

The 1954 Hague Cultural Property Convention addresses the protection of cultural property during international armed conflict.[183]It addresses personnel who are engaged in duties related to the protection of cultural property.[184]It also has provisions that apply during occupation.[185]Some provisions of the 1954 Hague Cultural Property Convention apply to noninternational armed conflict.[186]

The 1954 Hague Cultural Property Convention was adopted at a diplomatic conference on May 14, 1954.[187]The United States deposited its instrument of ratification to the Hague Cultural Property Convention on March 13, 2009, and stated four understandings.[188]

Two Protocols to the 1954 Hague Cultural Property Convention have been adopted.[189]

The United States has neither signed nor ratified either one.

19.17.1 Relationship Between the 1954 Hague Cultural Property Convention and Earlier Treaties.

19.17.1.1 *Relationship Between the 1954 Hague Cultural Property Convention and Certain 1899 and 1907 Hague Conventions*. In the relations between States that are bound by the Conventions of The Hague concerning the Laws and Customs of War on Land (IV) (Hague IV) and Concerning Naval Bombardment in Time of War (Hague IX), whether those of July 29, 1899, or those of October 18, 1907, and that are Parties to the 1954 Hague Cultural Property Convention, the 1954 Hague Cultural Property Convention shall be supplementary to the Hague IX Convention and to the Regulations annexed to the Hague IV Convention.[190]

19.17.1.2 *Relationship Between the 1935 Roerich Pact and the 1954 Hague Cultural Property Convention*. In the relations between States that are bound by the 1935 Roerich Pact and that are Parties to the 1954 Hague Cultural Property Convention, the 1954 Hague Cultural Property Convention shall be supplementary to the Roerich Pact and shall substitute for the distinguishing flag described in Article III of the Roerich Pact the emblem defined in Article 16 of the 1954 Hague Cultural Property Convention—in cases in which the 1954 Hague Cultural Property Convention and the Regulations for its execution provide for the use of this distinctive emblem.[191]

19.18 VIENNA CONVENTION ON THE LAW OF TREATIES

The Vienna Convention on the Law of Treaties (VCLT) was adopted on May 23, 1969, by the U.N. Conference on the Law of Treaties. The United States signed the VCLT on April 24, 1970. President Nixon transmitted the VCLT to the Senate for its advice and consent to ratification on November 22, 1971.[192]

The United States is not a Party to the VCLT but has viewed many of its provisions as reflecting customary international law.

19.19 BIOLOGICAL WEAPONS CONVENTION

After negotiations at the Conference of the Committee on Disarmament at Geneva and at the United Nations, the Biological Weapons Convention was opened for signature on April 10, 1972.[193]

The United States deposited its instrument of ratification on March 26, 1975.[194]

The Biological Weapons Convention followed the 1925 Geneva Gas and Bacteriological Protocol.[195]

In implementation of the Biological Weapons Convention, Congress enacted legislation prohibiting development, production, transfer, acquisition, or possession of a biological agent, toxin, or delivery system for use as a weapon.[196]

19.20 1977 ADDITIONAL PROTOCOLS TO THE 1949 GENEVA CONVENTIONS

A diplomatic conference between 1974 and 1977 negotiated two protocols to the 1949 Geneva Conventions that addressed international armed conflict (AP I) and non-international armed conflict (AP II).

Because these treaties were negotiated at the same time, they may provide information on States' views on differences between the rules applicable in international armed conflict and the rules applicable in non-international armed conflict.[197]

19.20.1 AP I. Protocol (I) Additional to the Geneva Conventions, and Relating to the Protection of Victims of International Armed Conflicts (AP I) contains rules on the conduct of hostilities during international armed conflict that are intended to supplement the 1949 Geneva Conventions.[198]For example, AP I provides additional protections for the wounded and sick, and provides for the establishment of an international humanitarian fact-finding commission.[199]

AP I, however, does not apply to the use of nuclear weapons.[200]

The United States signed AP I and stated two understandings.[201] As explained below, AP I is a significant law of war treaty that the United States has decided not to ratify. On January 28, 1987, President Reagan informed the Senate that AP I would not be submitted for the Senate's advice and consent to ratification because it is "fundamentally and irreconcilably flawed."[202] However, President Reagan noted that the United States would support "the positive provisions of Protocol I that could be of real humanitarian benefit if generally observed by parties to international armed conflicts."[203]

This manual references AP I provisions, some of which are consistent with DoD practice. Unless explicitly noted, no determinations are made about whether any of these AP I provisions reflect customary international law.

19.20.1.1 *Examples of AP I Provisions Incorporated Into Other Treaties That the United States Has Accepted.* Some AP I provisions have been incorporated into later treaties to which the United States is a Party. For example, AP I's definition of military objective in Article 51(2) is substantially similar to the definition in Article 2(6) of CCW Amended Mines Protocol and Article 1(3) of CCW Protocol III on Incendiary Weapons.2[204] Similarly, requirements under the Child Soldiers Protocol and U.S. law are comparable to AP I's requirements with respect to child soldiers.[205]

19.20.1.2 *Examples of AP I Provisions That Are Consistent With Longstanding U.S. Practice.* Certain provisions of AP I may not reflect customary international law, but may be consistent with longstanding U.S. practice. For example, AP I requires that Parties to AP I ensure that legal advisers are available to advise military commanders, and the United States has long employed legal advisers in this role.[206] Similarly, AP I requires that Parties to AP I undertake a legal review of, *inter alia*, new weapons, and the DoD policy and practice of conducting weapons reviews preceded this provision of AP I.[207]

19.20.1.3 *Examples of AP I Provisions That the United States Has Supported.* The United States has expressed support for certain AP I provisions. For example, United States has supported and sought to promote as a matter of national policy fundamental guarantees for the treatment of persons detained during international armed conflict that are reflected in Article 75 of AP I.[208]

19.20.1.4 *Examples of AP I Provisions Based on a Principle That the U.S. Supports, Even Though the Provision Is Not Necessarily Customary International Law Nor Militarily Acceptable in All Respects.* In some cases, the United States has supported a principle underlying an AP I provision, but the provision may not be customary international law or militarily acceptable in all respects. For example, the United States has supported the principle recognized in AP I that journalism is generally to be regarded as a civilian activity.[209]

Even where the United States has expressed support for an underlying customary principle that AP I provisions are based upon, the United States may disagree that the language of the provision reflects customary international law. For example, with regard to "direct participation in hostilities," the United States has supported the customary principle underlying Article 51(3) of AP I, but has disagreed that the language of that provision reflects customary international law.[210] Similarly, the intricacy of the provisions of AP I on objects indispensable to the survival of the civilian population make it doubtful that such provisions could be characterized as customary international law, although the United States has supported the principle that the starvation of civilians not be used as a method of warfare.[211] In addition, although the United States has supported the principle that civilian civil defense organizations and their personnel be respected and protected as civilians, a number of military operational problems have been identified with respect to the system of protection for civil defense

established by AP I.[212]

19.20.1.5 *Examples of AP I Provisions to Which the United States Has Objected.* The United States has objected to certain provisions of AP I. For example, the United States has objected to AP I provisions on:

- national liberation movements;[213]
- mercenaries;[214]
- works and installations containing dangerous forces;[215]
- criteria for lawful combatant status;[216]
- the obligation of combatants to distinguish themselves;[217]
- environmental protection;[218]
- presumptions in favor of civilian status in conducting attacks;[219]

• prohibiting the use of enemy flags, insignia, or uniforms to shield, favor, protect, or impede military operations;[220]and

• reprisals.[221]

19.20.2 AP II. Protocol (II) Additional to the Geneva Conventions, and Relating to the Protection of Victims of Non-International Armed Conflict (AP II) addresses certain types of non-international armed conflicts.[222]AP II reaffirmed, elaborated upon, and supplemented the protections contained in Common Article 3 of the 1949 Geneva Conventions.[223]

19.20.2.1 *The United States and AP II.* The United States is not a Party to AP II. The United States signed AP II and stated one understanding on December 12, 1977.[224]President Reagan transmitted AP II to the Senate for its advice and consent to ratification on January 29, 1987.[225]

After the Supreme Court's decision in *Hamdan v. Rumsfeld*, 548 U.S. 557 (2006), the Bush Administration moved AP II down on the Executive's treaty priority list, indicating that it did not seek Senate action on the treaty at that time, so that the Administration could study the treaty in light of the decision. On March 7, 2011, after interagency review, the Obama Administration urged the Senate to act as soon as practicable on AP II.[226]

Although the United States is not a Party to AP II, reviews have concluded that the provisions of AP II are consistent with U.S. practice, and that any issues could be addressed with reservations, understandings, and declarations.[227]

19.20.2.2 *AP II's Requirements for Its Application*. By its terms, AP II does not apply to all armed conflicts not of an international character. Instead, according to its terms, AP II only applies to armed conflicts that are not covered by Article 1 of AP I and that take place in the territory of a Party to AP II between its armed forces and dissident armed forces or other organized armed groups, which, under responsible command, exercise such control over a part of its territory as to enable them to carry out sustained and concerted military operations and to implement AP II.[228]

This threshold limits the application of AP II to a subset of armed conflicts that would be covered by Common Article 3 of the 1949 Geneva Conventions. For example, AP II would only cover conflicts in which dissident armed forces or other organized armed groups are organized and disciplined like regular armed forces.[229]In addition, non-State armed groups must be in a position to implement AP II before it is deemed to apply to all parties to the conflict.[230]

Although AP II specifies more strict requirements for its application, President Reagan, in submitting AP II to the Senate for its advice and consent to ratification, recommended that the United States take the position that it would apply AP II to all armed conflicts that would be covered by Common Article 3 of the 1949 Geneva Conventions.[231]

19.21 CCW, CCWAMENDED ARTICLE 1, AND CCWPROTOCOLS

19.21.1 CCW. The Convention on Certain Conventional Weapons (CCW) and Protocols I, II, and III were adopted at Geneva on October 10, 1980. Subsequent conferences adopted: (1) an amended Protocol II; (2) Protocols IV and V; and (3) an amendment to apply the CCW and Protocols I, II, III, and IV to non-international armed conflicts.

Protocols to the CCW are subject to ratification or other acceptance by a State before such protocols enter into force for that State. Additional protocols that have been adopted by a Conference of States Parties enter into force as provided in paragraphs 3 and 4 of Article 5 of the CCW.[232]Paragraphs 3 and 4 of Article 5 of the CCW provide, *inter alia*, that each of the CCW's Protocols may only enter into force for a State after that State has notified its consent to be bound by it.[233]Thus, a State's notification of its consent to be bound by a Protocol is a prerequisite to the Protocol's entry into force for that State.

Similarly, amendments to the CCW or its Protocols are subject to ratification or other acceptance by a State before such amendments enter into force for that State. Amendments to the CCW or its Protocols are adopted and enter into force in the same manner as the CCW and its Protocols.[234]As discussed above, a State must provide notification of its consent to be bound by a CCW Protocol before that Protocol's entry into force for that State. Similarly, the CCW is subject to ratification or other acceptance by States.[235]And, the CCW only enters into force for a State after that State has deposited its instrument of ratification or other acceptance.[236]Thus, a State's acceptance of any amendment to the CCW or its Protocols is a prerequisite to that amendment's entry into force for that State.

The United States is a Party to the: (1) CCW; (2) CCW Protocol I; (3) CCW Amended Mines Protocol; (4) CCW Protocol III on Incendiary Weapons; (5) CCW Protocol IV on Blinding Laser Weapons; and (6) CCW Protocol V on Explosive Remnants of War. The United States deposited its instrument of ratification to the CCW and accepted Protocols I and II on March 25, 1995. The United States provided notification of its consent to be bound by the CCW Amended Mines Protocol on May 24, 1999. The United States deposited its instrument of ratification to the Amendment to Article 1 of the CCW, and provided notification of its consent to be bound by Protocols III, IV, and V, on January 21, 2009.

19.21.1.1 *CCW Amended Scope of Application*. For the United States, the CCW and its Protocols apply to non-international armed conflict.[237]

On December 21, 2001, Parties to the CCW adopted an amendment to Article 1 of the CCW that modified the application of the CCW and Protocols I, II, III, and IV to apply to noninternational armed conflicts.[238]The CCW Amended Mines Protocol and CCW Protocol V on Explosive Remnants of War also apply to non-international armed conflicts.[239]The amended scope of the CCW does not prejudice additional protocols to the CCW adopted after January 1, 2002, which may apply, exclude, or modify the scope of their application in relation to Article 1 of the CCW.[240]

The United States deposited its instrument of ratification to the amendment to Article 1 of the CCW on January 21, 2009.[241]

U.S. policy before the entry into force for the United States of the amendment to Article 1 of the CCW was to apply the CCW, Protocol I, Protocol II, and CCW Protocol IV on Blinding Laser Weapons to non-international armed conflict. At the time of the deposit of U.S. instruments of ratification of the CCW, CCW Protocol I, and CCW Protocol II, the United States declared its intent to apply the provisions of those instruments to all armed conflicts referred to in Articles 2 and 3 of the GPW, *i.e.*, both international and non-international armed conflicts.[242]

Similarly, before the United States ratified the amendment to Article 1 of the CCW, U.S. policy was to apply the CCW Protocol on Blinding Laser Weapons to all armed conflicts and in peacetime.[243]

19.21.1.2 *U.S. Reservation to Article 7(4)(b) of the CCW*. The United States took a reservation to Article 7(4)(b) of the CCW.[244]

Article 7(4) of the CCW provides: This Convention, and the annexed Protocols by which a High Contracting Party is bound, shall apply with respect to an armed conflict against that High Contracting Party of the type referred to in Article 1, paragraph 4, of Additional Protocol I to the Geneva Conventions of 12 August 1949 for the Protection of War Victims: ...

(b) where the High Contracting Party is not a party to Additional Protocol I and an authority of the type referred to in subparagraph (a) above accepts and applies the obligations of the Geneva Conventions and of this Convention and the relevant annexed Protocols in relation to that conflict. Such an acceptance and application shall have in relation to that conflict the following effects: (i) the Geneva Conventions and this Convention and its relevant annexed Protocols are brought into force for the parties to the conflict with immediate effect;

(ii) the said authority assumes the same rights and obligations as those which have been assumed by a High Contracting Party to the Geneva Conventions, this Convention and its relevant annexed Protocols; and

(iii) the Geneva Conventions, this Convention and its relevant annexed Protocols are equally binding upon all parties to the conflict.

The High Contracting Party and the authority may also agree to accept and apply the obligations of Additional Protocol I to the Geneva Conventions on a reciprocal basis.

The U.S. reservation to Article 7(4)(b) of the CCW is consistent with the longstanding objections of the United States to Article 1(4) of AP I regarding national liberation movements.[245]

19.21.1.3 *CCW and Customary International Law*. The restrictions and prohibitions in the CCW and its Protocols on weapons generally have not been based on the premise that such restrictions or prohibitions are warranted because the weapons at issue are calculated to cause superfluous injury or are inherently indiscriminate.[246]Rather, the CCW and Protocols have been drafted on the basis that they are largely new contractual rules, and not a restatement of existing customary international law.[247]

19.21.2 CCW Protocol I. CCW Protocol I prohibits the use of weapons whose primary effect is to injure by fragments not detectable by X-rays.[248]

CCW Protocol I was adopted by the first CCW Conference in 1980. The United States accepted CCW Protocol I on March 24, 1995, when it deposited its instrument of ratification to the CCW.[249]

19.21.3 CCW Amended Mines Protocol. The CCW Amended Mines Protocol addresses landmines, booby-traps, and other devices.[250]

CCW Protocol II on Mines, Booby-traps and Other Devices was adopted by the first CCW Conference in 1980. The United States accepted CCW Protocol II on March 24, 1995, when it deposited its instrument of ratification to the CCW.[251]

CCW Protocol II was later substantially amended by the first Review Conference of CCW States Parties, which concluded its work on May 3, 1996. The CCW Amended Mines Protocol includes more detailed restrictions on the use of mines, booby-traps, and other devices, and prohibitions on specific types of such devices. The CCW Amended Mines Protocol also applies to non-international armed conflicts.[252]The CCW Amended Mines Protocol has provisions that are to be observed at all times, *e.g.*, provisions that are to be implemented in peace-time.[253]

Before the United States accepted the CCW Amended Mines Protocol, U.S. policy was to observe the requirements of the CCW Amended Mines Protocol to the fullest extent possible.[254]

The CCW Amended Mines Protocol entered into force on December 3, 1998.[255]The United States provided notification of its consent to be bound by the CCW Amended Mines Protocol with declarations and a reservation on May 24, 1999.[256]

19.21.4 CCW Protocol III on Incendiary Weapons. CCW Protocol III on Incendiary Weapons places certain restrictions on the use of incendiary weapons.[257]

CCW Protocol III on Incendiary Weapons was adopted by the first CCW Conference in 1980. The United States provided notification of its consent to be bound by CCW Protocol III on Incendiary Weapons on January 21, 2009, with a reservation and an understanding.[258]

19.21.5 CCW Protocol IV on Blinding Laser Weapons. CCW Protocol IV on Blinding Laser Weapons prohibits the use of blinding laser weapons and requires that feasible precautions be taken in the employment of laser systems to avoid causing permanent blindness.[259]

States Parties to the CCW adopted CCW Protocol IV on Blinding Laser Weapons at the first Review Conference of CCW States Parties, which concluded its work on May 3, 1996.[260]

This Protocol was consistent with DoD policy on blinding lasers before the Protocol was adopted.[261]

The United States provided notification of its consent to be bound by CCW Protocol IV on Blinding Laser Weapons on January 21, 2009, with an understanding.[262]

U.S. legal reviews before CCW Protocol IV was adopted concluded that the use of a laser for the purpose of blinding an enemy combatant would not constitute unnecessary suffering under the law of war.[263]Similarly, the restrictions in CCW Protocol IV on Blinding Laser Weapons may be characterized as arms control obligations in that these restrictions do not establish that the use of lasers as anti-personnel weapons, including for the purpose of blinding, is prohibited by the rule against weapons calculated to cause unnecessary suffering nor otherwise prohibited by the customary law of war.[264]

19.21.6 CCW Protocol V on Explosive Remnants of War. CCW Protocol V on Explosive Remnants of War addresses explosive ordnance and explosive remnants of war.[265]

CCW Protocol V on Explosive Remnants of War contains no restrictions or prohibitions on the use of these weapons; rather, it addresses what must be done with respect to unexploded munitions that threaten civilians and post-conflict reconstruction.[266]CCW Protocol V on Explosive Remnants of War addresses primarily the steps to be taken before or after hostilities, not during them, and it includes a Technical Annex of suggested best practices for greater munitions reliability.[267]

CCW Protocol V on Explosive Remnants of War was adopted on November 28, 2003.[268]

The United States provided notification of its consent to be bound by CCW Protocol V on Explosive Remnants of War on January 21, 2009.[269]The United States expressed its understanding relating to future arrangements in connection with the settlement of armed conflicts.[270]

19.22 CHEMICAL WEAPONS CONVENTION

The Chemical Weapons Convention creates extensive obligations for States with respect to the use and possession of chemical weapons.[271]

Article 22 of the Chemical Weapons Convention prohibits reservations to the Convention and prohibits reservations to the Convention's Annexes incompatible with its object and purpose.[272]

During the negotiation of the Chemical Weapons Convention, President George H.W. Bush on May 13, 1991, announced that the United States was formally forswearing the use of chemical weapons for any reason, including retaliation, against any State, effective when the Convention entered into force.[273]

The United States signed the Chemical Weapons Convention on January 13, 1993, and President Clinton submitted it to the Senate for its advice and consent to ratification on November 23, 1993.[274]In transmitting the Chemical Weapons Convention to the Senate for its advice and consent to ratification, President Clinton notified the Senate that the administration was reviewing the effect of the Convention's prohibition on the use of riot control agents as a method of warfare on Executive Order No. 11850, which specified the current policy of the United States with regard to the use of riot control agents in war, and would submit the results of that review separately to the Senate.[275]The review concluded that under then-current interpretations of the Chemical Weapons Convention, certain uses of riot control agents authorized under Executive Order 11850 would no longer be permissible and that a new Executive Order on riot control agents would be issued.[276]The Senate, as a condition of its advice and consent to ratification, required that the President certify to Congress that the United States is not restricted by the Chemical Weapons Convention in its use of riot control agents in certain cases and that Executive Order 11850 would not be modified.[277]President Clinton provided this certification on April 25, 1997.[278]

The United States ratified the Chemical Weapons Convention on April 25, 1997, with one condition that relates to the Annex on Implementation and Verification.[279]The treaty requires that sixty-five nations become Party to it before it enters into force, which occurred on April 29, 1997.[280]

In implementation of the Chemical Weapons Convention, Congress enacted laws prohibiting, *inter alia*, development, production, acquisition, transfer directly or indirectly, receipt, stockpiling, retention, possession, use, or threat of use of any chemical weapon.[281]

19.23 ROME STATUTE OF THE INTERNATIONAL CRIMINAL COURT

The Rome Statute of the International Criminal Court ("Rome Statute") establishes the International Criminal Court ("ICC") at The Hague, Netherlands.[282]Under the Rome Statute, the ICC may investigate and prosecute certain international crimes (*e.g.*, genocide, war crimes) in certain situations.[283]

The United States signed the Rome Statute on December 31, 2000, but President Clinton did not submit the treaty to the Senate for its advice and consent to ratification. Because of "significant flaws," President Clinton recommended that his successor not submit the Rome Statute to the Senate until the United States' fundamental concerns with the treaty were addressed.[284]In particular, the United States has expressed concern that the ICC would claim jurisdiction over personnel of States that have not ratified the Rome Statute.[285]In addition, for example, concerns have been raised that the ICC would not provide U.S. persons accused of war crimes with the procedural protections that they are afforded under U.S. domestic law.[286]

On May 6, 2002, the United States filed a letter with the U.N. Secretary General, the depositary for the Rome Statute, indicating that it "does not intend to become a party to the treaty," and that "[a]ccordingly, the United States has no legal obligations arising from its signature on December 31, 2000."[287]The United States participated as an observer in the Rome Statute's first Review conference in Kampala, Uganda, May 30-June 11, 2010, and has supported ICC efforts in certain cases.[288]The Department of State Legal Adviser has noted that U.S. policy is explicitly not to frustrate the object and purpose of the Rome Statute.[289]

19.24 1999 U.N. SECRETARY GENERAL'S BULLETIN FOR U.N. FORCES

The U.N. Secretary General has promulgated a bulletin for the purpose of setting out fundamental principles and rules of international humanitarian law applicable to United Nations forces conducting operations under United Nations command.[290]The bulletin is not a legally binding instrument, although, in many cases, the rules reflected in the bulletin reflect law of war rules that are binding upon the United States as a matter of treaty or customary international law. The bulletin has been the subject of criticism by States.[291]

19.25 2005 ICRC STUDY ON CUSTOMARY INTERNATIONAL HUMANITARIAN LAW There is no list of treaty provisions (or other codification) that the United States has agreed reflects the entirety of the customary international law of war.[292]

In 2005, the International Committee of the Red Cross published a study on customary international humanitarian law (ICRC CIHL Study), which has been criticized for its methodology and formulation of certain specific rules.[293]The United States likewise has expressed serious concerns about the methodology used in the ICRC CIHL Study and has stated that it is not in a position to accept without further analysis the Study's conclusions that certain rules related to the law of war in fact reflect customary international law.[294]

19.26 AP III

A third additional protocol to the 1949 Geneva Conventions was adopted on December 8, 2005.[295]AP III recognizes a red crystal as an additional distinctive emblem for the same purposes as the distinctive emblems of the Geneva Conventions.[296]

The United States deposited its instrument of ratification to AP III on March 8, 2007, without making any statements of reservation, declaration, or understanding.297[297]

List of Abbreviations

To make the manual easier to read, the use of abbreviations has often been avoided, especially in the main text. Nonetheless, the following abbreviations of the titles of documents have been used for frequently cited documents.

1899 Hague II: Convention with Respect to the Laws and Customs of War on Land, Jul. 29, 1899, 32 Stat. 1803

1899 Hague II Reg.:Regulations Respecting the Laws and Customs of War on Land, annexed to Convention with Respect to the Laws and Customs of War on Land, Jul. 29, 1899, 32 Stat. 1803, 1811.

1909 Declaration of London: Declaration Concerning the Laws of Maritime War, Feb. 26, 1909, *reprinted in* James Brown Scott, *The Declaration of London, February 26, 1909: A Collection of Official Papers and Documents Relating to the International Naval Conference Held in London, December, 1908—February, 1909* (1919)

1914 Rules of Land Warfare: War Department, Office of the Chief of Staff, *Rules of Land Warfare* (Apr. 25, 1914)

1925 Geneva Gas and Bacteriological Protocol: Protocol for the Prohibition of the Use in War of Asphyxiating, Poisonous, or Other Gases, and of Bacteriological Methods of Warfare, Jun. 17, 1925, 94 LNTS 65

1928 Pan American Maritime Neutrality Convention: Pan American Maritime Neutrality Convention, Feb. 20, 1928, 47 Stat. 1989

1929 GPW: Geneva Convention Relative to the Treatment of Prisoners of War, Jul. 27, 1929, 47 Stat. 2021

1929 GWS Geneva Convention for the Amelioration of the Condition of the Wounded and Sick of Armies in the Field, Jul. 27, 1929, 47 Stat. 2074

1940 Rules of Land Warfare: War Department, Field Manual 27-10, *Rules of Land Warfare* (1940)

1944 Chicago Convention: Convention on International Civil Aviation, Dec. 7, 1944, 61 Stat. 1180

1954 Hague Cultural Property Convention: Convention for the Protection of Cultural Property in the Event of Armed Conflict, 249 UNTS 240

1955 NWIP 10-2 : Department of the Navy, Office of the Chief of Naval Operations, Naval Warfare Information Publication 10-2, *Law of Naval Warfare* (Sept. 1955), *reprinted* as *Appendix in* Robert W. Tucker, The Law of War and Neutrality at Sea (U.S. Naval War College International Law Studies, Volume 50, 1955)

1956 FM 27-10 (Change No. 11976): Department of the Army Field Manual 27-10, *The Law of Land Warfare* (Jul. 18, 1956 with Change 1, Jul. 15, 1976)

1958 UK Manual: United Kingdom War Office, *Manual of Military Law, Part III: The Law of War on Land* (1958)

1976 Air Force Pamphlet 110-31: Department of the Air Force Pamphlet 110-31,: *International Law – The Conduct of Armed Conflict and Air Operations* (Nov. 19, 1976)

1989 NWP 9: Department of the Navy, Office of the Judge Advocate General, *Annotated Supplement to the Commander's Handbook on the Law of Naval Operations, NWP (Rev.A)/FMFM 1-10* (1989)

1997 Multi-Service Detention Regulation: Army Regulation 190-8 / Office of the Chief of Naval Operations Instruction 3461.6 / Air Force Joint Instruction 31-304 / Marine Corps Order 3461.1, *Enemy Prisoners of War, Retained Personnel, Civilian Internees and Other Detainees* (Oct. 1, 1997)

1997 NWP 9 :U.S. Naval War College, Center for Naval Warfare Studies, Oceans Law and Policy Department, *Annotated Supplement to the Commander's Handbook on the Law of Naval Operations* (1997)

2001 Canadian Manual: Canada, Department of National Defence, Joint Doctrine Manual B-GJ-005-104/FP-021, *Law of Armed Conflict at the Operational and Tactical Levels* (Aug. 13, 2001)

2004 UK Manual: United Kingdom Ministry of Defence, Joint Service Publication 383, *The Joint Service Manual of the Law of Armed Conflict* (2004)

2006 Australian Manual: Australian Defence Force, Australian Defence Doctrine Publication 06.4, *Law of Armed Conflict* (May 11, 2006)

2007 NWP 1-14M: Department of the Navy, Naval Warfare Publication 1-14M / Marine Corps Warfighting Publication 5-12.1 / Commandant Publication P5800.7A, *The Commander's Handbook on the Law of Naval Operations* (Jul. 2007)

2013 German Manual: Germany, Federal Ministry of Defence, Joint Service Regulation (ZDv) 15/2, *Law of Armed Conflict Manual* (May 1, 2013)

AJIL American Journal of International Law

AP I :Protocol (I) Additional to the Geneva Conventions of August 12, 1949, and Relating to the Protection of Victims of International Armed Conflicts, Jun. 8, 1977, 1125 UNTS 3

AP II Protocol (II) Additional to the Geneva Conventions of 12 August 1949, and Relating to the Protection of Victims of Non-International Armed Conflicts, Jun. 8, 1977, 1125 UNTS 609

AP III Protocol (III) Additional to the Geneva Conventions of 12 August 1949, and Relating to the Adoption of an Additional Distinctive Emblem, Dec. 8, 2005, 2404 UNTS 1

Appendix to 1985 CJCS Memo on AP I: Appendix to John W. Vessey, Jr., Chairman, Joint Chiefs of Staff, *Review of the 1977 First Additional Protocol to the Geneva Conventions of 1949*, May 3, 1985.

Bevans Charles I. Bevans, Assistant Legal Adviser, Department of State, *Treaties and Other International Agreements of the United States of America, 1776-1949* (13 Volumes, 1968- 1976)

Biological Weapons Convention: Convention on the Prohibition of the Development, Production and Stockpiling of Bacteriological (Biological) and Toxin Weapons and on Their Destruction, Apr. 10, 1972, 1015 UNTS 163

Bothe, Partsch, & Solf, New Rules: Michael Bothe, Karl Josef Partsch, & Waldemar A. Solf, *New Rules for Victims of Armed Conflicts* (1982)

CCW: Convention on Prohibitions or Restrictions on the Use of Certain Conventional Weapons Which May be Deemed to be Excessively Injurious or to Have Indiscriminate Effects, Oct. 10, 1980, 1342 UNTS 137

CCW Amended : Amendment to Article I of the Convention on Prohibitions or Restrictions on the Use of Certain Conventional Weapons Which May be Deemed to be Excessively Injurious or to Have Indiscriminate Effects, Dec. 21, 2001, 2260 UNTS 82

CCW Amended Mines Protocol: Protocol (II) on Prohibitions or Restrictions on the Use of Mines, Booby-Traps and Other Devices, as Amended on May 3, 1996, Annexed to the Convention on Prohibitions or Restrictions on the Use of Certain Conventional Weapons Which May be Deemed to be Excessively Injurious or to Have Indiscriminate Effects, May 3, 1996, 2048 UNTS 93

CCW Protocol I: Protocol (I) on Non-Detectable Fragments, Annexed to the Convention on Prohibitions or Restrictions on the Use of Certain Conventional Weapons Which May be Deemed to be Excessively Injurious or to Have Indiscriminate Effects, Oct. 10, 1980, 1342 UNTS 137

CCW Protocol III on Incendiary Weapons: Protocol (III) on Prohibitions or Restrictions on the Use of Incendiary Weapons, Annexed to the Convention on Prohibitions or Restrictions on the Use of Certain Conventional Weapons Which May be Deemed to be Excessively Injurious or to Have Indiscriminate Effects, Oct. 10, 1980, 1342 UNTS 137

CCW Protocol IV on Blinding Laser Weapons: Protocol (IV) on Blinding Laser Weapons, Annexed to the Convention on Prohibitions or Restrictions on the Use of Certain Conventional Weapons Which May be Deemed to be Excessively Injurious or to Have Indiscriminate Effects, Oct. 13, 1995, 1380 UNTS 163

CCW Protocol V on Explosive Remnants of War: Protocol (V) on Explosive Remnants of War, Annexed to the Convention on Prohibitions or Restrictions on the Use of Certain Conventional Weapons Which May be Deemed to be Excessively Injurious or to Have Indiscriminate Effects, Nov. 28, 2003, 2399 UNTS 100

Chairman's Commentary to the Copenhagen Process: Principles and Guidelines: The Copenhagen Process on the Handling of Detainees in International Military Operations, *Chairman's Commentary to the Copenhagen Process: Principles and Guidelines* (Denmark, Ministry of Foreign Affairs, Oct. 19, 2012)

Chemical Weapons Convention: Convention on the Prohibition of the Development, Production, Stockpiling and Use of Chemical Weapons and on Their Destruction, Jan. 13, 1993, 1974 UNTS 317

Copenhagen Process: Principles and Guidelines: The Copenhagen Process on the Handling of Detainees in International Military Operations, *The Copenhagen Process: Principles and Guidelines* (Denmark, Ministry of Foreign Affairs, Oct. 19, 2012)

Digest of United States Practice in International Law: Digest of United States Practice in International Law (Department of State, Office of the Legal Adviser)

ENMOD Convention: Convention on the Prohibition of Military or Any Other Hostile Use of Environmental Modification Techniques, Dec. 10, 1976, 1108 UNTS 151

Final Record of 1949 Geneva Diplomatic Conference: Diplomatic Conference of Geneva, *Final Record of the Diplomatic Conference of Geneva of 1949* (Switzerland, Federal Political Department, 1949)

Final Report on the Persian Gulf War: Department of Defense, *Conduct of the Persian Gulf War: Final Report to Congress* (1992)

GC : Geneva Convention Relative to the Protection of Civilian Persons in Time of War, Aug. 12, 1949, 75 UNTS 287

GC Commentary: Jean S. Pictet, *Geneva Convention Relative to the Protection of Civilian Persons in Time of War: Commentary* (International Committee of the Red Cross, 1958)

GPW:Geneva Convention Relative to the Treatment of Prisoners of War, Aug. 12, 1949, 75 UNTS 135

GPW Commentary: Jean S. Pictet, *Geneva Convention Relative to the Treatment of Prisoners of War: Commentary* (International Committee of the Red Cross, 1960)

Greenspan, Modern Law of Land Warfare: Morris Greenspan, *The Modern Law of Land Warfare* (1959)

Grotius, Law of War & Peace Hugo Grotius, *The Law of War and Peace: De Jure Belli ac Pacis Libri Tres* (translated by Francis W. Kelsey, 1925)

GWS : Geneva Convention for the Amelioration of the Condition of the Wounded and Sick in Armed Forces in the Field, Aug. 12, 1949, 75 UNTS 31

GWS Commentary: Jean S. Pictet, *Geneva Convention for the Amelioration of the Condition of the Wounded and Sick in Armed Forces in the Field: Commentary* (International Committee of the Red Cross, 1952)

GWS-Sea: Geneva Convention for the Amelioration of the Condition of the Wounded, Sick and Shipwrecked Members of Armed Forces at Sea, Aug. 12, 1949, 75 UNTS 85

GWS-Sea Commentary: Jean S. Pictet, *Geneva Convention for the Amelioration of the Condition of the Wounded, Sick and Shipwrecked Members of Armed Forces at Sea: Commentary* (International Committee of the Red Cross, 1960)

Hague IV: Convention (IV) Respecting the Laws and Customs of War on Land, Oct. 18, 1907, 36 Stat. 2277

Hague IV Reg.: Regulations Respecting the Laws and Customs of War on Land, Annex to Convention (IV) Respecting the Laws and Customs of War on Land, Oct. 18, 1907, 36 Stat. 2295

Hague IX: Convention (IX) Concerning Bombardment by Naval Forces in Time of War, Oct. 18, 1907, 36 Stat. 2351

Hague V:Convention (V) Respecting the Rights and Duties of Neutral Powers and Persons in Case of War on Land, Oct. 18, 1907, 36 Stat. 2310

Hague VIII: Convention (VIII) Relative to the Laying of Automatic Submarine Contact Mines, Oct. 18, 1907, 36 Stat. 2332

Hague X: Convention (X) for the Adaptation to Maritime Warfare of the Principles of the Geneva Convention, Oct. 18, 1907, 36 Stat. 2371

Hague XI: Convention (XI) Relative to Certain Restrictions with Regard to the Exercise of the Right of Capture in Naval War, Oct. 18, 1907, 36 Stat. 2396

Hague XIII: Convention (XIII) Concerning the Rights and Duties of Neutral Powers in Naval War, Oct. 18, 1907, 36 Stat. 2415

ICJ Statute: Statute of the International Court of Justice (Annex to the Charter of the United Nations), Jun. 26, 1945, 59 Stat. 1055

ICRC AP Commentary: Jean S. Pictet et al., *Commentary on the Additional Protocols of 8 June 1977 to the Geneva Conventions of 12 August 1949* (International Committee of the Red Cross, 1987)

ICTR: International Criminal Tribunal for the Prosecution of Persons Responsible for Genocide and Other Serious Violations of International Humanitarian Law Committed in the Territory of Rwanda and Rwandan Citizens Responsible for Genocide and Other Such Violations Committed in the Territory of Neighbouring States, between 1 January 1994 and 31 December 1994

ICTR Statute: Statute of the ICTR, Annex to U.N. Security Council Resolution 955 (1994), U.N. Doc. S/RES/955(1994) (Nov. 8, 1994).

ICTY: International Tribunal for the Prosecution of Persons Responsible for Serious Violations of International Humanitarian Law Committed in the Territory of the Former Yugoslavia since 1991

ICTY Statute: Statute of the ICTY, Annex to *Report of the Secretary-General Pursuant to Paragraph 2 of Security Council Resolution 808 (1993)*, U.N. Doc. S/25704 (May 3, 1993), adopted by U.N. Security Council Resolution 827 (1993), U.N. Doc. S/RES/827(1993) (May 25, 1993).

J.A.G.S. Text No. 11, Law of Belligerent Occupation: U.S. Army, The Judge Advocate General's School Text No. 11, *Law of Belligerent Occupation* (Jun. 1, 1944, reissued Jul. 2, 1945)

Lauterpacht, II Oppenheim's International Law: Lassa Oppenheim, *International Law, Volume II: Disputes, War and Neutrality* (edited by H. Lauterpacht, 7th ed., 1952)

Levie, Documents on POWs: Howard S. Levie, *Documents on Prisoners of War* (U.S. Naval War College International Law Studies, Volume 60, 1979)

Levie, POWs : Howard Levie, *Prisoners of War in International Armed Conflicts* (U.S. Naval War College International Law Studies, Volume 59, 1978)

Lieber Code: E. D. Townsend, Assistant Adjutant General, General Orders No. 100, *Instructions for the Government of Armies of the United States in the Field*, Apr. 24, 1863, *reprinted in Instructions for the Government of Armies of the United States in the Field* (Government Printing Office, 1898).

LNTS League of Nations Treaty Series

LOS Convention: United Nations Convention on the Law of the Sea, Dec.10, 1982, 1833 UNTS 396.

Message from the President Transmitting AP II

Message from the President of the United States Transmitting the Protocol II Additional to the Geneva Conventions of August 12, 1949, and Relating to the Protection of Victims of Noninternational Armed Conflicts, Concluded at Geneva on June 10, 1977, Treaty Doc. 100-2 (1987)

Message from the President Transmitting AP III, CCW Amended Article 1, and CCW Protocol V on Explosive Remnants of War: Message from the President of the United States transmitting Protocol Additional to the Geneva Conventions of 12 August 1949, and Relating to the Adoption of an Additional Distinctive Emblem (The "Geneva Protocol III"), Adopted at Geneva on December 8, 2005, and Signed by the United States on that date; The Amendment to Article 1 of the Convention on Prohibitions or Restrictions on the Use of Certain Conventional Weapons Which May Be Deemed to be Excessively Injurious or to Have Indiscriminate Effects ("The CCW Amendment"); and The CCW Protocol Explosive Remnants of War (The "CCW Protocol V"), Treaty Doc. 109-10 (2006)

Message from the President Transmitting LOS Convention: Message from the President of the United States transmitting the United Nations Convention on the Law of the Sea, with Annexes, and the Agreement Relating to the Implementation of Part XI of the United Nations Convention on the Law of the Sea, with Annex, Adopted at New York on July 28, 1994 and signed by the United States, Subject to Ratification on July 29, 1994, Treaty Doc. 103-09 (1994)

Message from the President Transmitting the 1925 Geneva Gas and Bacteriological Protocol: Message From the President of the United States Transmitting The Protocol for the Prohibition of the Use in War of Asphyxiating, Poisonous, or Other Gases, and of Bacteriological Methods of Warfare, Signed at Geneva June 17, 1925, Executive J (1970)

Message from the President Transmitting the 1954 Hague Cultural Property Convention: Message from the President of the United States Transmitting the Hague Convention for the Protection of Cultural Property in the Event of Armed Conflict (the Convention) and, for Accession, the Hague Protocol, Concluded on May 14, 1954, and Entered into Force on August 7, 1956 with Accompanying Report from the Department of State on the Convention and the Hague Protocol, Treaty Doc. 106-1 (1999)

Message from the President Transmitting the CCW Amended Mines Protocol, Protocol III on Incendiary Weapons, and Protocol IV on Blinding Laser Weapons: Message From the President of the United States Transmitting Protocols to the 1980 Convention on Prohibitions or Restrictions on the use of Certain Conventional Weapons which may be deemed to be excessively injurious or to Have Indiscriminate effects: The Amended Protocol on Prohibitions or Restrictions on the Use of Mines, Booby-traps or other devices (Protocol II or the Amended Mines Protocol); The Protocol On Prohibitions or Restrictions on the Use of Incendiary Weapons (Protocol III or the Incendiary Weapons Protocol); and the Protocol on Blinding Laser Weapons (Protocol IV), Treaty Doc. 105-1 (1997)

Message from the President Transmitting the Chemical Weapons Convention: Message from the President of the United States Transmitting the Convention on the Prohibition of Development, Production, Stockpiling, and use of Chemical Weapons and on their Destruction, Opened for Signature and Signed by the United States at Paris on Jan. 13, 1993, Treaty Doc. 103-21 (1993)

Message from the President Transmitting the VCLT: Message from the President of the United States Transmitting the Vienna Convention on the Law of Treaties Signed for the United States on April 24, 1970, Executive L (1971)

Moore's Digest: John Bassett Moore, *A Digest of International Law* (8 Volumes, 1906)

Official Records of the CDDH :Official Records of the Diplomatic Conference on the Reaffirmation and Development of International Humanitarian Law Applicable in Armed Conflicts, Geneva 1974-1977 (16 Volumes, 1978)

Outer Space Treaty: Treaty on Principles Governing the Activities of States in the Exploration and Use of Outer Space, Including the Moon and Other Celestial Bodies, Jan. 27, 1967, 610 UNTS 205

Public Papers of the Presidents: Public Papers of the Presidents of the United States (National Archives and Records Administration, Office of the Federal Register)

Regulations for the Execution of the 1954 Hague Cultural Property Convention: Regulations for the Execution of the Convention for the Protection of Cultural Property in the Event of Armed Conflict, Annexed to the Hague Cultural Property Convention, May 14, 1954, 249 UNTS 270

Roerich Pact: Treaty between the United States of America and other American Republics on the Protection of Artistic and Scientific Institutions and Historic Monuments, Apr. 15, 1935, 49 Stat. 3267

Rogers, Law on the Battlefield A.P.V. Rogers, *Law on the Battlefield* (1996)

Rome Statute: Rome Statute of the International Criminal Court, July 17, 1998, 2187 UNTS 90

Spaight, Air Power and War Rights: James Maloney Spaight, *Air Power and War Rights* (3rd ed., 1947)

Spaight, War Rights on Land:

James Maloney Spaight, *War Rights on Land* (1911) Stat. United States Statutes at Large

Technical Annex to CCW Amended Mines Protocol: Technical Annex to Protocol (II) on Prohibitions or Restrictions on the Use of Mines, Booby-Traps and Other Devices, as Amended on May 3, 1996, Annexed to the Convention on Prohibitions or Restrictions on the Use of Certain Conventional Weapons Which May be Deemed to be Excessively Injurious or to Have Indiscriminate Effects, May 3, 1996, 2048 UNTS 144

Technical Annex to CCW Protocol V On Explosive Remnants of War: Technical Annex to Protocol (V) on Explosive Remnants of War, Annexed to the Convention on Prohibitions or Restrictions on the Use of Certain Conventional Weapons Which May be Deemed to be Excessively Injurious or to Have Indiscriminate Effects, Nov. 28, 2003, 2399 UNTS 132

Trial of the Major War Criminals Before the IMT: Trial of the Major War Criminals Before the International Military Tribunal (42 Volumes, 1947-1949)

Trials of War Criminals Before the NMT: Trials of War Criminals Before the Nuernberg Military Tribunals Under Control Council Law No. 10 (15 Volumes, 1949-1953)

Tucker, The Law of War and Neutrality at Sea: Robert W. Tucker, *The Law of War and Neutrality at Sea* (U.S. Naval War College International Law Studies, Volume 50, 1955)

U.N. Charter: Charter of the United Nations, Jun. 26, 1945, 59 Stat. 1031

U.N. Law Reports: United Nations War Crimes Commission, Law Reports of the Trials of War Criminals (15 Volumes 1947-1949)

U.S. Response to ICRC CIHL Study: John B. Bellinger, III, Legal Adviser, Department of State, & William J. Haynes II, General Counsel, Department of Defense, *Letter to Dr. Jacob Kellenberger, President, International Committee of the Red Cross, Regarding Customary International Humanitarian Law Study*, Nov. 3, 2006, *reprinted in* 46 International Legal Materials 514 (2007).

U.S.C.: United States Code (as of the date of publication of this manual)

UNTS: United Nations Treaty Series

Vattel, The Law of Nations: Emer de Vattel, *The Law of Nations, or the Principles of the Law of Nature, Applied to the Conduct and Affairs of Nations and Sovereigns* (translated by Charles Fenwick, 1916)

VCLT: Vienna Convention on the Law of Treaties, May 23, 1969, 1155 UNTS 331

Von Glahn, The Occupation of Enemy Territory: Gerhard von Glahn, *The Occupation of Enemy Territory* (1957)

Whiteman's Digest: Marjorie M. Whiteman, Assistant Legal Adviser, Department of State, *Digest of International Law* (15 Volumes, 1963-1973)

Winthrop, Military Law & Precedents: William Winthrop, *Military Law and Precedents* (2nd. ed., 1920)

End

Made in the USA
Monee, IL
13 September 2026

63182433R00230